McGraw-Hill Networking and Telecommunications

Build Your Own

TRULOVE • BUILD YOUR OWN WIRELESS LAN (with projects)

Crash Course

Louis • BROADBAND CRASH COURSE
Vacca • I-MODE CRASH COURSE
Louis • M-COMMERCE CRASH COURSE
Shepard • TELECOM CONVERGENCE, SECOND EDITION
Shepard • TELECOM CRASH COURSE
Louis • TELECOM MANAGEMENT CRASH COURSE
Bedell • WIRELESS CRASH COURSE
Kikta/Fisher/Courtney • WIRELESS INTERNET CRASH COURSE

Demystified

Harte/Levine/Kikta • 3G WIRELESS DEMYSTIFIED
LaRocca • 802.11 DEMYSTIFIED
Muller • BLUETOOTH DEMYSTIFIED
Evans • CEBUS DEMYSTIFIED
Bayer • COMPUTER TELEPHONY DEMYSTIFIED
Hershey • CRYPTOGRAPHY DEMYSTIFIED
Taylor • DVD DEMYSTIFIED
Bates • GPRS DEMYSTIFIED
Symes • MPEG-4 DEMYSTIFIED
Camarillo • SIP DEMYSTIFIED
Shepard • SONET/SDH DEMYSTIFIED
Topic • STREAMING MEDIA DEMYSTIFIED
Symes • VIDEO COMPRESSION DEMYSTIFIED
Shepard • VIDEOCONFERENCING DEMYSTIFIED
Bhola • WIRELESS LANS DEMYSTIFIED

Developer Guides

Vacca • I-MODE CRASH COURSE
Guthery • MOBILE APPLICATION DEVELOPMENT WITH SMS
Richard • SERVICE AND DEVICE DISCOVERY: PROTOCOLS AND PROGRAMMING

Professional Telecom

Smith/Collins • 3G WIRELESS NETWORKS
Bates • BROADBAND TELECOM HANDBOOK, SECOND EDITION
Collins • CARRIER GRADE VOICE OVER IP
Harte • DELIVERING xDSL
Held • DEPLOYING OPTICAL NETWORKING COMPONENTS
Minoli/Johnson/Minoli • ETHERNET-BASED METRO AREA NETWORKS
Benner • FIBRE CHANNEL FOR SANS
Bates • GPRS
Minoli • HOTSPOT NEWORKS: WI-FI FOR PUBLIC ACCESS
Lee • LEE'S ESSENTIALS OF WIRELESS

Bates • OPTICAL SWITCHING AND NETWORKING HANDBOOK
Wetteroth • OSI REFERENCE MODEL FOR TELECOMMUNICATIONS
Sulkin • PBX SYSTEMS FOR IP TELEPHONY
Russell • SIGNALING SYSTEM #7, FOURTH EDITION
Saperia • SNMP ON THE EDGE: BUILDING SERVICE MANAGEMENT
Minoli/Johnson/Minoli • SONET-BASED METRO AREA NETWORKS
Nagar • TELECOM SERVICE ROLLOUTS
Louis • TELECOMMUNICATIONS INTERNETWORKING
Russell • TELECOMMUNICATIONS PROTOCOLS, SECOND EDITION
Minoli • VOICE OVER MPLS
Karim/Sarraf • W-CDMA AND CDMA2000 FOR 3G MOBILE NETWORKS
Bates • WIRELESS BROADBAND HANDBOOK
Faigen • WIRELESS DATA FOR THE ENTERPRISE

Reference

Muller • DESKTOP ENCYCLOPEDIA OF TELECOMMUNICATIONS, THIRD EDITION
Botto • ENCYCLOPEDIA OF WIRELESS TELECOMMUNICATIONS
Clayton • McGRAW-HILL ILLUSTRATED TELECOM DICTIONARY, THIRD EDITION
Radcom • TELECOM PROTOCOL FINDER
Pecar • TELECOMMUNICATIONS FACTBOOK, SECOND EDITION
Russell • TELECOMMUNICATIONS POCKET REFERENCE
Kobb • WIRELESS SPECTRUM FINDER
Smith • WIRELESS TELECOM FAQS

Security

Hershey • CRYTOPGRAPHY DEMYSTIFIED
Nichols/Lekkas • WIRELESS SECURITY

Telecom Engineering

Smith/Gervelis • CELLULAR SYSTEM DESIGN AND OPTIMIZATION
Rohde/Whitaker • COMMUNICATIONS RECEIVERS, THIRD EDITION
Sayre • COMPLETE WIRELESS DESIGN
OSA • FIBER OPTICS HANDBOOK
Lee • MOBILE CELLULAR TELECOMMUNICATIONS, SECOND EDITION
Bates • OPTIMIZING VOICE IN ATM/IP MOBILE NETWORKS
Roddy • SATELLITE COMMUNICATIONS, THIRD EDITION
Simon • SPREAD SPECTRUM COMMUNICATIONS HANDBOOK
Snyder • WIRELESS TELECOMMUNICATIONS NETWORKING WITH ANSI-41, SECOND EDITION

BICSI

NETWORK DESIGN BASICS FOR CABLING PROFESSIONALS
NETWORKING TECHNOLOGIES FOR CABLING PROFESSIONALS
RESIDENTIAL NETWORK CABLING
TELECOMMUNICATIONS CABLING INSTALLATION

NETWORK PROCESSORS

NETWORK PROCESSORS

Architectures, Protocols, and Platforms

Panos C. Lekkas

McGRAW-HILL

New York   Chicago   San Francisco   Lisbon
London   Madrid   Mexico City   Milan   New Delhi
San Juan   Seoul   Singapore   Sydney   Toronto

The McGraw-Hill Companies

Library of Congress Cataloging-in-Publication Data

Lekkas, Panos C.
    Network processors / Panos C. Lekkas.
        p. cm.
    ISBN 0-07-140986-6 (alk. paper)
        1. Routers (Computer networks)—Equipment and supplies.   2. Microprocessors.
    3. Application specific integrated circuits.   I. Title.

TK5105.543.L47 2003
004.6—dc21   2003051161

1 2 3 4 5 6 7 8 9 0   DOC/DOC   0 9 8 7 6 5 4 3

ISBN 0-07-140986-6

The sponsoring editor for this book was Stephen S. Chapman and the production supervisor was Pamela A. Pelton. It was set in Times by MacAllister Publishing Services, LLC.

Printed and bound by RR Donnelley.

 This book is printed on recycled, acid-free paper containing a minimum of 50 percent recycled de-inked fiber.

McGraw-Hill books are available at special quantity discounts to use as premiums and sales promotions, or for use in corporate training programs. For more information, please write to the Director of Special Sales, McGraw-Hill Professional, Two Penn Plaza, New York, NY 10121-2298. Or contact your local bookstore.

To my beloved wife and lifelong friend Régine,
and to our children
Jean-Chrysostome, Marie, Marina, and Nicholas
with love and gratitude.

CONTENTS

Preface   xvii

Acknowledgments   xxiii

Part 1   **Fundamentals**   1

**Chapter 1. The Evolution of Network Technology: Distributed Computing
                     and the Convergence of Networks**   3

In the Beginning   *3*
Departmental Machines Erode the Mainframe's Following   *4*
The First Local Area Network (LAN)   *5*
Moving Mainframes onto Desks: PC and Workstations   *6*
The Client-Server Model   *7*
Packet-Switched Versus Circuit-Switched Networks   *8*
The Internet, Routing, and Associated Web Technologies   *9*
Network Management   *11*
Switched LANs, Fast Ethernet, and Fiber-Distributed Data Interface (FDDI)   *12*
IP Networks: Intranets and Extranets   *13*
IP Telephony   *14*
ATM, LAN Emulation (LANE), Multiprotocol over Asynchronous Transfer Mode (MPOA), and IP over
ATM   *15*
Wireless Networks and Mobility   *16*
1 Gigabit and 10 Gigabit Ethernet   *17*
Storage Networks   *18*
The Convergence of Networks   *19*
Optical Networking Breakthroughs   *19*
Processors: RISC, Digital Signal Processor (DSP), and Integration Toward System-on-a-Chip (SOC)   *20*
The Quest for Bandwidth and QoS   *22*
Switching Evolution: From Layer 2 Switches to Routers to Layer 3 Switches   *23*
MPLS, Lambda Switching, and Wavelength Routers   *24*
VPNs   *25*
Security Co-processors   *26*
Traffic Engineering (TE)   *27*
QoS   *27*
Performance Constraints Imposed on Communications Network Equipment   *28*
Summary   *28*

Chapter 2. Network Processors: Justification   31

What Are Network Processors?   *31*
Functional Blocks in Networking Equipment   *32*
    The PHY Interface   *32*
    Switch Fabric   *33*
    Packet Processing   *33*
    Host Processing   *33*
A Closer Look at Packet Processing   *34*
Trade-offs When Designing with Standard Off-the-Shelf CPUs   *34*
Trade-offs When Designing with ASICs   *36*
The Network Processors' Breakthrough   *38*

The Value Proposition of Network Processors   *38*
Network Processors: Categories   *39*
Summary   *40*

Chapter 3. Packet Processing   43

Network Contexts: Client, Access, Edge, and Core   *43*
The Timing of the Network-Processing Evolution   *48*
The Overriding Requirements for Network Equipment   *49*
Data and Control Plane Processing   *50*
Packet-Processing Operations   *51*
    Packet framing   *51*
    Pattern search and packet classification   *52*
    CAM (Content-Addressable Memory)   *53*
    Search Engines   *54*
    Packet Parsing   *54*
    Packet Classification and Fast Forwarding   *55*
    Modification   *56*
    Switching   *57*
    Traffic Management and Other Operations   *57*
Summary   *57*

Part 2   Network Processor Architectures   59

Chapter 4. IBM PowerNP(tm)   61

IBM PowerNP: The Big Picture   *61*
    Architecture   *64*
Major Functional Blocks in the NP4GS3   *65*
Special Coprocessor and Assist Hardware   *68*
Software Architecture   *70*
Software and Systems Development Around the NP4GS3   *71*
    Performance   *72*
The NP4GX: IBM's Second-Generation OC-48 Network Processor   *73*
Trade-offs when Designing with NP4GS3   *73*
Summary   *74*
References   *75*

Chapter 5. Intel IXA™ Network Processors   77

Intel IXA: The Big Picture   *77*
Architecture   *79*
Software Architecture   *84*
Software and Systems Development Around IXA Architecture NPUs   *85*
Systems Considerations and Trade-offs When Designing with Intel NPUs   *86*
Summary   *92*
References   *92*

Chapter 6. AMCC nP(tm) Family of Network Processors   93

nP(tm) Architecture: The Big Picture   *93*
Developing Software for the nP Family of Network Processors   *96*
Traffic Management   *97*
Switch Fabric   *99*
Systems Considerations When Designing withAMCC nP Family NPUs   *101*
Fifth-generation Technology   *102*
Summary   *103*

Chapter 7. Agere PayloadPlus(r) Family of Network Processors

105

PayloadPlus(r) Architecture: The Big Picture   *105*
FPP   *107*
RSP   *109*
ASI   *112*
The DLB Algorithm   *114*
Agere's APP750NP (ex-NP10) and APP750TM (ex-TM10) Chipset   *114*
The APP550 (ex-INP5) Network Processor   *116*
Developing Systems and Software for the PayloadPlus Family of NPUs   *118*
Summary   *121*

Chapter 8. Motorola's C-Port(tm) Family of Network Processors

123

C-Port: The Big Picture   *123*
NPU Architecture   *124*
The Q-5 TMC   *127*
Developing Software for the C-Port Family of Network Processors   *131*
Systems Considerations When Designing with C-Port NPUS   *134*
Summary   *135*

Chapter 9. Other NPU Architectures   *137*

Silicon Access Networks' iFlow(tm) Chipset   *138*
Bay Microsystems' Montego(tm) and the InP(tm) Family   *141*
Cognigine   *145*
EZchip TOPcore(tm)   *148*
Vitesse IQ(tm) Family of Network Processors   *152*
Wintegra   *154*
Xelerated Packet Devices   *155*
Other Approaches   *156*
Summary   *156*

Chapter 10. Alternative Approaches to Network Processing: Net ASICs and Designing with IP Cores

159

Net ASICs   *159*
Designing with IP Cores   *160*
MIPS Technologies   *163*
ClearSpeed Technology   *165*
Tensilica   *173*
FLIX: Configurable VLIW   *182*
ARC Cores   *184*
Improv Systems   *186*
Summary   *187*
Suggested References   *188*

Part 3   Peripheral Chips Supporting Network Processors: Storage Processors, Classification Processors, Search Engines, Switch Fabrics, and Traffic Managers

191

Chapter 11. Storage Network Processors (SNPs)

193

Storage Network Processing: The Context   *193*
SAN-Enabling Technologies   *195*
Fibre Channel   *196*

IP Storage   *199*
  Network Interface Card (NIC)   *199*
  Storage HBAs   *200*
  iSCSI Adapters   *200*
Storage Virtualization   *200*
iSCSI   *202*
FCIP   *204*
Fibre-Channel-to-iSCSI Bridging   *204*
Typical Applications for an SNP   *204*
Requirements for an SNP   *205*
TCP Termination Engines or TCP-Offload Engines (TOEs)   *206*
Case Study 1: Trebia Networks' SAN Protocol Processor (SPP)   *210*
Case Study 2: Silverback Systems iSNAP(tm) Architecture   *211*
Security Issues in Storage Network Processing   *214*
Secure SNP Trends and Concerns Moving Forward   *215*
Summary   *216*
Suggested References   *216*

Chapter 12. Search Engines   **219**

The Packet Classification Context of a Search Engine   *219*
Content-Addressable Memory (CAM)   *220*
  Pros and Cons   *221*
CAM Structure   *221*
Management of Tables Inside a CAM   *226*
Systems Engineering Issues Surrounding the Use of CAMs   *227*
Reproaches Against CAM-Based Search Engines   *230*
Going Forward   *233*
Alternative Ways of Implementing a Search Engine   *234*
Summary   *235*
Suggested References   *235*

Chapter 13. Classification Processors   **237**

Two Types of Packet Classification   *237*
Lookup and Forwarding   *240*
Algorithms for Managing Lookup Table Updates   *244*
Algorithms and Data Structures to Support Lookup and Forwarding   *245*
Deep Packet Classification   *247*
Classification Based on Multiple Fields   *249*
Implementation   *251*
Classification Processors or CAMs?   *252*
Integrated Classification or Standalone?   *256*
Case Study: Raqia's Regular Expression Classification Coprocessor   *256*
Summary   *259*
Suggested References   *260*

Chapter 14. Switch Fabrics   **263**

The Definition of Switch Fabric   *264*
The Basics of Switching   *264*
Blocking   *268*
Basic Switching Elements   *269*
Generic Types of Switching Platforms   *271*
The Evolution of the Multiservice Router/Switch   *271*
Backplane Description   *275*
The Scalability of Switch Fabrics   *277*

The Redundancy of Switch Fabrics   *278*
Routing/Switching Systems Considerations   *279*
Switch Fabric Architectures   *281*
    Input-Buffered and Output-Buffered Switches   *283*
    Buffered Crossbar   *285*
    Arbitrated Crossbar   *285*
    Shared Memory Switches   *287*
Multistage switches   *288*
    Banyan-Based Switches   *288*
    Batcher-Banyan Switches   *289*
Other Examples   *290*
A Couple of Commercial Examples   *291*
    IBM PowerPRS(tm) Switch Fabrics   *291*
Agere Switch Fabrics   *298*
Summary   *302*
Suggested References   *302*
References   *304*

Chapter 15. Traffic Managers   311

The Definition and Purpose of a Traffic Manager   *311*
Traffic Managers as Standalone Chips   *311*
Fundamental Concepts in Traffic Management   *312*
QoS-Oriented Protocols   *317*
    RSVP   *317*
    IntServ   *317*
    DiffServ   *318*
Major Tasks and Algorithms   *321*
Statistics   *321*
Traffic Marking, Shaping, and Policing   *321*
Congestion Management   *322*
Scheduling and Buffer Management   *326*
Traffic Manager Case Studies   *329*
Summary   *329*
Suggested References   *330*

Part 4   Putting Everything Together   191

Chapter 16. Systems Engineering Issues   335

Memory Subsystems   *335*
    DRAM Flavors   *336*
    SRAM Flavors   *338*
    CAM   *338*
NPU Architecture Issues   *339*
Software Development Issues   *340*
Software Development Cost   *342*
A Real-Life Case Study: Design Issues with an MSR   *344*
    Task Definition   *345*
    Design Approach   *345*
    Preliminary Design Outlook   *346*
    Switch Fabric   *350*
    System Considerations   *350*
    Resources Budget   *353*
Summary   *354*
Suggested References   *354*

Part 5   Security Coprocessors 357

Chapter 17. Security Coprocessors 359

note   *359*
introduction   *359*
Secure Communications Applications in Network Processing   *360*
  VPNs   *361*
    Conducting Secure Electronic Transactions   *361*
    Wireless Security   *362*
Cryptography: Some Basic Notions   *363*
    Private- or Symmetric-Key Encryption   *364*
    Public-Key Cryptography   *364*
Block Ciphers, Stream Ciphers, and Cryptographic Modes   *367*
    Block Ciphers   *367*
    Stream Ciphers   *369*
    Cryptographic Modes   *369*
Important Cryptographic Considerations in Communications   *371*
    Weak Keys   *372*
    Protocol-Sensitive Encryption   *373*
    Hashing   *373*
    Message Authentication Codes (MACs)   *374*
    Digital Signatures   *374*
    Session Key Exchange   *375*
    Digital Certificates   *375*
    Embedded Sequence Counters   *375*
    Address Tunneling   *376*
    Timestamp (Nonrepudiation)   *376*
    Rekeying   *377*
    Security Associations (SAs)   *377*
Common Cryptographic Algorithms   *377*
    Diffie-Hellman (DH)   *380*
Common Public-Key Cryptography Algorithms   *382*
Standardized Security Protocols   *384*
  IPsec   *385*
  SSL   *388*
Security Coprocessors: A Classification   *389*
    Systems Considerations when Engaging a Security Coprocessor   *390*
Summary   *394*
Suggested References   *395*

List of Acronyms 397

Appendix I Overview of Network-Processor Products and Platforms 415

Appendix II Typical Traffic Load (in Millions of Packets per Second)
Correspondence at Various Link Speeds and Packet Sizes 419

Appendix III Standardization Efforts in Network Processing 421

Network Processing Forum (NPF)   *421*
    Hardware Working Group (HWG)   *422*
    Software Working Group (SWG)   *422*
    Benchmarking Working Group (BWG)   *422*
    Technical Education and Marketing Working Group (TEMWG)   *422*
    Implementation Agreements (IAs)   *422*

Optical Internetworking forum (oip)   *423*
ATM Forum   *424*
Institute of Electrical and Electronics Engineers (IEEE)   *424*
10 Gigabit Ethernet Alliance   *424*
Metro Ethernet Forum (MEF)   *424*
Internet Engineering Task Force (IETF)   *424*
InfiniBand   *425*
RapidIO   *425*
HyperTransport   *425*
Performance Benchmarking   *426*
Industry Forums   *426*
Industry Analysts   *427*

Index   **429**

PREFACE

The preface is usually the last part of the book an author has to write. The motivation, theme, and employed structure of the book must be clearly spelled out in it. As I sat down to write this preface, I looked for a good way to illustrate the confusion that often exists in many people's minds when someone talks about network processors. I remembered a joke that I had been told many years ago by one of my bosses back at IBM, who was also a highly decorated ex-Navy officer. I think it is highly appropriate.

> A fully armed and loaded nuclear warship is traveling through the notorious English Channel toward the North Sea on a typical foggy and rainy winter day under some pretty rough seas. The wind-swept mist from the towering waves is cluttering the radar and the crew up in the bridge is growing very agitated. The officer on duty at the bridge, already very nervous about the poor visibility and the heavy traffic of the dangerous straits, suddenly detects through his binoculars a small dim light right in front of the bow and immediately asks the captain for instructions. The captain orders that a radio message be sent on the international emergency frequency channel asking the vessel to move northward by one-tenth of a degree to avoid a collision. The warship's radio officer sends a message on the prescribed channel saying, "We are headed onto each other. Please set yourselves immediately one-tenth of a degree to the north. This is the captain."
>
> The warship immediately receives a response on the same frequency that says, "I see you. Please set yourselves immediately one-tenth of a degree to the south. This is an old sea wolf."
>
> The puzzled radio officer shows the message to the warship's captain who frowns and says, "The man is crazy. There will be a collision if he does not move out of the way immediately." He looks out again with his binoculars over the horizon and the small dim light in front of the bow seems to be getting closer.
>
> He then instructs the radio officer to send back a second message, this time sounding a little more threatening, so that they can intimidate the vessel out of the way. The radio officer sends a new message, "I repeat. Set yourselves immediately one-tenth of a degree to the north. This is the captain of the warship."
>
> A new answer comes back immediately on the same channel. "I repeat. Set yourselves immediately one-tenth of a degree to the south. This is an old sea wolf."
>
> The spellbound radio officer, by now at a loss, shows the cable to the captain who sees the approaching light in the fog and becomes impatient with the defying seaman out there who apparently seems to be ignoring the repeated warnings. "Send him a third message and make sure you sound stern this time," he barks at the radio officer. He immediately complies by sending a new message saying, "Set yourselves immediately one-tenth of a degree to the north. This is the captain of the nuclear-missile-delivery warship."
>
> An answer flies back on the same frequency. "Set yourselves immediately one-tenth of a degree to the south. This is the lighthouse."

I thought that this funny story captures the background and spirit of today's *network processor unit* (NPU) realm. Through my numerous contacts within the industry and the academic world, I found out not surprisingly that there is not a typical profile of a person who is interested in network processors. People from different professional or educational backgrounds, each with different interests and objectives, approach the subject from a completely different angle and perspective. When I tried to see how these various profiles feel about their level of attained knowledge and understanding of network processors as a new technology, I was surprised at what I found.

The exploding demand of traffic during the last decade has pushed the complexity and versatility of the indispensable very-high-speed routing and switching network gear to unprecedented heights of sophistication and capabilities. This demand was fed by an almost insatiable craze for online applications and connectivity. Although barely a dozen years ago it was mostly just a research curiosity, the Internet has now become a household name as well as a vital business and entertainment tool. In order to perform billions of times per second the mind-boggling multitude of functions that their users expect, special computing platforms and architectures are required to power switches and routers.

These platforms are now generically called *network processors.* This book explains, defines, and redefines these terms. They are special high-performance, highly programmable chips that can handle multigigabits-per-second wire-speed traffic and can be easily and efficiently programmed to reflect a new desired functionality, which is necessary with the expected evolution and upgrades of network gear. The idea, of course, is to handle the traffic using the *quality of service* (QoS) and *service level agreements* (SLAs) of this day and age, where one pays for what is used while allowing the customers who purchase network processors (such as the network equipment designers) to design flexible and modifiable functionality into their boxes. This allows their own customers (such as carriers, *Internet service providers* [ISPs], and enterprises) to preserve their investment, as the desired lifetime expectancy of new equipment becomes longer and longer.

Network-processing chips are here to stay, even after multiple generations of products in some cases. Even though many people have heard about them, many more have not. Some understand them partially, whereas others understand them more fully. Some people say they understand some things about them, but have more questions than answers. Some people admit they don't have any idea what network processors are and why they exist in the first place. A senior executive in a very large organization asked me sarcastically to explain to him what a network processor can do that an Intel Pentium cannot! Some people may understand network processors, but they may not know much about the other parts needed to make a network-processing system function. "A switch fabric? What is that?" "A traffic manager? Search engines? Does this have to do with Internet portals? The dot-com bubble is long over." The reactions in the industry are shocking. The concept seems to be surrounded by a dense fog.

However, this does not imply no sources of information are available about network processors. Many sources are available and some are very good. The problem, though, is that they are either hyper-fragmented in pieces of obscure product literature issued by numerous companies, in a disparate series of incoherent collections of unrelated white papers, or in a random collection of articles in the trade press, which sometimes does good and sometimes unfortunately peddles snake oil. Some crucial information is available in the research literature, but few working engineers have the time to research papers, scan publications, and absorb and digest it. Some technology analysis and market research organizations are doing a superb job covering the industry and compiling the information in massive reports that are usually published several times per year. However, the problem with their products is that at least a half dozen such reports are needed at one time and each one costs several thousand dollars. This creates a truly significant budget requirement that some companies may be able to afford in a limited number of copies, but practicing engineers or students most likely cannot.

I decided it was important that someone write a book to serve a general engineering audience that would like to learn more about network processors. As I looked around and realized that nothing like it existed on the professional bookshelves, I decided I would write a book that discusses the phenomenon of network processors for practicing, but not necessarily expert, engineers. The examination should be wide enough to cover the field, but it should also poke under the hood to show how things work. Additional references are provided for more information and to guide the reader in areas and channels through which he or she can attain deeper knowledge about the subject.

The main points are why network processors exist, how they appeared, what types of solutions they propose to materialize, to what type of problems they are applicable, what the trade-offs are, and what the real costs are when one tries to put together a real-world system based on this new technology. This last requirement implies that ancillary technologies in addition to network processors themselves, such as switch fabrics, traffic managers, and so on, must also be presented. This creates a well-rounded view of the realm.

While I was finishing this book, I realized that a couple of other books on network processors were just coming out. These include Douglas E. Comer's *Network Systems Design Using Network Processors* (Upper Saddle River, New Jersey: Prentice-Hall, 2003) (see www.npbook.cs.purdue.edu), Patrick Crowley's *Network Processor Design: Issues and Practices* (San Francisco: Morgan-Kaufmann, 2002), and Bill Carlson's *Intel Internet Exchange Architecture: A Practical Guide to Intel's Network Processors* (Intel Press, 2003). The arrival of a new textbook by a well-respected author only gives legitimacy to the effort. I vowed to ensure, as well as I could anticipate the other books' content, that not only would we not compete with each other, but we would actually end up having books that are complementary to each other. I was not disappointed when I saw that my expectation of Prof. Comer's book was indeed what his vast network-protocol experience would make it to be—designing systems (based on some specific network-processor architecture [Intel]) as a platform upon which he elaborates in his book to explain how to implement network-switching/routing protocols. The other two books mentioned have different audiences. One is an interesting work, but is only shown from Intel's point of view. The other is addressed to people who are interested in designing a network processor and want to get a better understanding of the research approaches taken so far.

My approach is very different. Intel, although undeniably an admirable company and technology platform, is definitely not the only available option to network-gear systems designers. Therefore, this book covers multiple industry platforms starting from typical off-the-shelf network-processor architectures from many vendors, some of which are established global players and some of which are promising startups. This book then discusses alternative techniques such as *net application-specific integrated circuits* (Net ASICs) as well as custom-designed system-on-a-chip (SOC) architectures that lend themselves to the design of powerful network-processing ASICs. Not all of these options are within the reach of everyone. Some companies want a quick and low-cost run to the market. Other companies want an optimized maximum-performance architecture and may have the skills, budget, and manpower to custom design their own super chips. These approaches are equally important, and in the scope of this book, the landscape is examined to show the reader all these available options. After discussing network-processor examples, this book covers all the necessary pieces in a switch/router realm, starting from switch fabrics and traffic managers and continuing to cover classification coprocessors, search engines, and *content-addressable memories* (CAMs). The book even covers *storage network processors* (SNPs) including *TCP Offload Engines* (TOEs) and security coprocessors in two separate chapters.

At every point, an overview of trade-offs and compromise issues is provided, and the costs, especially the hidden ones, that vendors will not easily talk about are revealed. Unless an individual is aware of the pitfalls, he or she stand a serious risk of making uneducated or emotional decisions that will end up being prohibitively costly to an organization. The book concludes with a real-case study that puts things into perspective by showing the back-of-an-envelope design of a large multiservice switch/router.

The process of actually coding protocols is not discussed, as this is not in the scope of the book. However, many cost issues associated with writing, maintaining, and upgrading code for NPU platforms are revealed. It is also shown that not all approaches are equal, no matter how jazzy they sound or how flashy they look.

A student who must learn how to program protocols on an NPU or a junior software engineer who wants to learn the Intel network-processing architecture should definitely read Comer's textbook as well as Carlson's monograph. However, a manager, consultant, or practicing engineer who is considering a move to networking/routing/switching systems should choose my book instead, as it will offer him or her a one-stop view of the landscape from where he or she can expand his or her own preferences accordingly.

In numerous industry discussions, I have encountered experienced software engineers who have implemented cutting-edge protocols, but have no idea what concepts such as scheduling, backpressure, switching fabrics, and classification mean. They complain they don't necessarily see the part of their contribution to a larger project and unfortunately do not necessarily understand the difficulties or challenges of their hardware and/or systems colleagues, and vice versa. Likewise, I have met top-notch ASIC hardware designers who understand fast path processing at multigigabit-per-second wire speeds, but have no idea how deep classification is performed on strings or how protocols are coded and allocated to multiple computational resources that are available in a chassis. I have met switch designers who are experts on switch fabric architectures and who eat, drink, live, and breathe scheduling and arbitration, but have little understanding of security coprocessing. Even fewer of these designers actually know what constitutes solid security and what real-time cryptography entails in a systems design.

All of these disparate backgrounds stand to gain from different parts of this book. Some readers will appreciate the handy and up-to-date review of the most promising architectures. Others will appreciate the introductory discussion of some mundane issues that can only be easily found if someone buys five or six different books and dares to ask in confidence some embarrassing questions. Some readers may want more information on a specific vendor's architecture or platform. My intention was not to replace a company's literature or to advocate one vendor over another. Trade-offs are presented in an objective way that I believe newcomers in the field will find useful.

I have tried to consolidate a lot of useful material in one handy book that can serve as an introductory guide to a young industry that perhaps grew faster than expected and then was stalled for multiple reasons that are not covered here but that now comes closer to the point where dire consolidation will be inevitable. Some vendors will prosper, whereas others will inevitably perish. However, some ideas will survive and continue to inspire. My intention is to increase the awareness about this fascinating technology. If I succeed even partially in this humble goal, I will consider myself happy because I will have not wasted my time.

The structure of the book flows as follows in five parts.

Part I is made up of Chapters 1 to 3. Chapter 1, "The Evolution of Network Technology: Distributed Computing and the Convergence of Networks," introduces the network technologies evolution historically. Chapter 2, "Network Processors: Justification," discusses the technical/business justification of network processors. Chapter 3, "Packet Processing," covers the fundamental concepts related to packet processing. It is addressed to newcomers in the field and to students who may need to put things into perspective and to understand fundamental terms and concepts in a friendly way.

Part II contains Chapters 4 to 10 and introduces some of the most important network-processor architectures. Chapter 4, "IBM PowerNP™," covers the IBM approach to network processing. Chapter 5, "Intel IXA™ Network Processors," discusses the Intel platform. Chapter 6, "AMCC nP™ Family of Network Processors," addresses the approach AMCC has taken regarding network processors. Chapter 7, "Agere PayloadPlus™ Family of Network Processors," provides an overview of the Agere network-processing platform. Chapter 8, "Motorola's C-Port™ Family of Network Processors," briefly covers Motorola's network processors. Chapter 9, "Other NPU Architectures," provides a brief discussion of other alternative and promising network-processor architectures offered by a series of companies such as Bay Microsystems, Cognigine, EZchip, Silicon Access, Vitesse, Wintegra, and Xelerated. Chapter 10, "Alternative Approaches to Network Processing: Net ASICs and Designing with IP Cores," discusses Net ASICs as a different way of designing network-processing systems and *intellectual property* (IP)-core-based custom-designed processors, where configurable technologies and massively parallel approaches are exposed. Concrete examples are given from the most promising technologies offered in this realm by companies such as MIPS, Tensilica, Improv Systems, ARC, and ClearSpeed Technologies.

Part II is therefore an overview of multiple platform approaches and is a quick reference source for someone who wants to compare approaches from established or startup vendors. It shows the different architectural approaches to solving the massive computational problem associated with wire-speed packet processing. Some companies use brute force to tackle it in a scalable way and some use sheer elegance.

Part III consists of Chapters 11 to 15 and examines the other components needed in order to build a complete cutting-edge multiservice switch/router around network-processing chips. Chapter 11, "Storage Network Processors (SNPs)," discusses SNPs, a breed of NPUs that combines protocol termination with most common NPU functionality in the context of rapidly evolving storage network technologies. Chapter 12, "Search Engines," discusses search engines and CAM. Chapter 13, "Classification Processors," covers classification coprocessors, which are separate components from the network processor in some architectures. Chapter 14, "Switch Fabrics," looks at switch fabrics and expands on two different representative architectures from IBM and Agere Systems. Finally, Chapter 15, "Traffic Managers," covers traffic managers, an indispensable component for the QoS era. A lot of thought has gone into these last two chapters—namely, as to whether it would be advisable to include information about most of the representative algorithms that are typically encountered in these two realms. I finally decided that it would significantly deviate from the focus of the book, in addition to massively expanding its size. Concentrating on actual chips, I just give an overview of the subject while ample references are provided from the research literature for readers who want to understand the switching internals in depth.

Part IV contains Chapter 16, "Systems Engineering Issues." This chapter wraps up the material of in previous chapters by taking a step back and looking at multiple architectures and how they measure up against the computational requirements involved in a network-processing environment. I also discuss some other topics that could not easily fit under any other category, such as memory subsystems. Different memory technologies are needed in a network-processing system, and each one addresses different performance and budget requirements. Aside from the few designers who know how to choose technologies and how to design systems memory for wire-speed applications, many people do not feel at ease with the concepts involved. Some readers may therefore appreciate the inclusion of an overview that clears up the topic and removes some ambiguities.

Chapter 16 also discusses the critical dimension of evolving software development by looking at efficiency and hidden costs. As a means of helping the reader digest the material that we have covered at that level and to show where each part fits in the large jigsaw puzzle of networking system design, this chapter concludes with a real-life case of designing a large scalable network-edge multiservice switch/router. I try to show the reader what requirements end up looking like and how the interaction of sometimes conflicting needs may dictate trade-offs, compromises, and constraints that require the architect to make specific system design decisions.

Part V contains Chapter 17, "Security Coprocessors." In this chapter, security coprocessors are discussed in extensive length, starting from cryptography fundamentals and going all the way to systems architectures, showing how this coprocessing can be engaged next to a traditional network-processing system. Because understanding these chips requires understanding cryptography and secure processing issues, it was necessary to include some fundamental material on cryptography and on communications security. I realize that this may be familiar to some readers. To avoid offending anyone, I decided to add this chapter as a standalone chapter at the end of the book. If someone needs to know about security coprocessing, the information is available and can be consulted at any time. If someone is already aware of the issues associated with the related security technologies, then the chapter is out of the way as he or she reads the book.

In addition to a handy list of the extensive number of acronyms used in the industry, a few useful appendices have also been provided at the end of the book to summarize the status of some network-processing platforms (arranged per vendor) in a tabular form and to give a clearer idea of the multiple efforts currently under way to standardize the network processor industry. As I mentioned earlier, this is a young industry still in search of consolidation, and some sense of overriding market discipline is needed and even expected. It is important to be aware of all these forums and efforts.

Although a few readers may try to read the book from cover to cover, this was not the model that I had in mind when writing it. The modular structure of the book in reality allows readers to skip the parts in which they have no interest and go directly to the modules that they want to tackle. It is my firm belief that such an approach increases the usefulness of a professional book. This way someone who is interested in switch fabrics, classification, hardware, *Transmission Control Protocol* (TCP) offloading, or security coprocessing can simply pick up the book for that specific chapter that covers their topic of choice. If the reader then realizes that terms are used that he or she does not understand, then the index will provide cues toward some more fundamental chapters earlier in the book. If the reader wants specific examples, then he or she can always turn his or her attention to specific platform chapters to see how things function together in that realm.

ACKNOWLEDGMENTS

In preparing such a book, it is obvious that I received help from several people in the industry and academia (too many to name here). Friends and colleagues helped me with their time as well as shared an abundance of information, expertise, and insight with me. I particularly appreciate the kindness of the many reviewers of multiple parts of my evolving manuscript and the persistent communication back and forth even at crazy times of the night. However, I want to single out some individuals to whom I am grateful for their kindness, help, and support.

First and foremost and through some very tough times, I want to thank Marjorie Spencer, my acquisitions editor at McGraw-Hill, and Jessica Hornick, her tireless assistant, as well as the rest of the top-notch editing and production crew. Special thanks go to Steve Chapman, Editor-in-Chief and Pamela Pelton, Production Supervisor, at McGraw-Hill. I also want to thank Beth Brown at MacAllister Publishing Services and her superb team that made this book possible. More specifically, my thanks go to Jeanne Henning, Angela Isner, Nonie Ratcliff, and Joann Woy.

From the industry, among many people, I want to especially acknowledge the help of Jeff Pauza, Gilles Garcia, and Rene Glaise at IBM; Rob Muñoz, Wei Li, and Venkata Krishna Mallampati at Agere Systems; George Taglieri from Tensilica and now back with Synopsys; Dr. Srinivasan Keshav from Ensim Corp.; Simon McIntosh-Smith at ClearSpeed; and Amir Eyal and Kosta Sidopoulos at EZchip. Despite all the help I have received, all mistakes or shortcomings of the work are mine.

I want to take the opportunity and express a special word of gratitude to my mother for all her support, as well as to pay tribute to my beloved father's memory. They have together injected into my brain an unquenchable thirst for knowledge and they have irrevocably shaped a fighting spirit in me that I hope to transmit even partially to my own children. In a few words they have made me who and what I am.

Last but not least, I want to thank my beloved wife Régine and our four children. Jean-Chrysostome helped me with my research and material collection, Marie helped me with proofreading and editing, Marina helped me with some of the intricate drawings, and young Nicholas kept me smiling and motivated whenever he decided that I must have had more than enough tranquility and estimated that I needed some change, which he offered in abundance by discussing hippos, sharks, walruses, Komodo dragons, tow trucks, and backhoe loaders while playing next to me. I especially thank my wife and lifelong friend and partner for her unwavering love and support during the writing of this book, which took place under some very difficult circumstances. Above all, I am grateful for the patience my whole family has shown me during endless nights and interminable weekends that I had to spend away from them. Thank you.

PANOS C. LEKKAS
Xstream Technologies, LLC
xstream@ieee.org
Boston, Massachusetts
June 2003

P · A · R · T · I

FUNDAMENTALS

CHAPTER 1

THE EVOLUTION OF NETWORK TECHNOLOGY: DISTRIBUTED COMPUTING AND THE CONVERGENCE OF NETWORKS

In this introductory chapter, we will review the unprecedented changes that have occurred in computing and telecommunications-related technologies over the last 30 years. We will also examine the chain of events that caused this extraordinary cascade of technical breakthroughs on multiple fronts. These breakthroughs ultimately helped generate the new high-speed broadband network requirements for which *network processors* will be indispensable.

The various subjects discussed in this book are documented extensively within the corresponding notes and references provided in this chapter. This chapter is more of an historical overview that intends to provide a context and background against which readers (especially recent college graduates) will be able to properly understand the macroscopic picture of how and why we arrived where we are. This background will enable readers to better view these complementary technologies in relation to each other and to appreciate and understand the main network-processing technologies discussed in this book.

IN THE BEGINNING

An explosion of *information technology* (IT) occurred predominantly in the last quarter of the twentieth century. Computers, which were exotic devices to previous generations, have by now become indispensable tools for our everyday work and leisure. Today all branches of industry, processes of workflow, channels and methods of education, manufacturing techniques, financial management tools, audio and video entertainment systems, transportation systems, electronics and engine control systems, and even humble video games have taken advantage of this unbelievable progress.

In the 1960s and early 1970s, when many of us were in college, working with a computer meant standing in line to use card punchers to write programs in primitive languages. A student programmer would have to wait until the following day to receive the printout results because the data-center staff had to feed numerous programs on a batch base daily into the university mainframe. The spooler was invented to manage the output for so many different people at different times of the day. This produced one single output point that would convey the results to the users who were expecting to see the fruit of their work. This all sounds unreal, yet it was still happening just 25 years ago.

Large mainframe computers were the solution for that era's IT problems. IBM was the leading paradigm for these computers. Companies that more or less emulated its business model, such as Amdahl, Burroughs, Control Data, and so on, also dominated the stage. Only universities, major organizations, and large (usually multinational) corporations could afford these machines. Some "enlightened" industry executives have even gone down in history affirming that there could not be any potential for more than two to three computers in the market!

Soon the card punchers disappeared and were replaced by alphanumeric terminals. People could sit in front of a computer screen and type in their code using a typewriter-like keyboard. The progress of compiling technology and operating systems facilitated interactive work sessions. Programmers no longer had to wait one day to get results. Once the programs were executed, the programmer could sit down and examine the results or reexamine the code and debug the program. Interactivity between man and machine started increasing.

The site topology and IT architectures of these machines were mostly based on an inverted tree structure. The mainframe, also affectionately known as the *big iron*, was at the top of the hierarchy (the root of the inverted tree). The structure contained a series of layers of controllers of variable performance. It had a capacity that would individually cluster several nearby or remote downstream devices. This would eventually create an array of terminals that enabled interactive users to use the mainframe's computing power on a time-shared basis.

IBM led the industry and the world by creating the first comprehensive and extremely powerful intercomputer communications architecture called the *Systems Network Architecture* (SNA).[1] This architecture was quite advanced for its time. SNA enabled mainframes to communicate with each other at different sites. Little by little, tasks that were previously tedious or impossible could be done in a complex but well-tested, documented, and straightforward way. Users could easily perform file transfers and log into other computers remotely. It would still take a few more years until SNA was developed enough to enable programs running on different systems to almost seamlessly communicate with each other, synchronize themselves, and exchange data in real time. This became possible in the late 1980s.

In the midst of all this change in the late twentieth century, semiconductor technology underwent a revolution. Because more powerful capabilities could be integrated into a silicon microchip, users could envision the ever-increasing possibilities in terms of the complexity, the integration of functions, the speed, and the accuracy. The commensurate progress that was made in software engineering, which was essentially driven forward by the ever-increasing requirements of new and more sophisticated IT applications, continued to try to use the available hardware capabilities. This formed an endless loop: Faster hardware was needed to run the more sophisticated software. The more sophisticated the software became, the more powerful the underlying hardware had to become. *Central processing units* (CPUs) became faster and more complex by first packing hundreds of thousands and then millions of transistors and even millions of logical gates on a chip (with typically four, six, or even eight transistors per logical gate).

It was only a matter of time before the centralized IT fabric changed. Computing power was essentially going to break up and would be physically distributed around corporate and organizational sites.

DEPARTMENTAL MACHINES ERODE THE MAINFRAME'S FOLLOWING

The organizational and political reasons why a corporate department, such as manufacturing or R&D, did not like to be connected to and controlled by a corporate IT center go beyond the subject of this book; however, they remain a fact of life. The founders of companies such as *Digital Equipment Corporation* (DEC), Hewlett-Packard, Prime, and Data General, which pioneered the so-called midrange systems or departmental machines, understood this problem.

With the advent of sleek interactive operating systems such as Digital's VAX/VMS and with the university world open-heartedly accepting the UNIX effort from Bell Labs, a new generation of com-

---

1. Atul Kapoor, *SNA: Architecture, Protocols, and Implementation*, J.Ranade IBM Series (New York: McGraw-Hill, 1992).

puter systems was developed. These systems were much more affordable than mainframes and were easy to run and manage with small teams of people. A plethora of these machines eventually appeared on academic and industrial campuses. People who used them were almost as enthusiastic about these machines as neophytes devoted to a cult.

THE FIRST LOCAL AREA NETWORK (LAN)

Around the early 1980s, *local area networks* (LANs) slowly moved out of the research community into the industrial world. Digital, Intel, and Xerox created the Ethernet based on research that was done at Xerox's *Palo Alto Research Center* (PARC). Technology suddenly became extremely interesting. For example, a user could be running a program on one VAX and interact with another system on the network to develop software code while choosing his or her own printer that was going to be shared among several users on the LAN. These users would quickly become indignant of the older and rigid mainframe technologies. In many cases, they would even look down on traditional data-center IT staff and qualify them as "nonenlightened." Two parallel popular cultures were created. At the risk of stereotyping, it seemed that one culture was dressed in a coat and tie, and the other was dressed in jeans and a T-shirt.

IBM followed suit with the introduction of the token ring, which was based on research that was mostly carried out at the IBM Research Lab in Rueschlikon, which is located outside of Zurich. The early introduction however of an open standard, coupled with the availability of off-the-shelf semiconductor chips that implemented the basic *Media Access Control* (MAC) and *physical layer* (PHY) interface functions, helped Ethernet keep its market lead. Several other manufacturers tried to come up with their own LAN approaches until the *Institute of Electrical and Electronics Engineers* (IEEE) stepped in and started standardizing the landscape. IEEE 802.3 covers the original Ethernet approach (*carrier sense multiple access with collision detection* [CSMA/CD])[2] and IEEE 802.5 covers the token ring. Vendors could now design adapters, also known as *printed circuit boards* (PCBs), that could be plugged into systems (for example, a departmental VAX computer) to connect devices on a LAN.

As IT managers realized that the proliferation of connected users was depleting the available network segment addresses, a wider structure was created. Gateways between LAN segments and bridges started appearing between token rings and/or Ethernets. By using a straightforward lookup table mechanism, they would remain two or more address spaces apart and steer traffic to and from the appropriate destinations and sources. If users were connected inside a building, it was only a matter of time before they would also require the appropriate levels of connectivity with the external world.

In the late 1970s and early 1980s, visionaries of the engineering community realized that the increasing complexity of design work in the mechanical as well as the electronic and civil engineering fields would require more sophisticated computer-based tools. Thus, the era of *computer-aided design/computer-aided manufacturing* (CAD/CAM) was born.

Very complex pieces of software were developed in the electronics arena to enable users to design sophisticated integrated circuits and multilayer PCBs. Similarly, in the mechanical area, advanced tools appeared in the market that would enable users to create two-dimensional and three-dimensional mechanical designs for car frames, ship hulls, airplane fuselages and wings, and even offshore drilling platforms. These tools were extremely computation oriented, especially when they combined mathematical techniques such as finite-element simulation modules. Special computing platforms were needed.

In addition to being too expensive for the average research and development lab, traditional IBM mainframes were not equipped with number-crunching capabilities. The IBM mainframe S/360 and

---

2. *http://standards.ieee.org/getieee802/*

S/370 architectures made their reputation as fast data-center machines due to the special IBM channel processor architecture, which could efficiently handle several *input/output* (I/O) requests from the CPU to and from the hard disks.

However, when an executed program was lean on I/O and heavy on computations, the IBM CPUs were weak. This gave rise to several new companies such as ComputerVision, Intergraph, and Applicon, which pioneered the field of CAD/CAM workstations and eventually *Electronic Design Automation* (EDA) for the electronics industry.

One of the reasons DEC was extremely successful at the time was because its VAX architecture was able to handle computationally heavy software better than the traditional IBM machines. As a result, DEC could capitalize on users with specific computing needs as opposed to the traditional IBM approach of "one architecture fits all." By the time IBM realized the pitfalls of their approach, DEC was an established global powerhouse. IBM responded by using channel-attached array processors, which were arranged by *original equipment manufacturers* (OEM), and by creating the 3090 mainframe, which had its own *vector facility* (VF). However, this was too little and too late. It would take one more IBM iteration, with offerings of really powerful *reduced instruction set computer* (RISC) workstations and departmental machines, before it would be able to compete in the new realm.

In the early 1980s, IBM sensed that the growth in the mainframe community would not be sustainable. It had to react to the emergence of departmental computing both as a defense against the erosion of its traditional IT dominance and as a new source of potential growth. If it could replace some of these departmental computer systems, it would increase its own market share. The question was how to go about doing this. IBM chose a three-pronged approach that enabled and ratified the client-server computing model:

- The creation of the *personal computer* (PC).
- The development of IBM's own midrange systems for scientific and engineering users.
- A wholehearted embrace of UNIX.

MOVING MAINFRAMES ONTO DESKS: PC AND WORKSTATIONS

While all of this was happening, other companies such as Apollo and Sun Microsystems appeared and introduced a new breed of machines: engineering workstations. These were powerful, beautifully packaged, sleek computers geared toward a single user. These workstations possessed a superb high-definition graphics display, a powerful computationally capable CPU with floating-point processing capabilities, lots of memory for heavy-duty computing, a big hard disk drive, and standard LAN interfaces. Most of these machines initially had a proprietary operating system (for example, Apollo had its own Aegis system); however, UNIX soon became the standard offering, although it was originally available in a palette of quasi-incompatible platforms. For example, UNIX versions were released from AT&T Bell Labs, Ultrix from DEC, UNIX BSD from the University of California at Berkeley, Xenix, and other less prominent industrial players. Less commercially successful versions were also released by various academia. These scientific and engineering workstations were not inexpensive devices for the average user, but they were absolutely essential in engineering organizations, where speed, performance, ergonomics, and the highest quality of comprehensive tools were imperative.

This new trend stalled the progress of traditional departmental machines, as epitomized by DEC's VAX. Manufacturers such as Prime Computer and Data General started feeling the pressure and several of them soon went out of business.

Around the same time, IBM introduced the PC. Several books and articles have been written about the success of the PC, the idea itself, the strategy, the pros and cons, and so on, so we will not dwell on this subject for long. However, it is important to understand that the arrival and phenomenal success of the PC sparked the explosion of decentralizing software applications even for ordinary data-center corporate computer users. People discovered it was more efficient to work at their desks rather than to go to a centrally located IT department and use the mainframe.

The PC was originally an underpowered piece of hardware that engineering workstation suppliers mocked. The atmosphere was bound to change, though. The more sophisticated the software applications became, the more powerful the hardware had to become. Once IBM opened up the architecture of the PC to cloning, a whole new industry was created. This not only drove the prices lower and made computing surprisingly affordable for ordinary consumers and startup companies, but it also enabled a humble PC to do unbelievable things. Intel developed and provided generations upon generations of microprocessor technologies on that same platform, whereas Microsoft and other software companies followed by developing more sophisticated operating systems and applications. An entire software industry was created, changing the method of computing.

THE CLIENT-SERVER MODEL

Huge armies of PCs in large corporations and organizations were soon connected to LANs, accessing information on larger machines. These included departmental machines and more traditional mainframes.

The idea had originated at IBM in the early 1980s and was dubbed *cooperative computing*. IBM wanted to put a network of industrial PCs in charge of *programmable logic controllers* (PLCs) on small manufacturing area LANs. The PC would control and feed the controllers with production data running on older Series/1 systems. These systems would in turn receive production planning and control information from mainframes mostly through traditional synchronous links such as SDLC/BSC protocols over coax connections supporting 3270 terminal emulation software and so on.

Connectivity between different computer systems became critical. For example, bridges allowed the interface between Unibus™ systems from DEC and IBM channels or between IEEE 802.4 *Manufacturing Automation Protocol* (MAP) industrial buses. At the time, MAP industrial buses were favored on the shop floor by the automotive manufacturing world, and Ethernet LANs were favored in the engineering realm, where VAXs and Apollo workstations lived and worked together.

The idea was simple: The individual PC (the client) would run applications locally, but whenever data was needed, it would have to be fetched from a server computer transparently to the user. The server would usually be a much more powerful machine that was situated upstream on the network hierarchy where databases were being kept around the clock. This model would ultimately require a radical rethinking of the programming methodology. New tools had to be developed, from programming languages all the way to the application structure and its development process. This was precisely the moment when the wave of object-oriented-language-based programming became widely embraced. Previously, this software approach flourished mostly in avant-garde academic research communities who knew about Smalltalk and *Common Lisp Object System* (CLOS). This was also one of the driving reasons C++ was subsequently created and then became well established. The Java paradigm was invented by Sun Microsystems, which like so many other UNIX vendors had been plagued by the UNIX flavors that bred incompatibilities. Sun Microsystems had the noble objective of achieving complete code portability over new architectures and operating systems. However, from a programmer's point of view, it was largely built on technology that C++ had already introduced to the world.

The IT architectural hierarchy by that time had been transformed into a community environment, where the mainframe was running central applications, such as payroll, while departmental machines were running their own applications. The lower one descended on this IT hierarchy tree, the more one was likely to run into client-server arrangements. Client-server arrangements fed data into PCs and engineering workstations on individual desks running a plethora of applications from accounting spreadsheets and general ledgers to CAD/CAM modeling and mathematical simulations.

As mentioned earlier, in addition to revolutionizing the world with the introduction of the PC, IBM responded to the IT decentralization trend by introducing its own series of midrange systems. These were powerful engineering workstations with RISC CPUs that soon gave birth to powerful decentralized servers such as the IBM RS/6000 supercomputer (better known worldwide by its prowess that eventually allowed it to beat the famous world champion Gary Kasparov in a game of chess).

IBM not only embraced UNIX, but it also created its own powerful version of it, which was dubbed *Advanced Interactive Executive* (AIX). That work was further compounded by the establishment of and support for the *Open Software Foundation* (OSF), an industry consortium that IBM helped set up together with other major vendors. Along with UNIX, programming legitimacy was now given to the C language, which was embedded inside the original UNIX offering. This became another deciding factor for the promotion and ultimate adoption of C++, which as we saw, strongly influenced the appearance of Java. With its sockets and inherent support of the *Transmission Control Protocol* (TCP), UNIX offered a very straightforward means to communicate with other computer systems, log in remotely, and activate file transfers. It was only natural to expect that because these UNIX machines could be connected on LANs and bridged networks, a different global connectivity paradigm was needed.

PACKET-SWITCHED VERSUS CIRCUIT-SWITCHED NETWORKS

In the 1970s, data communication was no longer just an item of curiosity and started becoming reality on a large scale. Modems were developed that enabled the transmission of digital information over analog telephone lines. For the first time, digital data could be superimposed onto an analog carrier wave that was transmitted on ordinary lines. At the time, it sounded like rocket science to the average person, even though we smile when we hear about it now. Organizations could transmit information from one site to another. Companies started realizing that they would need a certain level of guaranteed bandwidth per month for their data transfer operations between systems. The economics of buying or leasing a line (or a set of lines) became a typical business case study.

Carriers would block specific lines physically for customer A or B, while the capacity of other lines would be used on a time-shared basis among customers D, E, and F. Time multiplexing technologies and *pulse code modulation* (PCM) transmission techniques enabled such an arrangement. Time multiplexing was the first major carrier technology that enabled such an economic model. Time slots were created per units of time and a certain number of them were allocated to a specific customer. Traffic to and from this customer would be transmitted only inside the allocated slots and the carrier would charge the customer at the end of the month appropriately.

At the same time, two significant steps occurred almost simultaneously in the evolution of communications. One was the introduction and eventual global acceptance of the seven-layer *Open System Interface* (OSI) model, which profoundly shook the structure of systems development (although layering was not a new concept since IBM had established it with its SNA years before). The other was the invention of packetized transmission, a radical departure from the previously accepted model of sequential transmission and permanent connection.

This invention was going to become the beginning of all subsequent packet-based technologies, and it was originally epitomized in the introduction of the X.25 network. A permanent circuit would no longer need to be connected between two endpoints while a communication session was active. Routes (circuits) were switched at exchange locations, originally by giant racks of mechanical relays and then by solid-state electronics switches. With X.25, no precious switched resources would have to be reserved for a communications circuit that was only used part of the time.

The transmitted information would be broken up into structured chunks (also known as packets, frames, and messages). Then some meaningful tags would be generated and prefixed or suffixed to each packet—for example, the sender's address, destination address, *cyclic redundancy check* (CRC), the number of packets being sent, and the order of a specific packet in the transmitted sequence. As a result, the intermittent network gear would know where a packet was coming from and where it was going. The packet sequence could be transmitted through switched virtual circuits or permanent connections. If a switch ran into problems and went out of operation, for example, another link would be set up around the affected link to reestablish connectivity. This would enable the carrier to deliver the packets to their destinations reliably. X.25 was designed with reliability in mind.

The OSI model has been analyzed in depth in several publications (see, for example, Radia Perlman's book *Interconnections: Bridges, Routers, Switches, and Internetworking Protocols*),[3] so we will not elaborate on it here. However, we will use the numbering system of its layers in numerous places in this book, so the reader should be familiar with its fundamental premises.

X.25 was a success worldwide, but its performance limit of 64 Kbps quickly became a huge impediment for the improved transmission of data. As a result of the ongoing semiconductor technology evolution, computers, bridges, and switches became increasingly faster. It was impossible to accept that the global network infrastructure would keep things strapped down to low speeds. This was the impetus for the next step in the evolution of networks—*frame relay* (FR).

The reliability mechanisms of X.25 were stripped down and replaced by newer and less noisy transmission media (such as fiber optics). Clever bit-setting mechanisms in frames (a new formal name for the evolution of packets) were also introduced to signal advance congestion notification. These changes led to the creation of the newer technology of frame-relay networks.[4,5] This turned out to be a faster and higher-quality transmission technology. It continues to have many followers even today.

Both X.25 and frame-relay technologies correspond to the second layer of the OSI model (the data link layer), which means that essentially any layer 3 protocol could be transmitted over either one of them. IBM's SNA, *Transmission Control Protocol/Internet Protocol* (TCP/IP) (favored by the UNIX community), DECtalk, AppleTalk, and Novell's *Internet Packet Exchange* (IPX) were all options in a disparate layer 3 world at that time. It was only a matter of time until IP was going to rule the day and become the de facto standard. It became by far the greatest common denominator even among incompatible networks.

THE INTERNET, ROUTING, AND ASSOCIATED WEB TECHNOLOGIES

The introduction of the Internet in the late 1970s is the next spectacular stop in our fast-forward trip through the technology landscape of the last 30 years. The Internet is a one-of-a-kind phenomenon in history. The history of how the U.S. government through its *Defense Advanced Research Projects Agency* (DARPA) took the initiative to help connect initially specific university campuses and then some of its contractors and sister agencies has been well documented in multiple sources. Much has been written on how this originally small network of researchers grew exponentially to become the Internet. The interested reader can consult Prakash Ambegaonkar's book *Intranet Resource Kit with CD-ROM*,[6] Christian Huitema's book *Routing in the Internet*,[7] and Uyless Black's books *Internet Telephony: Call Processing Protocols*,[8] and *IP Routing Protocols: RIP, OSPF, BGP, PNNI, and Cisco Routing Protocols*[9] for more information.

3. Radia Perlman, *Interconnections: Bridges, Routers, Switches, and Internetworking Protocols*, 2nd ed. (Reading, Massachusetts: Addison-Wesley, 1999).

4. Jeff T. Buckwalter, *Frame Relay: Technology and Practice* (Reading, Massachusetts: Addison-Wesley, 1999).

5. Uyless Black, *Frame Relay Networks: Specifications and Implementations*, Computer Communications Series (New York: McGraw-Hill, 1995).

6. Prakash Ambegaonkar, *Intranet Resource Kit with CD-ROM* (Milwaukee, Wisconsin: Frontier Technologies, 1997).

7. Christian Huitema, *Routing in the Internet* (Upper-Saddle River, New Jersey: Prentice-Hall, 2000).

8. Uyless Black, *Internet Architecture: An Introduction to IP Protocols* (Upper-Saddle River, New Jersey: Prentice-Hall, 2000).

9. Uyless Black, *IP Routing Protocols: RIP, OSPF, BGP, PNNI, and Cisco Routing Protocols* (Upper-Saddle River, New Jersey: Prentice-Hall, 2000).

The three most important points about this rapid and spectacular evolution are

- The fact that IP became the uncontested link technology between computer sites all over the world.
- New sets of protocols were developed that reside and function on top of the IP layer. These proto-cols provide several services to communicating devices, from reliable end-to-end transmission to the reservation of network resources and the quantification of *quality of service* (QoS). These pro-tocols include some very well-known tools, such as the *Hypertext Transfer Protocol* (HTTP) or *File Transfer Protocol* (FTP) and the *Hypertext Markup Language* (HTML) family of languages, upon which the *World Wide Web* (WWW) has been based.
- The fact that numerous alternative routes could be calculated on-the-fly between points A and B on this globally deployed network thanks to advancements in routing technology.

We saw earlier how IP evolved to become the de facto communication link technology at layer 2. Now let us look at other WWW technologies that at first sight might appear unrelated to this evolu-tion of computer communications and networking.

The idea of using a markup language to encode web pages was truly brilliant. It would be unac-ceptable to eat up the available transmission bandwidth trying to transfer back and forth between com-puter systems large bit streams and bitmaps of graphics and pictures in order to create content that made sense in the current multimedia world. It would make much more sense to encode the structure of web pages in a new language (HTML) and send the encoding instead to the client computer that asked for a specific web page. As a result, web page text could be combined with graphics, pictures, sound, and even video. The web page would reside on a server that is connected to the Internet. A name server would know its address and broadcast it to anyone interested in communicating with it. When a computer user accessed this web page, a whole set of actions would take place transparently to the user whereby the HTML text of the page and its constituent components would be downloaded to the requesting computer. A special piece of software called a *browser*, residing on the requesting computer, would then interpret the incoming data on-the-fly and compose the content of the web page locally on the user screen. This turned out to be the basic mechanism for network users for the gen-eration of an insatiable demand for more bandwidth.

Routing was the third major factor of this tremendous explosion in operation efficiency. IBM had tried to contain this revolution by trying to squeeze SNA into every platform. This obviously had not worked at the departmental computing level (where IBM was not as powerful) as well as with the mainframes and originally even the PC. IBM was forced to accept the presence of IP as the common interconnectivity thread. In fact, it was forced to embrace it with its own departmental platform based on AIX running on RS/6000 offerings. The outbreak of an IP culture effectively isolated SNA into the IBM legacy world. While IBM was in a new painful state of denial (shocked at its loss of control to the clones of the PC market it single-handedly created), several small startups, among which was an unknown little entity at that time called Cisco Systems, started delivering small network machines called *routers*. They were simple microcomputers based on a bus architecture. I/O adapters for dif-ferent layer 1 and layer 2 protocols, such as RS-232, IEEE-488, SDLC/BSC, X.25, frame relay, and Ethernet, would be plugged into the fast backplane of the router chassis. A master CPU along with plenty of memory would route the traffic from any port to any port based on some forwarding poli-cies. These policies would associate addresses with end systems, and a lookup table would show from which port each address could be accessed and under what conditions or circumstances. The router was eventually sold with user-friendly configuration software, which would allow a network admin-istrator to easily configure the lookup tables and to install the router inside a network in a straight-forward way. A huge new multibillion-dollar industry was created.

The success of the router manufacturers enabled them to invest heavily in R&D. Carrying the torch of standardization bodies such as the *Internet Engineering Task Force* (IETF), a plethora of routing protocols were developed. They would enable adjacent routers to communicate automatically with each other. They would also notify their peers about the status of the network at every neigh-borhood, communicate route links toward specific target addresses, and so on. The *Routing Information Protocol* (RIP), *Open Shortest Path First* (OSPF), *Interior Gateway Protocol* (IGP),

Exterior Gateway Protocol (EGP), and *Border Gateway Protocol* (BGP) are now commonplace technologies for a networking professional, but less than a few years ago, they were truly breakthrough concepts.[10,11,12] A giant web of routers deployed on a worldwide scale and armed with the appropriate routing protocols and interface adapters could saddle the ever-increasing massive traffic of the Internet around the clock.

If a certain link was inaccessible, the routers would reroute a link around other less congested areas. The whole world would end up being a connected place. This new connectivity fabric would enable the realization of the original dream. From a circuit-switched world, which used old telephony network relay switches, traffic could now be completely packet switched. Even more striking is the observation that everything is digital in this transmission realm; therefore, the nature of the information semantics is irrelevant. All digital bits following the modulation stage of the transmission process are transformed into electromagnetic energy pulses. Regardless of whether the pulses are traveling down a fiber-optic cable as a bunch of light photons or down a coax cable as a collection of electrons, or whether they are transmitted over the airwaves as microwave photons, they will always be representing digitized and compressed voice, streaming audio/video, or alphanumeric data with the same likelihood. Voice and data were no longer distinguished from one another as they were in the past. It would not take a rocket scientist to realize that the Internet or IP telephony was now the logical outcome of such enabling technologies. Competition would be severe for the traditional voice communications providers.

Packetized transmissions would be generated by breaking up the information that was going to be sent into packets. The network would route these packets automatically and in an unsupervised manner through the optimal route that it calculated. Such an approach brought forth a new generation of problems. For example, some packets might arrive at their destination out of order, whereas others might get lost on their way for many reasons, such as looping around folded branches or timing out. They could also end up being misforwarded by an incorrectly configured router.

We will soon see how the industry started looking after these legitimate QoS concerns. However, first we will take a look at how the industry came to the (then) unbelievable point of being able to fully and reliably manage complex network gear from a distance.

NETWORK MANAGEMENT

The proliferation of interconnected devices would have created a nightmare of unprecedented proportions had the techniques that enable the remote management of network devices not been invented. One of the major breakthroughs that enhanced network management was the protocol analyzer, which allowed network engineers to tap onto problematic network segments and analyze the frames until the cause of the problem was identified and fixed.

The undisputable revolution in network management, however, has to be ascribed to the *Simple Network Management Protocol* (SNMP) protocol.[13,14] SNMP was developed by the IETF. It is a software system that is predominantly based inside a PC or a UNIX system in the network management station. This station is able to communicate automatically with the various devices deployed across a network to collect information and therefore detect problems or issues that may require attention.

10. Christian Huitema, *Routing in the Internet* (Upper-Saddle River, New Jersey: Prentice-Hall, 2000).

11. Uyless Black, *IP Routing Protocols: RIP, OSPF, BGP, PNNI, and Cisco Routing Protocols* (Upper-Saddle River, New Jersey: Prentice-Hall, 2000).

12. Radia Perlman, *Interconnections: Bridges, Routers, Switches, and Internetworking Protocols*, 2nd ed. (Reading, Massachusetts: Addison-Wesley, 1999).

13. William Stallings, *SNMP, SNMPv2, SNMPv3, and RMON 1 and 2* (Reading, Massachusetts: Addison-Wesley, 1999).

14. David Perkins and Evan McGinnis, *Understanding SNMP MIBs* (Upper Saddle River, New Jersey: Prentice-Hall, 1996).

When the network is running normally, SNMP collects and logs detailed statistics about numerous variables and returns them in easy-to-interpret displays and reports. All network-connected devices supporting SNMP contain and maintain a set of *management information bases* (MIBs) with network statistics.

In order to provide the network manager with meaningful information, the SNMP management station queries the MIBs of the network-attached devices. Based on the answers it obtains, it compiles a well-rounded more or less real-time picture of how the network behaves. SNMP is structured in a client-server model. The client model (also known as the *network manager*) establishes a virtual connection with a server program (also known as the *SNMP agent*), which runs on a remote network device. The local database maintained by the SNMP agent is known as the SNMP MIB. It contains a standardized set of statistics and values of specific control variables. Commands from the network manager (client) consist of identifiers of SNMP variables (also known as *MIB object identifiers* or *MIB variables*) along with instructions to either get the value of the corresponding identifier or set the identifier value to a new value. The network manager obtains the relevant information through queries issued to the agent's MIB. This is the traditional technique of *polling*. An alternative technique is used when unsolicited responses from the network-attached devices are sent to the SNMP management station. We are referring to "traps" that the agent is throwing at the manager to signal that something unusual has happened.

Beyond the standardized MIBs, network equipment vendors have also created private MIBs, which allow the remote management of several disparate devices.

SNMP turned out to be a large and heavy protocol; therefore, it was often implemented only on a limited scale by vendors who tried to minimize the computation and memory load that was allocated purely for SNMP processing inside a network device. In conjunction with private MIBs, this often created undesirable results with SNMP compatibility between devices from different vendors. SNMP also suffered from a lack of scalability. Polling generates significant network management traffic, which only exacerbates network congestion problems by eating away useful bandwidth.

To address this capacity concern, the IETF defined *Remote MONitoring* (RMON) as an addition to SNMP. RMON was intended to go beyond just using intelligent agents (something SNMP pioneered) and use these same agents (called *probes* in RMON jargon) to collect filtered data and information about a whole network segment for subsequent proactive transmission to the network manager when needed. RMON would reconstruct the data and the environment at the network management station, thereby enabling human operators to play back an incident to understand exactly what happened.

The introduction of RMON drastically reduced the problems associated with polling and extended the range of information it sent back to the SNMP manager. The interested reader can find more information in William Stallings's book *SNMP, SNMPv2, SNMPv3, and RMON 1 and 2*[15] and David Perkins and Evan McGinnis's book *Understanding SNMP MIBs*.[16]

SWITCHED LANS, FAST ETHERNET, AND FIBER-DISTRIBUTED DATA INTERFACE (FDDI)

As a result of the increase in desktop computing capabilities, the proliferation of the client-server computing model sparked a phenomenon. LAN bandwidth was being rapidly eaten away, and local congestion became a common problem. The frustration this caused among users put pressure on vendors to come up with a faster LAN. The most notable of the achievements that addressed this concern was the development of 100 Mbps Ethernet, which eventually became known as *Fast Ethernet*.

15. William Stallings, *SNMP, SNMPv2, SNMPv3, and RMON 1 and 2* (Reading, Massachusetts: Addison-Wesley, 1999).

16. avid Perkins and Evan McGinnis, *Understanding SNMP MIBs* (Upper Saddle River, New Jersey: Prentice-Hall, 1996).

Driving the cost down of Ethernet LANs was a process that had to go through at least a couple of evolutionary stages, from the original coax cable to the twisted pair (Cheapernet) to ultimately *unshielded twisted pair* (UTP). At the heart of the 10Base-T standard and in conjunction with the advent of switched LAN technology, UTP caused the explosive proliferation of LANs during the 1990s. The various segments of an Ethernet LAN were connected in a hub-and-spoke architecture that enabled easy deployment and scalability. The wide availability of hubs (more accurately called *LAN repeaters*) turned out to be an easy way for network management to allocate bandwidth and ensure easier physical connectivity, overall site management, and ultimately QoS to users. Small startups offering hubs such as 3Com, Cabletron, and Wellfleet/Bay Networks, soon became multibillion-dollar companies. The presence of repeaters on local networks, working in combination with routers when these networks were getting connected with large-scale *metropolitan area networks* (MANs) or *wide area networks* (WANs), made network management even more of an urgent and critical issue. This fact exacerbated the industry's efforts toward advancing and developing network management technology even further.

In campus networks, where periphery LANs often served many users with Fast Ethernet capabilities, the backbone that was feeding these periphery LANs started to show very serious problems of congestion. This is because fast LANs serving the desktop produced so much traffic that the campus backbone linking these LANs would choke. It was only a matter of time before some serious help was needed. The effort to control this problem led to the introduction and wide-scale acceptance of the *Fiber Distributed Data Interface* (FDDI) and the Gigabit Ethernet technologies along with the advent of *Asynchronous Transfer Mode* (ATM).

FDDI was based on a logical and physical ring structure that offered speed (like the original IBM token ring principle), high reliability (because the ring would logically fold back on itself in case of rupture or accident), and the avoidance of traffic congestion.[17] Due to their significance in this historical overview, we will discuss ATM and Gigabit Ethernet later in this chapter in separate sections.

IP NETWORKS: INTRANETS AND EXTRANETS

The arrival of the Internet signaled the beginning of the era of web technologies. Client-server models were being applied on a grand scale beyond campus- or site-wide deployed systems. Companies forced by deregulation better manage their resources started *restructuring* (a term that came in vogue during the late 1980s and early 1990s). This involved looking among other things at better streamlining their operations while cutting costs. In many cases, they radically changed the way they did business (processes) and ran their internal operations.

All of a sudden new words entered into everyday vocabulary, such as *e-business*, *e-commerce*, and so on. Companies started realizing that the use of these technologies could be applied toward improving their day-to-day operations. For example, corporate users could now dial into specific web sites and access their daily resources from anywhere on the planet. They could check with divisional associates and databases, and carry out their work efficiently from anywhere and at any time. These special internal networks that were deployed on top of the same physical Internet were called *intranets*.

It was only a matter of time until companies realized that some external users could also have legitimate access to parts of a corporate network. For instance, key suppliers could be granted access to their OEM customer's inventory status databases and help adapt the shipment dates to support a *just-in-time* (JIT) philosophy. Customers might need to log into specific customer support systems and probe for frequently asked questions or report problems. Companies called these networks *extranets*.

17. Amit Shah, G. Ramakrishnan, and Akrishan Ram, *FDDI: A High Speed Network* (Upper Saddle River, New Jersey: Prentice-Hall, 1993).

Information flowing back and forth suddenly made for a more efficient economy in a way that would have been absolutely unthinkable only 5 to 10 years ago.

IP TELEPHONY

In the case of telephony, the deregulation of the carriers first in North America and soon thereafter in other parts of the world enabled newcomers to enter the market. These were mostly startups that mastered all technological aspects of the new network fabric. They were poised to offer very competitive services. This placed a tremendous financial pressure on the traditional transmission technologies, as companies that had always deployed them in their business model could not be economically sustained without some sort of government intervention. If a new-generation carrier using IP technologies could offer connectivity for a fraction of the cost of the older guard carriers, why would someone continue doing business with the traditional telephony carriers?

In addition to the privileged capability of efficiently handling data transfers, the new network was also able to tackle the (then) lucrative voice transfer market. Of course, IP telephony was not going to materialize overnight.[18,19,20] Telephony, as dictated by the ergonomics and the sensitivities of the human ear, is a very demanding application in terms of the acceptable latency and quality required to satisfy a user. Even the term *satisfaction* is rather generic as voice applications have different levels of acceptable quality for different levels of cost; hence, terms such as *toll quality* are not always applicable (Bellamy).[21]

Besides the issue of audible quality, which could arguably be addressed with the advancements in low-bit-rate vocoders, users had to come to grips with the different statistics of the new traffic that mixes everything in the same digital bucket—voice, audio, video, and data. Traditional telephony statistics are extremely well understood and predictable. That fact was at the heart of the study and deployment of the public telephony network many decades ago. With the arrival of the Internet on the global communications market, however, everyone realized that this was a very unpredictable medium in terms of traffic load. Consequently, to be able to offer reliable telephony over an IP network, the new-generation carriers found out that they either had to have their own intranet, where they could more or less manage the allocation of bandwidth, or they had to have access to specific pieces of powerful transmission/routing equipment on the Internet with the appropriate resource reservation protocols, such as the *Resource Reservation Protocol* (RSVP) and *Real-Time Protocol* (RTP).[22,23] Whether this meant that alliances were needed with companies serving the backbone of the Internet or that only well-heeled players would have a chance to compete in this new business, only time would tell. (In many cases, these new carriers included some older guard carriers such as AT&T or Verizon, who shed their old skin and adapted themselves by reacting appropriately to the evolution of the industry.)

It could be argued that no matter what, the equipment or bandwidth investment would have to ultimately be passed onto the carriers' customers somehow. Therefore, the following reasoning should be considered: When communicating over the Internet, the connection cost itself has been shown to be negligible, even coming very close to zero (barring the nominal cost of an *Internet service provider* [ISP] connection and a modem). However, the QoS one receives for that link is sometimes equally

18. Uyless Black, *Voice over IP* (Upper Saddle River, New Jersey: Prentice-Hall, 1999).

19. Bill Douskalis, *IP* Telephony—The *Integration of Robust VoIP Services* (Upper Saddle River, New Jersey: Prentice-Hall, 1999).

20. Uyless Black, *Internet Telephony: Call Processing Protocols* (Upper Saddle River, New Jersey: Prentice-Hall, 2001).

21. John Bellamy, Digital Telephony (2nd Edition), Wiley, New York, NY, 1991.

22. Uyless Black, *Internet Architecture: An Introduction to IP Protocols* (Upper-Saddle River, New Jersey: Prentice-Hall, 2000).

23. Uyless Black, *Voice over IP* (Upper Saddle River, New Jersey: Prentice-Hall, 1999).

close to zero. As the quality requirements increase, some infrastructure cost will be required, which will ultimately reflect itself in increased costs for the customer.

Nevertheless, it should be clear that the advent of IP telephony and the deregulation of the telecom industry during the last 10 years have been deciding factors that contributed to the sharp decline of voice communication costs. The traditional local or long-distance carrier is now in serious danger of extinction if it does not adapt quickly to the realities of the new network.

ATM, LAN EMULATION (LANE), MULTIPROTOCOL OVER ASYNCHRONOUS TRANSFER MODE (MPOA), AND IP OVER ATM

The emergence of ATM in the 1990s as the promising successor of frame relay for reasons that have been widely documented is another factor that had to be taken under consideration.[24,25,26]

ATM was created as a versatile way for carriers and service providers to more flexibly allocate bandwidth and to provide different levels of QoS. The basic idea was to mesh together ATM switches on point-to-point ATM links or interfaces. These would usually be interfaces to the *Synchronous Optical Network/Synchronous Digital Hierarchy* [SONET/SDH] hierarchy. The transmission units of ATM are small fixed-length bit packets (53 bytes), which are called *cells*. ATM switches can indeed transmit traffic cells from one interface to another very fast (up to several gigabits per second), and traffic can be transmitted with a very small and predictable delay. This fundamental characteristic of ATM is the key enabling factor for the delivery of voice and data services with a certain QoS in terms of available bandwidth, delay, and jitter.

ATM was expected to become the solution to the backbone congestion problem we described earlier. With projections of sharply increasing sales, vendors of ATM products hoped that the costs of ATM products and more specifically adapters would drop significantly, thereby opening up the huge markets of desktops.

To facilitate acceptance of the technology, several standardization efforts were put forth by the ATM Forum, an industry consortium devoted to the promotion and advancement of ATM. These efforts led to the creation of protocols that allowed *LAN Emulation* (LANE) over ATM or the transmission of several network and transport protocols over ATM, *Multiprotocol Over Asynchronous Transfer Mode* (MPOA).

In retrospect, it is rather easy to state that ATM has failed to become the astounding success it had originally promised for a couple of reasons. The most important reasons are as follows:

- The establishment of IP running over ATM as the predominant realm, within which routing decisions were being taken by network equipment operating at a higher layer than where ATM was, left no room or need for an intelligent ATM switch under it.

- The bandwidth and QoS services that ATM was designed to offer in the WAN were going to be offered by the newer layer 3 switching techniques (such as *Multiprotocol Label Switching* [MPLS], which we will discuss in another section).

- At the campus level, other technologies appeared such as Gigabit Ethernet, which was not only faster than ATM's 622 Mbps transfer rate, but it was also completely compatible with legacy Ethernet applications and software written for 10Base-T era networks. ATM was left in a perpetually hopeful mood, only now without any real prospects.

24. David E. McDysan and Darren L. Spohn, *ATM Theory and Applications* (New York: McGraw-Hill, 1998).

25. Mohsen Gluzani and Ammars Rays, *Designing ATM Switching Networks* (New York: McGraw-Hill, 1999).

26. David McDysan, *QoS and Traffic Management in IP and ATM Networks* (New York: McGraw-Hill, 1999).

Today ATM is mostly confined in the backbone of some long-distance or metro carriers. As a result, it must be utilized for the efficient and billable transfer of pertinent layer 3 protocols such as IP. The IETF quickly established how IP should be transferred over an ATM network. It is one of the current techniques used for the transport of voice or video or data over such a fast layer 2 network arrangement. Of course, ATM itself was supposed to run over an appropriately supportive layer 1 such as SONET/SDH,[27] but this is beyond the scope of our discussion. There are ample references for the interested reader to pursue the subject.

In the evolution of the newly convergent networks, the concept of optical networking starts appearing often, and ATM does not seem to be part of the new backbone technology landscape that is taking shape for the longer run. Some industry insiders already envision the demise and elimination altogether of the ATM layer (for instance, running under IP and over SONET/SDH, which would run over optical *wavelength division multiplexing* [WDM]) in the effort to ultimately have IP run directly over the newer technologies of optical WDM.[28] One of the reasons for such a bleak outlook is ATM's inefficient transmission layer. This problem includes *ATM Adaptation Layer level 5* (AAL5) and ATM cell overhead, which when combined approaches 30 percent. As a result, it overrides the advantages of multiservice integration and QoS functionality that ATM purports to offer.[29]

WIRELESS NETWORKS AND MOBILITY

In the 1980s, the first analog cellular networks appeared timidly in the United States and Europe. They were an instant success with business people and the public at large. As the PC liberated the tormented corporate user from the need to be attached to the mainframe when he or she had some data-related work (IT) to accomplish, the arrival of the mobile telephone liberated users from the telephone jack on the wall. It enabled users to roam around while doing their business and leading their lives more productively and efficiently. Europeans embraced the wireless technologies much faster and to a larger extent than Americans so they moved quickly to the second generation of wireless networking—the digital *Global System for Mobile communication* (GSM) standard (helped by an intergovernment-guided standardization process). The United States kept its market unregulated for political, economic, and competitive reasons.

Digital wireless technologies brought a higher quality of voice to roaming users. The result was that several competing second-generation technologies appeared in the United States, such as *Time Division Multiple Access* (TDMA), *Code Division Multiple Access* (CDMA), and even GSM, along with the older analog *Advanced Mobile Phone System* (AMPS) networks. This is why U.S. mobile carriers never attained the same deployment economies of scale of GSM as European carriers and as carriers on other continents where European manufacturers exported it.

CDMA in its wideband varieties soon became accepted as the third-generation standard. It will be deployed in a couple of different standards in North America, Europe, and elsewhere, with the hope that some sort of compatibility of third-generation networks can be expected. Third-generation technologies promise to further enrich the lives of users by enabling high-speed interconnectivity, among other things, that can transmit images, compressed video, high-quality audio, and data onto multimedia-enabled handsets. Microbrowsers are already available in handsets equipped with a *liquid crystal display* (LCD) screen. This screen enables the mobile browsing of Internet web pages through technologies such as the *Wireless Application Protocol* (WAP). The m-commerce area that is enabled by such an infrastructure looks extremely promising.

27. Walter J. Goralski, *SONET* (New York: Osborne McGraw-Hill, 2000).

28. Peter Tomsu and Christian Schmutzer, *Next Generation Optical Networks: The Convergence of IP Intelligence and Optical Technologies* (Upper-Saddle River, New Jersey: Prentice-Hall, 2002).

29. Uyless D. Black, *Optical Networks: Third Generation Transport Systems* (Upper-Saddle River, New Jersey: Prentice-Hall, 2002).

To facilitate the gigantic investment needed to uproot older infrastructure and the massive deployment of new technologies for carriers, which is what the transition from second-generation technology to third generation implies, some intermediate solutions have been proposed by infrastructure equipment vendors and explored by carriers. Two popular and quite promising examples of this wave of technology include *General Packet Radio Service* (GPRS) and *Enhanced Data Rates for GSM Evolution* (EDGE). *Cellular Digital Packet Data* (CDPD) is commercially less successful. These 2.5-generation technologies provide enhanced transmission speeds, and they can be deployed for the most part on the current second-generation wireless infrastructures. This enables carriers to proceed with the delivery of third-generation-like services without having to foot the bill for the huge immediate investment that is required for the establishment of a full-fledged third-generation network. However, the sudden and explosive growth of wireless LANs (WLAN) and access technologies like IEEE 802.11, 802.16 etc. create an environment where the prospects of 3G cellular telephony may be endangered.

At the same time, the development of Mobile IP is leading to the possibility of having a unique IP address that will allow users' devices to be accessible no matter where they are. Clearly, we are moving toward a realm where the traditional phone number and the IP address of a computer are merged into the same sequence of digits. This trend is further supported by the fact that the traditional wireless handset has started embedding functionality that until recently was only available inside a *personal digital assistant* (PDA), an entertainment box such as an MP3 music player, or a portable video player most likely to be working along the MPEG4 lines. Today's wireless telephones make the handling of electronic transactions, such as purchases charged to one's credit card, instructions to one's stockbroker, and so on, relatively easy and secure.

The need for global and secure connectivity, coupled with ubiquitous computing capabilities, dictates that the flow of unprecedented communications traffic will need to be reliably and systematically managed between wired and wireless networks all over the planet, around the clock, and based on demand. The new global network is expected to be able to handle this type of demanding environment. This can largely be done with the advances in powerful microchips (network processors) that populate the motherboards of network switching equipment. These network processors are discussed in more detail in the following chapters.

1 GIGABIT AND 10 GIGABIT ETHERNET

One of the key technologies in the performance network arena is Gigabit Ethernet, which was developed as the result of the natural evolution of Fast Ethernet.[30] It preserves a very good compatibility with legacy software applications developed for and running on 10Base-T and Fast Ethernet networks (something that is always a good financial advantage). Above all, it offers a staggering bandwidth increase for campus networks. The ability to properly service heavy traffic and to interface Gigabit Ethernet networks with the rest of the world through switched equipment and routers is another dimension in the demand for fast network processing chips.[31] We will see this later in the book as a recurring phenomenon.

Although it sounded impossible a couple of years ago, the effort to further extend the Ethernet philosophy to a 10 Gbps network has already become a reality. The technology has become an IEEE standard (IEEE 802.3ae-2002). Several vendors have proposed components, subsystems, and systems that can function in this realm that promise to revolutionize the industry both on the LAN and MAN/WAN. This revolution will not only increase its speed, but it will also improve the software compatibility that it allows. As companies don't have to upgrade or change fundamental parts of their IT infrastructure, the business case becomes easier to justify. The effort in the 10 Gbps Ethernet dimension is

30. Jayant Kadambi, Ian Crawford, and Mohan Kalkunte, *Gigabit Ethernet: Migrating to High-Bandwidth LANs* (Upper-Saddle River, New Jersey: Prentice-Hall, 1998).

31. Radia Perlman, *Interconnections: Bridges, Routers, Switches, and Internetworking Protocols*, 2nd ed. (Reading, Massachusetts: Addison-Wesley, 1999).

further compounded by the work done by the *Metro Ethernet Forum* (MEF) and the 10 Gigabit Ethernet Alliance. More information about these groups can be found in the Appendix III, "Standardization Efforts in Network Processing."

STORAGE NETWORKS

With the establishment of the client-server computing model, IT managers realized that in order to cope with application growth and the demand for functionality on behalf of users, they would need to be able to attach storage space and devices onto an existing IT hierarchy. These devices include hard disks, tape drives, and so on. This storage attachment should enable several computer systems to gain access to the storage reliably and in a modular fashion. In order to do that, they had to adopt either the direct attachment model or the *network-attached storage* (NAS) model.

The direct attachment model meant that storage devices would hang from a server using the standard *Small Computer System Interface* (SCSI), which is currently at its Ultra3 level of iteration and is able to sustain a throughput of 160 Mbytes/sec. The NAS model required that the disk arrays and storage devices connect directly onto a traditional LAN using network adapters, such as Ethernet or Fast Ethernet cards or even hub connections.

NAS makes storage resources more readily available and helps alleviate bottlenecks associated with access to storage devices. It has proven more useful in areas where a relatively low volume of data traverses the links. In general, NAS has been shown to suffer from a couple of major drawbacks:

- As most NAS devices are coupled to the LAN through 10 Mbps Ethernet or 100 Mbps Fast Ethernet cards, a certain bandwidth shortage occurs when storage is accessed. This situation will continue to occur until Gigabit Ethernet and even 10 Gigabit Ethernet interfaces become commonplace in this area.

- A clear lack of cohesion exists among storage devices. If disk arrays and tape drives are on the LAN, managing the devices can be challenging because they are seen as separate entities and are not tied together logically.

As large enterprises want the ability to store and manage large amounts of information in a high-performance environment, a new technology has appeared in the landscape: *storage area networks* (SANs).

In a SAN environment, storage devices, such as *redundant array of inexpensive disk* (RAID) arrays, are connected to several kinds of servers through a high-speed interconnection, typically a Fibre Channel.[32] This provides fast access to storage from all types of servers. It also provides the convenience of alternative paths to storage through an alternative server, should the server of choice turn out to be unavailable or slow. Using a SAN, data can be easily mirrored and disaster recovery sites can be created, while storage access bandwidth can be added without burdening the main LAN. Online backups can take place on a SAN without causing any inconvenience to LAN users. When more storage is needed, it is directly attached to the SAN rather than being hooked up to one of the LAN servers. The greatest benefit this technology provides is that it is managed centrally as a single entity; each device is not managed individually. This makes it easier to manage very large "farms" of storage devices, which could potentially consist of dozens or even hundreds of servers and devices.

The Fibre Channel was developed by the *American National Standards Institute* (ANSI) in the early 1990s as a means to transfer very large amounts of data quickly. Fibre Channel is compatible with other legacy technologies such as SCSI, IP, IEEE 802.2, AAL, and Link Encapsulation. It can also be used over copper cabling or fiber-optic cable. Fibre Channel links usually offer a performance from 266 Mbps to over 4 Gbps. Devices can be distanced up to about 10 kilometers (6 miles), which

---

32. *http://www.fibrechannel.com*—the web site of the Fibre Channel Industry Association with tutorials, FAQs on the technology, and information on how it relates to SANs.

offers the possibility of convenient off-site connectivity for network managers. Fibre Channel supports several configurations, including point-to-point and switched topologies. A *Fibre Channel Arbitrated Loop* (FCAL) is usually used to create a reliable and high-speed environment where any-to-any connectivity is easily supported and where even simpler SCSI devices can be easily bridged onto and interfaced with a Fibre Channel.

The special functionality of the underlying sophisticated hardware, which must be able to identify, process, switch, and forward all transmitted packets quickly, is not found in ordinary CPUs; therefore, special architecture semiconductor chips are required that are classified among the greater family of network processors. We will examine these microchips in greater detail later in this book.

THE CONVERGENCE OF NETWORKS

Because of deregulation in the telecommunications industry combined with the technology revolution, conventional voice-based switching technology is being pushed out of commission. The infrastructure is being replaced by packet-based architectures using new hardware and software technologies. The deployment of these new technologies not only costs as little as 10 to 20 percent of the previous generation of systems, but it also enables the consolidation of multiservice voice and data transmission with much greater efficiency. Since 2000, data communication has overtaken traditional voice traffic (tomsu).[33] The explosive proliferation of Internet connectivity and corporate and organizational intranets and extranets is a new reality. Carriers have no other choice than to evolve their network to the new technologies.

This new type of consolidated network is invariably called the *new network* or the *converged network*. The gigantic process of uprooting the older network infrastructure and adding the newer transmission and switching systems has been dubbed as *the convergence of networks*. We will use this phrase throughout in our discussion.

OPTICAL NETWORKING BREAKTHROUGHS

The wide-scale deployment of fiber optics as the successor of the old and tried copper cable was one of the fundamental factors leading to the proliferation of high-speed networks.[34,35] Signals could be optically transmitted and the new technique produced a sharp decrease in transmission losses. It also provided higher security against passive eavesdroppers than copper cables, which usually generate radiation in their vicinity and can be easily tapped. Optical fibers allow the transmission of signals for many tens of miles without requiring traditional signal recovery, filtering, and reamplification.

The development of many generations of suitable integrated lasers and advanced doped-fiber optical amplifiers in the two major spectral windows of transmission in conjunction with WDM increased the capacity of the cable dramatically.[36] This meant that the sheer number of simultaneous transmission channels and the awesome speed of the transmission of digital data over these fiber-optic links would enable the extraordinary new capabilities that we have come to see in the infrastructure networks. These new broadband networks require equipment with remarkable computing power and intelligence in order to be able to process transmitted and received data at both ends of an optical link

33. Peter Tomsu and Gerhard Wieser, *MPLS-Based VPNs: Designing Advanced Virtual Networks* (Upper Saddle River, New Jersey: Prentice-Hall, 2002).

34. Ivan P. Kaminow and Thomas L. Koch, eds., *Optical Fiber Telecommunications IIIA* (New York: Academic Press, 1997).

35. Govind P. Agrawal, *Fiber-Optic Communication Systems,* Wiley Series in Microwave and Optical Engineering (New York: John Wiley, 1997).

36. Rajiv Ramaswami and Kumar Sivarajan, *Optical Networks: A Practical Perspective*, 2nd ed. (San Francisco: Morgan Kaufmann Publishers, 2001).

and at line speeds.[37] Therefore, from another point of view, we see the need for powerful network processors inside communications equipment. Until this processing is handled completely with optical technologies, fast microelectronics will play a key role; therefore, network processors enable this type of functionality at very high speeds of transmission.

PROCESSORS: RISC, DIGITAL SIGNAL PROCESSOR (DSP), AND INTEGRATION TOWARD SYSTEM-ON-A-CHIP (SOC)

Microprocessors were available in the 1970s, but they were simple 4- and 8-bit processors of small to medium levels of silicon integration. Given the very limited levels of integration of silicon that semiconductor technology allowed at that time, high-performance computers were based on complete CPU modules. These modules contained multiple specialized chips that handled all instruction fetching, decoding, and scheduling, as well as all arithmetic and logic processing functions and the necessary memory support and I/O interface logic.

However, the market for powerful microprocessors started taking off in the early 1980s with the arrival of the PC. The establishment of the IBM-compatible architecture as the de facto standard using the Intel platform (and later Advanced Micro Devices) dealt a severe blow to Motorola's then competing 68000 architecture. Motorola never really recovered in the PC market. Astronomical Intel sales funneled profits toward more R&D and plant/equipment investment. These sales were also profitable since PCs had not yet become a sales commodity item with razor-thin profit margins. New semiconductor fab lines were being built and existing ones expanded to meet demand. This economic cycle would further affect the improvement of the design and manufacture of more sophisticated, more complex, and less expensive semiconductor chips, due to the ever-increasing profits from larger, profitable, and enhanced operations. Microprocessors, dynamic and static memory, and I/O interface chips all profited from this progress. The computing landscape started changing dramatically.

Soon microprocessors were deemed so complex that new computing architecture paradigms had to be found. Research from academia (University of California Berkeley and Stanford) as well as from the industry (IBM) pioneered the concept of *reduced instruction set computers* (RISCs) as a means of shedding the unnecessary capabilities of the traditional microprocessors, which had come to be known as *complex instruction set computers* (CISCs).[38,39] The RISC CPUs used more optimized approaches that were heavily based on pipelines of multiple stages for fetching, decoding, and scheduling code instructions ahead of their time in a program. RISC CPUs would certainly offer simpler and faster hardware. However, software that was written for these new CPUs would run much faster if the novel RISC architectural schemes that the designers had developed were used.

A typical example would be loop and branching look-ahead in iterative code. Unfortunately, taking full advantage of the capabilities of a RISC CPU involved a deeper architectural understanding on behalf of the programmer, which he or she rarely had. Writing code in assembly was no longer an option (except for some minor optimization parts of an application) as the underlying CPU was designed to decode extremely simple operations. The programmer would prefer the opposite—that is, to compact as many different logical operations within the boundaries of one single instruction (a philosophy created in the minds of most computer science graduates largely by the CISC industry legacy). Therefore, the burden had to be shifted onto the compiler tool developers, who had to create new types of sophisticated development tools for these new processors, if these CPUs were to ever stand any chance of commercial success against the established market presence of CISC CPUs.

37. P. A. Perrier, "Position, Functions, Features and Enabling Technologies of Optical Cross-Connects in the Photonic Layer," Technical Paper, Alcatel, Nevada, September 1999.

38. David A. Patterson and John L. Hennessy, *Computer Organization and Design: The Hardware/Software Interface* (San Francisco: Morgan Kaufmann Publishers, 1997).

39. John Hennessy, David Goldberg, and David A. Patterson, *Computer Architecture: A Quantitative Approach* (San Francisco: Morgan Kaufmann Publishers, 1996).

Around the early to mid 1980s, the speed of electronics enabled the faster digitization of analog signals (voice, video, telemetry, speed, temperature, pressure, and so on). At the same time, the development of sophisticated digital-processing methods, algorithms, and mathematical formulation techniques that could take advantage of this progress had already made their way to the classrooms and laboratories in engineering colleges in the late 1970s. This resulted in a new army of signal-processing engineers in the industry and academia. These engineers would rather use digital-processing techniques to solve a problem than tinker with older analog, essentially nonrepetitive, complicated, and sometimes half-baked solutions, which may or may not provide reliable and consistent results.

Texas Instruments (TI), Motorola, and Analog Devices (and a plethora of less successful vendors) introduced multiple families and architectures for *digital signal processors* (DSPs).[40,41] These were sophisticated CPU-like chips that contained integrated circuitry to optimally and efficiently handle mathematical operations used in digital-processing algorithms in one single clock cycle—for example, the execution of *Multiply-And-aCcumulate* (MAC) operations like the ones used in digital filtering. DSPs and memory chips would now be integrated onto adapters and PCBs. A complete sophisticated DSP system could easily be developed, opening up horizons and possibilities for numerous new applications where classical CPUs could not have been envisioned.

Although Intel adopted RISC techniques relatively early in some of its embedded processor products (for example, i960), its bread-and-butter business involving CPUs (80286, 80386, Pentium, and so on) for the PC platform continued to evolve in the CISC dimension. The RISC flag, however, among several less well-known names, remained on the masts of IBM, Sun Microsystems, MIPS, ARM, and Motorola. IBM took the principle further to the supercomputer arena with the design of the famous RS/6000 family. Some of the CPUs developed for that realm, in variations on a theme, have also ended up powering IBM's networking equipment. IBM even proposed them as embedded CPUs in some network processing functions. However, ARM ended up becoming extremely successful in the 1990s as it was instrumental in establishing the RISC technology as the globally undisputed leading architecture for the implementation of main CPU components inside the upcoming *system-on-a-chip* (SOC) revolution.[42] We will talk more about this later in this book.

In embedded devices, where the volume of a projected solution allowed this approach, companies found out that by designing appropriately and by reusing available chunks of logic (sometimes very large and complicated ones), either by themselves or through third parties who were willing to license and support the developed intellectual property cores, one could patch together a whole integrated system inside a silicon die in a comparatively short amount of time. As a result, a new level of integration was created. Of course, it sounds much easier than it actually is. However, with the appropriate methodologies and a disciplined approach, it is now an undeniable fact that this new method of designing super chips is the only economically viable solution when striving for cost containment (the need to reuse components) and decreased the time to market. Until then, a certain system implementing a specific functionality would require an entire multilayer PCB with multiple CPUs, the memory of different types, and an I/O interface. Besides off-the-shelf components, it would also mean that one or more full- or semi-custom-designed chips would need to be designed. Now it finally became possible to combine the following:

- Large logic blocks called *megacells*, perhaps coming from unrelated in-house development teams.
- IP cores that were to be licensed from a third-party vendor, thereby keeping the proverbial lid over the erupting costs and gaining speed to market.

The main CPU in such a configuration is usually a RISC processor (very often but obviously not always from ARM). Other integrated modules are available that implement specific functions. One of these modules might be a powerful embedded DSP core (as offered now by several companies such

---

40. John G. Ackenhusen, *Real-Time Signal Processing; Design and Implementation of Signal Processing Systems* (Upper Saddle River, New Jersey: Prentice-Hall, 1999).

41. Lars Wanhammar, *DSP Integrated Circuits* (New York: Academic Press, 1999).

42. Stephen B. Furber, *ARM System-on-a-Chip Architecture* (Reading, Massachusetts: Addison-Wesley, 2000).

as TI, Infineon, and DSP Group) on which specialized DSP code runs along with the main SOC management/supervision software that runs on the main embedded CPU. The SOC die is completed with embedded *read-only memory* (ROM) for storing executable code, *programmable ROM* (PROM) for prototyping, flash memory for retaining something beyond the power constraints, and *random access memory* (RAM) for storing data during operation. RAM comes in various types and flavors.

The computing paradigm would then become as follows: The intended application would be partitioned into parts that would be implemented in hardware and parts whose behavior and functionality would be written in software. Special logic blocks or megacells would cover the hardware aspects. Some of them already existed in the company's logic block (cores) arsenal or would have to be developed. Some might have to be found outside the company among numerous third-party providers of IP cores. The rest of the application would be implemented in software, which would have to be running on the main embedded CPU or one of its adjacent peer CPUs or DSP inside the die. software engineers would then develop the code using high-level languages and *computer-assisted software engineering* (CASE) tools for higher productivity on traditional development stations (PCs or workstations). Cross-compilation, debugging, and linking with appropriate vendor libraries would eventually create the executable code that would be burned into ROM form. At mask preparation time, the semiconductor fab would personalize the ROM cell of the SOC with the binary executable ROMable code and the system would work (if it was properly debugged).

New methodologies and toolsets were developed for the joint co-development of hardware and software to minimize the risks of failure at silicon time (a very expensive problem).[43,44,45,46] With

For the most part, anything one desires is currently essentially available in the IP core market. With rather modest integrated systems design capabilities and with some handholding from a semiconductor manufacturer or a credible fabless design house, an SOC can be put together in a straightforward manner.

The word *fabless* has come about because these companies do not possess their own semiconductor manufacturing plant, which is known as a *fab*. Numerous fabless companies have appeared on the SOC horizon. This is changing the landscape and the industry forever since no one organization possesses the resources, skills, or specialization to come up with the optimal circuitry that implements a function.

The traditional make-or-buy debate has taken an altogether new dimension of importance in light of the shrinking product life spans, cut-throat competition, and an ever-changing market landscape where a new product becomes obsolete barely a few months after it is launched.

THE QUEST FOR BANDWIDTH AND QOS

The rapid evolution of technology for the desktop and mobile computing (PDAs and wireless handsets) has created a huge array of applications that until recently were unimaginable. These applications were developed for corporate and organizational users, as well as for casual consumers in their homes. The performance expectations are getting higher and higher, whether it is for the sales forces of companies who are able to consult and update secure corporate databases of inventories and orders in real time in front of their customers or for the excited Generation-Xer who engages in a multiuser video-game session with heavy animation involving three-dimensional graphics over the Internet. These new applications that provide exceptional local computing capabilities require additional

43. Henry Chang et. al, *Surviving the SOC Revolution—A Guide to Platform-Based Design* (Dordrecht, The Netherlands: Kluwer Academic Publishers, 1999).

44. Michael Keating and Pierre Bricaud, *Reuse Methodology Manual for System-on-a-Chip Designs* (Dordrecht, The Netherlands: Kluwer Academic Publishers, 1999).

45. Prakash Rashinkar, Peter Paterson, and Leena Singh, *System-on-a-Chip Verification—Methodology and Techniques* (Dordrecht, The Netherlands: Kluwer Academic Publishers, 2000).

46. Wayne Wolf, *Modern VLSI Design: System-on-Chip Design* (Upper Saddle River, New Jersey: Prentice-Hall, 2002).

transmission bandwidth compared to the past. This bandwidth was not previously available simply because the demand for it was not there. Applications drive the need.

In most of these new applications, the functional specification requirements for hardware and software designers are staggering for the underlying equipment. The communications landscape is no longer what it used to be: The multimedia transmission requirements in such a realm are combined with streaming audio and video, bringing in their own ergonomic levels of acceptability. In many cases, packets now cannot be lost or discarded, as it may not be possible to recover the traffic in case something inadvertent affects the transmitted bit stream.

Reconstructed voice from digitized and compressed data used to be an area where sophisticated vocoding would more than make up for the deficiencies of the transmission channel. The other party's voice might be distorted at times, but as long as it was intelligible, no one complained. In the worst case, if one party did not understand what the other party had said, the other party would just repeat what was just said. However, data is a different story. Transmitted data must arrive intact. The transmitter can resend the packet if it arrives corrupt. However, this affects the net throughput as it can be compared to the problem of taking three steps forward and then two steps backwards. So far, it had been the intelligence of the underlying protocol stacks and *forward error correction* (FEC) codes that would try to make it up for the users in case of trouble. If a user has to resort excessively to retransmitting corrupt frames or packets in order to achieve a reliable link, sooner or later the network capacity will be hampered down by redundant traffic chunks. As a result, the response time and latency as perceived by the user will be qualified at least as inadequate for several applications. This elevated the importance of also considering the QoS requirements. This time it had to be done in a thorough manner.

It goes without saying that in order to discriminate between what needs to be done on a bit stream, standard methods have had to be decided and agreed upon—namely, how to read, filter, inspect, parse, modify, store, and forward the frames and packets. The requirements for such local processing intelligence clearly point toward the need for specialized high-performance microchips for advanced and optimized architectures—the network processors, about which we will be talking in length in this book.

SWITCHING EVOLUTION: FROM LAYER 2 SWITCHES TO ROUTERS TO LAYER 3 SWITCHES

By cleverly replacing access to the shared media (for example, of the original coax cable for Ethernet) with dedicated bandwidth, switched LAN technology has greatly increased network performance. Users still have direct access to the network, but bottlenecks of shared Ethernet disappear as point-to-point switching is deployed.

Switched networks are generally flat domains that must be subnetted to alleviate broadcast overhead, spanning-tree loops, and inefficient addressing, and to provide some rudimentary security.[47,48] Standard IP network textbooks explain the concept and trade-offs of subnetting,[49–54] so we will not

47. Jayant Kadambi, Ian Crawford, and Mohan Kalkunte, *Gigabit Ethernet: Migrating to High-Bandwidth LANs* (Upper-Saddle River, New Jersey: Prentice-Hall, 1998).

48. Radia Perlman, *Interconnections: Bridges, Routers, Switches, and Internetworking Protocols*, 2nd ed. (Reading, Massachusetts: Addison-Wesley, 1999).

49. Douglas Comer, *Internetworking with TCP/IP, Vol. I: Principles, Protocols, and Architecture* (Upper Saddle River, New Jersey: Prentice-Hall, 1998).

50. _____ , *Internetworking with TCP/IP, Vol. II: ANSI C Version: Design, Implementation, and Internals* (Upper Saddle River, New Jersey: Prentice-Hall, 2000).

51. _____ , *Internetworking with TCP/IP, Vol. III: Client-Server Programming and Applications—Windows Sockets Version* (Upper Saddle River, New Jersey: Prentice-Hall, 1997).

52. Douglas E. Comer, David L. Stevens, Marshall T. Rose, and Michael Evangelista, *Internetworking with TCP/IP, Vol. III: Client-Server Programming, and Applications—Linux/Posix Sockets Version* (Upper Saddle River, New Jersey: Prentice-Hall, 2000).

53. W. Richard Stevens, *TCP/IP Illustrated: Volume 1, The Protocols* (Reading, Massachusetts: Addison-Wesley, 1994).

54. Gary R. Wright and W. Richard Stevens, *TCP/IP Illustrated: Volume 2, The Implementation* (Reading, Massachusetts: Addison-Wesley, 1995).

expand on it here. The important point to remember is that without subnetting, switched networks and LANs do not scale well. This issue was the fundamental reason routers were brought in during the 1980s to the switched networks to take connectivity beyond bridges and switches. Routing is an important function, but it remains a fact that typical routers installed in a LAN setting (for example, on a campus backbone) can handle around half a million packets per second. The high-performance LAN switches (serving the desktops) can produce millions of packets per second feeding the backbone, which can find itself incapable of handling the aggregate throughput. Routers are also expensive and relatively tedious to manage and configure compared to switches. Therefore, it has turned out that deploying a mix of switches and routers for local connectivity is not a wise solution. This is exactly where layer 3 switching came into play.

As documented in Radia Perlman's book *Interconnections: Bridges, Routers, Switches, and Internetworking Protocols*[55] and Kadambi, Crawford, and Kalkunte's book *Gigabit Ethernet: Migrating to High-Bandwidth LANs*,[56] switching is an inherently cheaper process than routing. It also removes the scalability and throughput restrictions that limit a network's growth. In March 1996, Ipsilon (which later became part of Nokia) introduced a technique for switching at the third layer called *IP switching*. The technique enabled the high-speed forwarding of IP packets onto underlying ATM networks. It claimed to be much less complicated than MPOA, which had been introduced by the ATM Forum.

About six months after that, Cisco introduced its *tag switching* approach, while IBM announced its *aggregate-route-based IP switching* (ARIS) technology and Toshiba launched its *cell-switched router* (CSR). The debate among these major vendors soon led to the formation of the MPLS working group at the IETF, which consolidated discussions and guided the industry into several new standards. These standards are referred to generically as *MPLS*.

These layer 3 switching techniques enable the introduction of many new interesting services. *Virtual LANs* (VLANs) and full-fledged *virtual private networks* (VPNs) became feasible.[57] *Traffic engineering* (TE), QoS, and the level of priorities are some of the issues that network equipment manufacturers can address while tailoring their offerings to their customers at easily justifiable costs.

MPLS, LAMBDA SWITCHING, AND WAVELENGTH ROUTERS

Traditional mesh-connected routing networks require any-to-any connectivity between all routers. This leads to the need for $n \times (n-1)/2$ virtual connections, for example, on an ATM network with n nodes. This obviously means that if a new router must be added, a virtual connection will be mandated with all the other routers. That is a problem.

Beyond this shortcoming, a network failure or topology change will provoke a massive amount of traffic that was generated by a routing protocol. Each router will have to communicate routing updates across each virtual connection to which it is connected in order to inform its neighbors about the new IP network reachability situation.

As if these problems were not enough, let us, for a moment, think about the following situation. A typical ISP network contains multiple routers at the edge of the ISP's network that have peer relationships with other ISP routers with which they exchange routing table information to provide global IP connectivity. In order to find the optimal path to any destination outside an ISP's network, the routers at the core of the ISP network must be made aware of all the network reachability information. Routers at the edge of the network can acquire this knowledge from the adjacent routers (which are outside the ISP network) that they are peering with. The result of this uncomfortable situation is

55. Radia Perlman, *Interconnections: Bridges, Routers, Switches, and Internetworking Protocols*, 2nd ed. (Reading, Massachusetts: Addison-Wesley, 1999).

56. Jayant Kadambi, Ian Crawford, and Mohan Kalkunte, *Gigabit Ethernet: Migrating to High-Bandwidth LANs* (Upper-Saddle River, New Jersey: Prentice-Hall, 1998).

57. Marina Smith, *Virtual LANs: Construction, Operation, and Utilization* (New York: McGraw-Hill, 1998).

that all the core routers of the ISP network must possess and maintain the entire Internet routing table, which requires an enormous amount of memory and leads to a very high degree of CPU utilization.

The MPLS standard introduced a fundamentally new approach in the deployment of IP networks. The control mechanism was supposed to be separate from the forwarding mechanism and the concept of label was supposed to be introduced for packet forwarding. MPLS can be deployed on router-only networks or in ATM environments that integrate the layer 2 and layer 3 infrastructures into one single consolidated IP+ATM network.[58,59,60]

An MPLS network has *label-switched routers* (LSRs) in the core of the provider's network and *edge label-switched routers* (Edge-LSRs) at the periphery of the provider's network. Within the MPLS network, traffic is forwarded using labels. The Edge-LSRs at the ingress side of the MPLS cloud (the MPLS network point from where an incoming packet is entering) assign the appropriate label to each packet and forward the packets onto their next-hop LSR along the path that the traffic has to follow in order to go through the MPLS cloud. The label's value is actually a pointer used by all LSRs on a table that points to the next hop and a new label. At each LSR, the old label is exchanged with a new one and the packet is forwarded onto the next hop. At the egress side of the MPLS cloud (the MPLS network point from where the forwarded packet must exit the MPLS network), the last LSR on the path will remove the label altogether and traffic will be forwarded using traditional IP-routing protocol mechanisms.

MPLS networks also use the concept of *Forwarding Equivalency Class* (FEC), which is a group of packets sharing the same attributes while traveling through the MPLS cloud. For example, these attributes can be the same destination address, some indication of QoS, or the identification of a specific VPN. All packets belonging to the same FEC receive the same label from the LSR. Different protocols such as the *Label Distribution Protocol* (LDP) exist that enable the LSRs to exchange the information that associates FECs with labels. The MPLS architecture enables carriers and service providers to offer new services, such as VPNs and *service-level agreements* (SLAs) with their customers, based on the sophisticated TE functions.

The TE-related MPLS-TE capabilities are important in order to understand the concept of the *Multiprotocol Lambda Switching* (MPLmS) architecture, which is being developed to provide dynamic wavelength provisioning in the optical transport network that starts to take shape as part of the converged network. In addition, when the new optical networks are implemented, wavelength routers are used. These routers are made up of wavelength switching cross-connect matrix fabric providing optical interfaces. Depending on the technology used for the switching backplane, the routers can be electrical wavelength routers, hybrid wavelength routers, and optical wavelength routers.[61,62] Electrical wavelength routers are usually deployed; however, hybrid wavelength routers are now appearing as a transition technology. All-optical wavelength routing seems to be the trend of the future technology, but many features and characteristics must still be researched and improved before this technology gains market acceptance.

VPNS

Even before intranets were invented, many corporations and organizations had already pushed the state-of-the-art connectivity toward VPNs. The need stemmed originally from a traditional precaution and demand for solid business privacy. However, it has evolved since the 1990s with the arrival of

58. Uyless Black, *Multiprotocol Label Switching* (Upper-Saddle River, New Jersey: Prentice-Hall, 2000).

59. Bruce Davie and Yakov Rekhter, *MPLS: Technology and Applications* (San Francisco: Morgan Kaufmann Publishers, 2000).

60. Peter Tomsu and Gerhard Wieser, *MPLS-Based VPNs: Designing Advanced Virtual Networks* (Upper Saddle River, New Jersey: Prentice-Hall, 2002).

61. Peter Tomsu and Christian Schmutzer, *Next Generation Optical Networks: The Convergence of IP Intelligence and Optical Technologies* (Upper-Saddle River, New Jersey: Prentice-Hall, 2002).

62. Uyless D. Black, *Optical Networks: Third Generation Transport Systems* (Upper-Saddle River, New Jersey: Prentice-Hall, 2002).

sophisticated hacking techniques and well-publicized cyberattacks employed by malicious intruders or eavesdroppers. Today intranets refer to closed-access private networks or networks designed to be inaccessible to unauthorized outsiders. In the 1980s, when corporations would lease X.25 lines from carriers, it was widely believed that these lines ensured that no other traffic could run on those lines simultaneously. Numerous cases (not in the United States, but almost invariably overseas) proved the contrary. Other people's traffic could run on the same physical lines and bandwidth slots that some-one else was paying for.

In the mid-1990s, secure communications companies designed layer 2 frame encryptors, which ensured that secure tunnels were created between equivalently equipped sites, regardless of the type or ownership of the public network between the sites (for example, X.25 or frame relay). Soon the effort was expanded to layer 3 devices, which would offer the same functionality on IP and/or IPX networks. These were the first true VPNs in the sense that communications were secure from eaves-droppers with access to the public network. The presence of these virtual tunnels ensured that traffic encrypted on-the-fly at the transmitting site was only going to be decrypted (again on-the-fly) by a similar piece of equipment upon arriving at the destination site. This intention for a sense of privacy, despite the fact that traffic was transmitted over the public and insecure network, was the basis for the name VPN.

In the second half of the 1990s, with the IETF's help, the *IP Security* (IPsec) consortium estab-lished similar types of VPN communications security at the network layer (layer 3) using strong encryption, tunneling, and potential encapsulation and authentication. IPsec became a standard set of techniques that had the noble goal of allowing secure intercommunication between pieces of equip-ment of different vendors.[63,64,65,66] IPsec intercompatability, of course, did not happen overnight, but it was gaining momentum and making progress. IPsec is a computationally very demanding environ-ment, especially if longer encryption key sizes are used. If it is executed on a main CPU of available systems, it can also tax the system's performance significantly or possibly bring the system to a com-plete halt, depending on the communication applications and their frequency of use. IPsec was orig-inally implemented in software for low-speed applications or where it made business sense, such as in a first-generation firewall. It was also implemented in hardware on acceleration systems that took the forms of plug-in adapter boards. Currently, it is becoming available in special security co-proces-sor chips, as we will see in the next section and in more detail later in this book. IPsec-compliant routers, IPsec-compliant firewalls, and IPsec-compliant switches are now available.

Although VPNs still have the same underlying principle of a certain degree of communications security, they acquired a different dimension altogether with the arrival of the layer 3 switching tech-niques. It especially changed after the concerted consolidation of major rival approaches from Ipsilon/Nokia (IP switching), Cisco (tag switching), IBM (ARIS), and Toshiba (CSR) into MPLS. MPLS-enabled carriers, as a direct result of the technology they possess, are able to offer VPNs as one of several value-added services they can provide to their customers.

SECURITY CO-PROCESSORS

In the mid-1990s, router and switch manufacturers realized that security was important. The competi-tion provided by traditional security companies with previous experience serving the military and intel-ligence markets was too intense. Network equipment manufacturers understood that they had to offer security inside their products or the solidity of their base would erode. The trend started as security

63. Naganand Doraswamy and Dan Harkins, *IPsec: The New Security Standard for the Internet, Intranets, and Virtual Private Networks* (Upper Saddle River, New Jersey: Prentice-Hall, 1999).

64. Elizabeth Kaufman and Andrew Neuman, *Implementing IPsec: Making Security Work on VPNs, Intranets, and Extranets* (New York: John Wiley, 1999).

65. Carlton Davis, *IPsec: Securing VPNs*, (New York: McGraw-Hill Professional Publishing, 2001).

66. Peter Loshin, *Big Book of IPsec RFCs: Internet Security Architecture* (San Francisco: Morgan Kaufmann Publishers, 1999).

software (encryption, authentication, firewall services, and so on) running on the main CPU of the router/switch. Given the performance penalty that such a piece of equipment would pay in a commensurate loss of switching capacity, they soon realized that hardware acceleration engines were required.

Alliances were formed between some security companies and some network gear vendors to initiate designs. In some cases, the network equipment manufacturers set up new specialized engineering teams to design their own in-house-developed add-on acceleration boards or *application-specific integrated circuits* (ASICs) in order to handle the heavy-duty mathematical processing required for encryption and authentication, which was to be the mandate of the *security co-processors*.

These are chips and/or sometimes whole subsystems that can handle predominantly cryptographic functionality quickly, something that ordinary CPUs were never designed to handle efficiently. With the arrival of the IPsec specifications and publications from the IETF, vendors started implementing IPsec first in software and then in hardware. The faster the network equipment became, the more programmers had to consider how to generate cryptographic keys and digital signatures as well as how to encapsulate traffic into new types of packets that provided at least the sense, if not the impression, of a secure tunnel.

Security co-processors are another relative in the family tree of network processors. We will discuss these in detail later in the book in Chapter 17.

TRAFFIC ENGINEERING (TE)

Another direct result of the introduction and acceptance of MPLS is the set of capabilities that it offers for TE. TE is geared toward decreasing the cost of network operations for carriers and service providers by enabling them to more efficiently allocate and manage the use of bandwidth resources. This prevents undesirable situations where some parts of the network are congested while other parts remain underutilized. Special intelligence and adequate processing speed are required to ensure the dynamic adaptation of the network to changing traffic patterns and loads. For instance, under these premises, the following would be required:

- The capability of fast rerouting.
- The possibility of calculating alternative routes.
- The facility of presignaling these new backup-plan routes, so that they can transparently pick up the workload from operating tunnels that are suddenly less efficient.

These capabilities directly increase the resilience and survivability of the network, while they indirectly improve its scalability. Currently, MPLS networks provide very powerful TE capabilities. This adds to the functionality and performance requirements of the router circuitry.

QoS

Customers are no longer interested in signing up with carriers or service providers for a number of communications channels at some aggregate data bit rate. Several applications that are tightly related with the customer's organizational needs require different levels of service, response time, bandwidth, delay, jitter, cost, and so on. Customers are not willing to pay the same rate for all their needs. New business models have been developed that bill the customers based on what they actually use and what the content is.

Service providers must now be able to treat different services that they provide with different criteria, which must be made to apply optimally to the customer's diverse requirements. In other words, not all bits are to be treated in the same way. Several new protocols appear from various standardization bodies, such as *Differentiated Services* (DiffServ), *Integrated Services* (IntServ), and RSVP.

Previously, frame relay and ATM treated issues of QoS at layer 2, whereas protocols such as IntServ and DiffServ now treat QoS concerns at layer 3.[67] As QoS ultimately becomes an end-to-end issue, the attention to it must eventually encompass all providers in a transparent way for the users.

Early generations of switching and routing equipment used simple *first-in first-out* (FIFO) queues for all traffic indiscriminately. When the forwarding rate exceeded the capabilities of these first routers and switches due to heavy traffic, packets would be dropped and the link reliability challenge was unloaded onto the shoulders of protocols that operated higher on the stack and that were more or less able to provide some recovery (for example, TCP). However, not all applications can afford this type of tinkering anymore. Services are now being differentiated according to their content.

Therefore, packets must be inspected in real time with special bits flagging higher- to lower-priority traffic. Lookup tables must be consulted to match services with forwarding policies as dictated in a *service level agreement* (SLA). This also requires fast and intelligent processing that goes beyond what a typical fast CPU can do. This is yet another angle from which we can look at the area of network processors.

PERFORMANCE CONSTRAINTS IMPOSED ON COMMUNICATIONS NETWORK EQUIPMENT

Figure 1.1 provides the historical overview of the last 30 years as it pertains to the evolution of computing and communications networks. From the top moving clockwise, the figure shows the evolving loop of applications requiring sophisticated software. This feeds a more complex hardware evolution that could justify more advanced software and so on. The two downward-pointing arrows show the effect that the progress of semiconductor technology in conjunction with networking and software technology advances has had. It is rather striking that both arrows converge on the need for higher capability in network equipment—thus creating the need for sophisticated network processors.

It must have become clear by now (at least qualitatively) why network processors are needed. In this introduction, we have seen how and why the sheer quantity of network-related data processing has rapidly evolved to unprecedented levels of sophistication and complexity from many angles. Today's network equipment must be able to parse packets, search lookup tables that document policies, resolve conflicting operations that seem necessary, potentially modify the packet's content by adding or removing bits, possibly encrypt payload and authenticate the other party, generate digital signatures and verify other parties' signatures, create secure tunnels (for example, IPsec stipulates building the so-called *Authentication Header* [AH] and *Encapsulating Security Payload* [ESP] headers), encapsulate traffic into tunnels, and engage to modular arithmetic (indispensable for encryption operations) and, more specifically, to modular multiplication and exponentiation.

The list of tasks for the hardware inside network equipment can go on and on. Ordinary CPUs simply cannot handle these tasks for many reasons, including software complexity, system throughput, and operations latency. Special hardware based on optimized architectures is needed together with the availability of cutting-edge development tools, which will help shrink the time-to-market nightmare that companies confront. In short, new types of advanced microchips are required for the timely and efficient handling of these requirements. These are called network processors, which are the subject of our study in the rest of this book.

SUMMARY

In this chapter, we provided a very short historical and qualitative overview of the evolution of computing and networking communications technologies over the last 30 years. This evolution has largely

---

67. Y. Berner et al. "A Framework for Differentiated Services," *draft-ietf-difserv-framework-02-.txt*, February 1999.

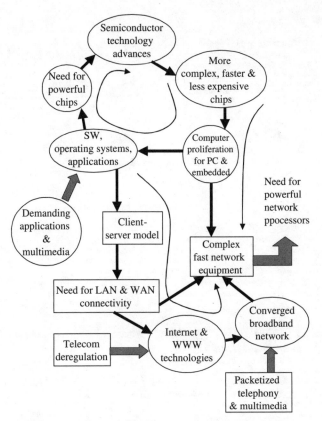

FIGURE 1.1   An overview of the historic evolution process in the computing and networking industries over the last 30 years. One can easily see the
major external and internal factors that converge to create the need for sophisticated and powerful network processors in the future.

been made possible thanks to the spectacular progress that we have witnessed over the same period
of time in semiconductor engineering and the commensurate advancement in operating systems and
software technologies. The insatiable demand for communications bandwidth and fast response time
in a real-time setting that many new applications require are at the heart of the unprecedented network
growth of the last several years. The convergence of these growing networks for the very high-speed
transmission of voice, multimedia, and data, coupled with the deregulation of telecommunications in
many parts of the world and the global de facto acceptance of packet-based technologies, requires a
new breed of extremely fast and efficient semiconductor devices. These devices, which are the basis
of network switching/routing equipment, will process the expected fast and at times very heavy traffic without compromising on the QoS expectations of network users. This new generation of advanced
microchips is now known under the generic name of *network processors*.

  In the next chapters, we will look under the hood of network processors. We will learn what they
are doing well, how they go about doing it, and why their performance is so superior to other architectural paradigms. We will also look at what differentiates each different approach taken by major
design houses and semiconductor manufacturers, and we will look at the trade-offs and position of
various predominant architectures.

CHAPTER 2
NETWORK PROCESSORS: JUSTIFICATION

In the previous chapter, we learned how the evolving landscapes in the computer, communications, and semiconductor industries have created a revolution that is built around the insatiable demand by users for global connectivity, applications portability, and user mobility. Users want technology that can be accessible anytime and anywhere. We also saw how these new demands translate into a convergence of the telephony and data networks with the Internet. Ever-increasing requirements for decreased costs, enhanced network performance and availability, and a new market framework where notions such as broadband speed, *quality of service* (QoS), and pay per use are now more important.

This remarkably rapid evolution has caused the arrival of *network processors*. We will elaborate on this evolution and explain why it occurred. First we will define and categorize network processors. Based on their classification, we will describe what functions they are able to perform, in which context, and why. Most importantly, we will explain the unique advantage that network processors bring to both the user and the developer communities. We will elaborate on their privileged cost/performance/flexibility positioning with respect to alternative design approaches—for example, architectures that rely heavily on the more traditional use of *application-specific integrated circuits* (ASICs) or *reduced instruction set computer/complex instruction set computer* (RISC/CISC) computing platforms to accomplish the same functions. By understanding the context in which network processors are revolutionizing the networking and communications industries, the reader will be properly equipped to tackle the fundamental technologies and internal technical intricacies that make up the heart and brain of the network processor microchips and the systems they enable.

WHAT ARE NETWORK PROCESSORS?

Network processors, also known in the industry and product literature of several vendors as *network processor units* (NPUs), are highly programmable specialized integrated circuits (processors). These circuits are used in the high-speed communications industry. They are used to optimize the performance of packet processing in the evolving functional framework of broadband network equipment.

Because of the unmistakable convergence of networks that we briefly discussed in the previous chapter, packet processing becomes the overriding function that is expected to be properly implemented in high-speed networking equipment such as routers, switches, and so on. Obtaining the appropriate performance and functionality in network devices is one of the key factors for determining the usefulness, desirability, and business potential of these devices within the corporate or service provider markets.

Throughout this book, we use the term *packet* in a general sense to describe a datagram unit, meaning either a cell, packet, or frame. If we are discussing something very network specific, such as the *Asynchronous Transfer Mode* (ATM) environment, we will call the datagram unit a cell.

FUNCTIONAL BLOCKS IN NETWORKING EQUIPMENT

Before we can examine alternative ways of implementing different packet-processing functions, let us take a look at the conceptual partitioning inside networking devices, as shown in Figure 2.1.

Functionality can be divided by the following four major blocks:

- The *physical layer* (PHY) interface.
- Switch fabric.
- Packet processing.
- Host processing.

The PHY Interface

The PHY interface is the first conceptual layer of functionality. It is currently compacted into one integrated component, such as a PHY chip. It is responsible for transmitting and receiving information. The bitstream, which must be *transmitted* by a networking device as part of being on a network, needs to be modified from its digital binary form into an analog form that can be efficiently transmitted over the communications channel medium. This must be done whether the medium is modulated light injected onto an optical fiber, an electrical current traversing a coax cable, or an electromagnetic wave radiated over the air. Similarly, in order to be *received*, the arriving light, electric current, or electromagnetic wave must have its content transformed from its analog transmission form (even when it carries digital information) back into a binary digital form that the rest of the receiver's logic can handle properly.

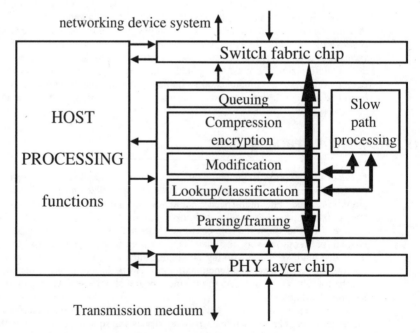

FIGURE 2.1   The conceptual functional partitioning of a network device.

The PHY interface chip is the component at the edge of the networking device closest to the physical medium and the bidirectional handling of traffic. PHY chips are designed for different transmission media. For example, 100Base-T networks, Gigabit Ethernet, and *Asymmetric Digital Subscriber Line* (ADSL) are some types of media that use PHY chips. Companies that offer PHY chips include Agere, Alcatel, AMCC, Broadcom, Conexant, Fujitsu, and IBM.

Switch Fabric

The networking device is physically structured in several ways. The most straightforward and modular way is to use either a bus or a backplane into which adapter modules or line cards are inserted. The switch fabric is a functional module that reads packets at an input (also known as the *ingress point*) and routes them to an appropriate output (also known as the *egress point*). The current switch fabric function is usually offered in a highly integrated standard off-the-shelf chip set, as proposed by several vendors, such as Agere, IBM, Vitesse, and Zetacom. Its speed is the most critical factor for defining the switching capacity of a network device. As an alternative, the designer/manufacturer of the network device sometimes proposes an in-house custom-designed *very large scale integration* (VLSI) chip that implements a tailor-made switch fabric implementation.

Packet Processing

The overall packet-processing set of operations is positioned between the PHY interface and the switch fabric. The industry usually categorizes these operations into two operation groups or two *processing paths*: a *fast* packet-processing path and a *slow* packet-processing path. A fast path refers to a data path that handles all operations that are executed in real time directly on a packet. These include, but are not limited to, the five fundamental operations of framing/parsing, classification, modification, compression/encryption, and queuing. A slow path refers to the required operations that are executed independently of the actual flow of packets. Some examples of slow-path operations are unknown address resolutions, new route calculations, updates of routing and forwarding tables.

Figure 2.1 serves more to clarify the structural context of the network-processing-based computing than to be a precise and rigid template of the sequence of events. For example, an external security co-processor is sometimes used to encrypt and authenticate packets. In that case, it may be advantageous in some applications to reverse the order of some operations and perform the modification function *after* the queuing function in the foreseen pipeline of events. This would allow some packets to be marked differently based on the congestion level that they encountered during queuing. It also facilitates a higher-performance multicast implementation where multicast packets/cells only need to be buffered once while being able to be read out multiple times, modifying each copy after it is retrieved from the packet buffer. This example reiterates why Figure 2.1 should be seen more as a generic representation of network computing and less as a necessarily fixed topology.

Host Processing

The term *host processing* refers to a number of generic processing tasks that do not reside on the flow path of the network packets. As a result, they are usually allocated to some *central processing unit* (CPU) that does not handle packets directly. Host-processing chores include implementing network management routines, configuring devices, running diagnostics, and managing internal communications between functional modules or subsystems of the network device. Host processing is usually implemented in software that runs on standard off-the-shelf RISC CPU processor chips such as IBM's PowerPC and various MIPS CPUs. In a few cases, network equipment manufacturers have chosen to implement their products' host processing on more common CISC processors, such as an Intel Pentium processor.

A CLOSER LOOK AT PACKET PROCESSING

Until network processors appeared, system architects had to choose between two ways of tackling the overall design of the packet-processing module in order to implement network devices.

One approach would entail using a standard off-the-shelf CPU, which would usually be a RISC processor. This choice is similar to choosing the CPU for the host-processing part of the design partition that we just discussed. However, in some cases, especially in network devices intended for low-end devices such as a small *wide area network* (WAN) router for the *small office/home office* (SOHO) market, it could also be a CISC processor. In fact, in some of these low-end cases, the packet-processing function and the host-processing function end up using the computational power of the same CPU chip on a time-sharing basis with the help of a real-time operating system kernel.

The other approach would be to design a specialized high-performance ASIC that would handle packet processing. Because most network device design houses are fabless companies, the designer would have to hand off the custom design of the ASIC to a semiconductor house (fab) to have it built. Of course, some major semiconductor powerhouses such as IBM and Intel happen to be both designers of networking chips and manufacturers of integrated circuits. Therefore, these vendors obviously enjoy a vertical integration that offers them a more robust market advantage.

However, this advantage is economically significant and sustainable only when the vertically integrated vendor has already been enjoying high levels of semiconductor manufacturing business that would allow the corresponding *complementary metal oxide semiconductor* (CMOS) processes at hand to achieve parity or overtake the economics of large silicon foundries such as TSMC. The availability of an in-house foundry is therefore not enough. The foundry must be already almost fully exploited from a capacity usage standpoint in order for this to be a real economic advantage.

TRADE-OFFS WHEN DESIGNING WITH STANDARD OFF-THE-SHELF CPUS

Packet processing would usually be implemented in software that runs on a standard off-the-shelf CPU because of the ease and flexibility with which such a CPU can be programmed. To obtain new functionality, a new software version with the appropriate additions or modifications is needed. Software can be easily downloaded into a system with the corresponding memory architecture (*read-only memory* [ROM], *erasable programmable read-only memory* [EPROM], flash, and so on). Bugs can also be easily fixed. When an entirely new functionality is required, implementing it is straightforward. The time needed to accomplish this kind of change is usually short, and this flexibility translates into a significant business advantage for the device vendor. This is also advantageous for the user, as the user does not need to invest in new hardware to obtain some enhanced or corrected functionality. From the user's perspective, an existing network device can be upgraded easily (many times directly on site) by updating its software, which costs much less than new hardware.

The downside to this approach is a decrease in performance, as off-the-shelf CPUs are designed for a general computing environment. Generally speaking, they will spend many clock cycles on tasks that are not directly related to packet processing; therefore, the percentage of their processing capacity that is used directly for packet processing is only going to be a small fraction of the network's requirements. For instance, the fastest off-the-shelf CPUs can currently only handle a throughput of around a couple of hundred megabits per second. This is far less than today's backbone networks, whose minimum requirement is easily tens of gigabits per second.

Figure 2.2 compares the growth of bandwidth requirements as dictated by Internet backbone connections (in megabits per second) and of the typical computational performance of off-the-shelf CPUs (in MIPS).

The demand for bandwidth is a combination of two unrelated events. On one side, web-based connectivity, e-commerce, Internet telephony, and multimedia on demand are combining with the deregulation of the traditional telecom access network and the arrival of many new players in the market.

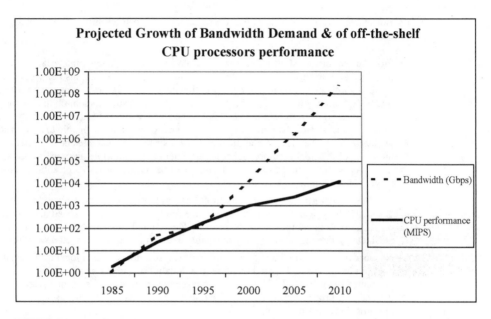

FIGURE 2.2   The historical and projected growth in bandwidth demand (as witnessed at the backbone of the Internet) and in computational power of typical off-the-shelf RISC CPU processors. *(Sources: Telstra and InStat/MDR, respectively)*

On the other side, a plethora of technological breakthroughs (such as *Digital Subscriber Line* [DSL], *dense wavelength division multiplexing* [DWDM], and broadband wireless local loop) enable the spread of faster connectivity to the converged network backbone. The ability to offer these types of services over large geographies and markets is becoming a matter of competitive edge and even survival for many service providers. The widening gap between the two curves in Figure 2.2 as time progresses is absolutely astounding. It shows that the functional requirements of evolving networking devices will simply not be able to be serviced by the expected evolution of CPU processors, where the progress of semiconductor capabilities has been more or less accurately predicted and charted for the last 20 years.

Another unrelated factor that contributes dramatically to the exhaustion of the computational capabilities of off-the-shelf standard CPUs in a networking environment is that we are witnessing a rapid move of the processing function upwards in the protocol stacks. Barely six to seven years ago, networking still used layer 2 processing. With the evolution of *Internet Protocol* (IP) and *Multiprotocol Label Switching* (MPLS), network computing started to involve layer 3 calculations. The recent trend has now climbed even higher by seeking, capturing, and exploiting information from the transport to the application layers (layers 4 through 7). In order to accomplish such a feat, network devices must be able to look deep inside packets to scan, parse, recognize, and extract features that reveal information from each packet about specific levels of QoS, service levels the user contractually has bought, or load balancing based on *uniform resource locators* (URLs). This type of intensive and intelligent packet processing implies that more bits per packet need to be examined and handled than before. It is estimated that a single standard CPU processor chip is incapable of performing deep-packet processing all the way to the application layer in real time faster than a couple of hundred megabits per second.

Beyond the shortcoming of off-the-shelf CPUs when it comes to network processing, one should also not underestimate the fact that processors that perform packet processing usually suffer from a serious memory bottleneck as well as from a suboptimal instruction set. The following bullets explain these factors:

- First, current off-the-shelf processors are clocked at rates of a few gigahertz. Due to their typically pipelined architecture, they are able to perform billions of instructions per second, thereby almost achieving the rate of executing one instruction per clock cycle. However, data must be fetched continuously from memory so that the processor can work at any moment on the instruction at hand. It also produces data from operations. This data needs to be stored back in memory before new instructions are tackled. However, memory read and write operations are unable to sustain activity at these gigahertz rates. Therefore, elaborate memory subsystem designs must be devised using a multilevel hierarchy of different memory technologies, interleaving multiple memory banks, and synchronizing memory pages and bus access—techniques that usually lead to a prohibitive cost and levels of power consumption.

  The lack of performance that results from this structural deficiency is an architectural paradox. On one hand, the typical processor pipeline stages end up being in high-speed networking applications often empty (a phenomenon called *pipeline bubble*) and consequently underutilized. On the other hand, the system remains squarely incapable of dealing with the expected workload.

- Network traffic obeying completely different statistics models than local traffic on a computer bus does not have the same spatial and temporal locality properties as regular desktop or client-server IT application workloads. The result is that the typical processor's cache systems are not effective in a network-processing environment. Without the benefits of their caches, conventional CPUs simply slow down to a proverbial crawl.

- The instructions that are needed to handle and modify live packets in a network-processing environment require specific bit-level operations that must be carried out at wire speed and they are not available as standard instructions in off-the-shelf processors. As ordinary off-the-shelf CPUs will need more than one of their standard instructions assembled into a microprogram that performs the intended functionality, these microprograms are executed over multiple clock cycles, stalling the pipeline and taking up time. This further negates the off-the-shelf CPU's capability to cope with the computational load associated with very high-speed traffic arriving and leaving in real time. This example illustrates the inadequacy of the instruction set associated with off-the-shelf CPUs. We will discuss this problem and its ramifications as well as ways to address it from a system designer's point of view in much depth in this book.

Some vendors have looked to allocate the necessary work to more than just one such CPU processor. However, in addition to the astronomical and direct increased cost of hardware with such an approach, the sheer complexity and cost (in time and money) of developing the intercommunicating multiprocessor real-time software that is needed to manage such a system efficiently, should not be underestimated.

It becomes clear that the use of off-the-shelf CPUs is not the solution to the problem and that something radically different is needed.

TRADE-OFFS WHEN DESIGNING WITH ASICS

Designers opt for the ASIC approach when the application requires the maximum performance. Typically, an ASIC in this environment delivers better performance than a typical off-the-shelf CPU with the same capacity properly programmed to handle the same packet-processing application. This performance edge has been instrumental in the wide-scale adoption of the ASIC approach in the high-speed network equipment design community. ASICs implementing efficient architectures can also be designed to operate extremely fast. As a result, they can certainly provide one path of evolution toward the ongoing satisfaction of the ever-evolving needs in the networking industry.

However, two negative factors of the design of ASICs in network gear should be considered:

- ASICs suffer from limited, if any, programmability, which causes them to be a rigid solution delivery platform. When new functionality is required, or when new protocols must be supported, the vendor does not have many other options than to drop the evolution of the product or to redesign the ASIC—a costly proposition for both the vendor and user. It is costly for the vendor because of the design and time. It is costly for the user because in order to benefit from the new functionality, the equipment must be upgraded. In the worst case, a user would have to buy a new system altogether. In the best case, a user would have to buy a new adapter (for example, a line card) with the new ASIC in order to replace the older one. This type of continuous quasi-forced upgrade is highly undesirable to the user community. In the long run, it hurts the relationship between the customer and the vendor.

    Seen from a different angle, the same lack of programmability is a serious impediment for ASICs. Consider the amount of protocols and data formats that are encountered at the different layers of a typical protocol stack. The higher a user goes on the stack, the more protocols a user is bound to run into. The device all of a sudden loses flexibility, despite the fact that it has improved in performance with the inclusion of specially designed ASICs.

- Designing a sophisticated ASIC requires more time than can be afforded. This is probably the most important downside of this approach, because the type of ASICs needed in high-speed networking devices usually requires a design cycle of somewhere between 12 and 18 months. Although the ASIC design process is now extremely well understood by many engineering organizations, it remains a fact of life that the process is not sensitive to ongoing changes. An organization starts by deciding on and fully specifying the ASIC functionality, and the engineers proceed with its implementation. Roughly 18 months down the road, a working product will come out of the production line and will hopefully operate as specified.

    What about the case where something must be either added to or modified in the originally specified ASIC functionality? In this case, the answer is easy because the vendor has to stop the design work and restart the development work to avoid wasting precious time and money. The exact point of retreat obviously depends on the individual case. A user may have to go back to the hardware coding language source level (VHDL or Verilog), carry out the needed modifications, and resynthesize the hardware encoding against the underlying technology library. A user may also have to go back and recode the whole design.

    Sometimes the extent of the work is so significant that it is easier to recode from scratch than it is to revamp obsolete or incorrect code. The specification disruption can be so significant that the design has to start again from the beginning, incurring extra cost and time to market. For example, say marketing did not fully understand what the market was looking for. In that case, all the interim work of scripting, synthesizing, verifying, simulating, documenting, and so on has probably vaporized. The precious time to market has been lost and the money needed for that work has essentially been thrown away, straining budgets and strangling metrics of profitability or *return on investment* (ROI). This is an unfortunate but all too real side effect of the ASIC design process, and its importance should not be underestimated.

    Many product line managers have lost their jobs because they had to go to their boss one day to announce the following:

    - There will be an extra *n*-month delay until the product actually hits the market due to new design requirements. This means that competitor A or B will be the first to acquire and build market share.

    - There will be a significant (often unbudgeted) extra cost that must be incurred due to all the wasted development work so far.

    - The reason for all this is that the product's content and functionality were not properly specified in the first place. For the boss, this usually means the market research work was not done thoroughly, and it is, of course, the product manager's fault.

Product lifetimes have shrunk dramatically. Launches of new products with enhanced features every six months make other recently launched products obsolete. The industry has become extremely competitive. This year's star is next year's casualty. These factors have created a cutthroat environment where the time to market is extremely important. ASICs do not fare well in this regard.

THE NETWORK PROCESSORS' BREAKTHROUGH

The reader must have realized that the argument so far between the two schools of thoughts, namely, the one favoring designs around programmable off-the-shelf CPUs and the one favoring high-performance ASICs inside network devices, is a classic engineering discussion about the trade-off between flexibility and performance. Engineers are trained to recognize these dilemmas early on. They manipulate the conceptual plane of variables by making appropriate design choices and compromises between conflicting requirements until they find the optimal combination of technologies that enable them to design and deliver products that meet performance and cost expectations.

Network processors have now entered the stage as the proposed solution to this debate. Network processors are state-of-the-art semiconductor chips that offer a powerful programmability similar to traditional off-the-shelf CPUs but with a performance level that approaches that of ASICs for packet-processing applications. By adopting network processors in their designs, network equipment manufacturers can obtain the sought-after high performance, while retaining their system flexibility and decreasing their development cycle.

So how do network processors do this? As we will find out when we examine various representative architectures later in the book, network processors provide specialized circuitry and appropriate architectural structures coupled with fine-tuned low-level instruction sets that coordinate a highly optimized performance for packet-processing functions compared to that expected from off-the-shelf CPUs. Network processors contain microengines that are wired to perform all the generic packet-processing functions exceedingly well at wire speed. In addition, they also usually embed a major programmable module, usually a tailor-made RISC CPUs (and sometimes more than just one) that allows the execution of real-time operating systems, handshake communications with other parts of a larger network device, and so on.

THE VALUE PROPOSITION OF NETWORK PROCESSORS

Based on what we have said, the benefits of adopting network processors in the new designs are as follows:

- **Shorter time to market** Instead of the 18 months it takes to design an ASIC, a vendor using a network processor platform can realistically expect to complete the development cycle of the packet-processing part of a major network device product within 6 months. Of course, a whole system development project will need more than 6 months to be developed. The actual length of time depends on the nature of the system and the engineering resources invested in tackling it. The network-processor-based product's performance will most likely be almost as fast as that of an ASIC-based one, while the programmability of the network processor will allow the flexibility of offering new features in the field without penalizing the customer.

- **Longer time in market** New functionality and enhanced features can be embedded into an network-processor-based product while it is deployed in the field without requiring the customer to buy a new product that uses a new ASIC design. This extends the product's time in market. Because it decreases the cost of product ownership over the life of the product, it creates more sales opportunities for the equipment vendor. The fact that the customer probably does not have to replace the product soon improves the quality of the relationship with the vendor. This is something that can also lead to repetitive and longer-term business prospects.

- *Just-in-time* (**JIT**) **delivery of new features**   As mentioned earlier, the rapidly changing reality of the market requires network gear vendors to provide new features and functional characteristics inside their products, such as the support of new protocols. Vendors who adopt the network-processor development path are able to embed this new functionality into their products without having to withdraw them from the market in order to replace them with something more recent or more updated.

- **Greater focus on other issues of business management**   The majority of the packet-processing functions in an network-processor-based environment are coded in a standard way, either by the network equipment vendors or third-party suppliers; therefore, the main core of software development is essentially available off the shelf. This decreases the overall time needed for software development. It also liberates resources so that vendors can concentrate their efforts on other aspects of the project that are equally important. They can focus on providing other necessary functionality such as network management, diagnostics, configuration, or different interfaces, as well as spend more time and money on the business management side of the equation, pursuing alliances and customer relations. Network processors are bound to revolutionize the industry by commoditizing the design of network devices, creating a phenomenon that is almost reminiscent of the PC industry in the 1980s.

NETWORK PROCESSORS: CATEGORIES

During the last few years, since its emergence, the network processors market has appeared to be a very "fizzy" environment. New startups are entering at a relentless pace, and major semiconductor houses as well as network device manufacturers are realizing that unless they participate in this process, they will miss the train of opportunity. Several startups have already left the field as the first casualties of a coming major shakeout. However, the consolidation process that has been taking place has started to show some underlying characteristics in this industry. Two major underlying classes of network processor chips can be identified within which essentially all network processor products can be categorized: *platform network processors* and *peripheral network processors*.

Platform network processors are usually complete chipsets that major vendors have designed to do the following:

- Handle all functions related to packet processing.
- Minimize the number of components needed and therefore the direct hardware cost in the final design.
- Optimize the trade-offs between performance and flexibility.
- Facilitate an accelerated and integrated software development cycle.
- Capture the largest possible number of design-wins by positioning themselves as the ideal source for one-stop shopping for the network gear designers.
- Attract third-party hardware and software players that will allow the build-up of an intertwined community that facilitates the easy and timely development of several modular products that share the common characteristic of being based on the vendor's network processor architecture (platform).

These chipsets are distinct with every vendor, but the overall partitioning of the platform architectures is quite similar among most of them. Their chipsets include PHY layer interface chips, NPUs, switch fabrics, traffic managers, and so on.

Peripheral network processors are microchips that have been designed to optimize a very specific function among the many that need to be handled in a packet-processing environment. Examples of a peripheral network processor include a compressor chip (such as the ones that HiFn offers) or an *IP Security* (IPsec) acceleration chip (such as the ones proposed by Broadcom, Cavium, or Corrent). These are highly specialized functions that require specific circuitry capabilities to handle the exceedingly heavy computational load efficiently in real time; therefore, it makes sense to offload them onto specialized co-processors. Specialized peripheral processor chips also conduct lookup/classification. In some cases, however, a full-fledged network processor conducts these operations instead. This processor operates in a sort of a slave-like mode adjacent to a master network processor that handles the live traffic flow. Other peripheral network processors are also available in the market to parse and frame specific layer 2 protocols, such as ATM cells, Gigabit Ethernet, and so on.

Yet another way of categorizing network-processing chips is based on whether they are implemented on configurable or unconfigurable hardware. *Field-programmable gate array* (FPGA) manufacturers have come up with very fast, highly integrated, and programmable chips over the last year or so. Network device vendors have also tried this alternative approach. We will not elaborate on this point, as it is unrelated to the topic of network processors.

To give the reader a full dose of reality from both sides, we should mention the other side of the argument about the network processors. More specifically, we should clarify that to a certain extent network processors did not succeed immediately or live up to the expectations that they had raised in the industry.

First, most network processor vendors needed to go through multiple generations of their designs just to get it right. To a large extent, this is an ongoing quest. NPU-chipset vendors have generally been characterized by a propensity toward responding impulsively in an affirmative fashion when customers, industry analysts, or even representatives from the trade press confront them with questions as to whether their chipset can accomplish specific tasks at wire speed. Customers, however, did not think about it and vendors conveniently never bothered to address what happens if other tasks must be performed at wire speed at the same time. Some of the major challenges confronting the industry include deciding the content of testing and agreeing upon how realistically performance-testing and benchmarking suites depict a traffic load, which can then be used as a satisfactory and truthful metric of anticipated or expected performance. The Network Processing Forum, an industry consortium, is actively working on these challenges; however, a lot of work still needs to be done before globally accepted and respected models and benchmarks are produced that emulate real-life network applications in realistic quantities and mixes of different types of traffic. The combination of these types of traffic can be used to obtain consistent, realistic, and meaningful ratings of performance.

The second challenge is that because many network processor chipsets turned out to be notoriously complex to program and fine-tune to achieve balanced wire-speed performance, customers found out that they do not have an easy metric to judge and compare the actual software-engineering development cost needed to develop upon a certain platform. We will see later in Chapter 16 how several unrelated factors, such as the sheer number of program lines needed for an application or the cost of licensing of application software from an NPU-chipset vendor, directly affect the choice of platform, the evolution of a product over multiple releases, and even the viability of a startup networking company that bets its future on a specific platform to deliver its product roadmap.

SUMMARY

In this chapter, we defined network processors and briefly discussed how they are structured and what they do. We reviewed macroscopically the two traditional methods of designing packet-processing network devices and communications equipment: using either software-based solutions that are based on off-the-shelf CPU processors or handling packet-processing operations in specialized hardware implemented as optimized high-performance ASICs. We identified the underlying trade-offs in flexibility and performance between the two approaches and introduced the idea of using network processors as a means for breaking free from the dilemmas of the two traditional schools of thought. We saw how network processors are optimized to handle packet processing with the flexibility of traditional CPUs and at the performance levels of networking ASICs.

In the next chapter, we will look inside the typical high-speed switching equipment and descend top-down into modules of functionality. This will enable us to see what types of operations typically occur in a network device and how they interact with each other. A solid understanding of such a functional breakdown and the corresponding modes of various operations is important because we will eventually look at network processor architectures and discuss the appropriate design choices by vendors. The arguments will only make sense if the reader can put them in the context of their applicability in a bottom-up approach, knowing what feature is useful in which context and what would actually be desirable (if something is missing) from a specific architecture.

CHAPTER 3
PACKET PROCESSING

In the previous chapter, we provided a general overview of the various switching technologies that possess evolving applicability, flexibility, and sophistication. The evolution of the concept and the technology were explained in relation to both time and complexity. We discussed how switching methods perform specific tasks. We also identified the engineering trade-offs and caveats that designers most often confront when using these approaches.

Most of these technologies are currently implemented in some of the most representative cutting-edge routing and switching gear in the world. However, these switching engines are not just suspended in thin air. They are invariably an integral part of an overall routing/switching system architecture. Within the architecture, a mind-boggling amount of complex operations takes place in real time in an orchestrated fashion. These operations usually become targets for *network processing units* (NPUs) in the most recent designs. We need to step back and examine several factors that must occur on the actual switching/routing system engine, so that the reader can understand the full scope of this discussion and appreciate the following challenges:

• The nature of the operations involved with fast packet processing.

• The way routers/switches are currently built.

• The types of physical modules one typically expects to find inside a router/switch chassis.

• How it all fits together with the latest trends in component integration—namely, chipsets of NPUs, search engines, classification and forwarding processors, switch fabrics, traffic managers, and security coprocessors. These trends address these combined requirements in the new design era, which we described in Chapter 2, "Network Processors: Justification."

NETWORK CONTEXTS: CLIENT, ACCESS, EDGE, AND CORE

Before we look more closely at the inner structures of systems and operations, we must clarify some common terms that will be used in this discussion. A reader who has had exposure to carrier-based services and products should be very familiar with this nomenclature. However, experience shows that many technical and business managers in the telecommunications industry either ignore many of the subtle, but nevertheless important, distinctions between these terms, or even worse hear them and use them without knowing exactly what they mean. The nuances become more important when the equipment requirements for the various realms are examined. A router designed for an enterprise network or a college campus where only moderate quantities of *Internet Protocol* (IP) traffic must be handled is quite different from a *multiservice router* (MSR) that is capable of multiple layer 2 and 3 protocols operating in the long-distance core of a *wide area network* (WAN) backbone between *Internet service providers* (ISPs) and legacy voice-based traffic carriers. However, most people refer to them with the same generic name of *router*.

We discussed earlier in Chapter 1 how the original switch and router concepts have slowly merged with each other to form a functionality that spans the multiple layers of protocol stacks. This is the main reason why we usually use these terms interchangeably throughout this book. In the specific cases where the two concepts must be distinguished from one another, however, we clarify which term will be used. In the industry, the switching/routing equipment is usually referred to by the physical place where it is installed, not by the corresponding stack layer at which it operates. Figure 3.1 shows the conceptual layering of multiple interconnected networks in the converged network that we are discussing.

The bottom of the hierarchy contains the *enterprise network*, which is also known as the *customer network* or *customer premises*. The term *customer premises equipment* (CPE) was created from this concept, although it was not originally coined in a packetized-data network concept. The enterprise network corresponds to the typical day-to-day user's Ethernet and Fast Ethernet networks that are located in companies, universities, and so on. The enterprise network contains one or more *local area networks* (LANs) on one side connecting ordinary user stations, such as PCs, with shared access to common resources, such as printers, faxes, and so on. In addition, this network contains faster local networks that enable an organization to connect its servers, its storage subsystems, and so on. Some of these faster enterprise LANs are Gigabit Ethernet networks.

A new type of switch has recently evolved that serves the latter community of servers. This switch may handle intelligent load balancing by switching traffic to and from specific servers, or it may merely manage a storage server farm or arrays of disk storage racks connected through a dedicated *storage area network* (SAN). The latter context shows the advantages associated with offloading the traditional LAN. A trend is formed toward creating newer IP-based techniques such as Fibre Channel over IP or *Small Computer System Interface* (SCSI) over IP. This trend may ultimately replace the reigning Fibre Channel.

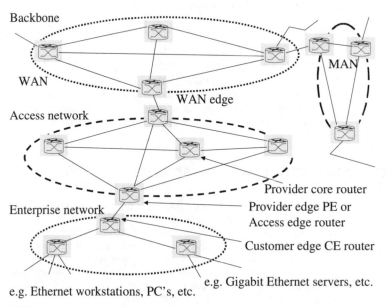

FIGURE 3.1   The conceptual hierarchy of networks.

Highly specialized and dedicated gateways are also usually provided on the landscape to bridge the locally connected systems to the rest of the world, depending on the applications and policies. For instance, telephony gateways translate the evolving *voice over IP* (VoIP) or video over IP realm back and forth and to and from *Public Switched Telephone Network* (PSTN) signaling and traffic. Firewalls and routers handle the normal traffic that enters and exits a site. At some point, some of the LAN-based systems require legitimate access to the rest of the world. This could require connectivity with other companies, sites of the same enterprise, suppliers, customers, partners, or classical web access for an organization's members.

Routers usually handle this access. Because these routers are situated at the periphery of the enterprise network, they are called *edge routers* (probably inappropriately as we will explain later). Edge routers are different from the other routers that operate in the heart of the enterprise network, which have different requirements for protocol support, port speeds, and so on. These routers are often called *core routers*. Edge and core routers should not be confused with one another and should be used appropriately in different contexts.

The next layer in the global network hierarchy is the *access network*, which is also known among many industry players as the *provider network*. An ISP's network is a typical example of an access network where the boundaries among a local telephone company, a long-distance company, and an ISP become more blurred. Everyone is stepping on everyone else's toes in a competitive stampede that is bound to reshape the industry landscape while optimizing the communications services and cost. Access networks consolidate (aggregate) customer traffic from the humble home-based PC modem users to the more sophisticated broadband cable clients. These networks prepare to feed the traffic through a larger pipe into the WAN. This could be done over the *Plain Old Telephone Service* (POTS), the Internet, or something else. Cable-based broadband access clients are multiplexed through the local cable-TV company's head-end equipment. The client might also use some sort of *Digital Subscriber Line* (DSL) modem or the latest wireless broadband last-mile access technology.

It does not take a rocket scientist to realize that the provider networks also comprise two types of routers: *provider network core routers* and *provider network edge routers*. These routers are illustrated in Figure 3.1, which provides a macroscopic view of reality. Once again, the terms *edge* and *core* are used loosely here so we will not follow this example. The typical speeds encountered inside an access network currently range between OC-3 and OC-48.

The top layer in this hierarchy is the WAN, which interconnects provider or carrier networks and is often referred to as the *backbone* (for example, of the Internet). The WAN also contains edge and core routers. The transmission technologies most often used at the WAN level are optical. The typical speeds achieved on a WAN currently range between OC-48 and OC-192. In some metropolitan areas, a trend is evolving to adopt the new emerging 10 Gigabit Ethernet as well.

Historically, the metropolitan network was largely used as a transport-layer medium (such as *Synchronous Optical Network/Synchronous Digital Hierarchy* [SONET/SDH] and *Plesiochronous Digital Hierarchy* [PDH]). The major innovation was that equipment designed for metro networks needed to be able to handle data traffic. As such, the core of the WAN still contains several *Asynchronous Transfer Mode* (ATM) switches. Although a move is being made toward handling fast-switched IP traffic on the WAN, the core and edge switches must be able to handle multiple protocols at wire speed, often including *time division multiplexing* (TDM) traffic and frame relay. IP traffic is still being transmitted as IP over ATM during this transitional era, while backbone service providers are adopting newer technologies to use on the more modern optical networks, such as terabit routers offered by companies such as Avici, Cisco, and Juniper.

To better depict the market situation of several companies competing from completely different angles of new data-driven business while supporting lucrative legacy business, it is more customary to use the four-layer approach shown in Figure 3.2. In this example, the WAN has been broken into the edge and core network, and the term *metropolitan network* denotes the combination of the access and edge networks. The terms *edge* and *core* are used correctly in Figure 3.2.

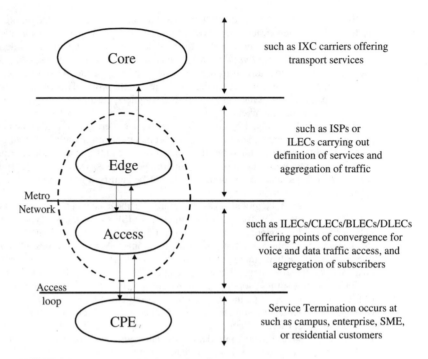

FIGURE 3.2 Four-layer model of network reality.

When examining the requirements of the switching equipment at these various levels of functionality, one should first look at the interesting variations on what a switch/router should be able to do in each situation.

LAN traffic is now mostly a mixed landscape of switched 10 and 100 Mbps Ethernet, while legacy token rings are also used in some cases and more and more Gigabit Ethernet is showing up on the LANs inside the enterprise network level. Traffic is generally switched at layers 2 and 3. *Workgroup switches* are network units that consolidate all the disparate users generating this traffic demand. This is done in a cost-effective way.

Web switches (load balancers) on top of traditional layer 3 switching must also be able to switch traffic at layer 4 (for example, at the *Transmission Control Protocol* [TCP] layer) all the way up to layer 7 (for example, for cookie detectors). In this case, traffic would need to be switched according to each application and based on the *Uniform Resource Locator* (URL).

LAN backbone switches aggregate all the enterprise workgroup switches and provide connectivity to the access network. Gigabit Ethernet is used in this situation, as switching multiple protocols at layer 3 has replaced the need for the cumbersome and expensive routers that originally handled this type of traffic.

Because each individual access line for routers at the edge of the access network has a speed of less than 10 Mbps, the switching requirements are well below the high-speed processing capabilities of the network-processing chip architectures that we discuss in this book.

We should mention some overall contextual differences between enterprise and service provider equipment:

- At times, multiple end customers who are typically indifferent toward each other's needs must be served.

- In many cases, the importance of managing bandwidth intelligently cannot be overemphasized. Bandwidth is typically not a scarce resource in the enterprise, but it may be scarce with a service provider.

- The differences between performance expectation and *quality of service* (QoS) treatment requirements must be considered.

- The operating environment requirements are very different in the two realms. For example, the requirements for a central office differ from those for a street cabinet/pole top where there is no forced air cooling and ambient temperatures can vary between great extremes.

- The two realms have different requirements in terms of reliability/availability and how easily the equipment can be upgraded in the field.

- The requirements for accommodating varying operations, administration, maintenance and provisioning requirements in both cases.

In the current WAN realm, both fast IP switching techniques and *Multiprotocol Label Switching* (MPLS) must be supported. This means that wire-speed IP routing at OC-192 is required. As mentioned previously, in addition to increased routing speed, MPLS offers carriers some unique capabilities for *virtual private networks* (VPNs) and other revenue-generating services based on its *traffic-engineering* (TE) advantages. Network processors are becoming useful for the timely design of advanced, but affordable, equipment that provides the manufacturer's clients with the possibility for such differentiation and potentially lucrative services.

The WAN edge routers must be able to consolidate multiple access network interfaces. A typical example of this environment is a Cisco Edge Service Router that can provide the equivalent sustained throughput of 43 T3 lines. A modular design allows multiple combinations of throughput—for instance, in multiples of T1 or even fractional T1. The uplinks from the access network can be found in Gigabit Ethernet or more often in OC-12 links. These WAN edge routers must be extremely reliable to ensure around-the-clock functionality. This implies that they must be designed to be field serviceable. In other words, critical components must be fault tolerant (or even redundant in some cases) and line cards must be *hot pluggable*, meaning that they can be replaced without bringing down all of the network equipment. These are very different requirements from a customer premise router.

Another type of WAN edge router is the MSR out of which the *Multiservice Providing Platform* (MSPP) has evolved. MSPPs are essentially SONET *add/drop multiplexer* (ADM) equipment that combines IP routing and ATM switching. Due to their versatility and performance, network processors are expected to be especially useful in the implementation and delivery of MSPP products.

Some companies use the term *metropolitan area network* (MAN) as another context. Although the technical environment of the MAN is essentially similar to the one encountered in the WAN, the cost of deployment and the economics of justifying product investment are different. This enables less expensive technologies to be adopted for the implementation of similar solutions. MAN equipment is an interesting business context, but for our technical purposes, we will consider it just as another example of a WAN technology that consists of the traditional access and edge networks, as shown in Figure 3.2. We will no longer single it out specifically. This is because the design requirements of network equipment for MAN applicability (especially pertaining to network processors, switch fabrics, and traffic managers) are easily deduced by the edge and access network equipment requirements.

In the midst of all these combinations of functionality, protocols, line interfaces, and wire speeds, one should not forget the ever-present need to consolidate along all the more recent techniques the legacy voice traffic, although at some point this need is bound to become more or less obsolete. This traffic is the most lucrative service a carrier could offer at this point. It is usually delivered on

TDM-multiplexed voice channels. This only exacerbates the need for flexibility and upgradeability in network switching equipment, something the network processors are ideally suited to handle.

It is assumed that the reader has a basic background on traditional PSTN-type telephony-inspired link platforms known as T1 and T3 (in North America) and E1 and E3 in Europe. These platforms allow transmission based on several protocols such as voice over TDM, frame relay, IP, and ATM. Table 3.1 summarizes the bandwidths of the more recent SONET-based links in the WAN.

THE TIMING OF THE NETWORK-PROCESSING EVOLUTION

As of the late spring of 2003, the accepted state-of-the-art speed in deployed network processing is OC-192 although there are devices that are able to function in OC-768 links; therefore, much of our attention will be focused on chips and architectures functioning at that speed. However, network equipment designers are already extremely anxious about the scalability of their architectures and designs in relation to the next logical performance step of OC-768. This step requires processors capable of processing traffic at 40 Gbps wire speed. We will discuss some of the issues and trade-offs that architects will sooner or later have to confront. The economic downturn following the stock market collapse in 2001 and the general slowdown that resulted following the tragic events on September 11, 2001, have also significantly delayed numerous investment plans and deployment schedules for 10 and 40 Gbps projects. The naturally induced financial conservatism has consequently affected the rate of market adoption for many of these new technologies. For the next couple of years, the main emphasis of the network-processor market will most likely be in the OC-48 and multiple Gigabit Ethernet realms.

The result has been mixed. On one hand, it has been negative for some vendors of cutting-edge network-processing technology (component or system) who were hoping to launch new platforms. On the other hand, it has been positive for others who needed some more breathing space to conclude their race against the clock designing sophisticated chips and putting the final touches on their development work. The vendors hoped that by doing their work more thoroughly, they would be able to weather the financial storm and be ready with real and stable products when the market would finally be ready to talk to them.

The same context seen from the market's viewpoint has also had a double effect. Carriers realized that they were not under the gun to deploy products such as VoIP in lieu of traditional legacy voice technologies. The pressure therefore was sent back to the network equipment manufacturers to accommodate TDM traffic along with their more targeted packetized future network. The purported demise

TABLE 3.1 Bandwidth of Typical WAN Links

Physical Layer	Bandwidth
T1	1.5 Mbps
E1	2.0 bps
T3	45.0 Mbps
E3	34.0 Mbps
OC-3	155.0 Mbps
OC-12	622.0 Mbps
OC-48	2.5 Gbps
OC-192	10.0 Gbps
OC-768	40.0 Gbps

of ATM was also delayed. Interesting new concepts appeared to take advantage of this window of opportunity, such as technology from Litchfield Communications, whose chips packetize TDM traffic and feed it into network-processing systems and switch fabrics that have been designed to handle packetized flows.

Of course, the decelerating economy of 2001 and 2002, the corresponding market repercussions, and the ensuing delay that affected the deployment of faster technologies had adverse effects on the financing and development pace of startups. This created a domino effect that few companies could avoid. We will discuss these issues in later chapters when we identify the trends of attrition, consolidation, and inevitable evolution in this rapid market.

THE OVERRIDING REQUIREMENTS FOR NETWORK EQUIPMENT

The four different network realms illustrated in Figure 3.2 are characterized by various requirements, which are summarized in this section. We intend to highlight the following two points:[1]

- How the systems designer has to cope with some very specific constraints imposed on him or her by the context within which the equipment will be called to operate.
- How to trigger the imagination of the reader as to how the network-processing platforms are called to deliver solutions for the different problems that are encountered at each level in this conceptual hierarchy.

For CPE equipment, which will require more packet-based services in the future as new services are introduced into this market space, the systems designer wants to ensure that the equipment has the following characteristics:

- Is interoperable with the access network's behavior at layers 1 and 2 and potentially with the edge equipment's expectations at layer 3 and above in the service provider's network (for example, applications like e-mail protocols such as *Simple Mail Transfer Protocol* [SMTP] and *Post Office Protocol* [POP] or widely used layer 4 protocols such as TCP).
- Can handle the WAN wire speed.
- Is designed and proposed economically.
- Does not require large physical space for its deployment so that it can be offered to multiple environments without undue customer resistance.
- Is easy to manage and configure remotely.
- Is highly integrated so multiple services can be offered through it.
- Allows room for modular future expansion.

Access network equipment, which is the first layer of consolidation and the last layer of distribution by the service provider of traffic to and from multiple users, has a different set of essential requirements. The systems designer focuses primarily on the following:

- A large-scale and (as much as possible) low-cost aggregation of multiple physical connections to subscribers. This is usually accomplished with rack-mounted devices and modem banks.
- A small footprint of these rack-mounted and chassis devices.
- Low power consumption as racks at carrier premises have tight constraints that must be respected.
- The ability to communicate in multiple physical interfaces and layer 2 protocols.

---

1. See, for instance, "The Role of Network Processors in Next Generation Networks: Defining the New Network Processor Landscape from CPE to Core," a white paper from Intel Corporation, Network Processor Division, August 2001.

For edge network equipment designers, the problem is not how to collect low-speed traffic from users or how to distribute it to them. It is how to aggregate multiple traffic streams (flows) into traffic classes that reflect specific characteristics of differentiated services.

Therefore, the designer of edge equipment is interested in ensuring that

- Services can be easily provisioned when and where they are needed.
- Both the performance and functionality of the equipment are scalable.
- The density of the design must be maximized.
- The reliability and availability of the design must be optimized for this context.
- The network equipment can be serviced, maintained, and upgraded easily.
- The design is modular so that it can be expanded and upgraded with new protocols, standards, and required functionality such as customized billing.

The core network is comprised of optical fibers connecting hundreds, if not thousands, of edge routers and switches, each requiring the ability to handle hundreds of gigabits-per-second traffic and often having a terabits-per-second switching capacity. The designer of core equipment is preoccupied with the following characteristics:

- Scalable performance when switching or routing.
- High reliability.
- High availability.
- Fault tolerance and, in most cases, sheer redundancy.
- Field serviceability, which means modular design and hot-pluggable cards or modules.
- An industry-standard design that is *Network Equipment Building Standards* (NEBS) compliant in terms of cards and chassis size, power consumption (maximum and typical), MTTF, etc.

Of course, NEBS-type requirements apply to many types of service provider equipment, not just core boxes. In fact, the environmental requirements on access systems that sit in street cabinets are arguably even more stringent.

The combination of these fundamental requirements and some obvious market dynamics create a new set of challenges for the market and an ensuing set of opportunities for network-processor manufacturers. For example, new packet classification requirements will appear with the widespread adoption of MPLS. Like windows capabilities originally reserved for fancy and expensive engineering workstations slowly but surely arrived on the humble PC when it became equipped with a powerful microprocessor and *input/output* (I/O) bus, functionality that was previously only available or expected inside edge and/or core equipment will continue to expand its presence toward the access points, as the CPE devices increase their sophistication (due to application-driven demand) and speed. This increase will enable these devices to take advantage of 10 Gigabit Ethernet networks and to be deployed in the metropolitan areas.

At the same time, a noticeable trend is that sophistication and service granularity migrate from lower-speed equipment to higher-speed equipment. This is possible because technology enables more work to be done in a given power/cost/space envelope and more work enables higher-value services to be delivered.

DATA AND CONTROL PLANE PROCESSING

We learned in Chapter 2 how a typical switching/routing system is structured architecturally in functional units that combine to compose two parallel processing paths. These are called the *slow* and *fast processing paths*, respectively. We also learned how these paths received their names. The former deals with processing operations *about* packets, such as network management, routing protocol handling and routing table updating, and traffic regulation. The latter deals with operations that are directly performed *on* packets, such as header modification, filtering based on content, classification, and the encryption of fields.

The slow processing path takes place on a parallel data path implemented by slower *central processing units* (CPUs) without the tremendous pressure of having to keep up with many packets that require special and different attention arriving in real time at wire speed. This is the responsibility of the fast processing path. In order for the switch/router to be able to cope with the high-speed packet-processing requirements during packet parsing, classification, forwarding, field processing, potential encapsulation, scheduling, and switching, blazing-fast circuitry and a smart and efficient architecture are required. This is why we say that on the fast processing path, the system operations are exercised directly on packets.

Two other mainstream terms that are equally used by the network equipment community to describe this processing reality are *control plane processing*, which is another way of referring to the slow processing path, and *data plane processing*, which is a synonym for what occurs along the fast processing path. Some people even extend these terms further and adapt them to the actual hardware choices for the implementation of the two processing data paths. For example, a control processor could refer to a CPU or an *application-specific integrated circuit* (ASIC) that is used to handle the slow processing path functions, whereas a data plane processor could refer to either an NPU or a fast specialized network ASIC. Data plane processor could also even refer to a *reduced instruction set computer* (RISC) or a *complex instruction set computer* (CISC) CPU, but this is less common.

It is worth mentioning that some vendors logically split the control plane into two adjacent and complementary computational slices, which they dub the *control* and *management plane*, respectively. This is more of a cosmetic implementation-dependent characterization, which simply delineates the host CPU (usually a processor based on the PCI bus) that oversees the macroscopic management of the line card or system from another control plane processor, which may be handling much narrower day-to-day control responsibilities.

We will not make this hypergranular distinction in this book and will continue using the predominant model of thinking in terms of two planes.

It is also worth mentioning that NPUs may integrate in Ethernet MACs, but they rarely integrate in Ethernet PHYs. In addition to the complications of the mixed signal design (containing analog and digital inside the same silicon die) and the additional power dissipation, the pins used for SMII and GMII interfaces are electrically compatible with SPI-3/*Universal Test and Operations PHY Interface for ATM Level 3* (UTOPIA 3). Therefore, no additional package pins are required when only MACs are integrated in. I/O pins are often a scarce resource for massively packaged NPU chips.

PACKET-PROCESSING OPERATIONS

So far we have used several generic terms to describe handling operations applied by the high-speed router/switch onto packets. We will now take a closer look at these operations:

Packet Framing

In the Ethernet environment, the MAC and PHY layers implemented in the transceiver are sometimes implemented in different chips. In high-speed links, however, such as SONET where either ATM or *Packet over SONET* (POS) is being transmitted, a separate *framing* unit is used to map the ATM cells or *Point-to-Point Protocol* (PPP) packets into the SONET frames for transmission. These frames are then passed through a *serialization and deserialization* (serdes) module before they are handed over to the transceiver. The inverse occurs at the reception point. Network processors can obviously handle the framing function in the high-speed links of the future.

Pattern Search and Packet Classification

The generic classification task (as the term itself implies) means that some rules and conditions must be applied in real time to every incoming packet and, more specifically, to its headers or to parts of its overall content in order for the switch/router to assign the packet to one among several logical outcome options. This is usually associated with specific QoS or forwarding decisions. Typically, a table of addresses and, more recently, the whole database of rules and policies must be searched in real time so a context-sensitive decision can be made, according to which packet will be classified and forwarded.

As a result of classification, the incoming stream of packets gets partitioned into multiple logically separated output streams, which will then need to be handled appropriately. For instance, one stream of packets may need to be forwarded to its destination port with a higher priority, whereas another stream may have to be relegated to a lower priority because of other more urgent tasks. One of these output streams might also be the subject of a special billing procedure, whereas another may not. Traditionally, especially in older routers where the line speed was quite low compared to the more recent generations of very fast switching/routing gear where the high wire speed requires hardware implementations of the classification work, packet classification algorithms were implemented in software that was running on a standard off-the-shelf CPU. The more recent hardware implementations of packet content search and classification are completely focused toward supporting designs realized with network processors.

The classification algorithms themselves depend on the application at hand. Generally, when one looks for match, several criteria actually constitute the required degree of matching. To give an analogy, if one needs to assert whether a specific phone number is from the Boston area, one does not need to exhaustively list all the numbers in Boston among telephone users in the United States and then check where each number is located in a more elaborate way. One just has to check the area code and ensure it is the number 617 in this example. On the other hand, if one wants to sort out the numbers that are in Boston and belong to the same local exchange, say, 754, one simply has to match all numbers against the area code 617 and the prefix 754 using wildcard characters for the rest.

The same principle works with IP addresses. Depending on the application itself, one may require an exact match for a specific address search or may just need a prefix match. Mask bits can be applied to select whatever bit positions one decides based on the appropriate criteria. The system will then find the most suitable entry by looking it up in a *content-addressable memory* (CAM), which will yield the necessary address.

In the most rudimentary setting, today's routers use the *Classless InterDomain Routing* (CIDR) routing protocol to calculate the address to which a packet must be forwarded. For example, CIDR uses the *longest-prefix match* (LPM) algorithm for the calculation of the next-hop address. We will discuss the internals of this algorithm in more detail later in Chapters 12 and 13. For the moment, we will just mention that in order to implement this type of classification environment, until very recently, switching/routing systems designers in conjunction with an ASIC or a RISC/CISC CPU would involve a CAM module, which allows a fast and efficient implementation of the classification scheme in several configurations.

Some instances of classification occur at layer 7—for example, looking up specific URL strings associated with the *Hypertext Transfer Protocol* (HTTP) protocol. However, classification usually occurs at layers 3 and 4.

A typical example of a need for such a sophisticated classification would be in a *Differentiated Services* (DiffServ) environment. In this environment, the lookup must be executed based on multiple fields from the TCP and IP headers. This is where the classifier must apply the *five-tuple* lookup in order to extract the appropriate forwarding information based on data provided from a joint TCP/IP set of headers and, more specifically, from the following five distinct fields of data therein:

- The IP source address (32 bits).
- The IP destination address (32 bits).

- The specific IP protocol used (8 bits).
- The TCP source port (16 bits).
- The TCP destination port (16 bits).

We will revisit this case in more depth in Chapter 12, "Search Engines," and Chapter 13, "Classification Processors."

Returning to our DiffServ example, in this five-tuple lookup operation, the classifier will need to locate and extract 104 very specific bits from the combination of the IP and TCP header. It will then look into a CAM using these 104 bits as index in order to find a new bit field from the CAM, which will then be used as index to an associated data memory bank from where the exact result will be extracted. Based on that final result, the switch/router will decide which DiffServ flow it must allocate to the packet. To make a long story short, this is accomplished by the generation of the *DiffServ Code Point* (DSCP) bit pattern that is written by the switch into the *type of service* (TOS) field of the IP header. The DSCP code will notify all routers/switches in the DiffServ domain as to what type of treatment should be reserved for this specific packet at each hop of its trip.

A similar operation occurs at the ingress points of MPLS networks when label tags must be swapped or stripped on-the-fly based on specific rules.

As another example, in the case of a simple address filtering, pick a bit mask among the several stored that reflects the desired filtering and then use modulo-2 to add it to the search destination (that is, use *exclusive OR* [XOR] on it, ignoring the last carry) and throw it to the CAM as the so-called key. The output of the CAM should then provide the bit sequence that should be used as an index to an external memory bank that determines which yields the intended and desired destination.

CAM (Content-Addressable Memory)

For the unfamiliar reader, CAM is a specialized memory bank used in switching/routing environments in what has come to be known as *search engines*. These search engines have nothing to do with web-browser-based Internet search engines that look for web pages. Traditional memory lookups ask the following question: What content is stored in address X? However, in CAM, the question is inverted: In which address is content Y stored? CAM memory is arranged in such a way that when a specific entry is looked up, the memory bank will rapidly compare the specific request with all its contents. If a match is found, the corresponding address (where it is stored) will be returned. When a table is stored inside a CAM, the CAM is said to be *initialized*. When we want to look up something in that table, we *write a search key* to the CAM. In reality, no one is writing anything to anyone in this case. A bit pattern is simply presented (also known as the *key*) to the CAM. The CAM will try to match it with one or more of the entries it contains. If it succeeds, it returns the address where the match was found.

For example, in switching/routing systems where the next-hop address must be found, the specific address obtained as a result of the CAM lookup operation is used as a pointer to a specific address located inside some external *static random access memory* (SRAM) bank that is known as *user data memory* or, more appropriately, *associated data memory*. That is exactly where an IP address or a MAC address (depending on the application) will be found that satisfies the packet's destination requirements that the system was seeking when it initiated the lookup.

The lookup operation described so far is based on a *binary CAM* because the bit positions either match the key content or fail (0 or 1 in every bit position). Most CAM products used in current search engines offer the possibility of *ternary CAM* (TCAM) operations, which enable the creation of masks for every entry word using 0, 1, or don't-care values. This is required for several of the currently used search algorithms, including the LPM algorithm. If more than one match is found, the lower address is usually returned, although TCAM chips are present, which are structured with an embedded priority-encoding mechanism that returns multiple matches in a certain priority. This is obviously a mechanism that taxes the real-time performance of the search engine severely, so it is only used when it absolutely and undoubtedly makes sense.

Although we will be discussing CAMs in length later in Chapters 12 and 13, it is worth mentioning here that the brute-force hardware lookup capabilities that CAMs provide are expensive and often require a lot of power. CAMs are attractive for the following reasons:

- They recognize large bit patterns (they do a lot of work per trip across the I/O pins where approaches using conventional memory typically need to make many trips as the bit pattern to be classified gets bigger).
- They are useful where table sizes are small (storing large lookup tables in CAMs is prohibitively expensive in terms of cost and power).
- They are helpful where lookup latency is critical (although latency can often be hidden with a suitable use of pipelining and threads with memory-based approaches).

Search Engines

When wire speed becomes so high that an external SRAM is not the best approach (OC-48 and above), an integrated search engine must be considered. This is either a TCAM implementation or part of a dedicated classification processor with the appropriate system speed design. The newer switching/routing systems are based on network processors; therefore, the interface between NPUs and search engines poses a real challenge. Some current designs are implemented around cumbersome *field-programmable gate arrays* (FPGAs) that are meant to replace a sea of glue logic. Several standardization efforts are currently under way to address this problem, as we will see later in Appendix III in the discussion about standardization. Some search engine vendors interface their engines to the NPU through a memory interface so that the NPU is essentially fooled to believe that it communicates with an ordinary memory bank. The leading providers of search engines include IDT, NetLogic Microsystems, SiberCore, and *Kawasaki LSI* (KLSI).

Typical search engines offer the possibility of about 100 *millions of searches per second* (Msps), but this number is rapidly increasing. If a certain piece of switching/routing gear requires bandwidth that is higher than 100 Msps, multiple search engines may be used. They can either be centrally located in the switch/router (a demanding proposition for high-speed links), or the designers can include a search engine on each line card, reducing the performance constraints on the search engine, which is most often the case. The key size is nowadays usually 72 bits; however, with the appropriate soft configuration at half the clock speed, search engines will usually also work with a search key of 36 bits.

Therefore, capacity and speed are the two most important rating parameters for the specification of a search engine's performance. Because the word size is 36 bits (and not 32 bits as in normal memory) when using the convention of calling a search engine with 1Mb or 2Mb capacity, we obviously mean they contain 1.125Mb or 2.25Mb, respectively. One can cascade search engines and increase the capacity, although this may adversely affect the latency of the overall system, as more cycles may be needed from the moment that a key is presented to the search engine until the moment when a result has been produced at the output port of the engine. The reader is referred to Chapters 12 and 13 for more details.

We also expand on other interesting topics associated with this subject in Chapter 15, "Traffic Managers."

Packet Parsing

Unlike the traditional ATM environment, where all cells are of equal length, deep packet inspection is not a trivial operation when incoming packets are of variable length. Special architectural capabilities must be designed in order to ensure a flexible handling of the field alignment for subsequent processing. Network processors are very well suited for this type of function. Some contain embedded functionality that can do this, whereas others must be augmented by ancillary chips (preferably from the same chipset family and vendor) to implement the desired scheme.

Packet Classification and Fast Forwarding

The two terms *packet forwarding* and *packet routing* may confuse some readers at this point. On one hand, forwarding refers to the selection of an output port by a switch/router based on the destination address of the packet and in conjunction with a routing table that stipulates what goes where. On the other hand, routing, microscopically speaking, in many contexts refers to the process of actually building the table itself, although macroscopically it is associated in people's minds mostly as implying the forwarding function.

This is the right time to clarify two common terms that may already be familiar to many readers and are used quite frequently in the industry in order to characterize specific design philosophies that lead to specific architectures.

A *store-and-forward* architecture, as the term implies, stores the incoming packet temporarily and then decides what to do with it. On one hand, it gives the architect more flexibility and wider applicability for the final outcome of the packet handling process. On the other hand, it also implies a higher implementation cost, as buffering facilities must be provided and as an overall longer end-to-end delay is incurred due to increased latency from the ingress point to the egress point. This is obviously the case even if the storage time is shrunk down to minimum acceptable levels for a specific application. For example, this is certainly applicable in a typical low-end router and/or Ethernet hub. An incoming packet is first written into memory and then the switching device's CPU decides on which port to output it.

A *cut-through* design eliminates both the cost of the extra buffering and the longer latency associated with store-and-forward architectures by making the forwarding decision on-the-fly, based on specific bit fields that it parses in real-time on the incoming packet headers. In many cases, given the high wire speed, this decision must be made even before the incoming serialized packet has completely entered the switch/router. A typical example would be the latest MPLS switches on the WAN backbone, where the small label tag that has been prefixed to the arriving packet already signals to the switch the switch output port from which the packet must egress. It is clear that this approach only lends itself to some very fast implementations.

A certain risk exists that the fate of a packet may already have been decided upon and that the packet may already have been forwarded onto a certain egress port before some other functionality of the switch had the time to step in and decide about some other overriding alternative that precludes the forwarding decision that was previously made. To further illustrate the point, cut-through architectures are also well suited for low-end Ethernet switches since for instance they cannot manage traffic congestion with a QoS framework, they cannot check *cyclic redundancy check* (CRC) before actually forwarding, and so on.

For example, say that an incoming packet associated with a VPN enters the switch from one of its ingress points and with the appropriate prefixed label tag in an MPLS backbone network. The switch proudly decides on-the-fly that based on this label tag and the internal forwarding-associations table, the egress point for this specific packet is going to be its output port called X. The packet is now switched by the switch fabric onto the port X and bits start to exit the switch from that port while bits are still coming in at the ingress point. A traffic manager in conjunction with specific QoS requirements that the switch must satisfy may then intercept a specific bit field of the incoming packet at the ingress point, which may for all practical purposes be located deep inside the parading packet. The switch/router all of a sudden may realize that this specific application class requires priority handling over a separate link. It may also require some exceptional treatment that is guaranteed through reserved bandwidth resources (as a result of *Resource Reservation Protocol* [RSVP] and DiffServ actions) associated with another egress port called Y. It will flag the event as an exception. The switch may then realize that the packet should not have been forwarded through egress point X in the first place. It will try to block the output, but it may already be too late for the next hop station, as some bits and maybe whole packet headers most likely have already exited the switch and continued their trip downstream. The reader should be able now to see the trade-offs involved with the two design approaches.

All incoming packets must undergo a deep examination by the switch that goes beyond the traditional header inspection. Based on the results of such an inspection, the packet will be immediately classified to some class or category for subsequent processing. Therefore, the correct classification of packets may be implied by various needs.

For example, a user may have to deal with making a specific routing decision (layer 3) for some packet based on some specific routing protocol, which may use the LPM algorithm based on a handy data structure called a *trie*. We will discuss this in more detail in Chapters 12 and 13. In the implementation of address-prefix matching algorithms, the forwarding database that must be consulted generally contains a *dictionary* of address prefixes. The algorithm is used to find the longest initial substring of the destination address that is included in the forwarding database. During a classification operation, the network processor (or ASIC, or other CPU for that matter, chartered to take care of the task) will traverse the trie looking for the LPM. We look at several approaches to improving this technique in Chapters 12 and 13.

As we briefly discussed in Chapter 1, "The Evolution of Network Technology: Distributed Computing and the Convergence of Networks," with the arrival of MPLS, data flows are tagged by each router with a small route-specific label that is extremely reminiscent of ATM headers on top of IP traffic. It might be tempting to conceptualize about MPLS traffic as merely network load that must be routed/switched at wire speed as ATM but on real IP packets. These are notorious about their varying lengths; hence, switching has to occur at layer 3 without the nuisance (in this case) of the fixed-cell length that ATM is imposing. In fact, some researchers[2] openly admit that MPLS has borrowed the good design attributes of ATM without the need to set up calls and without the need for a fixed-length cell. These factors were once perceived as the two major drawbacks of ATM. The classification issue we just discussed becomes absolutely critical for the performance of the MPLS networks, as switching must occur in extremely high speeds at the backbone of the Internet (where wire speed attains at least several tens of gigabits per second) based on the content of a small prefix (MPLS label tag) attached to the IP packet. Similar concerns are found when implementing other applications such as load balancers for server networks. We also discussed relevant issues in the section "Packet Classification and Fast Forwarding" earlier in this chapter.

Modification

Modification is a generic term that can be applied to several contexts. In a typical modification, a packet must be encapsulated. This means that new headers must be calculated. In some instances, new trailers and CRC checksums must also be calculated. This is done for example when IP over ATM is running, where an extra overhead of 8 bytes is created and added to an IP packet. *ATM Adaptation Layer level 5* (AAL5) is subsequently used to carry the encapsulated packet, which now carries the original content along with the appropriate ATM headers, the required AAL5 padding, and the necessary trailers. Another example that we will see in the security coprocessor Chapter 17 is the encapsulation of encrypted and authenticated packets in an *IP Security* (IPsec) environment. This involves the creation or removal of the *Authentication Header* (AH) and *Encapsulating Security Payload* (ESP) headers, and especially the dressing (or undressing for that matter) of packets depending on whether the link operates in the tunneled or encapsulated mode. Network processors are more than up to the task.

---

2. Radia Perlman, Interconnections: Bridges, Routers, Switches, and Internetworking Protocols, 2nd ed. (Reading, Massachusetts: Addison-Wesley, 1999).

Switching

Once a decision has been made on what should happen to an incoming packet and once any relevant processing on it has been concluded, the packet will go through the switch fabric. Inside an MSR, for instance, a handful of critical architectural contexts are available. These include the backplane that connects everything inside the chassis, the actual switch fabric, and, of course, the various line cards where the network-processing chips are situated. Sometimes packets are broken transparently to the user inside the switch fabric into smaller manageable chunks called *cells* (which have nothing to do with ATM cells). These cells facilitate their transition from the input to the output of the fabric and are reassembled as packets at the output of the switch fabric. We will look at the context of these fundamental categories of hardware later and will also dive deep into the heart of the actual switching/routing device—the switch fabric itself, which we discuss in Chapter 14.

Traffic Management and Other Operations

When the packet is ready to be transmitted to the subsequent stage in the chain of processing, scheduling needs to be applied to it. Two types of scheduling must be performed: scheduling before the packet is handed over to the switch fabric and scheduling when the switched fabric is launched on the output port. This falls more generically under the category of traffic management, which takes care of handling queues and flows based on the various *classes of service* (CoSs), generating the appropriate billing information and ensuring that traffic abides by the applicable *service level agreements* (SLAs) and levels of QoS.

Traffic management is the major category of functionality where the problem of traffic congestion is handled along with traffic shaping in environments such as the one that the latest trend for DiffServ requires. Traffic management includes queuing, buffer management (including the application of sophisticated algorithms such as *Weighted Random Early Detection* [WRED], RIO, *Early Packet Discard* [EPD]/PPD, which we discuss later, and ideally with multiple buffer pools for better traffic isolation), and scheduling/shaping. Shaping is this context refers to effective non-work-conserving scheduling. Interestingly, bandwidth can be guaranteed even with a simple *first-in first-out* (FIFO) scheduling by carefully managing the buffer space that each flow is allowed to occupy (although it is certainly preferable to also use a differential scheduling treatment as part of the overall QoS toolkit). We discuss these issues in greater length in Chapter 14, "Switch Fabrics," and Chapter 15.

Such a system is usually integrated inside one shelf. For larger router/switch designs where multiple shelves are involved, the intersystem communication is handled by optical fiber interconnect. In terms of implementation, we will present snapshots of reality through various vendor case studies in several chapters in which we cover representative products for each category.

SUMMARY

In this chapter, we continued looking at fundamental concepts. We first defined the multiple contexts of network realms from CPE to the WAN core. We clarified the different design requirements that drive the network equipment manufacturers' thought processes in each network context. We briefly discussed the most important operations that need to happen in real time and at wire speed inside switching/routing gear that is designed to handle packets. We also introduced many of the fundamental concepts, which we will review in depth in the subsequent chapters that discuss the techniques and typical products that implement them.

NETWORK PROCESSOR ARCHITECTURES

CHAPTER 4
IBM POWERNP™

We will now take a closer look at some of the most advanced *network processing unit* (NPU) architectures that have been proposed by several vendors. Some of these vendors are solidly established and some are promising startups. Our coverage will not be exhaustive in detail for two reasons. First, this book does not take a cookbook approach. Second, the subject is massive and only so much can be packed in one single book. Detailed information can be found in each vendor's product datasheets and chipset documentation. We relied on these items as the main sources for compiling these overview chapters.

Our approach is to explain the fundamentals of each architecture by not only showing the breakthroughs, but also by highlighting the techniques, modules, analogies, and paradigms that we may have already reviewed in earlier chapters. We will look at how a complete network gear solution can be implemented for tackling design problems through the various NPU architectures and will pinpoint the strong and weak points of each approach.

In this chapter, we look at IBM's PowerNP™ family of network processors. More specifically, we look at the architectural structure of the NP4GS3 network processor, the capabilities, and the complementary peripheral chips (queue managers, switch fabrics, interface converters, and so on) that are required to produce a working system based on the IBM platform. This requires an overview of the systems model that IBM NPUs favor. Finally, we will examine the tools that enable and support development of this IBM NPU and will discuss the design trade-offs that these network processors impose on the designer of switching/routing equipment.

IBM POWERNP: THE BIG PICTURE

IBM is one of the uncontested leading vendors in the global information industry. It combines advanced networking expertise with unparalleled microelectronics technology, deep submicron design, and semiconductor manufacturing process capabilities. Through its IBM Microelectronics unit, a very large engineering group has been put together to design and bring to market the various IBM families of network processors. In addition to being the leading captive semiconductor producer on the planet, IBM has been one of the leading network equipment manufacturers for many years, as evidenced by their NPUs. IBM tangibly implements pertinent sophisticated know-how, which continues to pour out of its world-famous research teams and, more specifically, teams that are engaged in fast networking development in Yorktown Heights, New York, in Rueschlikon, Switzerland, and in Haifa, Israel.

If we step back and look macroscopically at the PowerNP family, we find IBM's NPU flagship—NP4GS3—at the top of the line. NP4GS3 is also known in the industry as *Rainier*.

The two variants of the NPe405 are available at the low end of the spectrum of IBM's NPU offering. The NPe405 (dubbed H and L, respectively) network processors have embedded support for various interfaces, such as Fast Ethernet, *High-level Data Link Control* (HDLC), and so on. They are focused primarily on the access equipment market and are not capable of handling very-high-speed routing/switching in a multiservice routing/switching environment like their more powerful sibling. Because the underlying architectures are different, the executable code for the NPe405 and the code for the NP4GS3 are incompatible. IBM customers using the e405 family will have a certain migration path inside the e405 family as equipment performance requirements increase. This, of course, helps preserve the customer's software investment. However, we will not be expanding on this low-end product here. Interested readers can find more information about it directly from the IBM Microelectronics web site at *www.chips.ibm.com.*

Below the NP4GS3 and above the NPe405, IBM introduced the NP2G in 2002. This network processor is based on the same powerful architecture, but has fewer resources than the NP4GS3. More specifically, it offers 12 picoprocessors. Interested readers can find more information once again from the IBM Microelectronics web site.

The NP4GS3 is one of the most powerful NPUs currently in the market. It contains 16 so-called picoprocessors that handle packet manipulation operation. It also contains a powerful PowerPC 405 *central processing unit* (CPU) core that handles control functions. Each picoprocessor is a full-fledged 32-bit *reduced instruction set computer* (RISC) CPU running at 133 MHz with a 1-cycle *arithmetic logic unit* (ALU) and with arithmetic, logical, compare, shift/rotate, and bit test/set/clear instructions. It also contains a scalar read-only register bank that provides interrupt vector management, a timestamp, *pseudorandom number generation* (PRNG), processor status, and work queue status, namely whether the information at hand refers to an ingress or egress queue. Each picoprocessor supports 2 threads in hardware (for a total of 32 threads per NPU) and includes 9 hardwired function units for common tasks such as copying string, checking bandwidth policy, and generating and verifying checksums.

Besides a switching engine, each NP4GS3 also contains *tree-search engines* (TSEs), one of which is shared with each pair of coprocessors. A TSE is based on 3 different algorithms. There are also frame processors, Ethernet *Media Access Control* (MAC) controllers, four 1 Gbps media access ports

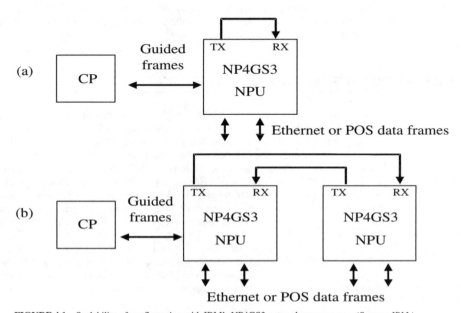

FIGURE 4.1   Scalability of configuration with IBM's NP4GS3 network processors.   *(Source: IBM.)*

(given the fact that the aggregate bandwidth of the NP4GS3 is 4 Gbps), 2 full-duplex switch fabric interfaces, and separate interfaces to 10 external memory banks. These interfaces can support up to eight *double data rate* (DDR) *synchronous dynamic random access memory* (SDRAM) ports and two *zero bus turnaround* (ZBT) *static random access memory* (SRAM) ports. The NP4GS3 offers sophisticated capabilities in terms of hardware-based scheduling, policing, and flow control, including the *Shock-Absorber Random Early Detection* (S-RED) algorithm. IBM claims that S-RED is more elegant, dynamic, and efficient in its ability to self-adjust to different traffic rates as well as handle peak traffic flows than the traditional *weighted random early detection* (WRED) algorithm. Therefore, IBM has been pushing for the industry-wide acceptance of this algorithm through the *Institute of Electrical and Electronics Engineers* (IEEE) standardization process.

The NP4GS3 can easily cope with a single OC-48 channel or up to 40 Fast Ethernet 100 Mbps ports. Alternatively, it can be configured to handle a *fat pipe* in an OC-48c configuration. In each NP4GS3 NPU, the 16 parallel coprocessors (picocode processors) and the 9 available hardware-assisted coprocessors (one for each of the 16 parallel picocode processors) provide in total a staggering 2,128 *millions of instructions per second* (MIPS) of processing power with 32K words of internal instruction memory. Its flexibility is driven by picocode and application software rather than any *application-specific integrated circuit* (ASIC) components.

Scalability was one of the highest priorities for the IBM designers. As a result, the NP4GS3 processor can be connected in series with another NP4GS3 chip as shown in Figure 4.1 using the switch fabric interfaces under the control of the PowerPC 405 core in one of the two NPUs. This scheme effectively doubles the bandwidth of the system. This brings the NP4GS3 extremely close to 10 Gbps, which is the next expected equipment performance milestone. IBM is working on new single-chip products that will be able to comfortably handle that speed.

To further emphasize the scalability of the architecture, we must mention that up to 64 NP4GS3 NPU chips can communicate with each other through an external switch fabric under the control of an external CPU in order to provide massive scalability to higher bandwidths. In such an arrangement, each NPU will only handle a portion of packet-processing operations. Figure 4.2 depicts this scheme.

IBM sees the NP4GS3 as a candidate for customer premises network equipment, edge network devices, or even core network gear. Due to its performance and scalability, it is targeted to *local/metropolitan/wide area network* (LAN/MAN/WAN) routers, *Multiprotocol Label Switching* (MPLS) routers, *Internet Protocol* (IP) over *Synchronous Optical Network* (SONET), SONET Transport, *digital subscriber line access multiplexer* (DSLAM), *Internet service provider* (ISP) access boxes (*quality of service* [QoS]), firewalls, server adapters, *storage area network* (SAN) and LAN adapters, and so on.

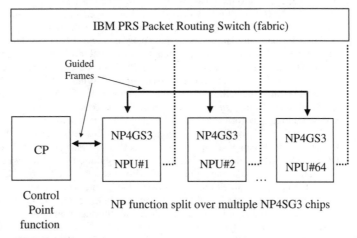

FIGURE 4.2   Up to 64 NP4GS3 network processors can be connected via an external switch fabric. *(Source: IBM)*

In order to help consolidate a data plane network-processing idea into a complete product design, IBM also provides a wide selection of other necessary components. The components include the leading switch fabric chips, SONET/*Synchronous Digital Hierarchy* (SDH) framers, optical transceivers, backplanes, and interface converters. Almost all types of dynamic and static memory needed for packet buffering and lookup tables can be added to this impressive list. Traditional PowerPC CPUs, which are used on the control plane to complete a systems design, should also not be forgotten.

Architecture

The NP4GS3 architecture combines an array of eight so-called *dyadic processor units* (DPPUs) next to the embedded PowerPC 405 CPU core. These offer a combined total of 16 active threads and 16 inactive threads. This means that a single NPU can process up to 32 frames at the same time with zero context-switching overhead when switching between threads. In other words, absolutely no cycles are lost when switching from one thread to another. All incoming packet data reside in system memory on the NPU and do not need to be copied to and from some *working, register,* or *user area* for processing, which is usually the case in a computing environment. The data are processed right where they are stored, which definitely improves the performance of the architecture. Support for large lookup tables for layers 2, 3, 4, and other higher-layer functions are performed by hardware-assisted programmable picocode processors using specialized coprocessors for tree searching and updating.

The packet-processing prowess of the NPU is distributed among its picoprocessors, coprocessors, and hardware-assisted units. The NPU system design minimizes contention for access to the coprocessor engines. Forwarding and filtering is done without retaining any data copy by hardware-implemented mechanisms, which ensures the wire-speed performance of the chip. Common layer 2, 3, 4, and higher functions can be implemented in extremely fast schemes. For example, support is available for the on-the-fly alteration of frames on well-known protocol elements, such as the *Time to Live* (TTL) field in the IP header. Tag deletion for *virtual LANs* (VLANs) and MPLS label manipulation (such as delete or swap) can be implemented efficiently and quickly in IBM's picocode.

As mentioned earlier, in order to ensure scalability with high performance, the NP4GS3 enables different connectivity schemes that distribute the necessary functionality in *steady state* and *nonsteady state processing*. By executing NPU picocode, the NPU itself performs all steady state operations. These operations include filtering, frame forwarding, frame alterations, protocol layer 2, 3, and 4 processing, classification, QoS, traffic management, and accounting. At the same time, the so-called *control point* (CP) processor performs nonsteady state functions. These include route discovery, updates to the tree, updates to the *Open Shortest Path First* (OSPF) database, *Simple Network Management Protocol* (SNMP) agent processing, debug/diagnostics, configuration management, and deep frame processing, as well as executing applications that the *network equipment vendor* (NEV) develops.

The CP is an external processor that supervises and serves a system comprised of several NPUs. The NP4GS3 is designed to accommodate many vendor designs with various CP-NP configurations. Refer to Figure 4.1 for an example of two NPUs and one CP. Either of the internal PowerPC CPUs or an external one can be used as the configuration CP. When traffic requires nonsteady state operations in such an arrangement, the NPU communicates with the CP by special frames of a special EtherType. IBM calls these frames *guided frames*, and they can contain data and one or more commands. The CP uses them to update forwarding tables in the NPUs (that is, trees).

This two-NPU configuration can support 80 Fast Ethernet (10/100 Mbps) ports or 8 Gigabit Ethernet ports. It can also support eight OC-3/OC-12 *Packet over SONET* (POS) ports or even two OC-48 POS ports. In contrast, a single-NPU configuration where no switch fabric is required can support half as many of the same ports in an NP-CP scheme. Up to 16 NPUs can be controlled by 1 CP. When the design requires more than two NP4GS3 network processors, as shown in Figure 4.2, a switch fabric is required for the data movement. In the configuration shown in Figure 4.2, the NPUs split the set of chores of layer 2 forwarding and filtering (frame repository and queuing), layer 3 forwarding and filtering (flow control and frame alteration), and layer 4 flow classification based on priority and multicast handling. They also maintain network management counters. The CP in

Figure 4.2 handles layer 2 support (spanning tree), layer 3 support (OSPF, the *Routing Information Protocol* [RIP], and the *Border Gateway Protocol* [BGP]), networking management (the *Remote Network Monitoring* [RMON] agent), configuration, diagnostics, and other box-related functions. Up to 64 NPUs can be connected with a switch fabric in one system. This scheme supports up to 1,024 Fast Ethernet ports or multiple POS configuration possibilities.

The NP4GS3 is built using a 0.18μ copper-interconnect process. It is housed in a 1,088-pin package (with 815 signal *input/output* [I/O] lines) using a 24-pin debug bus. Its core is powered by a 1.8 voltage supply, whereas the DDR and ZBT RAMs are powered by 2.5V and the so-called *data mover units* (DMUs) as well as the PCI interfaces are supplied by a 3.3V supply. Power dissipation is estimated at 14 watts.

MAJOR FUNCTIONAL BLOCKS IN THE NP4GS3

Figure 4.3 shows an architectural view of the NP4GS3 system. This illustration shows its major functional components with the abbreviations that IBM uses to describe them in its technical literature. These blocks are as follows:

- *Physical MAC Multiplexer* (PMM).
- *Ingress Enqueue/Dequeue Scheduler* (I-EDS).
- *Switch Interface* (SWI).
  - *Switch Data Mover* (SDM).
  - *Switch Cell Interface* (SCI).
  - *Data-Aligned Serial Links* (DASLs).
- *Egress Enqueue/Dequeue Scheduler* (E-EDS).
- Traffic Shaper.
- *Embedded Processor Complex* (EPC).
- *Embedded PowerPC Complex* (ePPC).

Various storage areas are also deployed throughout the system.

Imagine that the data flow on the ingress side proceeds from the bottom of the drawing (the network) upward, toward the left-hand side of the drawing (where the I-EDS block is), and then upward toward the top center to the output (switch fabric). In an egress flow, data enters the chip from the top of the drawing and proceeds toward the right-hand side of the drawing (where the E-EDS block is) and then downward toward the network interface. The center of the drawing contains the processor complex that acts on the frames while on ingress or egress flows. The various types of storage are also shown macroscopically. We will now look at each of these blocks.

The PMM provides interfaces from POS framers and Ethernet *physical layer* (PHY) chips to the NPU's four flexible external ports. It contains two banks of five DMUs each for the ingress and the egress ports. One pair of DMUs is reserved for internal wraparound communications from egress to ingress inside the NPU. The rest can be configured to support either 10 Fast Ethernet 10/100 Mbps ports per DMU, a 1 Gigabit Ethernet per DMU, 4 OC-3 POS per DMU, 1 OC-12 POS per DMU, or 1 OC-48 per 4 DMUs. Each port contains an Ethernet MAC, which can support 1 full-duplex Gigabit Ethernet link or, with *time division multiplexing* (TDM), 10 full-duplex Fast Ethernet connections. All RMON groups are supported by special hardware counters in each MAC for remote monitoring in network management. The MAC controllers support 802.3ad link aggregation, 802.1q VLAN detection, flow control, and even jumbo frames. The NP4GS3 through its standard *Gigabit Media-Independent Interface* (GMII) interface supports Gigabit Ethernet PHY chips that are directly attached. Alternatively, the available SMII interface can be used to support any mix of 10 Fast Ethernet ports running in combinations of 10 and 100 Mbps.

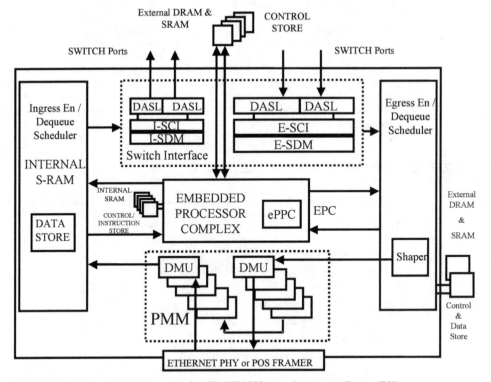

FIGURE 4.3 The internal block structure of the IBM NP4GS3 network processor. *(Source: IBM)*

The I-EDS stores frames from the DMUs into the data store. It performs some filtering decisions and frame alterations, such as VLAN tags. It then dequeues the frames from the data store and schedules them to be forwarded or discarded. This happens when the target NPU or the switch fabric indicates to this NPU that they are running low on resources.

The SWI provides a data cell-based interface between NPUs either via switching fabric (for three or more NPUs) or direct wire connections (for one or two NPUs). The SDM and the SCI for both the ingress and the egress path convert the output of the EDS logically into a cell flow, and vice versa. They also provide/receive cells to/from the PHY.

The DASL is IBM's fast method for implementing the physical interface between the NPU and the switch fabric, between the ingress and egress sides of one NPU, or between the ingress and egress sides of two NPUs.

The E-EDS receives frames through the switch interface. It reassembles them because they arrive in a cell flow. It then enqueues the resulting frames into its data store where extensive frame processing is provided. It finally dequeues the frames from the data store and schedules them to be forwarded.

The Traffic Shaper manages bandwidth on a per-frame basis for all egress DMU ports. It is an optional NPU component and can be configured by software. It implements *weighted fair queuing* (WFQ) regulation for up to 2K queues, which sustains a good performance in a *Differentiated Services* (DiffServ) environment. The Shaper discards traffic depending on its configuration and based on several algorithms such as RED and WRED.

The heart and brains of the NP4GS3 are formed with the combination of the EPC and the ePPC. Figure 4.4 shows the EPC in more detail. It contains eight DPPUs and nine Hardware-Assist coprocessors. It determines what must be done with the frames received in the data store on either the ingress or the egress side of the NPU. It provides the overall steady state control and programmability of the NPU—in other words, the code that makes the NPU equivalent to a programmable ASIC. The ePPC is a specialized PowerPC CPU core with 16KB of instruction cache and 16KB of data cache, which can be used to provide CP functionality—in other words, the nonsteady state processing for packets.

As mentioned earlier, each NP4GS3 has 8 DPPUs (16 programmable protocol processors) and each DPPU has 9 Hardware-Assist coprocessors. These packet processors share 128K of the local control store memory; for more space, external memory is needed. Incoming packets/frames are allocated and assigned to specific threads. When processing is completed, they are de-allocated and passed over to the corresponding scheduler.

NEVs can modify the code that runs in the NPU or develop their own software that runs on the CP processor. IBM provides both high-level (C *application programming interfaces* [APIs]) and low-level APIs to facilitate the interface with the network-processor system for software developers.

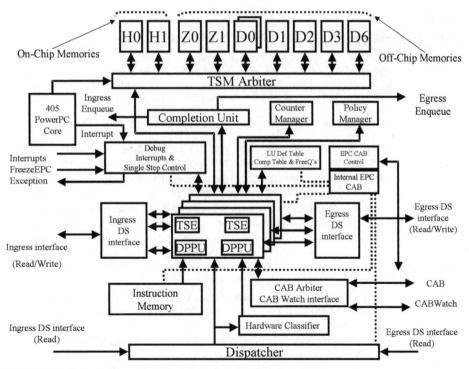

FIGURE 4.4 The internal structure of the NP4GS3 chip's EPC. *(Source: IBM)*

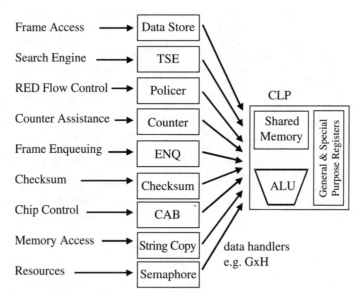

FIGURE 4.5   Nine Hardware-Assist coprocessors per DPPU. *(Source: IBM)*

SPECIAL COPROCESSOR AND ASSIST HARDWARE

Figure 4.5 shows the block structure of a DPPU where the picocode is executed. The figure also includes the nine Hardware-Assist coprocessors. These coprocessors are associated with each DPPU and function in parallel with the data movement by accessing and maintaining internal registers per thread.

The suite of the nine Hardware-Assist coprocessors comprises the following units:

- **Data Store Coprocessor**   This handles all data transfers (read/write) between ingress and egress data stores and the shared memory data pool. It is structured to handle 128 bits per transfer.

- **CAB Interface Coprocessor**   This provides all DPPUs with access to internal registers, counters, and memory for debug or statistics gathering.

- **Enqueue Coprocessor**   This interfaces with the Completion Unit (discussed later in this section) from the special hardware units to enqueue frames to the switch and to the target port queues.

- **Checksum Coprocessor**   This deals with half-word data in order to generate half-word header checksums based on RFC 1071 for the computation of Internet checksums. It works based on two instructions: *generate checksum* and *verify checksum*. All checksum calculation results are stored in a special accumulation scalar register.

- **String Copy Coprocessor**   This moves multibyte data within the shared memory pool. The commands it understands pass the source address, the destination address, and the number of bytes needed to encode the string.

- **Policy Coprocessor**   This examines the flow control and information, and checks to make sure everything conforms to preallocated bandwidth.

- **Counter Coprocessor**   This interfaces threads with the Counter Manager. It updates counts and manages an eight-level command queue.

- **Semaphore Coprocessor**   This controls access to shared resources such as tables. It grants access based on a handshake mode that issues Request Order and Dispatch Order pairs of signals.

- **TSE Coprocessor**   This handles all table searches and updates. Almost every frame that is processed by the system uses this coprocessor. The search engine retrieves forwarding decisions from the local routing tables in each NPU. If these local tables need to be updated, the CP processor will do it through the use of guided frames. The TSE Coprocessor provides tree search and modification functions for requests issued by picocode threads. As two coprocessor locations are used, every thread can execute two searches simultaneously. The NP4GS3 relies heavily on searching tree structures for issues such as layer 3 IP address routing tables, layer 3 and higher frame filtering, layer 2 MAC address port mapping, flow control, and so on. It supports three types of tree search algorithms: *full match* (FM), *longest prefix match* (LPM), and *software-managed trees* (SMTs), which is an IBM algorithm invention that allows multiple leaves that can be chained in a linked list. The TSEs can perform 8.5 million searches per second for layer 3 routing (using the LPM algorithm) and 12 million searches per second for layer 4 classification (using the five-tuple approach). These numbers can be improved with the external use of a *content-addressable memory* (CAM).

Beyond the coprocessors, the NP4GS3 contains special hardware units that are also shown inside the EPC block, as depicted in Figure 4.4. These units offer the following functionality:

- A Dispatcher tracks the use of threads. It is engaged right at the beginning of processing as it is the unit that fetches the initial frame data before thread assignment occurs.

- A Completion Unit is responsible for maintaining the order of frames, which are enqueued, so that both ingress/egress flow control and the overall scheduler can function properly.

- A Policy Manager performs policy management based on four management algorithms as specified in *Internet Engineering Task Force* (IETF) RFCs 2697 and 2698. They are the *single-rate three-color marker* (srTCM) (in color-blind or color-aware modes) and the *two-rate three-color marker* (trTCM) (also in color-blind or color-aware modes) algorithms.

- A Hardware Classifier is engaged in the classification of frames from various realms, such as Ethernet (802.3 and DIX), layer 3 (IP), VLAN header detection, and guided traffic.

- A Counter Manager is used by the EPC to control several counts used by the picocode for various purposes, such as statistics, policy management, and flow control.

IBM considers the last two of these units particularly critical for the robustness of network equipment designs built around the NP4GS3 and for the predictable delivery of desired functionality.

The NP4GS3 supports different types of memory that are connected to the chip in different locations and used for different purposes. Memory for the NP4GS3 can be internal (on-chip) and SRAM, or it can be external (off-chip). In the latter case, it is either ZBT SRAM or DDR SDRAM. Internally, the NP4GS3 contains 384KB of memory that is used for internal NPU control information or for storing frame data. The large amount of supported memory enables a large size of forwarding tables to be used in the local NPU. The NP4GS3 can easily sustain 500,000 table updates per second.

Figure 4.6 shows what types of external memory are used and for what type of storage. The abbreviation Z stands for ZBT SRAM memory. The *pattern search control blocks* (PSCBs) are structures that define trees. They are used by the TSEs to locate or update tree data. Since trees are used extensively in the NP4GS3, the PSCBs are set up deliberately in ZBT SRAM memory where very fast tree searches can occur. The abbreviation S stands for SRAM, and the abbreviation D denotes DDR SDRAM.

The two high-speed 7 Gbps switch fabric interfaces can be used to connect the NP4GS3 to two different switch fabrics (such as IBM PowerPRS™ chips) to provide redundancy and fault tolerance. The use of an IBM-provided DASL-to-CSIX converter chip in conjunction with a fabric interface chip enables the use of other non-IBM switch fabrics. If both interfaces are used, these two extra chips must be doubled.

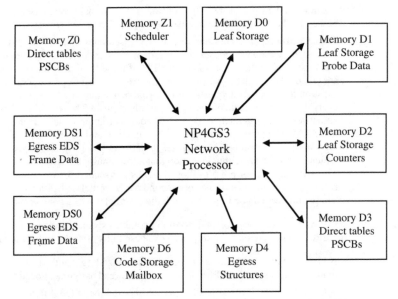

FIGURE 4.6 Types of external memory used in systems built with the IBM NP4GS3 .
(Source: IBM)

An external PCI bus operating at either standard 33 MHz or 66 MHz is provided for the interface of the NP4GS3 with a host CPU or an external CP processor.

SOFTWARE ARCHITECTURE

Picocode is designed for the NPU's EPC part called *General-Purpose Processors* (GPPs). These processors contain array registers, scalar registers, and general-purpose registers. The picocode threads execute in the EPC's DPPUs, which contain what IBM calls *Core Language Processor* (CLP) engines. The CLP in general is a nonpreemptive, event-driven processor accessible in IBM NPU Assembler. Each CLP can execute up to two threads. IBM NPU Assembler language predictably contains integer operators, built-in functions, string operators, and string expressions. IBM has also developed a native C compiler. Before that, the implication of the lack of high-level language support was that the architectural incompatibility between the NP4GS3 and the lower members of IBM's NP family—the e405—meant that assembly code for one cannot be used for the other. This could be a problem for some users. It has been resolved with the arrival of IBM's NPU C compiler.

To briefly address the computing model, NP4GS3 consists of four types of what IBM calls data handlers. Picocode executes when threads are dispatched using the appropriate handlers inside the CLP engine. Thirty-two handlers are available (the same as the number of threads):

- *General Table Handler* (**GTH**)   A GTH handles control frames, which require access to tree memory. There is only one GTH per NPU chip, and it operates only on the egress side of the network processor.
- *Guided Frame Handler* (**GFH**)   The GFH handles control frames that are coming from or going to the CP or other NPU chips. A GFH can forward frames to the GTH by re-enqueueing frames to

the internal GTH queue. There is only one GFH per NPU, and it operates on either the ingress or the egress side of the network processor.

- *General PowerPC Handler* **(GPH)**   The GPH handles control and data frames transmitted to or received from the CP processor. One thread receives flows and the other one transmits them. Each NP4GS3 network processor has two GPH.

- *General Data Handler* **(GDH)**   The GDH handles data frames that enter from the network through the PHY ports. Each NP4GS3 network processor has 28 GDHs.

SOFTWARE AND SYSTEMS DEVELOPMENT AROUND THE NP4GS3

Figure 4.7 shows how the various software components are combined in a system that comprises multiple NPUs to produce a modular and flexible solution. The customer's applications can communicate through special facilities. The NPU runs control picocode, management picocode, and forwarding picocode. Through special low-level APIs, the CP interfaces its NPAS environment with the network-processor realm. NPAS with high-level APIs interfaces for instance a local SNMP agent or exception forwarding code with other vendor applications, such as routing table management, an OSPF routing protocol, and so on.

Customers' applications, which must execute on the CP processor, can communicate through the *Network Processor Application Services* (NPAS) (application services) high-level C-language API. IBM supports Linux and WindRiver's VxWorks. Customers can develop and test their application code under various versions of Windows, Sun Solaris and Red Hat Linux. IBM offers a developer's toolkit, which provides a series of development tools that start from a Core Simulation Model, a *network-processor-specialized assembler* (NPASM), a C compiler, an interpreter of the picocode binary image file, a debugger, a network simulator, a performance profiler, a test-case generator, and scripts

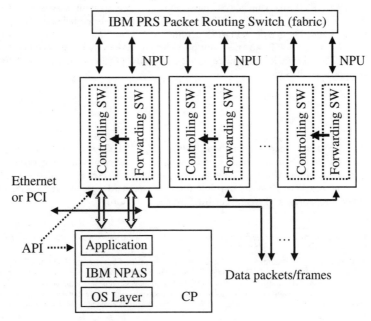

FIGURE 4.7   Software structure for a system based on multiple NP4GS3 chips. *(Source: IBM)*

that can be extended by engineers via the use of *Tool Control Language/Toolkit* (TCL/Tk). Network-processor code can be developed and tested without even having access to hardware prototypes of a new system.

The C compiler, which was lacking from the initial product launches in this family, is a valuable addition that can dramatically simplify PowerNP application code development. It is especially useful for creating prototypes of an application quickly. This optimizing C compiler implements a subset of ANSI C and provides a set of APIs to access onboard coprocessors. It also supports inline assembly to allow hand optimization of critical sections of the code during the fine-tuning process of a finished product.

An important factor from a business standpoint is that IBM actually delivers code that the customers can do two things with. They can keep it as is, concentrating their efforts on developing their application or supervising software (which is the most typical case) in cases where their software will be meant to run on the CP while leaving the internals of the NPU software intact. On the other hand, in more elaborate cases where fine-tuning is required, they can modify the picocode to produce the desired and intended behavior and performance. IBM provides handholding from authoring device drivers all the way to full-fledged hardware/software design validation and consultation. With NPAS, IBM's customers can license production-quality infrastructure, control, protocol, or forwarding software. NPAS contains numerous components that vary from MPLS to IPv4 over SONET, from 802.1D bridging and 802.1Q VLANs all the way to *File Transfer Protocol* (FTP), *Transmission Control Protocol* (TCP), and *Point-to-Point Protocol* (PPP) implementations, from full-fledged DiffServ to simply handling jumbo frames on Gigabit Ethernet, from *management information base* (MIB) support to unicast/multicast filtering/forwarding of IPv4 on Ethernet, etc.

Besides the data plane processing, IBM's basic and advanced software offerings provide strong support for control plane development for both internal PowerPC and external choices of a CP processor. Code is readily available for boot, system management, diagnostic services, interface management, protocol services, memory management, GxH (with *x* as a wild character here) frame handler formatting, *traffic-engineering* (TE) management, physical transport services, exceptions, and so on.

After simulations, code can be executed and debugged on physical hardware by using IBM's Reference Platform. This is a 5U rack-mountable chassis with integrated power, cooling, and backplane assemblies. It contains a Packet Routing Switch Fabric blade (target) option. Up to four blades can be stacked with external DASL cabling. A PCI card implements a CP processor with a PowerPC 750. An optional 4GS3 carrier card provides an NP4GS3 with its own embedded PowerPC 405. It offers 22 sockets with the choice of a 2-port GBIC Gigabit card, a 20-port 10/100 TX card, or a 1-port OC-48c POS card.

Performance

It is important to mention that the IBM PowerNP NP4GS3 was the first network processor to pass all the required tests in the OC-48c configuration for the new LinleyBench 2002 benchmark.[1] In addition, the NP4GS3 was the first chip in the industry objectively verified to operate at 10 Gbps while running the new IPv4 forwarding industry standard benchmark[2] established by the *Network Processing Forum* (NPF) and certified by The Tolly Group.

---

1. The details about this benchmark can be found at the Linley Group's web site at *http://www.linleygroup.com/benchmark/linleybench.html.*

2. More specifically, regarding the OC-48c configuration of the LinleyBench 2002, the NP4GS3 passed all required IPv4, DiffServ-with-30K-routes and DiffServ-with-100K-routes tests. The NP4GS3 passed all the IPv4 forwarding tests by forwarding all the frames at all the frame sizes with zero frame loss in an environment that included the generation of Internet-like traffic, which was sent to the NP4GS3-based system and then successfully routed the entire data stream to its next destination without any errors. A full disclosure of the results can be found at *http://www.ibm.com/chips/techlib/techlib.nsf/pages/linleybench.*

IBM is an active member of the NPF. We discuss this organization in more detail in Appendix III, "Standardization Efforts in Network Processing." Among other things, the NPF has created an industry standard IPv4 forwarding benchmark. IBM's results, along with interfaces, configuration parameters, and test setup, have been independently certified by The Tolly Group[3] and released by the NPF.[4] IBM has been reported to achieve greater than 10 Gbps of throughput by employing three PowerNP NP4GS3 network processors in the data path.

THE NP4GX: IBM'S SECOND-GENERATION OC-48 NETWORK PROCESSOR

At the Network Processors Conference West in October 2002, IBM Microelectronics announced the arrival of the NP4GX, its second-generation OC-48 processor. The impressive characteristic of the new NPU is that it enhances the performance of the NP4GS3 by offering almost an instant tripling of computational "lung" capacity while preserving full compatibility with the NP4GS3 processor's software environment.

The NP4GX is built using the IBM 0.13μ Cu-metal *complementary metal oxide semiconductor* (CMOS) process technology, and it has been targeted to operate with a 500 MHz clock. The die contains 16 packet processors and several coprocessors like the NP4GS3, but the instruction memory has now been doubled to contain 64K instructions. Given the fact that its predecessor was more than capable of handling sophisticated DiffServ types of applications, this should now enable more applications that can utilize the significant computational headroom that the new processor offers. The PowerPC 405 core previously used in the NP4GS3 network processors has been replaced in the NP4GX by a PowerPC 440 core, which is a dual-issue superscalar RISC processor offering 1,000 MIPS capabilities that runs at 333 MHz or 500 MHz.

In terms of interfaces, the previously integrated DASL ports of the NP4GS3 are now replaced by a CSIX-L1 for the interface with a switch fabric, whereas a couple of look-aside interfaces implemented according to the NPF LA-1 specification allow the support of either external coprocessors or *quad data rate* (QDR) SRAM memory. Cleverly, the memory controllers of the multiple DRAM channels of the NP4GX have been designed to also support *fast cycle RAM* (FCRAM), in addition to the native DDR SDRAM.

The NP4GX network processor's package will be a HyperBGA replacing the ceramic package of the NP4GS3. It is estimated that it will consume around 10 watts. IBM released the first samples of this network processor in early 2003.

TRADE-OFFS WHEN DESIGNING WITH NP4GS3

Designing high-speed network equipment with the IBM NP4GS3 network processor brings some clearly discernible advantages to the designer, but he or she must face some trade-offs as well.

On the positive side, the performance afforded by the NPU architecture is flexible, fast, and scalable. The behavior choices implemented in picocode are endless. Customers can easily implement differentiating features into their products by simply developing the appropriate picocode in the NPU(s) they use. Beyond that, however, IBM's fine-tuning of the robust internal systems design

---

3. *http://www.tollygroup.com.*

4. *http://www.npforum.org.*

makes the need for the customers to design their own optimized networking ASICs a problem of the past. The NP4GS3 is so integrated with ancillary functionality, such as traffic management, MAC layers, and switch fabric interfaces, that a whole Gigabit Ethernet line card can be produced simply by adding PHY chips and memory. IBM's software is not only tested, but it is also fully validated. This means that customers can simply plug it into their own system (if they don't require any modifications of the picocode) and it will work, thus saving themselves precious time to market. All future specification changes of network equipment designed around the NP4GS3 can be done in software offering flexibility and further preservation of the customer's software development investment.

The throughput speed, which results from the wide-range of optimized hardware-assisted functionality in conjunction with the distributed-computing platform of the NP4GS3, is undisputed. Since scalability is a major concern for NEVs, IBM offers multiple ways of drastically and easily expanding the bandwidth of a system built around its NPUs while preserving software compatibility and investment. One cannot ignore the fact that the NP4GS3 is coming from a globally successful giant with highly diversified and deep technology know-how. IBM backs a product with ancillary product offerings, tremendous technical support on numerous fronts, presence around the clock worldwide, and a unique commitment to the industry.

On the less positive side, this NPU performs traffic management only on the egress side; therefore, if traffic management for some customers must be done on the ingress side, then an external traffic manager must be used. This will significantly complicate the overall system design. It is easier to integrate the NP4GS3 with IBM PowerPRS switch fabrics. The extra flexibility gained comes at the price of extra hardware if a non-IBM switch fabric is used. This is an extremely complex product. Programming it in picocode represents significant challenges. Developing picocode in assembler, fine-tuning the overall system, and deciding what lies in which memory bank and which of the numerous coprocessors needs to be invoked at what time in order for the application to achieve optimum performance is a rather complicated task. No one should underestimate it.

We will conclude this chapter by saying that in order to appreciate the full impact of the IBM technology and the trade-offs involved in designing a fast network-processing system using IBM network-processing components, it is obvious that using an IBM switch fabric is a less tedious and more straightforward approach. The leading IBM switch fabrics, as well as the corresponding IBM chips that handle the sophisticated interface of the switch fabric and backplane with a network processor inside a complete fast-switching/routing system, are extensively discussed as one of the leading-vendor-technology case studies in Chapter 14, "Switch Fabrics." Interested readers may want to consult that chapter in order to obtain a more rounded view of the IBM approach. This chapter may also be of interest to readers who want to take a closer look at the intensity and breadth of network technology research that IBM has been conducting at its world-famous lab in Rueschlikon, Switzerland.[5]

SUMMARY

In this chapter, we reviewed the architecture of IBM's PowerNP by taking a close look at NP4GS3, IBM's flagship network processor. We reviewed its structure as well as its many advantages and pinpointed a couple of potential shortcomings. We identified design issues with which a systems architect must be familiar, reviewed software development tools and approaches for the IBM NPU platform, and pointed out several trade-offs that should be considered when deciding whether this is the right platform to use for a new design. For a complete view of the IBM network-processing product-line landscape, however, interested readers are referred to Chapter 14. Chapter 14 is dedicated to switch fabric technologies and provides extensive coverage of IBM's leading switch fabrics and fabric-NPU interface chips that accompany what has been described in this chapter.

---

5. *http://www.zurich.ibm.com/cs/index.html.*

REFERENCES

High-quality technical documentation for IBM's network processors, switch fabrics, queue manager, and interface conversion chips offered by IBM Microelectronics can be found at their web site at *www.chips.ibm.com*, where an elaborate technical library is available with detailed datasheets, presentations, and application notes.

CHAPTER 5
INTEL IXA™ NETWORK PROCESSORS

In this chapter, we will look at Intel's approach to network processing. At the time of this writing, Intel had announced three new *network processing unit* (NPU) chips as part of its second generation of network processors. These are all part of its evolving *Internet Exchange Architecture* (IXA) architecture family.

Compared to IBM's approach, which as we learned in the previous chapter is characterized by the ability to offer very high performance and to offer systems designers a complete one-stop shopping solution, Intel has taken a different route to tackle the network-processing challenge. It originally started with NPUs that performed modestly (namely, the IXP1200 family). These NPUs solidified the company's grip on the *local area network* (LAN) market, consisting of mostly *customer premises equipment* (CPE) and access equipment. So far, these have proven to be the most commercially successful network processors based on the number of market designs, according to Intel's claims in the trade press. Intel has capitalized on the ease of systems hardware and software design around its NPUs, especially given its outstanding software development environments and support (also from third parties). As it continues to improve in performance with its second-generation processors, Intel is clearly setting its sights on the faster, more lucrative edge and core equipment markets.

INTEL IXA: THE BIG PICTURE

Intel IXA is an end-to-end family of high-performance, flexible, and scalable hardware and software development building blocks that have been designed to satisfy the growing performance requirements in today's networks. The architecture is based on programmable silicon and software building blocks.

At the low end of its offering, Intel has positioned its IXP220, 225, and 425 NPUs as integrated solutions that are suited for *small office/home office* (SOHO) and *small medium enterprise* (SME) equipment in a CPE premise. However, the cornerstone of Intel's IXA family is the IXP1200 network processor and its variants IXP1240, 1250, and so on. These NPUs run at different clock frequencies and with or without added features such as embedded *cyclic redundancy check* (CRC) and *error correction code* (ECC) memory access. On top of the 1200 family, Intel has recently brought a couple of powerful additions into the market. For the OC-12 to OC-48 (2.5 Gbps) realms, Intel introduced the IXP2400 in 2002. For the OC-48 to OC-192 (10 Gbps) realms, Intel's flagship NPU is the IXP2800 network processor.

Unlike IBM, Intel does not yet offer switch fabrics; therefore, standard interfaces must be provided to connect to fabrics provided by other vendors. Intel's first-generation NPUs were already designed to enable the high-speed manipulation of packets across several media types and forward packets efficiently with the appropriate modification of packet headers while reserving sufficient compute cycles (headroom) for network management and other analytical tasks. In the latest entries, performance can be scaled from OC-3 (155 Mbps) links all the way to OC-192 (10 Gbps).

Intel IXA is a systems architecture that is used for network-processing purposes. It can be characterized by three predominant traits:

- **Intel's Microengine technology** A subsystem of programmable, multithreaded 32-bit *reduced instruction set computer* (RISC) microengines that have hardware multithread support. When these traits are combined, they provide over 1 giga-operations per second (more than 1,000 mega-operations per second). This combination enables high-performance packet processing in the data plane through Intel's *Hyper Task Chaining*, a high-speed multiprocessing data plane technology that features software pipelining and low-latency sequence management hardware. Hyper Task Chaining is discussed in further detail later in this section.

- **Intel's XScale™ technology** As of this writing, this provides the highest performance-to-power ratio in the industry. It can perform up to 1,000 *millions of instructions per second* (MIPS), and its power consumption can be as low as 10mW for the low-power, high-density processing of control plane applications.

- **The Intel IXA Portability Framework** An easy-to-use modular programming framework providing several advantages. It provides software investment protection through code portability and reuse across hardware and software development or operating system platforms between network-processor-based projects. It also enables a faster time to market and compatibility with future generations of Intel IXA network processors.

Microengines are essentially packet processors that are characterized by flexibility and customizability that is similar to *application-specific integrated circuits* (ASICs). New functions or modifications of older ones can be easily implemented with little cost and engineering effort. Costly equipment upgrades are eliminated, and new service capabilities can be added to network equipment merely through software. Microengine technology capabilities span a wide range of speed and functionality requirements from layer 2 through layer 7. They can deliver deep packet inspection (as required by the latest intelligent applications) at wire speeds up to OC-192 and beyond.

XScale is a new Intel microarchitecture that provides a high-performance, ultra-low power environment that is compliant with the ARM™ Version 5TE ISA instruction set (excluding the floating-point instruction set). The microarchitecture surrounds the ARM-compliant execution core with instruction and data memory management units, and instruction, data and mini-data caches. It also has other features such as write, fill, pend, and branch target buffers; power management, performance monitoring, debug, and *Joint Test Action Group* (JTAG) units; a coprocessor interface; a *Media Access Control* (MAC) coprocessor; and a core memory bus. Although it is obviously targeted to control plane applications, this microarchitecture can take care of communicating with a backplane, managing and updating data structures that are shared with microengines (such as routing tables), and setting up and controlling media and switch fabrics. It can also handle exception packets that require complex additional processing.

At OC-192 speeds, if carriers and network service providers are to provide new services and bill their customers accordingly, Intel estimates that deep packet inspection must occur within a short time window of around 35 nanoseconds. Within this interval, the network processor must execute all the pertinent and relevant layer 3 through layer 7 applications on these packets and then transmit them in the correct sequence (not to mention at the correct speed rate) and without bit losses to their destination. Intel uses a store-and-forward architecture that lends itself well to this model.

The speed of the second-generation NPUs is more than enough to handle the 10 Gbps wire speed. The highly parallel processing afforded by the multiple microengines allows the segmentation and partitioning of a single-stream packet analysis, such as routing into a set of multiple, sequential tasks including packet receive, route table lookup, and packet classification.

The microengine design of Intel's second-generation network processors constitutes the first implementation of Intel's Hyper Task Chaining, as shown in Figure 5.3. This approach provides hardware support for managing data-dependent operations among multiple parallel processing stages with low latency.

Intel has also introduced a series of patented techniques of register technologies that enable data and event signals to be shared among threads and microengines with virtually zero latency while maintaining coherency. We discuss several of them in the following section.

ARCHITECTURE

Figure 5.1 shows the internal block diagram of the Intel IXP1200 network processor. The architecture combines an embedded Intel StrongARM™ processor, which is targeted for control plane applications and is supported by a 8KB data cache and a 16KB instruction cache with a set of 6 microengines that are used for packet processing.

Other important features in the IXP1200 architecture include the IX bus unit (which we discuss later in this section) along with the *hash unit* that expedites address table lookup by performing polynomial hash on several values simultaneously. It also contains *scratch pad memory* (used to exchange data back and forth between microengines), a *Peripheral Computer Interface* (PCI) unit (used to interface with an external host *central processing unit* [CPU] or other PCI-compatible peripherals), and separate *static random access memory* (SRAM) and *synchronous dynamic random access memory* (SDRAM) controllers. Each microengine supports multithreading by maintaining four copies of the program counter. Zero overhead occurs when switching contexts between threads. Each thread uses 32 general-purpose registers as well as 32 *transfer registers*. The 128 transfer registers are used for the temporary retention of data that happens to be in transition to or from memory. An internal *direct memory access* (DMA) engine, which automatically steps in after software has loaded the registers, accomplishes the actual data transfer.

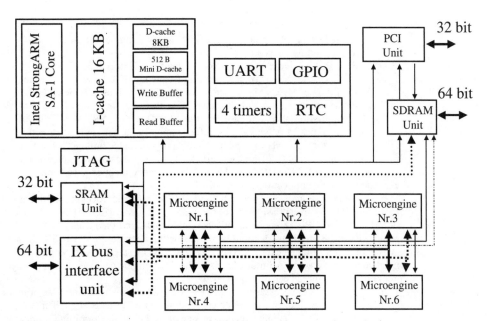

FIGURE 5.1 Internal block structure of the Intel IXP1200 network processor. *(Source: Intel)*

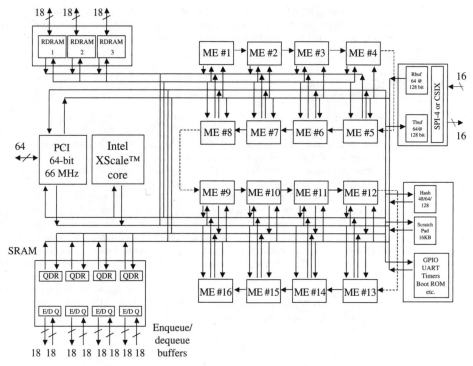

FIGURE 5.2 Internal architecture of the Intel IXP2800 network processor. *(Source: Intel)*

Whereas the StrongARM processor core and the microengines are clocked at 166 MHz, 200 MHz, or 232 MHz (depending on exactly which member of the 12x0 network processor family is used), the IX bus and the PCI bus have their own clock domains. PCI runs at 33 or 66 MHz point to point. The IX bus on the IXP1200 has a typical operating frequency of 33 to 85 MHz. In many designs, if the Intel IXF440 Ethernet MAC chip is used, the clock speed will usually be 66 MHz.

The memory interfaces run at half the speed of the core, thus 100 MHz SDRAM and 100 MHz SRAM are required on a system based on the 200 MHz core. SRAM is typically used for lookup tables, whereas SDRAM is typically used for temporary packet payload storage. The SRAM interface actually has three signals with independently programmable timings: SRAM, flash, and the memory-mapped *input/output* (I/O) device interface. It provides the common interface with different types of memory besides SRAM (flash) and even other memory-mapped peripherals. This feature may be convenient in some applications.

A typical boot sequence begins with the IXP1200 network processor booting a *real-time operating system* (RTOS) off its flash memory (or *read-only memory* [ROM]) that is connected through the SRAM port. The NPU resets its main functional blocks and then transfers from the flash or ROM memory bank the programs that will be run inside the microengines. The SRAM port handles up to 8MB of program storage next to 8MB of SRAM data storage. Each microengine has a 2K×32 RAM-based code control store. All four threads in the microengine can use the same program. A separate program for each thread can also be loaded.

In addition, it is not necessary to utilize every thread in a microengine. One or more microengines can be set up in which only one thread could be run or no threads at all. Threads in a microengine share control registers, a context enable register, and other context arbitration functions. Each thread

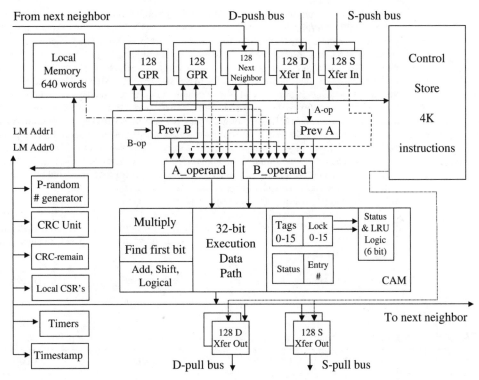

FIGURE 5.3 The internal architecture of Intel's 2nd generation microengines. *(Source: Intel)*

in a microengine has its own program counter, signal events registers, wake-up events register, and segmented storage among the 256 transfer and general-purpose registers within the microengine.

The microengines are programmable using a symbolic microcode instruction set optimized for bit stream manipulation. It offers bit, byte, word, and double-word instructions, as well as a variety of optimization tokens. A key feature of the IXP1200 is its ability to swap contexts from one thread to another without affecting performance. The key benefit of multithreading is that each microengine can do useful work even while other threads are waiting for memory transactions to complete. This feature makes the IXP1200 rare, if not outright unique. Software engineers working on the embedded code will have a vested interest in taking advantage of this ability to tune code for maximum parallelism and performance. The architecture of the IXP is clearly based on *symmetric multiprocessing* (SMP). As a result, it is very flexible. However, this flexibility comes at a price.

The IX bus is a 64-bit-wide bus with a bandwidth of 4.2 Gbps at 66 MHz, 5.1 Gbps at 80 MHz, and 6.26 Gbps at 104 MHz. It works in a demultiplexed fashion (unlike PCI), so it allows easy external device interfacing. In its *split mode of operation*, it can be configured as two separate 32-bit buses.

From the newer Intel NPUs, the 2400 offers 2 unidirectional 32-bit media interfaces (*receive signal* [Rx] and *transmit signal* [Tx]) programmable to be *System Packet Interface version 3* (SPI-3), Utopia 1/2/3, or CSIX-L1. Each path is configurable for 4×8-bit, 2×16-bit, 1×32-bit, or combinations of 8- and 16-bit data paths. We do not intend to present an exhaustive inventory of the 2400 NPU capabilities. Rather, we show what can be expected from its specifications. This flexibility provides industry-standard cell and packet interfaces to media and fabric devices that deliver a performance rate of 4 Gbps. Therefore, the 2400 can support OC-48 plus fabric encapsulation overhead or even four channels of 1 GbE. The standard interface also simplifies the design and interface to custom ASIC devices that a customer may decide to connect.

On the other hand, the 2800 offers SPI-4 Phase 2 operation based on a transfer clock of 311 to 500 MHz using 16-bit *Low-Voltage Differential Signaling* (LVDS) dual-edge signaling. Figure 5.2 shows the internal architecture of the Intel IXP2800 network processor. The switch fabric can also be interfaced using a CSIX interface with the same clock rating and LVDS dual-edge signaling. In terms of memory banks, the 4 channels of *quad data rate* (QDR) SRAM offer the IXP2800 a peak bandwidth of 1.6 GBytes/sec per channel using 200 MHz SRAMs (800 MBytes/sec read and 800 MBytes/sec write). The 3 channels of RDRAM offer a peak bandwidth of 1.6 GBps (12.8 Gbps) per channel, supporting 800 to 1066 MHz RDRAM. Notice that bandwidth on memory interfaces is quoted in *megabytes per second* (MBps) or in gigabytes per second (GBps) (corresponding to stored capacity measurement units, file sizes, and so on), whereas transfer rates on serial links are rated in *megabits per second* (Mbps) or *gigabits per second* (Gbps). The QDR SRAM interface is used for lookup tables, access lists, *content-addressable memory* (CAM) or *ternary CAM* (TCAM) associative memories, the connection of *Internet Protocol Security* (IPsec) coprocessors, and other coprocessors standardized by the *Network Processing Forum* (NPF). The *double data rate* (DDR) DRAM memory subsystem supports the nuts and bolts of the network processor's store-and-forward processing model.

Table 5.1 provides a very raw comparison between the capabilities of the IXP1200 and the more recent IXP2400 and IXP2800. For a more detailed description and comparison, see the Intel product literature available from the company's networking products web site at *www.intel.com/design/ network/ixa.htm*.

Intel incorporated several second-generation enhancements into the IXP2400 and IXP2800 network processors in order to handle packet-processing operations flexibly and powerfully. One of these

TABLE 5.1 Comparison of the Major Characteristics between the Most Prominent Intel Network Processors

FEATURE	IXP1200	IXP2400	IXP2800
Speed realm of applicability	OC-3 to OC-12	OC-48	OC-192
Number of microengines	6	8	16
Instruction store for each microengine	2K	4K	4K
Giga-operations per second	>1	>5.4	>25.2
Packet-processing performance in numbers of enqueue/dequeue packet operations per second		14 million	60 million
Integrated memory controllers	SRAM and SDRAM	DDR DRAM and 2QDR SRAM	3 RDRAM and 4 32-bit QDR SRAM
Processor core frequency	166 MHz with other family chips at 200 and 232 MHz	400/600 MHz	700 MHz
Microengine operating frequency	166 MHz	400/600 MHz	1.4/1.0 GHz
Peak bandwidth of I/O bus	6.26 Gbps		
Package		1356 Ball FCBGA	1356 Ball FCBGA
Power consumption	3.8 watts at 166 MHz	10 watts at 600 MHz	
Standard interfaces beyond PCI	104 MHz IX bus	2 unidirectional 32-bit media interfaces, which can become SPI-3, Utopia 1/2/3, or CSIX-L1, all at 25 to 125 MHz	2 unidirectional 16-bit LVDS data interfaces programmable as SPI-4 Phase 2 or CSIX

enhancements is local memory (refer to Figure 5.3). Local memory is now available in each micro-engine to improve performance, built-in resources for tasks such as *Asynchronous Transfer Mode* (ATM) *segmentation and reassembly* (SAR), *pseudorandom number generation* (PRNG) for table lookups, timestamps for supporting flow metering, and a multiply function for performing complex algorithm calculations such as those encountered in *quality of service* (QoS) environments. These latest network processors also automatically align code and data bytes for better code streamlining, thus enhancing the productivity of software engineering.

The following are other interesting and innovative features of this architecture:

- *Next-neighbor registers,* which enable the rapid transfer of data and state information from one microengine to an adjacent one.

- *Reflector mode pathways,* which ensure that data and global event signals can be shared by multiple microengines using 32-bit-wide unidirectional buses (called the *D* and *S bus*) that connect the IXP2800 network processor's internal processing and memory resources.

- *Ring buffers,* which establish producer-consumer relationships between microengines, thereby providing a very efficient mechanism for the flexible cascading of linked tasks among multiple software pipelines.

This combination of flexible software pipelining and fast interprocess communication accounts for a large part of the suitability of the IXA architecture NPUs in core, edge, and access applications.

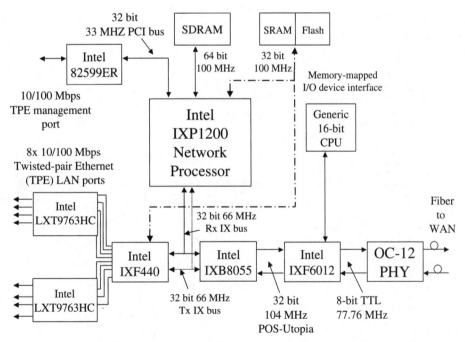

FIGURE 5.4 A systems design based on the IXP1200 network processor for an enterprise IP router connecting fast Ethernet with SONET over OC-12. (*Source: Intel*)

SOFTWARE ARCHITECTURE

Optimized microengine libraries and tools provide continuity between changes in the microengine instruction set and architecture. The libraries include a hardware abstraction library that provides interoperability across multiple hardware configurations, a protocol library, and a utility library for hardware-optimized operations on protocol-created packet headers and data structures in general. Figure 5.6 shows the model. Microblock code can be easily developed using the high-level Microengine C language environment. The Portability Framework is an integral part of the Intel IXA *Software Developer's Kit* (SDK).

A modular programming model, which is also part of the IXA Portability Framework, enables optimal partitioning of an application across the microengines and threads. Therefore, it facilitates the integration of customer-written code along with microblocks, which can be supplied by Intel or third parties. These *microblocks* are independent building blocks of software that are specifically written for the microengines. These blocks perform a clearly defined set of functions. This modular model enables software reuse—that is, the flexible mixing and matching of software components. Intel's microblock library is also designed to support the pipelined architecture of the network processor microengines by providing the flexible connection of these microblocks.

Intel's XScale microarchitecture source code libraries enable modular core component development. They also enhance portability between multiple operating environments. Third parties provide several compilers, assemblers, linkers, and debuggers to support Intel's XScale architecture. Of course, programming the embedded StrongARM core can be done with an equally wide array of tools and software development platforms that are provided from third parties that support work for ARM CPUs.

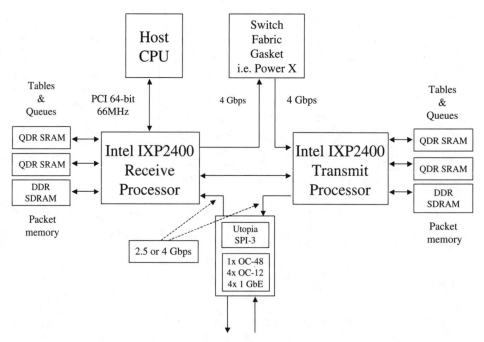

FIGURE 5.5 Typical architecture of an OC-48 system showing two IXP2400 network processors that are needed to handle the transmit and receive paths respectively. *(Source: Intel)*

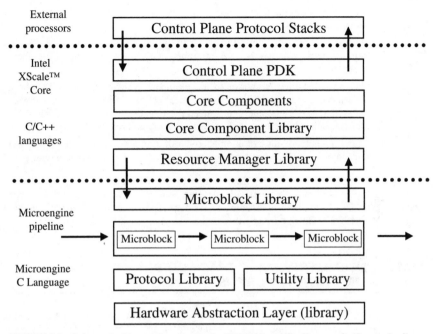

External processors

Intel XScale™ Core

C/C++ languages

Microengine pipeline

Microengine C Language

Control Plane Protocol Stacks

Control Plane PDK

Core Components

Core Component Library

Resource Manager Library

Microblock Library

Microblock → Microblock → Microblock

Protocol Library — Utility Library

Hardware Abstraction Layer (library)

FIGURE 5.6   Software Architecture based on the Intel IXA™ Portability Framework. *(Source: Intel)*

Intel also provides a core-control plane *Platform Development Kit* (PDK), which offers a common interface and interconnect protocol for control plane stacks that may be running on external processors.

SOFTWARE AND SYSTEMS DEVELOPMENT AROUND IXA ARCHITECTURE NPUS

Intel IXA SDK offers an integrated environment with functionality that enables rapid code development and simulation for both control and data plane applications, with a choice of embedded operating systems. It is supported by a comprehensive hardware platform. More specifically, the SDK contains several interesting tools:

• The Integrated Microengine Development Environment provides an integrated environment for the advanced graphical simulation, profiling, and debugging of a system working exclusively in software. It enables development engineers to create prototypes quickly, and intuitively optimize and support data for both data and control plane applications. The *transactor* from this tool resolves concurrency issues by simulating packets going in and out of the network processor. It can be used to gather statistics. It can also aid in creating and verifying the architectural design and by providing a fine level of internal detail, including pipeline execution stages. In other words, it can pinpoint things and situations that would not be visible otherwise.

- Intel's Microengine C compiler facilitates code development for the microengines and improves time to market.
- The SDK is provided with support for the Wind River™ VxWorks and MontaVista™ Linux operating systems, whereas the IXA environment also provides support for other third-party embedded operating systems.
- The provided libraries shorten the development cycle as part of the IXA Portability Framework by offering the systems designer some critical chunks of infrastructure software that is pretested and validated. Intel's customers can embed these blocks of quality code into their own software flow to deliver their intended application more quickly and reliably.
- A comprehensive suite of completed building blocks and sample applications further improve the customer's software development through the use of common networking building blocks.

In order to complement the development environment, Intel also provides several hardware development platforms for the parallel development of hardware simultaneously with the software. These standard-form platforms enable processing performance among other realms at OC-48 (2.5 Gbps) and OC-192 (10 Gbps) wire speeds.

SYSTEMS CONSIDERATIONS AND TRADE-OFFS WHEN DESIGNING WITH INTEL NPUS

Intel IXA network processors work together with several other Intel families of chipsets in various complementary technologies to produce working systems that are straightforward to design because they all essentially share common interfaces:

- Embedded Intel architecture control processors improve the scalability of the design while providing broad software support in communications environments.
- Intel media signal processors can be used in conjunction with NPUs for applications such as *voice over IP* (VoIP), as shown in Figure 5.8 and discussed in this section.
- Intel I/O processors are extensively used for networked storage applications.
- Intel provides a very broad line of framers, media access controllers, and even *physical* (PHY) layer devices. These features significantly facilitate the overall systems design process.

Designing systems with Intel's network processors implies that in high-speed links, one NPU is required for the ingress (receive) path and another is required for the egress (transmit) path. This is a characteristic of the whole family and not just of one of the network processor chips that Intel proposes. In certain applications, however, a single Intel network processor may be adequate for the available traffic load. For example, a single network processor is adequate for a VoIP gateway that works up to an OC-3 (155 Mbps) capacity, as shown in Figure 5.8. This gateway system is connected on one side on multiple Gigabit Ethernet (1000Base-T) and Fast Ethernet (10/100 Base-T) media and on the other side on the *Public Switched Telephone Network* (PSTN) through a *time division multiplexing* (TDM) backplane that transfers voice channels.

In this example, based on an Intel reference design, voice is carried over IP packets coming in from Ethernet and Gigabit Ethernet links. After the respective PHY and MAC stages of their reception (which is handled by other convenient Intel chips, as shown in the Figure 5.8, and require no other glue logic around them), the packets are forwarded through the split IX bus to the IXP1200 network processor for subsequent processing. Deep packet-processing applications are partitioned among the NPU's microengines. All supervisory systems control functions that will be exercised onto the NPU are dispatched by a host CPU externally through the PCI bus. A *field-programmable gate array* (FPGA) is required to handle the application-specific glue logic translating the IX bus cycles into VX bus cycles. This is required because on the *time slot interchange* (TSI) side the data are coming in and going out serially in real time, whereas on the IX side the NPU prefers to handle data in burst mode.

The shown IXS1000 chip is the Intel media processor responsible for translating the VX bus traffic to and from TSI slots for the TDM-multiplexed H.110 backplane used to interface with the telephony world. The IXS1000 media processor is a good choice for many reasons. It can handle 240 voice channels split over 512 full-duplex TDM channels; mix and match call configurations with all classical vocoding schemes such as G.711, G.726, and so on; take care of G.168-compliant echo cancellation; adopt fax modem or fax relay behavior based on V.17, V.29, and so on; and handle typical A-law and/or μ-law *pulse code modulation* (PCM) interfaces. In short, it can implement all the necessary signaling context of a typical PSTN network interface with functions such as *Dual Tone Multiple Frequency* (DTMF) detection and generation.

This hardware design along with the appropriate software can manage the TSI slots. It can easily process all signaling messages for the call setup and teardown. It can also manage the combination of *Real-Time Protocol* (RTP)/*User Datagram Protocol* (UDP)/IP for the handling of the voice traffic itself and the combination of *Real-Time Control Protocol* (RTCP)/*Transmission Control Protocol* (TCP)/IP for the associated control packet traffic. Although the design approach is clean-cut and straightforward, in several cases, significant help will be offered to customers either from Intel or from third parties in the form of Verilog code or even a complete FPGA design (at a price, of course). However, in some cases, the need and the associated cost to design and include a special FPGA for the implementation of glue logic or interfaces from one realm to another may discourage some potential users, who could choose to approach a network-processor vendor that offers a more integrated and seamless solution.

Another example of a single IXP1200's ability to handle a traditional enterprise/campus routing system for modest performance proportions is shown in Figure 5.4. The router of this example connects eight 10/100 Mbps Fast Ethernet RJ-45 ports on one side with a *Synchronous Optical Network* (SONET) OC-12 optics backbone pipe to the *wide area network* (WAN) handling layer 3 IP switching and routing functions along with key routing protocol support. *Simple Network Management Protocol* (SNMP) network management can be handled via a specially assigned Fast Ethernet port.

The Intel IXF6012 SONET Framer properly encapsulates IP packets coming into the router from the Ethernet realm, as it is capable of both SONET and *Synchronous Digital Hierarchy* (SDH) encapsulation of ATM or *High-level Data Link Control* (HDLC) frames. It offers either a *Packet over SONET PHY Level 3* (POS-PL3) or a standard Utopia interface to higher-level protocols. It can operate in single OC-12c or quad OC-3c mode on the line side. A generic 16-bit processor interface is provided for configuration and network management.

To explain the other shown parts of the design, we will briefly say that the IXB8055 is a POS-to-Utopia bridge—an implementation in Verilog that Intel can provide to its customers. Customers will then have to implement it by themselves in an FPGA. The 104 MHz clock rate of the bridge operation in this Intel reference design example can only be realized with a specialized ASIC, as FPGA implementations will have to function at a smaller clock rate. The LXT9763HX (Hex PHY) provides six standard *media independent interface* (MII) ports for various Ethernet media. Only four of them are used in this example to match the number of MAC units. The IXF440 is an octal MAC. It provides eight standard MII 10/100 Mbps Ethernet ports without requiring glue logic to connect with the IXP1200 network processor. The 82599ER is an Ethernet controller that handles the interface with a 10/100 Mbps twisted-pair Fast Ethernet port, which is used here for network management and the overall configuration.

In this design example, if it was implemented in real life, layer 3 routing across the optical network would also require other more complex protocols implemented in software and running on the IXP1200 itself. In addition, in such an environment, the IXP1200 can also run other gateway-type software. As a result, this system can ultimately serve as the front-end network interface in a CPE environment connecting to the WAN and LAN with substantial local traffic.

The combined ingress traffic in this example of 1.422 Gbps is within the measured performance for the IXP1200. These network processors can drive 16 Fast Ethernet ports at wire speed while at the same time perform layer 3 routing (with 1.6 Gbps unidirectional traffic as its theoretical maximum).

The system buses used in the design shown in Figure 5.4 are summarized as follows:

- **IX bus** This consists of two separate 32-bit paths for transmit and receive flows operating at 66 MHz each. It offers 2.1 Gbps bandwidth, which is well above the 1.422 Gbps ingress requirement mentioned previously. The total ingress and egress IX bus bandwidth in this example is 4.2 Gbps.

- **Ready bus** The ready bus is an 8-bit bus that runs parallel to the IX bus and provides sideband messaging between IX bus devices. The IXP1200, as the IX bus master, manages the collection of ready flags from IX bus peripherals/slaves through this ready bus. The ready bus can also perform other functions, including flow control.

- **Memory-mapped I/O interface** Sharing the SRAM interface, this bus offers the possibility of independently programmable timing. It can also serve as the third connection between the IXP1200 network processor and another peripheral processor sitting on the IX bus. This bus behaves like a slow port. As a result, it can be used for configuring Ethernet MAC controllers, managing an attached device, and even collecting statistics in the context of *Remote Network Monitoring* (RMON) and/or SNMP.

- **POS-PL3** This is a *first in/first out* (FIFO) interface that is 32 bits wide. It works at a rate of 104 MHz for each transmit and receive path. This amounts to a consolidated bandwidth of 3.3 Gbps paths on this interface.

- **MII bus** This is a standard MII, and it forms the link between the Ethernet MAC ports in the Intel IXF440 MAC and the Intel LXT9763 PHYs.

- **PCI bus** In Figure 5.4, the 32-bit 33 MHz PCI bus provides a point-to-point connection from the IXP1200 network processor to the 82559ER Fast Ethernet management port.

In yet another case in a much higher-performance environment, Figure 5.5 shows a typical block structure for an OC-48 line card that is built around the IXP2400 NPU. A strikingly similar approach is taken with IXP2800 in a core network application, as shown in the LAN/WAN example of Figure 5.7. The scalability of the Intel architecture at this point should be quite obvious. On the ingress path of this example, the first IXP2800 is responsible for issues like SAR, classification, metering, pricing, and initial congestion management. On the egress path of the example, the second IXP2800 handles flexible traffic shaping, *Differentiated Services* (DiffServ) for IP traffic, traffic management such as TM 4.1 for ATM networks, or custom traffic shaping.

Regarding systems design and connection with coprocessors from other vendors, such as IPsec security coprocessor chips in a *virtual private network* (VPN) system, Intel recommends the use of either the SRAM interface bus or the IX bus to attach an IPsec coprocessor that will offload the network processor. In the case of the former, it can be done directly, if the IPsec coprocessor is compatible with the bus signals. In the worst-case scenario, it can also be done through using glue logic that must be implemented in an FPGA. In the case of the IX bus attachment, an IX bus bridge is required to interface the security coprocessor bus signals with the IX bus itself. If two network processors are available for the ingress and egress paths, the traffic load should be considered so important that potentially two IPsec coprocessors must be used to support the computational load of calculating in real time and creating or stripping IPsec-encapsulated packets while still providing headroom to the NPUs for other fundamental networking packet processing. We will discuss these issues in more detail in Chapter 17, "Security Coprocessors."

Another issue to keep in mind is that the SMP-based architecture, which offers a potential parallelism and software-based pipelining (as microengine threads can be cascaded essentially in any desired chained-link configuration), is essentially an environment that is more difficult to program than other NPUs that offer a single run time image environment. The high quality of the software development tools and, more specifically, of application software profiling tools and application partitioning and fine-tuning tools, that Intel and its partners offer becomes a very critical consideration in such a context. Intel's vast relationships with third-party developers seem to affect this issue. However, the major problem with this distributed approach is that in very-high-speed heavy-traffic-load contexts, the performance of an application cannot be gauged before the application has actually been developed.

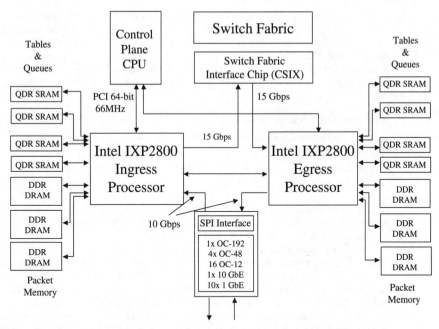

FIGURE 5.7 Configuration of a typical LAN/WAN interface using the IXP2800 network processor. *(Source: Intel)*

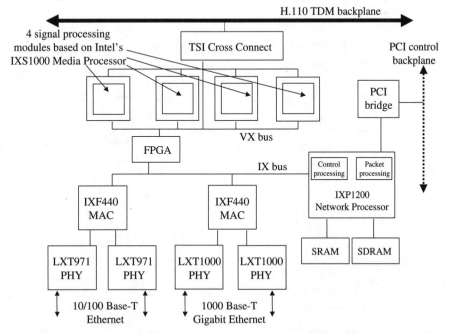

FIGURE 5.8 Design example of a Voice-over-IP gateway linking Fast and Gigabit Ethernet LANs with the TDM-multiplexed PSTN telephony network. *(Source: Intel)*

In order to better understand the principle of allocating parts of the packet processing to different microengines, we must also look at a real-life application and, more specifically, at how Intel recommends the application be logically partitioned over the available microengines in order to optimize performance. The example design is of a simple router that is implemented as a full-duplex ATM-to-Fast-Ethernet conversion engine handling IP packets and working over a dual OC-3 (155 Mbps) link. The router design example in real life obviously requires software to properly handle the following tasks:

- SAR of ATM cells and IP packets
- IP over ATM encapsulation based on *Subnetwork Access Protocol* (SNAP)/*Logical Link Control* (LLC)
- *ATM Adaptation Layer* (AAL-5) as unspecified bit rate (UBR) traffic
- CRC-32 for reliable transmission

As a reference design, the complete software can be licensed from Intel. It can be modified by Intel clients who are eager to shorten their time to market and who want to create their own version of a similar design but cannot afford to start from scratch.

Figure 5.10 shows macroscopically and conceptually the protocol conversion that needs to happen in both directions—namely, from Ethernet to ATM and vice versa. In this generic approach, Ethernet *Institute of Electrical and Electronics Engineers* (IEEE) 802.3 packets go through LLC/SNAP encapsulation and are then followed by segmentation into AAL-5 cells. The opposite process is applied onto ATM cells, which are stripped from their ATM headers and finally reassembled into Ethernet packets.

Figure 5.9 gives an overview of the control flow and an idea of how to apportion the packet processing needed over the available (in the case of an IXP1200 network processor) six microengines. In this case, three of the available six microengines are tasked to handle the ATM-to-Ethernet data flow, whereas the other three are assigned to the reverse direction from Ethernet to ATM. Multiple queues are used by the microengines to send data from one stage to the next. Details as to how this

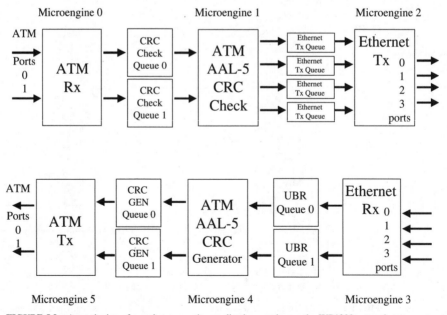

FIGURE 5.9 Apportioning of a packet processing application running on the IXP1200 network processor over the 6 available microengines. *(Source: Intel)*

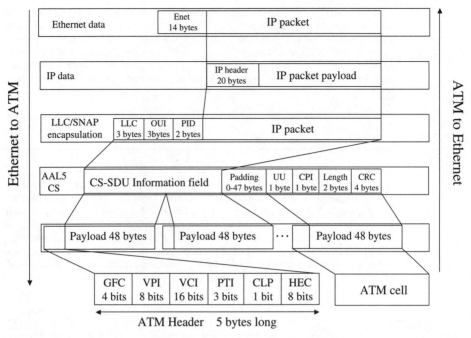

FIGURE 5.10   Apportioning a packet-processing application running on the IXP1200 network processor over the six available microengines. *(Source: Intel)*

can be done are beyond the scope of this book. The corresponding code structure, interprocess signaling, data structures, initialization and startup, and so on can be found in a detailed application note that Intel provides called "IXP1200 network processor ATM OC-3/Ethernet IP Router Example Design." It is available from Intel's web site at *www.intel.com/design/network/ixa.htm.*

Right before this chapter went to press, Intel announced a 1.4 GHz follow-on device to the 2800 —the IXP2850 network processor. This is a simplex 10 Gbps processor and is scheduled to be sampled by mid-2003. The interesting feature of this NPU is that it embeds encryption capabilities. More specifically, it contains two *crypto engines* as modules. Each crypto engine contains special hardware (some of them in multiple instances) for the implementation of the *Advanced Encryption Standard* (AES)/Rijndael, Triple DES, and SHA-1 cryptographic algorithms, which we discuss in length in Chapter 17. The 2850 is also capable of calculating TCP checksums. Interested readers can learn more details about this TCP termination-engine functionality in Chapter 11, "Storage Network Processors." Other hashing algorithms that are often needed such as MD5 or encryption algorithms such as RC4 are to be implemented in software on the microengines. However, the 2850 clearly positions Intel NPUs to handle multigigabit-per-second IPsec types of VPNs in a powerful way. Again, unfamiliar readers are referred to Chapter 17, where these concepts are discussed in more detail.

The important message with this announcement is that a major NPU vendor like Intel, with a truly dominant position in market share, takes the proactive step of integrating critical security functionality inside some of its network processors. This movement, which is bound to be copied by some of Intel's competitors such as Broadcom, is expected to have a major impact in many designs against the perceived need for an external security coprocessor, which is attached either in band or in a look-aside configuration. It will definitely tilt the market tendencies significantly away from the previous need to incorporate an external stand-alone security coprocessor. The IXP2850 costs a couple of hundred dollars more than the 2800 and consumes about 2 watts more.

This means that in some designs requiring a security coprocessor, the chip count of the system becomes smaller with the use of the 2850. The direct cost of purchase is also less, as a security coprocessor costs much more than the difference we just mentioned, and it probably needs extra memory and interface logic. The power consumption is less than that of stand-alone coprocessors.

This concept will also probably add significant market pressure against stand-alone security coprocessor vendors in the long run. Some of them may survive, but they will remain in a shaky position.

SUMMARY

In this chapter, we reviewed Intel's IXA architecture of network processors and looked more specifically at its IXP1200, 2400, and 2800 models. We also provided some information on its more recent 2850 chip, which integrates sophisticated security functions. We identified their underlying characteristics and looked at the advantages they offer as well as some of the few associated inconveniences for a systems designer. We finally described a few typical applications using various configurations implemented along a common architectural theme that is characteristic of this family of NPUs. Intel has a powerful and wide family of network processors. Combining these processors with an exceptional array of software tools and third-party development platforms will most likely further consolidate Intel's leading position in this market.

REFERENCES

Extensive literature with detailed product datasheets, technology white papers, and application notes, along with links to other related Intel communications and networking sites, can be found at Intel's network processing web site at *www.intel.com/design/network/ixa.htm.*

Information a bout the building blocks needed in networking applications around Intel's offerings can be found at the web site *http://developer.intel.com/design/network.*

Intel's technical literature center can be found at *http://developer.intel.com/design/litcentr/.*

CHAPTER 6

AMCC nP™ FAMILY OF NETWORK PROCESSORS

Applied Micro Circuits Corporation (AMCC) has become one of the leaders in the field of network processing. Its acquisition of a few companies with state-of-the-art technology and products in the *network processing unit* (NPU) and switch fabric fields, as well as the consequent breadth of its offerings, has positioned AMCC as one of the leading contenders. AMCC is now able to offer the advantage of one-stop shopping to its customers. It covers the entire spectrum of a network equipment designer's needs from scalable OC-192 switch fabrics and NPUs all the way to transceivers and framer chips for *Synchronous Optical Network* (SONET) and Gigabit Ethernet realms.

In this chapter, we review AMCC's nP network-processing architecture. We briefly look inside some of the company's most powerful network processors to form an impression of how AMCC's approach compares to that of other leading vendors. Finally, we discuss some of the company's other associated chips that facilitate the integration of a complete switching/routing system design by efficiently handling major technical challenges such as traffic management, scheduling, and the actual switching process.

nP™ ARCHITECTURE: THE BIG PICTURE

AMCC[1] has been consistently expanding its NPU offerings by building on an underlying scalable architecture called nP™. Although the company offers several network-processor products, we will look at only a few of their most recent and powerful ones: the nP7250, which is a network processor rated for the OC-48c realm, and the more recent nP7510, which is AMCC's flagship OC-192c network processor.

The *network-optimized instruction set computing* (NISC) architecture is at the heart of AMCC's network processors. This architecture is implemented in the company's patented nPcore™, the fundamental engine replicating which dramatically scales the performance and bandwidth of a network processor based on the nP architecture. The company's NISC model was already developed at MMC Networks (before the company was acquired by AMCC) in response to the performance shortcomings of traditional *reduced instruction set computer* (RISC) processors in the late 1990s. These shortcomings were especially apparent as link speeds exponentially increased and traffic loads exploded due to increased bandwidth demand. The company estimates that with the implementation of its NISC instruction set in a multitasking environment and its inherent zero-cost task switching, the nPcore engines achieve 4 to 12 times the network-processing capacity of typical RISC *central processing units* (CPUs).

---

1. Data sheets, application notes, and white papers on AMCC products and technologies can be found at the company's web site at *www.amcc.com*.

AMCC deduced the instruction set after studying the most typical routing and switching algorithms and understanding the kind of operations involved. The result of the analysis was a highly specialized instruction set that optimizes the parsing, search, and modification of packets. An example based on RFC 1812 routing shows it can be implemented in just 50 NISC instructions, where each instruction takes 1 clock cycle. AMCC estimates that a typical RISC-based NPU implementation uses 200 to 800 instructions to accomplish the same task. If layer 2 and layer 4 classification were added to the RFC 1812 routing, the nPcore engine implementation would only need 5 more instructions for a total of 55 instructions. At the same time, a RISC-based NPU would need between 350 and 1,200 instructions (and clock cycles). By implementing this NISC model in the nPcores without implementing unusable instructions (such as arithmetic operations), AMCC eliminated the waste of silicon. The company further improved the efficiency of the design by adding features that allowed for future expansion, performance scalability, and the attachment of specialized coprocessors either internally or externally.

As shown in Figure 6.1, the architecture of the nP is straightforward. The NPU is positioned between the switch fabric on one side and an array of multiple *physical* (PHY) interfaces on the other side. Several nPcores are used depending on the link speeds that the device is expected to sustain. For instance, the nP7250 designed for the OC-48c realm uses two nPcores inside the die, whereas the nP7510 designed for OC-192c links uses just six nPcores and does not require any major architectural changes. Figure 6.2 illustrates the block structure of the OC-192c-capable nP7510. They both provide significant extra headroom for other features or additional computational loads beyond what a typical application such as layer 3 switching or routing on multiple gigabit streams provides. In

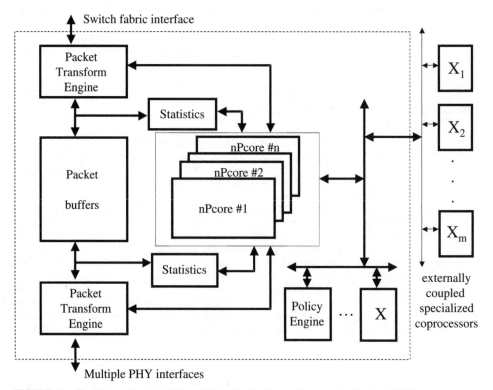

FIGURE 6.1 The block architecture of the AMCC nP family of network processors. *(Source: AMCC)*

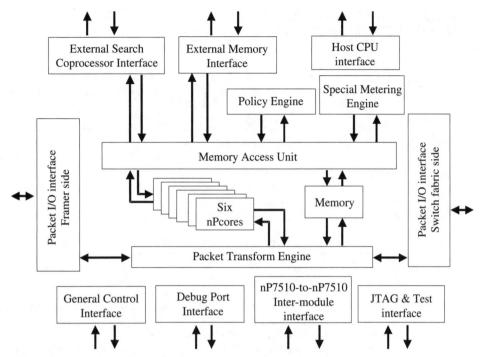

FIGURE 6.2 The block structure of the OC-192c-capable nP7510 network processor. *(Source: AMCC)*

Figure 6.1, X denotes any generic coprocessors. These coprocessors could be internally integrated on the same die as the network processor or externally coupled.

AMCC's network processors include an innovative embedded on-chip engine called the *policy engine*. This engine is an example of an on-chip coprocessor that supports a single-clock-cycle simultaneous lookup of layer 2, 3, and 4 packet header components. A software-configurable database supports configurations that have access to multiple logical tables using 32- to 512-bit-wide keys, support Best Match searches, and even possess a patented feature called *weight array* that allows easier table management and, more specifically, the handling of low-cost insertions. The policy engine can be used to implement layer 4 switching, such as packet prioritization based on some layer 4 information. It allows functionality as dynamic port assignment in applications such as *voice over IP* (VoIP). It can also be used to expedite the mainline network-processor packet examination and classification code. The coprocessor interconnection bus can be extended off-chip, thereby facilitating a broader spectrum of products with potentially different search requirements such as web switching via *Uniform Resource Locator* (URL) matching or Internet core routing.

The AMCC approach involves two other important characteristics: the single programming image that the architecture provides to the designer and the company's ability to offer ancillary chips such as traffic managers, switch fabrics, and so on, which create an almost complete design of a whole system with minimal hardware effort.

We discuss the single programming image in the section "Developing Software for the nP Family of Network Processors." We cover the topic of the company's ability to offer ancillary chips in a separate section, "Systems Considerations When Designing with AMCC nP Family NPUs."

DEVELOPING SOFTWARE FOR THE NP
FAMILY OF NETWORK PROCESSORS

Figure 6.3 shows the cleanly layered structure of nPsoft™ Services, which is essentially a software services architecture. The company's approach has the advantage of only requiring the addition of parts necessary for the overall desired design functionality, without anything superfluous. This stream-lined software architecture is comprised of the following:

- An open *applications programming interface* (API) with custom-written application-specific code or other third-party software packages
- Transparent access to other coprocessors available from other vendors, such as search engines, encryption acceleration chips, and so on
- Traffic management engine interactions and switch fabric configuration and management
- A library of common networking functions
- A modular interface for customer-developed NPU software

Customers write their application software, without loss of efficiency, as if it was intended to run on one single CPU. The system will automatically repartition it over the available nPcores. From the beginning of its development efforts, AMCC was extremely sensitive to the fact that embedded software written for a high-speed switching system must be fine-tuned for true wire speed so that hardware-computing resources would not remain idle even for small amounts of time. A typical situation where this occurs is with the phenomenon of a *pipeline bubble*. In a pipeline bubble, inactivity at some point in time propagates down the pipeline stages, further promulgating the effect of temporary idleness and multiplying the effect of efficiency loss.

Supercomputer designers have found out the hard way that scheduling multiprocessor-based computing tasks for the time-sensitive execution of software is a difficult task. The unpredictable nature of network traffic, coupled with the extreme high speeds involved in today's links, can cause interdependency situations and force undesirable wait states on some processors. As a result, partial idleness can be incurred pending the completion of an intermediate and necessary task that runs on another

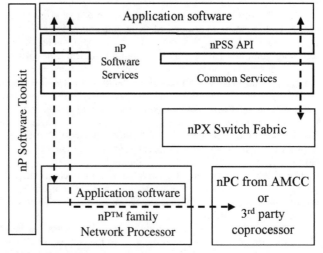

FIGURE 6.3 AMCC's nPsoft, a layered software services architecture. *(Source: AMCC)*

processor inside the same network processor. Writing task distribution algorithms in such a computing model remains tedious. It also does not offer any guarantee of performance. In addition, even if a designer experiments with a certain traffic load context and creates superbly crafted code that implements such a fine-tuned task distribution, the code will still need to be radically rewritten as soon as some new feature or functionality is introduced into the overall application code. This can happen at any time as part of mere upgrading or maintaining the code.

In AMCC's single-image computing model, software engineers write software in one logical block of code as if they were programming one single logical CPU. They do not worry about allocating tasks or scheduling. As long as the clock cycle budget allows more tasks to be executed, the model, which is based in zero-cycle task switching overhead, guarantees that the written code will be executed at wire speed without any further tweaking and tinkering. Perpetual load balancing is no longer necessary among multiple cores.

In addition to the fully functional preintegrated hardware development systems that enable the parallel development and testing of hardware and systems code in real-life networks, AMCC also offers a C/C++ compiler, an assembler, and a debugger, which facilitate the software development cycle. However, compared to the extent and quality of the development tools offered by some other vendors, this set of tools may be considered insufficient for enabling the wider-scale adoption of the company's platform by many more *network equipment vendors* (NEVs).

TRAFFIC MANAGEMENT

To scale performance eventually above 40 Gbps, AMCC realized early on that traffic management (a key foundation upon which a carrier can offer *quality of service* [QoS] and guarantees) cannot be fully integrated into one and the same silicon die with the network processor. Therefore, it adopted a chipset architecture, which is based on separate chips for the traffic manager as well as for the switch fabric. This physical separation allowed the company to pursue the optimization of these functions. AMCC realized early on that provisioning per-subscriber services requires many thousands of separate logical queues and the ability to schedule these queues on an individual basis in order to provide guaranteed access to network resources such as bandwidth. To illustrate the magnitude of the problem, consider, for example, the number of the queues required to handle the number of *Digital Subscriber Line* (DSL) connections that can be aggregated into an OC-192c trunk. For the sake of argument, assume that an average connection load has a rating of 0.5 Mbps per subscriber:

$$10.96 \times 10^9 / 0.5 \times 10^6 \approx 22,000 \ logical \ queues$$

In order to provide these bandwidth guarantees, the traffic management engine must implement individual queues for each subscriber. AMCC has implemented a feature called *per-flow queuing*. This feature ensures that each traffic flow is managed as a separate entity. In other words, it is queued and scheduled independently from the other flows. It is impossible to integrate such a granular level of traffic management inside a network processor in hardware or software. However, service providers who must implement QoS contexts with different services and features as demanded by the market require such a granular level of traffic management. Congestion experienced by one flow is prevented from interfering with the traffic conditions of another flow. As a result, QoS is maintained. Traffic scheduling enables the hardware scheduling of traffic on a per-flow basis through the support of cell- and packet-based algorithms such as rate, strict priority, *weighted fair queuing* (WFQ), and *weighted round robin* (WRR).

AMCC also refers to a feature called *virtual SAR*. This means that expensive external *segmentation and reassembly* (SAR) devices are not required when the nPX5700 is used. Instead, the SARing function is inherent in the chipset and is a natural result of the way in which the nPX5700 accomplishes per-flow queuing and scheduling. This explains the term *virtual SAR*.

Another interesting feature is its ability to support point-to-point multicast connections. This indicates that traffic that is received on one input flow can be sent to one or more output flows, either on separate output ports (physical multicast) or on the same output port (logical multicast).

The nPX5700 can also operate in *snooping* mode. This means that it can send a duplicate flow originally meant for another port to an output. This is useful if someone tracks items with an attached network protocol analyzer, eliminating the need to move the analyzer from one switch port to another.

One of the useful capabilities of the 5700 chipset is that it enables packets entering on separate ports to be merged to exit from a single port, as is required in *Multiprotocol Label Switching* (MPLS).

In addition to standard OC-3/OC-12 *Asynchronous Transfer Mode* (ATM) and 10/100 Ethernet ports, the nPX5700 can handle multiple slower speed pipes, such as T1, fractional T1, and DS-0, aggregated into a single physical port. Conversely, multiple ports can be aggregated into a single high-speed pipe. For example, up to 16 OC-3 ATM ports can be combined into a single OC-48 ATM port.

In very high-speed applications, two separate traffic managers will be needed: one on the ingress path of the switch/router and one on the egress path. AMCC traffic managers support thousands of queues, and sort and queue traffic by flow.

The nP5700 traffic manager is one of AMCC's promising products that enables the company to develop an integrated solution. The nPX5700 is a chipset that consists of the nPX5710 control logic chip (which is responsible for tasks such as admission control, scheduling, and queuing functions) and the nPX5720 buffering chip (which is responsible for managing payload memory). The 5710 is packaged in a 601-pin PBGA, whereas the 5720 is presented in a 1125-pin PBGA form. Figure 6.4 illustrates their block structures.

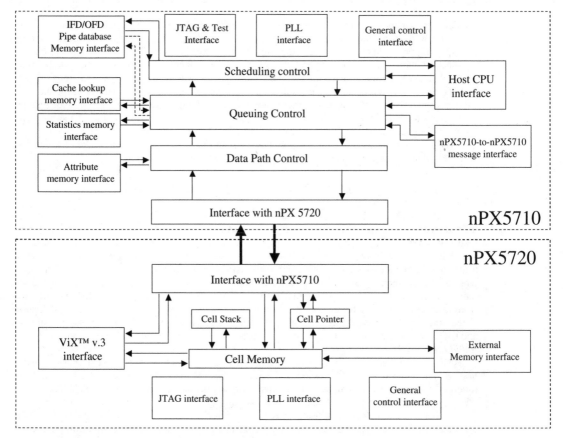

FIGURE 6.4   Architecture of AMCC's nPX5700 traffic management chipset. *(Source: AMCC)*

Many of today's intelligent carrier, service provider, and *customer premises equipment* (CPE) platforms require a feature-rich 10 Gbps traffic management context for the provision of subscriber bandwidth, the flexible scheduling of capabilities, and the exercise of rigorous admission control. The nPX5700 per-flow queuing mechanism offers very high levels of granularity and supports tens of thousands of subscribers and hundreds of thousands of queues. More specifically, the nPX5710 control logic chip can easily support up to OC-192 bandwidth scheduling in fine-grain 256 subports, 64,000 virtual pipes (aggregates), and 256,000 input flows. Similarly, the nPX5720 memory management device, which can support up to four OC-48 channels or one OC-192 channel, has its own embedded *dynamic random access memory* (DRAM). Therefore, it can provide local storage for up to 8 million cells of payload storage.

NEVs who are designing network equipment can use the chipset to implement a variety of sophisticated admission control techniques. These techniques include dynamic marking and discard threshold levels, *Random Early Detection* (RED), *Weighted RED* (WRED), *Early Packet Discard* (EPD), and Partial Packet Timeout to manage and control potential congestion and enforce programmed service levels. Maximum flexibility is also preserved in the sense that the systems designer is free to implement policy-based QoS features that support strict priority, WFQ, *round robin* (RR), WRR, *constant bit rate* (CBR), *variable bit rate* (VBR), and minimum and maximum bandwidth control among several intrinsically supported and available possibilities.

SWITCH FABRIC

The switch fabric function further augments the model based on which the designer must physically separate the network-processor chip from the traffic managers and then both of these functions from the switch fabric chipset. The switch fabric does this by maintaining local logical queues that are built upon the concept of classes and are further sorted per output port. Figure 6.5 illustrates this concept.

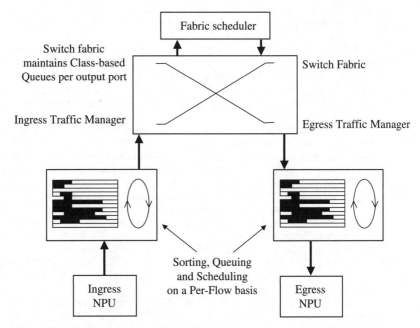

FIGURE 6.5 An example of switching and managing traffic with the nP family of products. *(Source: AMCC)*

AMCC offers several products in this realm, but we will focus on the nPX5800 switch fabric targeted for the area of OC-48 and OC-192 systems with a desired throughput of up to 160 Gbps.

The nPX5800 switch fabric is a high-speed, scalable switching element that along with the traffic manager completes AMCC's network-processing platform. The nPX5800 implements nonblocking virtual output queuing technology to achieve 40 Gbps (20 Gbps full duplex) to 320 Gbps (160 Gbps full duplex) switching capacity. It is scalable to support up to 16 full-duplex 10 Gbps OC-192c *Packet over SONET* (POS), ATM, or 10-Gigabit Ethernet interfaces with a significantly lower chip count than other existing solutions. AMCC's nPX switching family offers additional future architectural scalability over 1.2 Tbps.

For seamless platform implementations, AMCC uses its proprietary, nonblocking, QoS-enabled ViX™ interconnect bus, which eliminates the need for a high-speed memory bus and replaces it with much simpler, cheaper, point-to-point connections. This means that the switch's cost increases linearly with the number of ports. This is unlike non-ViX architectures, where the cost increases exponentially. Despite its use of a proprietary in-house-developed interconnect bus, AMCC is an active participant in the *Network Processing Forum* (NPF) (formerly CSIX) and it is contributing toward the definition and adoption of next-generation, standard 10 Gbps and QoS-enabled interfaces. The maximum allowed payload on the ViX bus is 64 bytes plus a 16-byte header that is full of special bit fields used for specifying the destination port, parity, priority, credit, flow control, and so on. The 5700 traffic manager chipset and the 5800 switch fabric communicate via *serialization and deserialization* (serdes) devices over the ViX bus by sending special ViX-bus-formatted cells over multiple 16-bit sub-buses. These sub-buses operate at 125 MHz. An aggregation of eight sub-buses can handle an OC-192 link, leaving plenty of overspeed for other system functionality.

Internally, the nPX5800 is based on a shared-memory architecture with a centralized scheduler. The chip is built with 16 input ports and 16 output ports, which are interconnected through 256 internal queues. Incoming traffic cells destined for one of the output ports are stored in the appropriate logical output queue. They will be authorized to exit by the centralized scheduling logic based on the highest priority among cells with the same output destination. When a conflict arises for access to the same output port by cells that are rated at the same priority level, the scheduler simply cycles through the same priority queues. Multicast cells are assigned to one of four traffic classes. They are queued at the input port before they can be sent to the output for which they have been earmarked. Multiple multicast requests are scheduled based on an RR fashion and multicast cells receive priority over unicast cells of the same priority level.

In order to operate with performance in systems that require a higher throughput than 20 Gbps, multiple nPX5800 chips must be connected in a master-slave configuration. In this configuration, an incoming cell gets sliced into several pieces (slices), which are then switched in a distributed fashion by the group of interconnected nPX5800 chips. This is done according to the master chip's scheduling decision instructions. It takes place over multiple serial links simultaneously and in perfect synchronization among slices.

The attached switch fabric devices exchange control messages over a 4-bit ring bus that helps them remain coordinated. The master chip manages an in-band back-pressure mechanism using Xon/Xoff signals or credits. The credit system works in AMCC's nP family in the following way: Every time a cell in the fabric leaves its queue for an output port, the nPX5800 sends a credit, which the traffic manager nPX5700 uses as a grant to send a new cell to the fabric. The traffic manager stops sending new cells when the credit balance available becomes zero.

Another interesting AMCC switch fabric that we must mention is the nPX8005, which is a terabit-class fabric that is based on a three-dimensional crossbar architecture with a large number of virtual output queues and distributed scheduling. As this fabric is using fixed-cell switching, it can handle *time-division multiplexing* (TDM) traffic on top of *Internet Protocol* (IP) and ATM flows. This is very significant as the tight requirements that traditional TDM traffic places on delay jitter and latency can be extremely hard to handle (if at all possible) for an average switch fabric chip that was designed only for IP and ATM traffic switching.

The nPX8005, which is positioned by AMCC for metro access network, metro core network, and *storage area network* (SAN) switching applications, is actually a chipset comprised of a memory subsystem (S8905), a scheduling device (S8805 or S8505), and a crossbar with an integrated arbiter

(S8605). It is designed to work seamlessly with AMCC's 7510 and 7250 network processors as well as with the company's nPX5700 traffic manager chipset. It features an integrated 2.5 Gbps serdes, high-speed terminations, and memory, so it is poised to provide strong QoS support, combined with a low-power, and a high-capacity switch fabric all packaged in a small form factor.

The nPX8005 provides eight *classes of service* (CoSs), thereby enabling greater granularity when handling traffic subject to *service level agreements* (SLAs) that require improved handling and reliability for time-sensitive realms such as VoIP or other system-critical data transfers as opposed to some types of data transfers, such as web page downloads. In general, these can be characterized as lower-priority tasks. For additional flexibility, the nPX8005 offers several robust scheduling algorithms. These include WRR, which is appropriate for fixed-length cell traffic; DRR, which is a wiser choice for variable-length packet traffic such as IP-over-Ethernet; WFQ, which is suited for egress traffic shaping and finer granularity scheduling; and maximal matching RR for connecting ingress to egress.

SYSTEMS CONSIDERATIONS WHEN DESIGNING WITH AMCC NP FAMILY NPUS

A systems designer should consider several factors when designing with AMCC nP family NPUs. First, to partition the logic into logical parts of a chassis-based design, the traffic manager 5700 chipset must be implemented on the line card, whereas the switch fabric 5800 must be integrated on the fabric card. As no serdes controllers are integrated in either of these products, unless a very low-speed single-board system is being designed (when the traffic manager and switch fabric can be connected directly), the chassis-based systems designer must use separate serdes components. More specifically, he or she must use four of them for each 5800 fabric chip. AMCC offers serdes devices (such as S2512, which provides four full-duplex 2.5 Gbps serial links) that are seamlessly compatible for such an application. Figure 6.6 shows a configuration of the scalability of the solution for OC-48 or OC-192.

An OC-192 or 10 Gigabit Ethernet configuration based on the newer nP7510 network processor uses two NPUs: one for ingress and one for egress connected with their respective nPX5720. Both NPUs would share a search engine or have their own engine (a much more expensive proposition). They would also be connected toward the line side through a ViX-to-SPI-4.1 bridge to an OC-192 framer or a 10 Gigabit Ethernet *Media Access Control* (MAC), which offer SPI-4.1 interfaces. As the nPX5800 switch fabric is a single-chip product, if a designer wants to combine chips for a 16-port fabric solution, then up to eight of them can be connected. Each of these fabric ports can support a quad (4x) OC-48c line card; therefore, a system can be put together with up to 64 OC-48c ports.

Looking at compromises in chip count, in a quad OC-48 line card, one nP7250 would be required per OC-48 link connected with the framer through a POS-PHY or *Universal Test and Operations* PHY *Interface for ATM* (UTOPIA) interface. With 10 Gbps line rates, a pair of nP7510s will replace four 7250 chips.

The interface of the 7520 with the search engine is a request/response type of interface that can be configured as dual 8-bit ports or as a single 16-bit wide port. A systems designer can connect AMCC's nPC2110 search engine or other devices without any further glue logic as recently announced by vendors such as IDT and NetLogic. Typical search engine devices will require glue logic implementation using a *field-programmable gate array* (FPGA).

The nP7520 has two symmetric ports that are used on the switch and on the line side, respectively. These ports can be configured in any one of five modes: UTOPIA 3, *POS PHY Level 3* (POS-PL3), FlexBus 3, dual RGGI, and AMCC's own ViX v.3. The line port where a framer is connected is usually configured as UTOPIA, POS-PHY, or FlexBus. The dual RGGI is used to connect Gigabit Ethernet MAC controllers. The switch side is configured as AMCC's ViX bus. If the switch port of an nP7520 is connected to the line port of another similar NPU, the system bandwidth is effectively doubled by processing the packets in a pipeline fashion. The *synchronous static random access memory* (SSRAM) interface is 64 bits wide and runs up to 104 MHz. It can be configured to support externally connected coprocessors such as classification chips from other vendors.

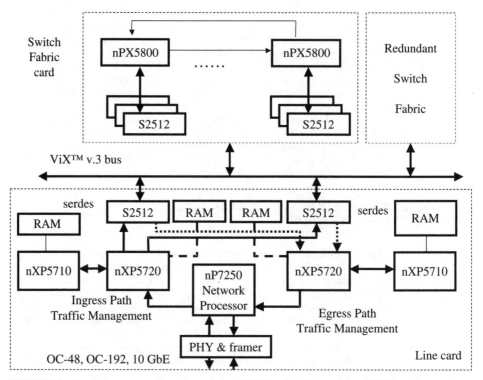

FIGURE 6.6 A typical systems configuration with the nPX5800 switch fabric and the nPX5700 traffic management chipset. *(Source: AMCC)*

AMCC documentation says that the nPX8005 family must be used to operate a switch fabric at a combined throughput above 160 to 320 Gbps. If we look at an example in a 16×10 Gbps switch fabric, then 5 chips need to be used for the switch fabric function, one in master, 4 chips need to be used in slave mode, and 48 chips need to be used for the queue management function. In addition, 16 serdes must be used for the switch interface and 16 FPGAs must be used for the line interface. Without counting memory, such a system requires a minimum of 85 chips if it is implemented with current AMCC technology. It will consume above 300 watts.

FIFTH-GENERATION TECHNOLOGY

We will conclude this chapter by adding a few comments on the company's fifth-generation technology, which AMCC introduced in late 2002 under the name nP5TM.

In addition to pursuing highly integrated products that efficiently offer headroom and flexibility to customers who need to come up with economic complete design solutions, AMCC is now offering the possibility of designing products that can handle multiple protocols and services at a lower cost, power, and size than before. The following added features accompany the main features of this new technology generation:

- The company's hardware-based functionality that was previously available in its nPX5700 fine-grained traffic management coprocessor is now integrated into the new platform. This tight integration enables designers to take advantage of the flexibility that can be afforded by software programmability. At the same time, the actual delivery of feature-rich subscriber services can be completed at high wire speeds.

- A richer programming model and the associated process flow inside the company's nPcore-based network processors allow a more extensive range of application coding without the programming complexity that is associated with another on-chip control plane CPU, which would obviously also impose its own extra power consumption and silicon real-estate requirements.

- While differentiating applications and services, customers require equipment that is designed around NPUs with significant "lung" capacity. AMCC's fifth-generation technology offers a respectable fivefold increase in performance over previous generations; therefore, it offers a significant amount of headroom to pursue sophisticated and differentiable applications.

- The previous on-chip coprocessors are now enhanced to allow simultaneous operations with the embedded traffic manager. This enables layers 2 to 7 packet processing together with a wire-speed OC-48 ATM SAR within one and the same device.

- The adoption of the latest NPF and *Optical Internetworking Forum* (OIF) interface standards allows a flexible and low-cost integration of memory subsystems along standardized ways, thereby enabling low-cost system solutions and creating a shorter time to market.

- Compatibility with the company's existing 100 Mbps to 10 Gbps network processors including the nPsoft Development Environment, in conjunction with support from the company's partners, enables customers to further leverage their existing investments in systems design and software programming.

The company has announced that its first priority with this new technology will be a next-generation, services-oriented 5 Gbps integrated NPU-traffic-management MAC solution. The intention is to enable designers to produce highly modular system designs that can support any service on any port, multiple concurrent high-value services, multiple technology capabilities, high subscriber density, and revenue-generating, per-subscriber statistics. The result should be products that enable carriers and service providers (who are the customers of the company's customers) to dramatically decrease both capital expenditures and operational expenses.

SUMMARY

In this chapter, we briefly reviewed AMCC's nP family of scalable network processors and discussed the main characteristics of the architecture. We also looked at other associated AMCC chips that handle traffic management and switch fabric issues in the framework of this complete family of interconnecting products. AMCC has a powerful combination of having the scalability of its architecture and the extremely advantageous feature of being able to offer multiple chips to the designer of networking equipment enabling the development of a complete solution quickly. Its solid business performance and robust financial health are important additional gauges of stability for customers who consider employing the company's network-processing technology into network equipment that they design.

CHAPTER 7

AGERE PAYLOADPLUS® FAMILY OF NETWORK PROCESSORS

Agere Systems is a recent spin-off from Lucent Technologies. It was formed after Lucent's acquisition a couple of years ago of a network-processing startup with the same name and the actual business of the former Microelectronics Division of Lucent. Agere Systems is now one of the world leaders in the sale of communications semiconductors. The company designs, develops, and manufactures integrated circuits for use in a broad range of communications and computer equipment. It recently announced its exit from the industry of optoelectronic components for communications networks. Its full line of communications chips includes network processors, switch fabrics, framers, *Synchronous Optical Network* (SONET), *Synchronous Digital Hierarchy* (SDH), *Plesiochronous Digital Hierarchy* (PDH), high-speed physical-layer-related products, and even *digital signal processor* (DSP) products.

In this chapter, we will only be looking at the most advanced members of the company's PayloadPlus family of network processors in both the OC-48c and OC-192 realms. This product family is geared toward the implementation of intelligent communication equipment with processing capabilities that span layers 2 through 7. These products focus on the wire-speed data stream. They work in conjunction with physical interface devices, traditional lower-speed microprocessors, and backplane fabric offerings to provide a complete solution for networking and communication applications. We will conclude our review of Agere's approach after also taking a brief look at other associated chips from Agere that provide the advantage of a complete systems solution.

PAYLOADPLUS® ARCHITECTURE: THE BIG PICTURE

Agere System's PayloadPlus is a comprehensive network-processing solution used in the OC-48c realm. It has been recently expanded to the OC-192 realm through the NP10/TM10 chipset (the two were recently renamed APP750NP and APP750TM, respectively). Until recently, this was basically a three-chip solution that handled all of the classification, policing, traffic management, *quality of service* (QoS)/*class of service* (CoS), traffic shaping, and packet modification functions required for a carrier-class network platform.

This network-processor family includes the *Fast Pattern Processor* (FPP), the *Routing Switch Processor* (RSP), and the *Agere System Interface* (ASI). The FPP and RSP process the wire-speed data stream. The ASI provides an industry-standard *Peripheral Component Interconnect* (PCI) interface between a host processor and other high-speed processors from Agere that are responsible for control and management functions, including routing table and virtual circuit updates, hardware configuration, and exception handling. The ASI also helps the FPP police *Asynchronous Transfer Mode* (ATM) and frame-relay traffic at rates up to OC-48c while maintaining state information on data flows and even capturing statistics.

In midsummer 2002, Agere announced a new integrated version of its 2.5 Gbps network-processor solution in the form of a new superchip called the *APP550* (previously known as the *INP5*). The APP550 integrates the FPP, RSP, and ASI; doubles the performance; and reduces the power, cost, and space required for supporting external memory. The goal is to drastically cut down the chip count of an integrated solution, improving the customer's time to market and system cost, performance, and density. In fact, a single APP550 can replace a six-chip configuration of the first-generation PayloadPlus chipset. Agere Systems has announced two members of the APP550 family: a 266 MHz version supporting 2 to 4 *Gigabit Ethernet* (GbE) or full-duplex 2.5 Gbps *Packet over SONET* (POS)/ATM processing capacity and a 133 MHz version supporting 1 to 2 GbE or full-duplex 622 Mbps POS/ATM processing capacity.

The entire network-processing solution rotates around the capabilities of the FPP, which can be called to action by programming the FPP chip through a high-level language that Agere has developed called *Functional Programming Language* (FPL). Through FPL code, the FPP can analyze and classify patterns based on the bit content of every byte of the payload or the headers of packets and/or frames. Agere's patented search and pattern-matching technology enables the buildup of very large lists. The search time is also deterministically limited. You can search for any length of data pattern, and the search time is only limited by the pattern length, not by the number of entries in the search table.

On top of these three fundamental chips, Agere has also introduced another member of the PayloadPlus family known as the *Voice Packet Processor* (VPP). This coprocessor chip is capable of *ATM Adaptation Layer 2* (AAL2) *segmentation and reassembly* (SAR) and switching functions supporting up to 32,767 conversations.

Figure 7.1 shows the block structure of the PayloadPlus architecture. It is based on a patented search technology called *Pattern-Matching Optimization*. According to Agere, this architecture enables the company's network processor to achieve a performance more than five times greater than network processors based on advanced *reduced instruction set computer* (RISC) cores. This performance attains the level of fixed-function *application-specific integrated circuits* (ASICs) while providing the flexibility and programmability of RISC. The architecture achieves this by using less

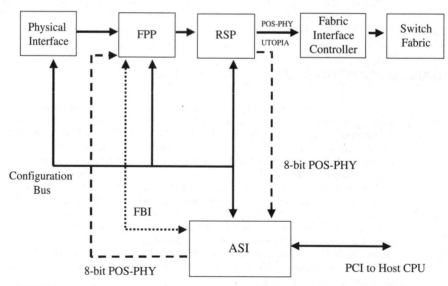

FIGURE 7.1   The block architecture and an overview of Agere PayloadPlus. *(Source: Agere)*

overhead, fewer clock cycles, and more data processing per clock cycle than enhanced RISC-based processors.

As shown in Figure 7.1, the FPP takes packets or frames from the PHY chip over an industry-standard interface that can be either a *POS PHY Level 3* (POS-PL3) or a UTOPIA 2 or 3 interface. Then it performs protocol recognition and classification as well as reassembly. The FPP can classify traffic based on information contained at layers 2 through 7. Once this is done, the FPP sends the packets and its classification results via a POS-PL3 interface over to the RSP. The RSP is responsible for handling queuing, packet modification, traffic shaping, the application of QoS tagging, and segmentation.

The FPP and RSP chips interface with the ASI chip. The ASI chip handles exceptions, maintains state information, and is responsible for the interface with a host *central processing unit* (CPU) over a PCI bus. The FPP and the RSP are configured and updated via the ASI chip over the *Configuration Bus Interface* (CBI). A special 8-bit asynchronous bus called the *Management-Path Interface* (MPI) enables the FPP to receive management frames from the local host CPU through the ASI. A third system bus called the *Functional Bus Interface* (FBI) connects the FPP to an ASI and/or other application-specific custom logic that is used to externally process function calls.

All memory interfaces are 64 bits wide either to standard PC-133 *synchronous dynamic random access memory* (SDRAM) or 133 MHz pipelined *zero bus turnaround* (ZBT) *synchronous static random access memory* (SSRAM). This is a significant advantage as the FPP stores all pattern-matching data in standard memory rather than in expensive and power-hungry *content-addressable memory* (CAM) devices.

If the arrows of the data flow shown in Figure 7.1 are inverted, the egress path can be determined; therefore, it explains how the same chipset can operate in a full-duplex line card as in OC-48c. If packets on the egress side require further classification, a new FPP needs to be inserted into the egress path. If packets need queuing at the egress path, another RSP chip will be needed. Finally, if separate statistics gathering is required at the egress path, a separate ASI chip is needed. In the worst case, the configuration of Figure 7.1 should also be replicated on the egress path, as well.

For systems that are based on the use of the VPP, the VPP is inserted in the structure shown in Figure 7.1 between the FPP and the RSP. It connects both upstream and downstream with 32-bit POS-PL3 interfaces. It can be configured by the ASI over the CBI bus, and it supports a 64-bit SSRAM interface for maintaining state and statistics. The VPP chip cannot handle speeds of above OC-40, (broken down as a maximum of OC-12 of AAL cells and a maximum of OC-12 of CPS packets). As a result, we do not intend to cover it in more detail here. Interested readers can refer to technical documentation from the Agere web site for more details on the VPP.[1]

In terms of physical presence and power consumption, both the FPP and RSP are available in *ball grid array* (BGA) packages that have 655 pins each. The ASI comes in a 448-pin BGA. The maximum total consumption of the set of three chips is 9 watts when it operates at 13 MHz.

FPP

The FPP is a pipelined, multithreaded processor that can simultaneously analyze and classify up to 64 *protocol data units* (PDUs). Each incoming PDU is assigned its own processing thread, which is called a *context*. The context is essentially a processing path that keeps track of all the blocks of a PDU, the number of the input port through which the PDU arrived, the data offset for the PDU, the last-block information, any potential program variables that are associated with the PDU, and, of course, the classification information that is related to the PDU. The FPP does not suffer from the *speculative execution* of instructions that cannot be followed up by the rest of the executable code—a situation that all too often stalls pipelines in RISC processing environments. It also does not suffer

---

1. Technical documentation with white papers, application notes, and data sheets is available at the Agere web site at *www.agere.com* and directly from the company.

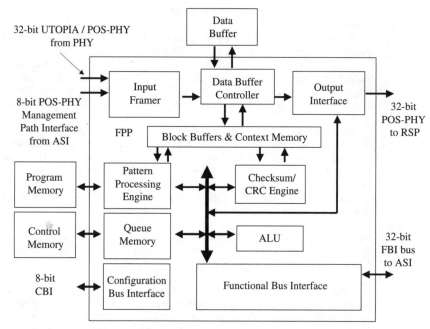

FIGURE 7.2 The internal block structure of the FPP chip. *(Source: Agere)*

from the undesirable switching-context overhead that is typical in most architectures that process data sequentially.

Figure 7.2 shows the internal structure of the FPP. Some blocks have an identifiable function such as the *arithmetic logic unit* (ALU) or the checksum/*cyclic redundancy check* (CRC) engine. The purpose of the other major blocks is as follows: The input framer frames the incoming stream into 64-byte blocks. Then it writes these blocks into the data buffer and into the block buffers and context memory. The latter temporarily stores blocks that are being processed as well as other associated context data for the execution of the FPP operations on the incoming data. The output interface strips the payload away from PDUs, such as packets or frames, according to block offsets, and forwards them along with their classification conclusions to the next processing stage downstream, which is usually the RSP chip.

The *Pattern Processing Engine* (PPE) of the FPP performs pattern matching to determine how the incoming PDUs are classified. This will decide how they must eventually be processed. The *Queue Engine* manages FPP replay contexts, provides addresses for block buffers, and maintains information on blocks, PDUs, and connection queues.

The FPP processes bit-stream data in two passes: first it processes the PDUs as separate 64-byte blocks and more specifically, the data offsets of the various blocks are stored and printer links are established between the blocks out of which the PDU is composed.

In the replay phase (second pass) the PDU is processed as a whole entity. Pattern matching is executed at the same time as integral transmission is handled of the PDU toward the output interface. The latter will reassemble the PDU and if needed it will strip a certain amount of data away from the blocks of the PDU, of course according to the data offsets, which were defined during the first pass.

Agere's architecture distinguishes the allocation of computational resources into a *fast* processing path and *slow* processing path. These paths were discussed in Chapter 2, "Network Processors: Justification." This logical partitioning is strongly reminiscent of the data versus control plane processing debate. With the PayloadPlus approach, the FPP, the RSP, the FBI bus, and part of the ASI are considered the fast processing path elements because they have to perform their tasks at wire speed directly on the traffic bitstream. The rest of the ASI, the MPI bus, and the PCI-based host, along with the host CPU itself, are the elements of the slow processing path, which is computationally responsible for handling exceptions, configuration, management, system updates, and so on.

RSP

The RSP handles the classification and analysis results of the FPP's work on the incoming PDUs. This happens over 64 logical input ports. In addition to the PDU, it comes in the form of a transmit command from the FPP that essentially instructs the RSP as to how to handle the specific PDU. The latter proceeds by identifying the necessary processing for each PDU. The PDU is added to a queue and stored into the PDU SDRAM. The transmit command determines the QoS, the CoS, and the required PDU modifications for the RSP.

The RSP supports up to 65,535 (64K) programmable queues. Each queue is based on programmable QoS and CoS criteria for processing and routing. It can schedule independently up to 256 logical output channels mapped onto 32 physical output ports. It can also connect to an external overriding scheduler that can monitor and schedule all RSP queues. It interfaces downstream with a potential fabric interface controller over a configurable industry-standard 32-bit POS-PL3 or UTOPIA 3 interface. This output can be configured to be one 32-bit interface, two 16-bit interfaces, or four 8-bit interfaces.

The RSP has fully programmable packet-discard policies (including *Random Early Detection* [RED], *Weighted RED* [WRED], and *Early Packet Discard* [EPD] algorithms) and outgoing packet data modification capabilities. It is also equipped with intrinsic support for multicast packets and virtual paths and has the native ability to segment (which is handy for interfacing with cell-based fabrics or ATM/POS-PHYs) and cope with real-time traffic such as *variable first-rate-real-time* (VBR-rt).

The RSP has the following four major areas of functionality:

- Queuing.
- Traffic management.
- Traffic shaping.
- Packet modification.

Figure 7.3 shows the hierarchy of criteria applied for the scheduling the RSP. Up to 16 CoS queues feed a single QoS queue to support PDU-based shaping policies. Each QoS queue is assigned to a single scheduler that is configured by connection rate type, such as *constant bit rate* (CBR), *variable bit rate* (VBR), or *unspecified bit rate* (UBR). A set of schedulers is defined for each logical port. Each scheduler supports a single type of traffic (such as CBR, VBR, or UBR).

Figure 7.4 shows the extremely efficient data flow inside the RSP. As we mentioned earlier, the systems designer has the extra flexibility to connect an external scheduler. This opens up the possibility of custom-written algorithms beyond the ones that the RSP offers. This feature is useful when processing priorities need to be changed based on live traffic conditions. In some cases, it is even imperative. This may be the case in situations where a switch fabric is used that makes global decisions about the overall scheduling of traffic.

Figure 7.5 shows the RSP chip's internal block structure. Three powerful compute engines based on *very long instruction word* (VLIW) architecture are cascaded in a pipelined fashion that allows heavy-duty computing performance while maintaining wire speed compatibility. These three engines are a Traffic Management Compute Engine, which enforces packet-discard policies and keeps queue statistics; a Traffic Shaper Compute Engine, which ensures QoS and CoS for each queue; and a Stream Editor Compute Engine, which performs all potentially necessary PDU modifications. In each queue definition, the RSP includes a destination, scheduling information, and pointers to programs for each

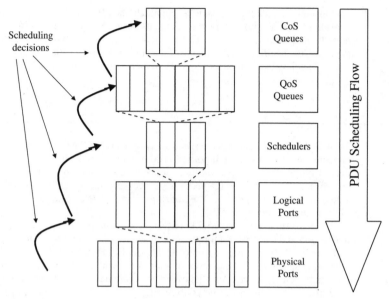

FIGURE 7.3 Scheduling hierarchy for each PDU. *(Source: Agere)*

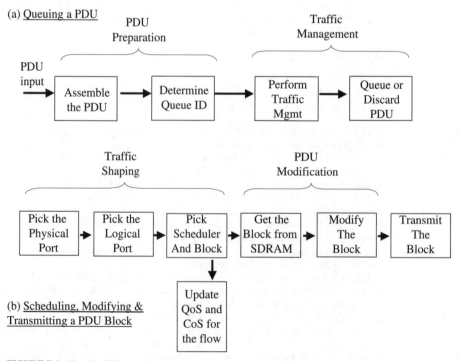

FIGURE 7.4 Queuing PDUs and block scheduling. *(Source: Agere)*

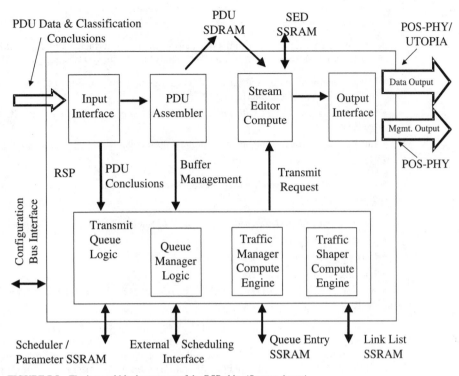

FIGURE 7.5   The internal block structure of the RSP chip. *(Source: Agere)*

of the three VLIW compute engines that we just mentioned. By selecting a queue definition that performs the desired processing, the RSP can execute multiple protocols. The external host CPU can also be used to dynamically add queue definitions, as needed, to set up ATM virtual circuits, for example.

To execute code, the compute engines must be properly configured. This means that a program, along with the necessary parameters, must be loaded at configuration time or dynamically during operation. The number of compute engines configured depends on the operation of the system, the size of the engine code, and the available internal RAM. Channels and physical ports are configured first. Then logical ports are configured and assigned to the physical ones. After these steps are completed, the desired compute engine program is loaded. The next step is the creation of schedulers for each logical port. The definition of each logical port includes the program selection that will handle traffic management, policy, and shaping, as desired. The compute engine programs are loaded at configuration time, but they can be selected for queues dynamically.

For the definition of queues, the queue must first be added to a data structure called the *stream editor destination ID table*. This table includes a pointer to the Stream Editor Compute Engine's modification instructions for the queue. The compute engine program parameters must then be defined. These are used to set thresholds for the discard policies or to define bytes to add or replace when modifying a PDU. Finally, the queue must be assigned to a scheduler. By doing that, the actual programming of the Traffic Management Compute Engine and Traffic Shaper Compute Engine are chosen, as well as both the physical and logical ports that will need to be used for the queue. Again, all these steps can occur at configuration time or dynamically during operation.

In terms of memory interfacing, the RSP comes equipped with a 64-bit interface that can be clocked up to 133 MHz for queuing PDUs in SDRAM and with four 32-bit-wide interfaces that offer point-to-point memory access up to 133 MHz.

ASI

As mentioned earlier, the ASI chip's role is to seamlessly interface the FPP and RSP to a supervising host processor. More specifically, it makes it possible for the systems designer to do the following:

- Create a method for the centralized initialization and configuration of the network-processing system and all its physical interfaces.
- Send routing and *Virtual Path Identifier/Virtual Connection Identifier* (VPI/VCI) table updates to the RSP.
- Implement various routing and management protocols.
- Handle any occurring exceptions.

The ASI also enables other high-speed, flow-oriented state maintenance tasks for the FPP, which include the following:

- Gathering *Remote Network Monitoring* (RMON) statistics needed for remote network management
- Timestamping packets.
- Checking packet sequence.
- Policing ATM and frame relay at up to OC-48c rates.
- An 8-bit POS-PHY interface over which the ASI sends packets to the FPP and receives packets from the RSP.

The ASI is connected to the host CPU by a PCI interface, which is a 64-bit, 66 MHz bus designed in a full master-slave implementation with full interrupt and *direct memory access* (DMA) support. Its support for SSRAM is based on two industry-standard, 32-bit-wide memory interfaces.

The ASI's 8-bit CBI bus enables the initialization and configuration not only of the FPP and RSP, but also of six additional devices. It is interesting to note that it has been designed deliberately to be compatible with both Intel and Motorola bus formats, so it enables the configuration of third-party devices such as framers or PHY interfaces. The CBI also loads the FPP and RSP chips with their corresponding programs and the dynamic updates to the FPP tables and RSP queues, respectively.

The FBI is a 32-bit bus that extends the capabilities of the FPP by enabling the FPP to make function calls that are executed by the ASI itself. These function calls can involve requiring the use of an ALU for a calculation and looking for access to data that is stored in SRAM, or it can be as all encompassing as taking control of the FBI bus itself.

Through several configurations of the *leaky bucket* (LB) algorithm, the ASI performs high-speed policing of ATM and frame-relay traffic. Its default configuration, for instance, uses the *generic cell rate algorithm* (GCRA) as defined by the ATM Traffic Management Specification, version 4.0. This works as follows: We saw earlier that the FPP is programmed in FPL. It is important that the FPL code can invoke functions that are sometimes executed on external hardware, thereby extending the capabilities of the FPP. At the same time, the ASI contains an ALU and an SSRAM interface state buffer, which are used to implement functions invoked by the FPL code. This means that when the FPL code, for example, invokes the policing function for a PDU, the ASI checks whether the PDU is compliant and returns an appropriate flag. The FPL program then determines what exactly must happen. For example, it can choose to just flag all noncompliant PDUs or it can discard them altogether, depending on the application.

Figure 7.6 shows the internal structure of the ASI chip. We have already discussed the role of most of its blocks. It is interesting, however, to note a couple of points. Two ALUs are available for processing FPP external function requests. One is for policing and the other is for maintaining state-related information and calculating statistics. Likewise, the two SSRAM interfaces, which were

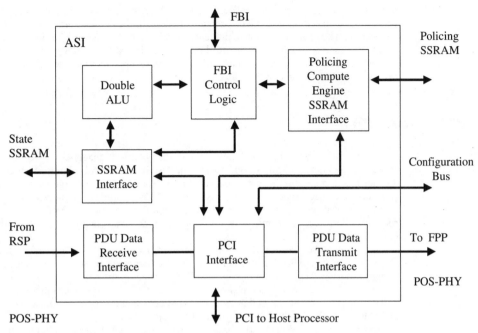

FIGURE 7.6 The internal block structure of the ASI chip. *(Source: Agere)*

intended to handle memory access without contention, are used to simultaneously access two banks of SSRAM memory: one with policing information and one with state information.

Transfer of management frames and statistics to a host CPU application is supported over the ASI's PCI bus. More specifically, through its *direct memory access* (DMA) master capabilities, the ASI forwards this information to host memory. Likewise, if the host wants to generate specific PDUs, it will do so and download them to the ASI over their PCI connection, and the ASI will then send them out through its 8-bit POS-PL3 interface. In terms of management information, the ASI maintains a very large database where it stores the state-related information and statistics it gathers. This information can be updated by FPL function calls invoked by the FPP and sent over the FBI bus. The code can run ALU operations to modify or compare values in the database and the ASI can return values to the FPL code. The ASI also maintains a second database that contains information used to determine compliance with the imposed traffic control constraints.

In its several variations, the *dual leaky bucket* (DLB) algorithm (whose one subset is the ATM-standard specified GCRA) is implemented on a programmable compute engine. When the FPP makes the appropriate function call to the ASI regarding a specific PDU, the ASI starts running the corresponding policing algorithm. When the algorithm execution is finished, the ASI flags the PDU (frame or cell) as compliant or not by returning a pass/fail value to the FPP. In the case of a DLB implementation, it will also stipulate from which bucket it identified the PDU's nonconformance.

It is important to realize that when we say that the ASI performs its policing by checking the conformance for up to 64K connections, flows, or aggregates at up to OC-48 rates, it does not mean that it schedules or shapes any traffic. It just identifies the cells or frames that do not comply. Also, when the LB algorithm is applied, numerous options in the GCRA parameters are chosen for each connection. The only constraint is that each PDU's arrival time must be measured with the same degree of granularity across the board. For instance, if ATM and frame-relay connections will be policed at the same time, the timeout counter must be set up to measure the smaller between the ATM cell rate and the byte time of the frame relay connection.

THE DLB ALGORITHM

In Chapter 14, "Switch Fabrics," and Chapter 15, "Traffic Managers," we cover issues related to scheduling and flow control. Among these issues, we discuss the LB algorithm and how it applies to a policy that decides how and when to discard packets. In the ASI chip, Agere has implemented a very flexible model that serves the traffic constraints in ATM networks extremely well.

In a classical *single* LB implementation, the algorithm uses two parameters: the *Limit* (L) and the *Increment* (I) value. The Limit value corresponds to the bucket depth, whereas the Increment value corresponds to the leak rate of the bucket.

In a *dual leaky bucket* (DLB) implementation, two buckets are applied to each connection. Depending on the application, each of the Limit and Increment parameters of the two buckets can be assigned to several connection parameters. For instance, in the context of an ATM connection, one bucket may be made to leak at the *sustained cell rate* (SCR), whereas the other may be made to leak at the *peak cell rate* (PCR). In that case, the ATM cells that do not conform can be tagged appropriately by setting their *Cell Loss Priority* (CLP) bit equal to one.

Several variations of the DLB, including how to use the CLP bit as a policing parameter, are stipulated in the ATM Forum TM 4.0 specification. In Agere's approach, both cells with CLP = 0 or CLP = 1 are added to both buckets. All discarded cells are marked as either SCR or PCR discards. All action that will be taken is determined ultimately by the FPP and RSP programming, thereby giving tremendous flexibility to the systems designer. More specifically, it enables systems to be implemented that can answer the following questions for each connection:

- Which algorithm will be used?
- What will the negotiated cell rates be, including the SCR and the PCR?
- What will the ATM tolerance parameters be, including the *maximum burst size* (MBS), the *burst tolerance* (BT), and the *cell delay variation tolerance* (CDVT)?
- What are the supported access line rates for frame-relay connections, such as the *committed information rate* (CIR)?

AGERE'S APP750NP (EX-NP10) AND APP750TM (EX-TM10) CHIPSET

Agere had originally targeted the PayloadPlus family to the OC-48c (2.5 Gbps) market. It has recently introduced a new chipset (originally called PP10G) that scales the architecture up for the OC-192 (10 Gbps) realm and offers carrier-class performance in edge and core networks. The NP10 network processor and the TM10 traffic manager chips (recently renamed APP750NP and APP750TM, respectively) comprise the new chipset, which can handle complex multifield packet classification, policing, queuing, statistics, scheduling, shaping, buffer management, and, of course, cell or packet modification.

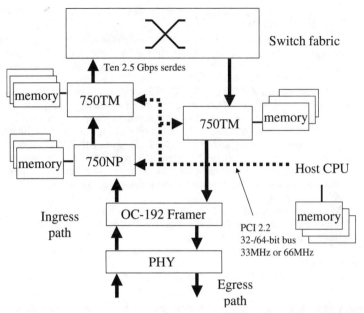

FIGURE 7.7 A block diagram of a typical OC-192 line card based on the APP750NP/APP750TM chipset. *(Source: Agere)*

Figure 7.7 shows the block structure of a typical 10 Gbps system based on these new chips. The three-chip configuration can easily handle full-duplex 10 Gbps, supporting wire-speed processing based on *access control lists* (ACLs) with thousands of ACL rules; however, an additional APP750NP *network processing unit* (NPU) may have to be used if the intended system design requires egress classification.

One of the major advantages of the new chipset is that it works with inexpensive external DRAM. It requires very little SRAM to provide high-performance functionality. As the classification rule database is stored in *fast cycle RAM* (FCRAM), which is also referred to as *network DRAM*, no external CAM is needed. For instance, 1 million *Internet Protocol version 4* (IPv4) routes can be kept in DRAM with separate information for each *virtual private network* (VPN) supported. Statistics and policing databases are kept in *quad data rate* (QDR) SRAM.

In terms of traffic management, the APP750NP/APP750TM chipset is extremely powerful and flexible at the OC-192 realm. For example, VPNs are supported with traffic isolation and *service level agreements* (SLAs). Dynamic service provisioning is ensured through dynamic bandwidth and QoS/CoS modifications in real time. Two million different packet-handling behaviors with three buffer management profiles per behavior type are available to guarantee a fine granularity in service differentiation. External packet buffer memory can be expanded to 256MB or more per direction.

As its predecessor, the APP750NP/APP750TM chipset is predominantly programmed using Agere's FPL. Complex classification policies such as IPv4/IPv6, *Point-to-Point Protocol over Ethernet* (PPPoE), *Layer 2 Tunneling Protocol* (L2TP), and *Multiprotocol Label Switching* (MPLS) can be implemented in FPL. Even when they are executed, they will still leave plenty of headroom for other packet computing work.

Statistics, policing, and several other modification functions can be implemented in Agere's C-like scripting language called *Agere's Scripting Language* (ASL). This preserves investment in software engineering for the implementation of queuing, policing, statistics gathering, as well as packet classification and code modification.

Although the APP750NP/APP750TM chipset can be directly connected to Agere's PI40 switch fabric through redundant integrated *serialization and deserialization* (serdes), it also provides support for both cell- and frame-based switch fabrics, given its programmable classification and SAR capabilities. This means that minimal if any at all glue logic is needed to interface third-party fabrics, which can be connected using an *Netw*ork *Processing Forum* (NPF)-like streaming interface based on *System Packet Interface 4.2* (SPI-4.2). Agere also provides a system reference design with full software support that can be extremely useful for *network equipment vendors* (NEVs) trying to minimize their time to market. A connection with the framer is also made via an industry standard SPI-4 Phase 2 frame interface.

Port-based rate shaping is programmable for up to 256 media ports and various configurations are supported, such as one OC-192c, four OC-48c, mixtures of 1 Gbps or one Gigabit Ethernet, 192 DS-3 links, and so on.

The chipset is accessible by a supervising host CPU over a PCI-2.2-compliant, 66 MHz, 32- or 64-bit bus.

THE APP550 (EX-INP5) NETWORK PROCESSOR

As mentioned in the beginning of this chapter, at the end of July 2002, Agere announced the APP550 (originally introduced in the market as INP5). APP550 is an integrated network processor that further optimizes the position of the product family for the OC-48 realm. It has also been designed to minimize the chip count (an issue that was perceived as the Achilles heel of the architecture previously offered by the company) and offer significantly decreased power consumption and a reduced overall systems cost.

A comparison of a typical OC-48 solution based on the company's previous three chips and the APP550 single-chip solution, along with associated memory as well as PHY and fabric interface chips in both cases, shows some impressive results. More specifically, the APP550-based system costs less than half the cost of the three-chip solution. It takes only about 60 percent of the printed-board space needed for the three-chip solution and consumes 19 watts (including all of the associated memories) as opposed to 43 watts for the three-chip implementation. The company introduced the first APP550 chip samples by the end of 2002.

Figure 7.8 shows how the APP550 fits between the PHY/framer and the switch fabric. A full-function classifier, a policing engine, and a traffic manager are integrated into the APP550, along with Ethernet *Media Access Control* (MAC) controllers and 3MB of on-chip DRAM. The APP550 interfaces to the line and to the fabric side through standard GMII/SMII or POS-PHY/UTOPIA interfaces. At the same time, it can be interfaced with a supervising host CPU through a PCI bus and with external optional coprocessors through a standard POS-PHY interface. Figure 7.8 also shows the data path through the APP550 and the internal architecture of this highly integrated network processor.

For fast table lookup, the APP550 uses FCRAM, which is a fast-cycle DRAM and which offers SRAM-like performance at DRAM prices. This means that for memory clock rates of 200 to 400 MHz, the network DRAM can achieve data rates equivalent to 400 to 800 MHz. Agere already has several large memory suppliers (such as Samsung, Fujitsu, and Toshiba) signed up and committed to the FCRAM used by its APP550 and APP750NP/APP750TM chips. The use of DRAM for the table lookup function saves significant cost and power and greatly increases the capacity compared to the

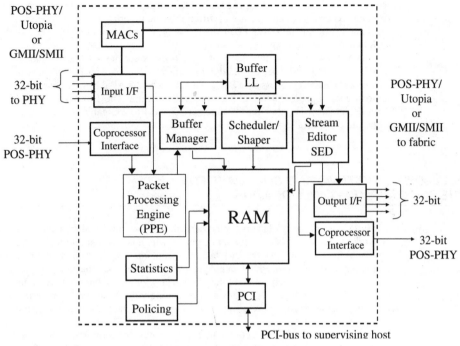

FIGURE 7.8   The internal architecture and data path of the APP550 network processor. *(Source: Agere)*

use of CAM or SRAM. FCRAM is used to provide a system with storage capabilities for high-density interfaces, such as tree memory, packet buffering, and data modification parameters. It is characterized by less power consumption than conventional DRAM. It is also optimized for small bursts of activity and random access, such as that needed in graphics and network applications (web content).

In terms of memory *input/output* (I/O) paths, the APP550 supports multiple types of memory usage:

- In *double data rate* (DDR) SRAM, it maintains a 32-bit-wide interface with linked-list memory, a 32-bit-wide optional *stream editor* (SED) context memory, an optional 32-bit-wide interface with memory that contains policing and statistics-related information, an optional 32-bit-wide interface with queue memory, and another optional 32-bit-wide memory bank that stores scheduler parameters.

- In FCRAM, APP550 maintains a 32-bit-wide interface with packet-buffer memory, an optional 32-bit-wide interface with memory that stores reassembly-related information, a 16-bit-wide interface with SED parameter memory, a 1 or 2 16-bit-wide interface with FPP program memory, and an optional 16-bit-wide interface with FPP control memory.

By its ability to perform 128K simultaneous reassembles, the APP550 can support a large number of virtual circuits, while the chip's integrated capacity of 256K queues enables the per-flow queuing for a large number of queues. Programmable data segmentation and modification allow the support for tunneled protocols and the use of different switch fabrics, whereas sophisticated buffer management and traffic shaping over 1,024 shapeable ports enable a more efficient use of bandwidth and a high-density system design.

The APP550 has been announced at two clock frequencies—133 MHz and 266 MHz. It is offered in a 1,413-pin FCBGA package and is manufactured in a 0.13μ *complementary metal oxide semiconductor* (CMOS) process by TSMC. The 266 MHz version has a throughput of 5 Gbps (or 2 to 4 bidirectional Gigabit Ethernets) and consumes 9 watts. The 133 MHz version (targeted by Agere toward the realm of applications between OC-3 and 1 to 2 Gigabit Ethernets) has a throughput of 2.5 Gbps and consumes less than 6 watts.

DEVELOPING SYSTEMS AND SOFTWARE FOR THE PAYLOADPLUS FAMILY OF NPUS

The FPL is one of the key factors for the flexibility and versatility of the PayloadPlus family of Agere's network processors. It is a functional language, which is a computing model that is somewhat reminiscent of the approach that the Lisp language implemented. It has nothing in common with the programming model of a procedural language, such as C. In a functional language, the programmer writes code that tells the underlying computer resources what to do, but not how to do it. Getting the code to do the latter is usually very tedious, excruciatingly detailed, and highly error prone. Worst of all, the code must be rewritten every time a slight modification of a protocol or operational procedure has to be implemented.

As an illustrative example, contemplate the difficulty of coding the task of sorting a list of long bit patterns according to some criteria and reordering them accordingly. In the functional programming approach, the task is specified as the sorting of the original list. In a procedural language realm, however, the programmer has to correctly code bit per bit all the manipulations that must occur in the appropriate order by properly monitoring and managing buffer usage. If the list of bit patterns changes, the procedural code must be rewritten. In the functional language, the same sorting code must be rerun, but this time it is simply applied on a different list of bit patterns.

FPL provides an order of magnitude of reduction in the number of instructions needed to carry out a task compared to C/C++; hence, it offers a significant improvement in productivity of software engineering. It also eliminates the need to hand-optimize assembly or microcode in order to achieve wire-speed performance. We revisit this context and the language's advantages in Chapter 16 where we discuss systems engineering considerations and trade-offs regarding the cost of development over the entire lifetime of a project or product.

Communication protocols are described in FPL, and the processor ends up "learning" pattern-matching processes. The software engineer does not have to write exhaustive code that explains how to seek out specific bits and what to do with them.

In the case of Agere's network-processing solution, code must be written in FPL to create a program in order to handle the PDUs. The code is then compiled and an image (executable) is loaded into the FPP. Every time a PDU arrives at the FPP input, a program must run. Typical examples of code written in FPL would perform operations such as layer 2 and above protocol processing, SARing of ATM cells, checking the size of programmable PDUs, performing timeout checks on ATM cells, handling CRC and checksum processing, and determining the PDU output queue and the PDU's corresponding CoS.

Code written in FPL must start from one of two possible entry points (program statements) called *roots*. These actually stipulate which FPL function should be invoked first. For example, the ROOT function will receive a data stream either from the framer or from the internal queue inside the FPP. We commonly say, "A PDU is being replayed from queue." The principle of replaying a PDU manifests itself in the FPL computing model. This requires a two-pass process when handling a PDU:

1. An initial processing pass must be performed, while the PDU data stream is read into the queue engine memory in blocks of 64 bytes at a time. For instance, this occurs when identifying the type of PDU, reading specific packet values, and assembling cells (in the case of ATM).

2. A second processing pass is performed, while the PDU is replayed from the queue. For instance, the program may decide to simply forward the PDU to its next-stop application engine destination, or some operations stipulated by a higher-level protocol may need to be performed on the replayed PDU.

The FPP Queue Engine (programmed by parts of the FPL code) enables the programmer to process a PDU that may be embedded in a higher-level protocol, and then send it back to the queue. It may even process it again for another protocol.

It is also important to note that through the use of an *application programming interface* (API), the software engineer can add or delete certain types of FPL statements to and from the image dynamically. Two types of pattern-matching statements are available: *single-rule* pattern statements, where a single pattern must be matched with one or more functions to perform, and *multiple-rule* pattern statements, which allow the definition of tables (for example, IP routing tables) to process a pattern with many variations. The former can only be changed slowly, whereas the latter can be updated very rapidly. The latter multiple-rule statements are called *trees* by Agere.

FPL offers the capability of specially tagging a PDU, which provides the definition of special processing paths for functions to handle the different types of data. All PDU processing ends with the option of either aborting and halting processing (in which case perhaps the application at hand dictates that an exception must be initiated and handled under the auspices of the host CPU) or sending the PDU to the downstream application logic waiting for it.

In addition to FPL, Agere is offering its ASL, a C-like scripting language, which can be used to program procedural tasks that can be associated with the workload typically executed by the RSP and the ASI chip. It can be compiled by Agere's VLIW compiler into VLIW engine code. In order to ensure that freshly written code executes within the available number of clock cycles, the programmer also has access to the VLIW instruction simulator. The effort customers put forth to write their own code from scratch to implement various common functions or protocols is further minimized by Agere's library of code blocks that provide reference implementations of protocols. These include protocols such as IP over AAL5, IP over SONET, and POS/*Point-to-Point Protocol* (PPP) as well as raw switch functionality such as the implementation of the WRED algorithm, aspects of ATM policing, or traffic shaping.

The array of available tools inside Agere's Festino™ *Software Development Environment* (SDE) in its latest version 3.0 includes a full-fledged performance and functional simulator of individual chips from the product family and of systems with multiple-chip topologies and configurations. This enables the offline analysis and simulation of switch designs that even include external custom logic. The latter is depicted in the SDE environment by using an extended model based on *eXtensible Markup Language* (XML). A source-level debugger for FPL completes the toolset along with a traffic-generation module, a throughput-accurate software simulator, and one common environment that offers support for both the OC-48c and the OC-192 realms. In addition to a convenient *graphical user interface* (GUI) approach, the environment has the following:

- A tracer tool, which keeps track of an individual packet during its lifetime inside a system and logs all functions and subsequent actions taking place on it.

- A profiler, which can help by throwing the proverbial spotlight on performance bottlenecks through the identification of the number of clock cycles spent on a particular context or on a specific "tree" (in Agere's meaning of the word, as we have seen).

An available *Software Development Kit* (SDK) enables the designer to write C- or Java-code models to describe other systems hardware that interacts with Agere's chipsets in a larger configuration. As a result, their behavior is brought into a global simulation run. SDE runs under Sun Solaris, Linux, or Windows NT.

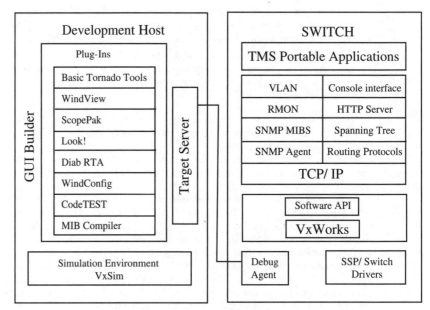

FIGURE 7.9 The TMS architecture for the development of software on Agere's network-processing platform. *(Source: Agere)*

It is also important to note that Agere provides strong support for the development of routing and switching applications that are meant to run on the PayloadPlus family of network processors. One of the preintegrated supported software options is based on WindRiver's TMS system that contains the very well known Tornado environment.[2] The latter is now the de facto development environment for embedded software systems in the infrastructure network community. It is coupled with software that addresses essentially all aspects of layers 2, 3, and above of communications protocols, management, and so on in the Internet world. The TMS protocol stack runs under VxWorks and communicates with Agere's reference boards. Driver support is available from the company, along with software support to interface with a PCI-based chassis system, which is called *Switch Support Package* (SSP). Figure 7.9 shows the concept.

A chassis-based hardware development system built around a Pentium- or PowerPC-hosted system that is operating under either the Linux or VxWorks operating system is also available for the development of systems based on Agere's network-processing solution.

We conclude the discussion of Agere's network-processor technology by referring you to Chapter 14, "Switch Fabrics," where we cover switch fabric technologies and where Agere's 40 Gbps switch fabric chipset is covered in more detail as a leading-vendor technology case study.

---

2. More information about the TMS and Tornado development systems can be found at WindRiver's web site at *www.windriver.com.*

SUMMARY

In this chapter, we reviewed Agere's network-processor family known as PayloadPlus as well as the company's latest 10 Gbps chipset and the most recently announced APP550 network processor, which is a highly integrated OC-48 realm solution and the latest entry into the family. We discussed the unusual partition of packet processing and switching tasks that the original Agere approach dictated and identified its interesting characteristics. We reviewed the programming model for the Agere NPU platform, which is based mainly on the company's FPL programming language. FPL allows tremendously shorter and efficient code writing compared to traditional C language coding, thereby minimizing development time. We will expand on these issues in Chapter 16 where we review systems considerations and trade-offs. Agere's 40 Gbps switch fabric chipset is discussed in Chapter 14 as a case study.

CHAPTER 8
MOTOROLA'S C-PORT™ FAMILY OF NETWORK PROCESSORS

Motorola has followed a two-pronged approach into the network-processing arena. At the top of their line, they offer the C-Port family of network processors and traffic managers, which we will review in this chapter. At the lower end of the spectrum, Motorola offers the PowerQUICC™ architecture, which is based on the company's original and very successful product recipe of including a very common *central processing unit* (CPU) in the same chip die (such as a member of the 68000 or the PowerPC families) with Ethernet or other networking and communications interfaces. The latter family has earned a tremendous amount of business for the company in the *local area network* (LAN) and access equipment industry, effectively propelling the company to an undisputed leadership position for communications processors; however, this same family cannot technically approach the requirements of the high-speed, heavy-duty-performance network processing that we study in this book. Therefore, we will not cover it here.

Of course, Motorola quickly realized the limitations that its PowerQUICC architecture would experience when it dealt with edge and especially core networks. This is why they decided to acquire a promising Massachusetts startup called C-Port a few years ago. Since then, the company has been developing and introducing new products in the network-processing market. They have preserved the same brand name.

C-PORT: THE BIG PICTURE

The C-Port family is composed of mainly three network-processor chips: the C-3e, the C-5, and the C-5e. The C-3e is a fully programmable 3 Gbps throughput *network processing unit* (NPU) with programmable interfaces along with integrated Ethernet *Media Access Control* (MAC) controllers (10/100/1000) and *Synchronous Optical Network/Synchronous Digital Hierarchy* (SONET/SDH) framers (155/622 Mbps). Integrated coprocessors handle classification and traffic management, but an externally connected Q-3 chip from Motorola can handle traffic management, offering multilevel hierarchy scheduling and support for up to 64K queues.

Motorola is positioning the next-step-up product—C-5—for a wide range of network applications around the OC-12 level. The latest product—C-5e—is geared for the OC-48 realm. The potential applications include *multiservice access platforms* (MSAPs), edge routers, *digital subscriber loop access multiplexers* (DSLAMs), wireless base stations, cable head ends, load balancers, web switches, and so on. The company's publicized product roadmap indicates that the C-10 and Q-10 chipsets will be introduced in 2003 to handle 10 Gbps of sustained throughput. Motorola is not present in the 40 Gbps realm yet.

The family contains two additional chips: the Q-5 (a traffic manager) and the M-5 (an interface-adapter chip that enables full-duplex and channelized OC-48 applications for the C-5e). The Q-5 provides fine-grained traffic management by handling relevant issues such as traffic policing, shaping, and scheduling.

The C-5 has a throughput of 5 Gbps and is available in the following clock frequencies: 166, 200, and 233 MHz. The more recent C-5e is clocked at 266 MHz. The C-5 is offered in an 840-pin *high thermal coefficient of expansion ceramic ball grid array* (HiTCE CBGA) package. Typically, it consumes 15, 17.5, and 20 watts respectively with its three available clock frequencies. The HiTCE material out of which the package is built has the unique characteristic of expanding thermally at the same rate as a typical *printed circuit board* (PCB). This accounts for the exceptional reliability levels attained by the Motorola C-5 and C-5e processor packages over a wide temperature range. The C-5e, on which Motorola is pinning lots of hope, is offered in a slightly different 840-pin HiTCE CBGA package, but it consumes only 9 watts as it operates from a 1.2V supply. The Q-5 and M-5 chips are presented in a 600-pin EBGA and a 352-pin TBGA package. They typically consume 4.5 and 2 watts, respectively.

NPU ARCHITECTURE

Figure 8.1 shows the basic architecture of the C-5e. The network processor combines 17 programmable *reduced instruction set computer* (RISC) cores for packet and cell forwarding, along with 32 *very long instruction word* (VLIW) engines called *serial data processors* (SDPs) for processing data

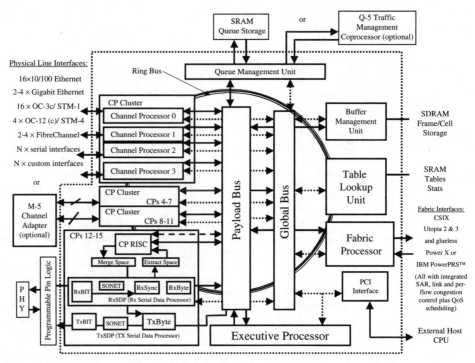

FIGURE 8.1 The internal architecture of Motorola's C-5e network processor. *(Source: Motorola)*

streams. Several powerful embedded coprocessors that handle other functions are located next to these. These include a *buffer management unit* (BMU), a *table lookup unit* (TLU), a *queue management unit* (QMU), a *fabric processor* (FP), and a supervising CPU that Motorola affectionately calls *executive processor* (XP).

In addition to this heavy artillery inside the NPU die, intrinsic provisions are available to interface externally with an optional *traffic management coprocessor* (TMC). This role is ideally fulfilled by the company's powerful Q-5 chip. The CPU combination of what is available inside the C-5e leaves more than 4,500 *millions of instructions per second* (MIPS) of computing power for a switching/routing systems designer who may be confronted with the task of adding services throughout the proverbial protocol stack.

Sixteen *channel processors* (CPs) are at the heart of the C-5e design. These are extremely flexible computing engines that can be individually programmed. Their flexibility means that each engine can be programmed to play different roles depending on the application at hand. Therefore, they can be made to easily support *Asynchronous Transfer Mode* (ATM), *Internet Protocol* (IP) over Ethernet IP, IP over *Point-to-Point Protocol* (PPP), SONET/SDH, frame relay, and even proprietary protocols.

Each CP consists of a dedicated RISC core and dual SDPs: one for ingress and one for egress computing in each CP. The CPs can be assigned to physical interfaces that the network processor is called to support. They can be combined into aggregates that support *input/output* (I/O) bitstreams of higher bandwidth, or they can be assigned to other computational tasks internally as dedicated coprocessors.

The SDPs handle all data encoding/decoding, framing, formatting, parsing, *cyclic redundancy check* (CRC)-based error checking, and data movement. As the SDPs can also control an external programmable pin logic block, they enable systems designers to implement almost any layer 1 interface. This flexibility includes connecting with T/E carrier framers, Ethernet PHY (RMII), Gigabit Ethernet PHY (GMII or TBI), OC-3/STM-1 PHY, and OC-12/STM-4 PHY through the M-5 Channel Adapter, and a *Universal Test and Operations PHY Interface for ATM Level 3* (UTOPIA 3)/*Packet over SONET/physical* (POS-PHY) interface, which can support OC-48/OC-48c/STM-16 MPHY capabilities. Also note that OC-3/STM-1, OC-12/STM-4, and OC-12c/STM-4 framers are built into the architecture of the SDPs.

Moving up one level to layer 2, the SDPs can be independently configured to support Ethernet, *High-level Data Link Control* (HDLC) streams, POS, frame relay, ATM, and Fibre Channel, as well as almost any other required format, including *Multiprotocol Label Switching* (MPLS) and other encapsulations. The SDPs are highly programmable; therefore, they support a whole array of diverse MAC interfaces and data-parsing requirements to the extent that each port can be made to implement a different protocol. Programming the SDP must be done in microcode. Motorola provides the microcode for a vast spectrum of applications (such as all flavors of Ethernet, IP and ATM over SONET, T/E carrier serial data streams, and so on). Interestingly, no coding is required on behalf of the user for the support of the diverse MAC interfaces.

The RISC core of each CP is clocked at the same frequency as the core clock rate of the C-5e. It possesses its own instruction and data memory of 32KB and 48KB per cluster (that is, a group of four CPs). The RISC core engine's instruction set is a subset of the widely known and used MIPS instruction set, so Motorola judiciously capitalizes on using a de facto industry standard. The RISC core is programmable in C or C++. This feature lends the computing power of the RISC core of each of the CPs to tasks that can be best implemented in a high-level language. These tasks include the decision making for forwarding, scheduling, statistics gathering, and so on. The natural result is that bit-level operations can be offloaded to the specialized SDPs; therefore, RISC core capacity is preserved for applications that require it.

In order to maximize the impact of any combination among the main parameters of processing power, throughput, and bandwidth, the systems designer can easily combine the CPs of the C-5 network processor. For instance, to scale the bandwidth, multiple CPs can be clustered in parallel logical aggregates for wider data streams while maintaining the same simple and straightforward software model. Likewise, to increase the processing power for a particular application, the CPs can be cascaded in a pipelined fashion to enable higher-performance processing on the same bitstream. This is an interesting way of applying processing power to a set of tasks independently of the actual data rate. Sophisticated hardware mechanisms allow one or both of these techniques to be engaged without placing a further burden on the overall software complexity.

The C-5e can be used as a stand-alone device with the possibility of supporting up to OC-48 line rates or four OC-12 streams in full duplex. However, for higher-speed applications that may require OC-48c full-duplex capabilities and channelized applications, the Motorola's M-5 Channel Adapter will be used. The M-5 Channel Adapter can seamlessly connect the external world onto the physical interfaces or the fabric interface of the C-5e in various user-defined configurations. The M-5 Channel Adapter accepts both *Packet over SONET PHY Level 3* (POS-PL3) and UTOPIA 3 framer interfaces into the C-5e network processor's 16 clustered CPs, as well as its FP interface at up to OC-48c/STM-16 wire speeds. Both SPHY and MPHY framers are supported on the C-5e CPs, and the FP also supports SPHY framers. Up to 48 logical interfaces can connect through the MPHY, thereby enabling virtual channelization down to the *Synchronous Transfer Signal, Level 1* (STS-1) level of granularity within an OC-48/STM-16 bit stream.

We mentioned earlier that the C-5e contains a set of powerful and highly specialized coprocessors. We will now take a closer look at them:

- **TLU** The TLU is a flexible and high-speed classification engine. It allows the implementation of a broad spectrum of traffic classification functions and supports the execution of multiple and different search algorithms. These search algorithms are executed simultaneously with the lookup operations. The performance afforded enables you to handle OC-48c/STM-16 class applications while leaving plenty of extra headroom for other needed computing chores. The TLU speed is certified by Motorola to achieve more than 46 million IPv4 lookups per second and more than 133 million index lookups per second. This impressive performance is a result of its highly pipelined architecture.

  Typical lookups that the TLU is called to perform include IPv4/IPv6 *longest prefix match* (LPM), ATM *Virtual Path Identifier/Virtual Connection Identifier* (VPI/VCI), Ethernet MAC/*virtual LANs* (VLANs), and MPLS. In addition to table lookups, the TLU can also be configured to perform integrated real-time statistics counting. Among the multiple search algorithms that the TLU can execute, support is available for the indexed pointer, hash, LPM, trie, key, as well as data, chained index, and chained hash tables. The TLU can be configured with up to 32 unique tables, which can each contain up to 16 million entries. Each entry in these tables ranges from 8 to 1,024 bytes.

  An interesting feature of the TLU architecture is that to prevent table updates from interfering with ongoing lookups, the TLU can support shadow table capabilities through its interface to 64-bit-wide 133 MHz *zero bus turnaround* (ZBT) *static random access memory* (SRAM). On top of that, if even further classification capabilities are required in a system application, the C-5e makes it possible to attach an external classification coprocessor to the SRAM interface, in which case the TLU will simply act as a proxy to the external coprocessor. The TLU can handle up to 64MB of external memory (arranged as 128Mb×32 pins).

- **QMU** The integrated QMU (working in internal mode) can support up to 512 queues, which is considered adequate to satisfy the requirements of most applications. However, this queue-management performance can be scaled by engaging the QMU in its external mode. By attaching the Q-5 TMC (a task that does not require glue logic), which we discuss in the following section, a very powerful *quality of service* (QoS) management platform can be achieved across the spectrum and over both IP and ATM applications.

- **FP** Through its programmability, the highly configurable FP offers the possibility of implementing a wide range of fabric parameters, such as cell size and self-routing headers, enabling control to be applied on a per-flow basis. It can also handle *segmentation and reassembly* (SAR) and integrated scheduling of up to 128 queues. The FP can run at 125 MHz with movement that is 64 bits wide (32-bit *transmit* [Tx]/32-bit *receive* [Rx]). It can support a bandwidth of up to 3.2 Gbps full duplex.

It offers the flexibility of a broad spectrum of standard interfaces such as UTOPIA 2, UTOPIA 3, and 32-bit 125 MHz CSIX-L1. Without any glue logic, it interfaces to the Power X TeraChannel®[1] fabric architecture and the IBM PowerPRS™ switch fabric family, which we discussed in Chapter 4, "IBM PowerNP™." It can be further configured to support other proprietary fabrics. Interestingly, multiple C-5e network processors can be connected through their fabric interfaces to a common switch fabric. As a result, aggregate bandwidth performance can reach a rate of terabits per second.

- **BMU** The BMU is 139 bits wide based on 128 bits of data, 9 bits for *error correction coding* (ECC), and 2 control bits. The size of buffer memory under its supervision can be up to 128MB.

- **XP** The XP handles supervisory tasks and is also a 32-bit RISC CPU core. It is equally programmable in C/C++ with the same instruction set as the RISC cores that are inside the 16 CPs. Externally, it provides support for a 32-bit 33/66 MHz *Peripheral Computer Interconnect* (PCI) bus and a serial *programmable read-only memory* (PROM) interface, along with a two-wire serial bus interface that supports 400 Kbps links.

As shown in the architectural structure of Figure 8.1, several internal communications buses can be found in the C-5e network-processor chip:

- The *payload bus* is 128 bits wide, transfers 64 bytes at a time, and can handle a throughput of up to 34.1 Gbps.

- The *ring bus* is 64 bits wide, transfers anything from 8 bytes to 32 bytes, and can handle a throughput of up to 21.1 Gbps.

- The *global bus* is a 32-bit bus that can transfer 4 bytes at a time with a maximum bandwidth of 4.2 Gbps.

The M-5 Channel Adapter supports a 5 Gbps aggregate and can be configured in 1 to 48 ingress channels. It essentially maps external links onto C-5e channels, and vice versa. For instance, an OC-1 link maps as three M-5 ingress channels to one C-5e CP channel, whereas an OC-3c maps as one M-5 ingress channel to one C-5e CP channel, an OC-12c link maps as one M-5 ingress channel to one C-5e CP cluster (four CP channels), and an OC-48c link maps as one M-5 ingress channel to four C-5e CP clusters (16 CP channels). An OC-48c can also map as one M-5 ingress channel to one C-5e FP channel, if it is connected onto the FP instead. The M-5 handles *packet data units* (PDUs) that are 52 bytes long for ATM cells. For POS, the packet length can vary from 28 bytes to 9,216 bytes. Figure 8.2 shows a typical configuration of a router system based on the Motorola C-5e network processor in conjunction with the company's M-5 Channel Adapter chip.

THE Q-5 TMC

As discussed previously, the sheer variety of applications that service providers must deliver while doing so under a diversified set of requirements and customer-imposed end-to-end QoS levels spans the whole spectrum from *voice over IP* (VoIP) and streaming video all the way to web casting, without forgetting, of course, mundane data transfers. These diverse services are characterized by different traffic patterns and rates. As a result, building networking systems that implement these next-generation services requires active and sophisticated traffic management. Motorola has introduced the Q-5 TMC to address this need. The Q-5 performs its mission by being coupled without glue logic to the company's flagship network processor C-5e in order to provide QoS management into the data-forwarding path (data plane).

---

1. Information about Power X TeraChannel fabric can be found at *www.powerxnetworks.com/products/*.

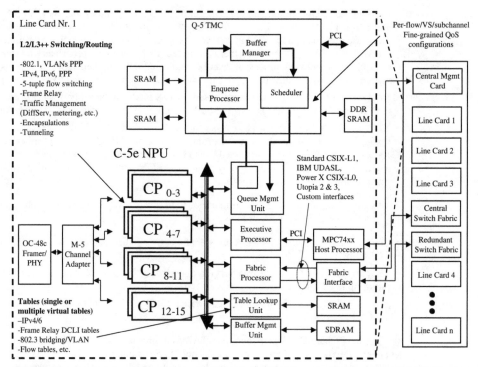

FIGURE 8.2 A typical line-card architecture based on Motorola's C-5e network processor and Q-5 TMC in a high-function edge router. The backplane is implied on the right side of the drawing running vertically across all the cards. *(Source: Motorola)*

Due to the flexibility of the Q-5 TMC interfaces, the company targets it to different markets of network equipment, such as Internet access routers, optical edge multiservice platforms, *virtual private network* (VPN) access devices, packet/ATM internetworking devices, IP/ATM access/aggregation devices, and even devices for wireless network infrastructure, base stations, and so on.

The Q-5 TMC offers the following interface possibilities with a network processor (or special *application-specific integrated circuit* [ASIC]), a host processor, or memory:

- A PCI host interface that is 32 bits wide, is clocked at 66 MHz, and can be used for system configuration and statistics gathering.

- An external *traffic management interface* (TMI) that is 58 bits wide and works at 100 MHz between the Q-5 TMC and a network processor or ASIC. The TMI is used to pass descriptors and control information. The definition and role of the descriptors are described later in this section with a real-life example of a high-performance edge router. For the moment, think of this as simply a data structure associated with the internal description of traffic payloads. In the Motorola C-5e network processor, the TMI replaces the QMU's external SRAM.

- A *double data rate* (DDR) *synchronous dynamic random access memory* (SDRAM) interface for descriptor storage. This interface is 72 bits wide, is clocked at 133 MHz, and can address a maximum of 64MB of storage.

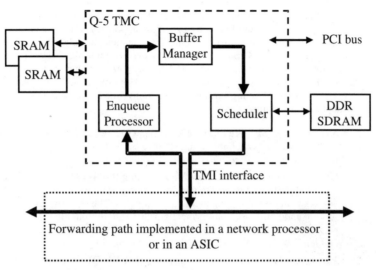

FIGURE 8.3   The logical flow of operations with the Q-5 TMC. *(Source: Motorola)*

- Two ZBT SRAM interfaces—namely, one for parameter storage, which is 72 bits wide, is clocked at 133 MHz, and can address a maximum of 8MB of storage, and one for queue-link storage, which is 18 bits wide, is clocked at 133 MHz, and can address a maximum of 10MB of memory space.

Typical QoS configurations such as policy-based *active queue management* (AQM) with fair buffer sharing, statistics collection parameters, and traffic-monitoring policing and shaping. Even policy-based priority and fair bandwidth allocation to flows, along with the scheduling of flows, can be easily implemented by software engineers working on switching/routing systems through the use of QoS *application programming interfaces* (APIs). These same APIs also enable the rapid modification of the QoS configurations so the user can provide real-time service provisioning and reprovisioning.

With its 5 Gbps throughput, the Q-5 provides multiprotocol support for virtually any type of link, enabling the implementation of QoS management up to OC-48c wire speeds in protocol environments, which can be anything among IP, ATM, frame relay, Ethernet, and POS. With the Q-5 TMC, the user can implement high-density per-flow and/or per-VCI queuing and very fine-grained traffic shaping for a broad range of packet- and cell-based applications. A three-level scheduling hierarchy, which provides support for up to 4,000 *virtual channels* (VCs), enables the implementation of a vast array of services including deep channelization and even integrated multicasting.

The Q-5 TMC is designed as a look-aside traffic manager, which enables it to provide both ingress and egress traffic management. Ideally, it should be combined with Motorola's C-5e network processor, but it can function equally well in a system as a stand-alone TMC. Figure 8.3 shows the flexibility with which the Q-5 TMC and its enqueue processor, buffer manager, and scheduler can implement advanced QoS.

In order to provide robust scheduling and ensure that *service level agreement* (SLA) stipulations are met for priority, fairness, and data rate, the Q-5 TMC offers a three-level scheduling hierarchy depending on the level of aggregation required. The schedulers at any of these three levels, as shown in Figure 8.5, can be configured with an assortment of algorithms to perform integrated shaping/ scheduling on different traffic types depending on the exact traffic requirements. The base element in the scheduling hierarchy of the Q-5 is the traffic queue, which represents an individual connection, a collection of connections, or a flow. Traffic queues can aggregate through up to three scheduling levels. This scheduling hierarchy provides support for priority and multiprotocol (ATM, IP, frame relay, MPLS, or a mixture) fair scheduling and shaping algorithms. The schedulers at level 3 can aggregate up to 128K traffic queues into a class or multiple classes. Level 2 schedulers can consolidate up to 32 level 3 schedulers. Level 1 schedulers can cluster up to 32 level 2 schedulers.

Offering a wide selection of algorithms, Motorola's Q-5 TMC enables customized implementations of QoS by allowing various combinations used by the schedulers. The following are among the supported algorithms:

- *Strict priority* (**SP**)   In this case, each input to a scheduler is statistically assigned one of 32 priority levels without any minimum guarantees. All of the nonempty inputs within each level of priority are served on a *first-in first-out* (FIFO) basis.
- *Guaranteed bandwidth weighted fair queuing* (**GBWFQ**)   This is a non-work-conserving WFQ-type of algorithm. It is used to provide guaranteed (*constant bit rate* [CBR]) bandwidth to inputs of any scheduler by assigning them 22-bit weights. The concepts and distinction between work-conserving and non-work-conserving algorithms are thoroughly discussed in several good computer-network theory books, such as *An Engineering Approach to Computer Networking: ATM Networks and the Telephone Network* by Srinivasan Keshav.[2]
- *Excess bandwidth weighted fair queuing* (**EB-WFQ**)   With this algorithm, each input to a scheduler is assigned one of 32 possible 22-bit weights. Bandwidth is served to the nonempty inputs relative to these weights. The WFQ algorithm distributes bandwidth proportionally to the weights, even in the presence of variable-length packets.
- *Frame-based deficit round robin* (**FBDRR**)   This algorithm, which is only available for use with the level 3 schedulers, apportions the bandwidth according to the weights that have been assigned to traffic queues. The FBDRR variant of the well-known *deficit round robin* (DRR) algorithm uses a configurable service quantum to reduce the latency and jitter, which are intrinsic to the fundamental DRR approach.

The Motorola Q-5 TMC practices what one would call *Active Queue Management* (AQM). The combination of a flexible buffer-sharing scheme at flow, class, and interface levels enables a wider regime of operating conditions when confronted with traffic congestion without any significant degradation of QoS levels associated with flows or connections. The traffic-payload descriptors are stored once they are received. The Q-5 TMC forwards them to the appropriate destination only when it must transmit them—something that it does as part of the scheduling operation. This information is stored internally in a *descriptor buffer*. The Q-5 TMC supports up to 2 million descriptor buffers, and each one is configurable from 8, 16, 24, to 32 bytes in size. This flexibility enables the dynamic allocation of buffer space and the easy maximization of buffers, which are allocated to active traffic queues. This is why the scheme is called *active queue management*.

To further complete the AQM picture of the traffic management capabilities within the Q-5 TMC, it is worthwhile to note that *Random Early Detection* (RED) and *Weighted RED* (WRED) AQM schemes are supported and are mapped onto the chip's shared hierarchical buffer model. All packet/cell-discard models are parameterized and configurable, and all PDUs are either tagged or discarded based on the corresponding congestion schemes, which the user may have chosen to configure in the Q-5 TMC.

For the sake of illustration, we will discuss an example of how the implementation of a typical QoS solution flows through a system that is based on the Q-5 TMC. The example is illustrated in Figure 8.2, which shows the implementation of a real-life high-performance routing system.

In this design example, as soon as packets/cells enter the system, which is composed of the C-5e network processor and the Q-5 TMC, the ingress processor sends the actual data over the internal pay-

2. Srinivasan Keshav, *An Engineering Approach to Computer Networking: ATM Networks and the Telephone Network* (Reading, Massachusetts: Addison-Wesley, 1997).

load bus to the BMU for temporary storage and it does so after parsing and classifying the incoming bitstream. The ingress processor can be one or more of the 16 CPs (depending on the application and the point in time when the system functionality is looked at), the XP, or even the FP.

Simultaneously, the internal processors of the C-5e network processor (one of the CPs, the XP, or even the FP) create an application-specific control packet called a *descriptor*, which is then enqueued into the Q-5 TMC through the auspices of the C-5e network processor's QMU. When packets or cells must be routed to different embedded processors (one of the CPs, the XP, or the FP), this is effectively done through the Q-5 TMC, which is always using the descriptors as proxies for the corresponding individual packets. Descriptors are transferred as part of the enqueue operation (both unicast and multicast) and are returned as part of the dequeue operation.

As mentioned earlier, the Q-5 TMC stores the payload descriptors when they are received. It then forwards these descriptors individually to the appropriate processor for subsequent payload-related processing through the network processor's QMU. This means that when a descriptor reaches its destination processor (the CP, the XP, or the FP), the payload data that is associated with this descriptor will be pulled from the temporary storage under the supervision of the BMU and forwarded to the corresponding destination, which is now the processor that possesses the descriptor. The following section discusses how to program QoS-related services with C-Ware APIs.

DEVELOPING SOFTWARE FOR THE C-PORT FAMILY OF NETWORK PROCESSORS

Motorola is offering a powerful toolset and development system for the overall development of software in conjunction with new hardware engineering. The C-Ware Applications Library and the C-Ware API enable the timely development of rich NPU source code that can be tested and analyzed by the toolset, simulation, and performance analysis environments. The C-Ware Simulation Environment enables the fast and performance-accurate simulation of all aspects of hardware in the C-Port family of NPUs, traffic managers, and even adapters. The environment further provides open interfaces for system simulation creation (including the host CPU, the control plane, the fabric, and any potential coprocessors). The C-Ware iPerformance® Analyzer offers an advanced integrated *graphical user interface* (GUI) with capabilities for monitoring per CP or per thread, and it enables graphical C-language-level debugging. The compiler and debugger are solid and GNU based, offering both performance and code-size optimization capabilities. The big picture of the development environment is completed with performance-analysis and traffic-scripting tools.

To interface the main network application with specialized network-processing code, which handles data parsing, classification and table management, traffic management, data modification, control plane management, and buffer management, independent of whether the functions occur at the forwarding or control planes, a series of APIs provide the peace of mind associated with code compatibility and the preservation of investment.

Figures 8.4 and 8.6 illustrate this point. These APIs, which act in a similar way as APIs found in the traditional computing world, abstract the underlying hardware architecture of the C-5e network processor and its associated Q-5 TMC. They offer support for the most common among network task-building blocks, such as physical interface management, data forwarding, table lookups, buffer management, and queuing operations. Writing code that interfaces with these APIs is a good way to ensure software compatibility and scalability from generation to generation of Motorola's C-Port family of network processors.

More specifically, in terms of QoS requirements, the combination of Motorola's APIs and standard C language is more than enough to configure the Q-5 TMC to perform its QoS-related tasks along with a main application, which runs on the C-5 or C-5e network processor itself. The APIs allow the coding of software that implements the QoS service from as low as the physical-level functions all the way to host-based supervisory and billing functions. If the Q-5 TMC is used independently of Motorola's network processors as a stand-alone traffic manager, the same APIs enable the correct configuration of the chip.

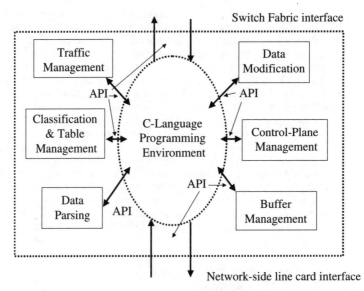

FIGURE 8.4 Conceptual use of APIs to engage all hardware functions of the C-5e. *(Source: Motorola)*

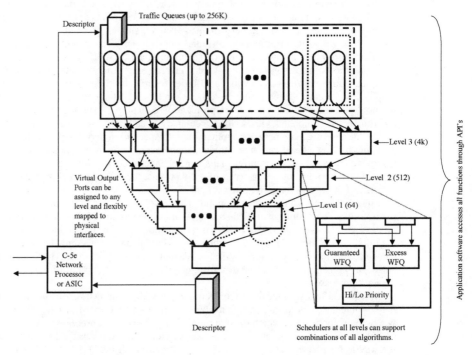

FIGURE 8.5 Organization of the data flow through the Q-5 Traffic Management Coprocessor. *(Source: Motorola)*

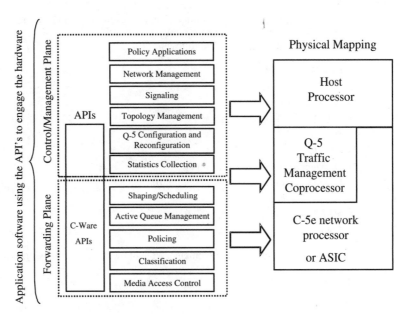

FIGURE 8.6 The use of APIs to address all functionality in both the data and control/ management planes. *(Source: Motorola)*

At the data plane, the AQM can be programmed as well as all aspects of traffic policing and shaping, statistics collection parameters, and the scheduling of flows. This type of modular functionality is needed to implement higher levels of QoS features that are required in network equipment so the service providers can provision special services and exercise policy management. In addition to the configuration capabilities of the Q-5 TMC that we have discussed so far, the following features are supported:

- **Multicast enqueue elaboration** A predefined table of multicast groups is used in order to determine the number and destination of traffic queues for multicast traffic. When a multicast enqueue is created, the corresponding descriptor references one of these multicast groups.

- **Acceleration of ATM SARing** For the support of ATM SAR and, more specifically, for AAL5 and AAL2 protocols, the Q-5 TMC has an interesting ability to enqueue a single descriptor on a per-packet basis. It can then leak that descriptor out n times (n corresponds to the number of the smaller segments of a large packet) at a rate that matches the required traffic specifications. It also obviously has the inverse ability to reassemble packets.

- **Collection of statistics** Not surprisingly, the Q-5 TMC can collect statistics on common objects such as queue lengths, queue discards, and so on. However, it can also gather statistics on buffer pools (enqueues, dequeues, and discards). Based on the relevant work that it compiles, the information it produces can be conveyed either over the PCI bus to a supervising host CPU or through the external TMI that exists between the C-5e network processor and the Q-5 TMC to other processors active in a system—that is, the CPs, the XP, or the FP.

The C-Ware Applications Library contains several implementations of protocols and interfaces that facilitate an overall switching/routing systems design for Motorola's customers. Among its implemented protocols, it contains the following:

- POS layer 2/3 switch.
- ATM AAL-5 SAR.
- ATM aggregation.
- AAL2 for two OC-3c ports.
- 802.1p
- 802.1Q
- *Differentiated Services* (DiffServ).
- Frame relay to DS-3 clear channel interface.
- Fibre Channel MAC.
- MPLS *label-switched router* (LSR).
- IPv6

Among the interfaces it implements, we will mention the following:

- 10/100 Ethernet.
- Gigabit Ethernet.
- OC-3c
- OC-12c
- OC-48c

Motorola is also offering an integrated C-Ware Development System. This is a joint hardware-software systems-engineering platform, which in conjunction with the availability of pre-existent hardware reference designs can definitely accelerate the overall development cycle. It is based upon a compact-PCI chassis into which you can plug one or multiple C-5e switching modules, a Q-5 TMC-based daughter board, a supervising computer board such as Motorola's MPC7400 Series Host Application Module, various other *physical interface modules* (PIMs), and several hardware reference designs that can facilitate the time to market for Motorola's customers. More detailed information about this development system can be found at Motorola's web sites.[3]

SYSTEMS CONSIDERATIONS WHEN DESIGNING WITH C-PORT NPUS

Unlike other NPU vendors, Motorola is not offering one-stop shopping. However, the company has documented compatibility with several vendors of complementary hardware as well as with both software and hardware development systems.

The security acceleration area, for instance, directly supports Corrent's 7120 Hurricane™ IPsec accelerator at above 2 Gbps throughput by interfacing with the C-5e to provide fast-path security solutions such as VPNs. In terms of search engines, the Network Database Search Engines (CYNSE70032) from Cypress as well as the Cypress coprocessor (CYNCP80192) can connect with

---

3. High-quality technical documentation with tutorials, white papers, application notes, users guides, and data sheets is available at the web site of Motorola's network and communications processing group at *http://e-www.motorola.com/webapp/sps/ site/overview.jsp?nodeId=03M0ylgx1KsM0yrfgP8S* and at Motorola's documentation library at *http://e-www.motorola.com/ webapp/sps/library/docu_lib.jsp*. The company's design resources for their business can be found at *www.motorola.com/ networkprocessors*. The support web site for Motorola's network-processor group is at *http://motorola.cport.com/support*.

the C-Port C-5 NPU over the latter engine's ZBT SRAM port. The combination delivers significantly higher levels of search performance and throughput for mission-critical applications. To provide OC-48 rate classification of content based on layers 2 through 7 processing, the C-5 NPU can be interfaced with the PM2329 ClassiPI™ high-performance content processor from PMC-Sierra, with a specialized *Software Development Kit* (SDK) that is available from the latter engine as well. The PAX.port™ 2500 classification processor from Solidum is another classification processor that has been announced to be connectable to the C-5e in order to enable multigigabit processing (up to 2.5 Gbps).

One of the most important elements in this teaming or alliance approach is the switch fabric. Motorola does not offer switch fabrics; It relies on relationships with other vendors. For example, IBM's PowerPRS fabric connects to the C-5 network processor through the IBM U-DASL interface, whereas the more recent IBM PowerPRS fabrics can connect to the C-5e directly through the CSIX-L1 interface.

Besides the popular *Software Development Environment* (SDE) *Tornado for Managed Switches* (TMS 2.0), which is tightly integrated with the C-Port family development environment and is offered by WindRiver as the C-5 *Switch Support Package* (SSP),[4] Netplane's MPLS routing stack is also supported.[5] HCL[6] and Tality[7] are two examples of other companies that offer expert design services for the C-Port network processor family, including hardware design and embedded networking and telecom software development. Tality specializes in extending the spectrum of C-5 NPU interfaces and offers a POS-PHY/UTOPIA interface adapter among other things.

SUMMARY

In this chapter, we reviewed Motorola's C-Port network processor family. We discussed in quite some detail the architecture of the family and looked at the C-5e as well as the company's Q-5 TMC and M-5 Channel Adapter. Motorola is the current market-share leader in network-processing sales. Regardless of what the rest of the market will do, it is a formidable player that combines world-class semiconductor expertise in both design and manufacturing as well as deep networking and communications know-how, along with tremendous financial and engineering resources. Therefore, it is more than safe to bet that the company and its products will remain key players in the network-processing field for years to come.

---

4. WindRiver's web site offers information about their support of the SDE *Tornado for Managed Switches* (TMS) in a C-Port network-processing context at *www.windriver.com/products/html/maswitch.html.*

5. Netplane (now a Conexant company since its recent acquisition) and its products are described at the company's web site at *www.netplane.com.*

6. HCL's web site describes their offerings at *www.hcltechnologies.com.*

7. Tality's web site can be found at *www.tality.com.*

CHAPTER 9
OTHER NPU ARCHITECTURES

Up to this point, we have discussed some of the most established architectures in the network-processing realm that have been developed by a few of the leading and most entrenched vendors. However, the field of network processors is extremely fertile and involves more than a few highly active participants. These participants range from global powerhouse corporations, which are mostly captive semiconductor manufacturers and/or communications equipment providers, all the way to small and fabless companies, which are mostly promising startups that often develop exciting technology. The network-processing field is extremely dynamic, but it must be put into the context of the overall economic situation. Because we are discussing technology developed in startup companies, it is prudent to consider the risk and reality of these products.

An extremely hostile environment is created when the economic rigors of a highly competitive market where companies struggle for differentiation are coupled with the general sluggish economy following the collapse of the amazing technology craze of the 1990s, which provided entrepreneurs with easy access to venture capital funds. Startup companies in this field now vie for acceptance through design wins and market share while confronting the day-to-day struggle to survive financially. This overall context sketches the background of an extremely competitive industry where the stakes are very high. The natural result will be the time-proven template of markets that sooner or later consolidate around a few major players. In other words, the market will ultimately only have room for no more than a half dozen significant players.

As this chapter is being written, major players with deep pockets and powerful vertically integrated market positions are acquiring some of the startups that we just discussed. Meanwhile, some promising startups, such as Clearwater Networks, simply vanish from the radar screen, having slowly laid off their engineering staff and used up their last pennies of funding. In some of these cases, such as Terago, the ailing companies have actually delivered a cutting-edge product to the market. Nevertheless, some of them fail to secure funding and are forced to cease operations.

Nowadays, a network-processing startup must do more than just possess technology, have a product and revenue, and execute a predetermined business plan. It must secure operational funds on time and obtain actual design wins from customers who are established market players in their own markets. This is difficult to accomplish since customers want to see a working product with differentiable characteristics that mean something for the customer along with a support structure, development tools, and so on. Many customers justifiably worry whether their key suppliers will be around next year or three years down the road; therefore, they require financial robustness from the network-processing vendors in order to make a favorable business decision.

As many of these young companies have taken their last breath, some of the technical material that was originally planned to be included in this chapter suddenly became nonapplicable and was omitted. This book intends to leave the job of passing final judgment as to who is a viable player and who is not to the rigors of the market. Consequently, we are taking the approach that we should cover much alternative material as the scope of a textbook allows. However, it has been our intention to abreast of the rapid evolution of this market in order to ensure that the material is kept up-to-date until the book goes to press.

In this chapter, we take a look over the landscape of other network-processor vendors. Some vendors offer interesting and innovative approaches, whereas others combine their products with other ancillary chips they have designed, such as traffic managers, classification processors, and switch fabrics, to propose a more or less integrated solution. Some vendors, such as EZchip or Silicon Access Networks, are funded by major industry players (in this case, IBM and Intel, respectively). In addition to being investors, these industry players have a brute interest in the startup's success—for example, IBM is EZchip's silicon foundry. On the other hand, they seemingly compete for network-processing business against the very startup they support.

We will try to cover these multidimensional relationships in the appropriate chapters of the book, although we may have to mention some of these issues in this chapter. The material is organized this way because some vendors have come up with nonspectacular or nondifferentiable network-processor chips, whereas others have also come up with powerful traffic managers or switch fabrics. These chips are so potent that they can be used as standalone traffic managers or as switch fabric solutions in systems that may end up being built with network processors from another competing vendor.

SILICON ACCESS NETWORKS' iFLOW™ CHIPSET

The iFlow chipset from Silicon Access Networks (*www.siliconaccess.com*) has been designed to operate at speeds between 10 and 40 Gbps. The company advertises it as a 20 Gbps solution to indicate that it can handle duplex OC-192 links, unlike several other products advertised as 10 Gbps *network processing units* (NPUs). The iFlow chipset is made up of several products: a packet processor called iPP, a traffic manager (to be formally announced) called iTM, an accountant chip that handles statistics and policing called iAC, and two search engines known as the *address processor* (iAP), and the *classifier* (iCL). The family does not contain *Media Access Control* (MAC) controllers, framers, or switch fabrics, but industry-standard interfaces ensure the connectivity between these products from other vendors and the heart of a network-processing system that is designed around the iFlow architecture.

Figure 9.1 shows how the chipset can be used to design a full-duplex OC-192 line card (or 2×10 Gigabit Ethernet card). The company specifies that the iFlow chipset is capable of handling layer 3 processing and forwarding at a rate of 50 *million packets per second* (MPPS). The figure shows the two search engines on the ingress path; however, depending on the application, classification capabilities may or may not be required on the egress path. The pair of iPPs and iTMs on the egress path in Figure 9.1 can be completely skipped for lower-speed applications, thereby saving two chips from the overall chip count.

Although the number of chips needed to develop an integrated solution may seem daunting, the network-processing solution from Silicon Access Networks has an interesting advantage. The extensive embedded memory eliminates the need for external *static random access memory* (SRAM) or even *content-addressable memory* (CAM). It even reduces the need for external *dynamic random access memory* (DRAM).

The iCL is used essentially for applications such as *access control lists* (ACLs), *Differentiated Services* (DiffServ) flow classifications, and controlled flow management based on *quality of service* (QoS) and *class of service* (CoS). It contains a 5Mb CAM that is rated for 100 *millions of searches per second* (Msps) plus 4.5Mb of 128-bit-wide associated data memory. The iCL can handle multiple 216-bit searches per minimum length packet at 10 Gbps wire speeds. It also supports large multiple-field classification tables with additional features such as range matching, per-entry masking, and/or per-lookup masking.

Interestingly enough, in addition to the traditionally required discrete *ternary CAM* (TCAM) that it displaces, the iCL also contains the associated data memory (we describe the use of this memory

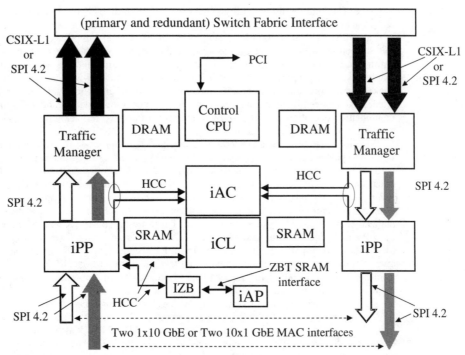

FIGURE 9.1 An example of an OC-192 line card based on the Silicon Access iFlow NPU architecture. *(Source: Silicon Access.)*

in the context of CAM in Chapters 12 and 13). Therefore, it actually saves the external SRAM that is normally required when such an external CAM is used. The iCL can handle classification tasks for layers 4 through 7 with 36K entries up to 144 bits each providing both per-entry and per-hop associated data in a single access.

Multiple iCL and iAP chip pairs can be combined to support larger tables. It is important to note that both iCL and iAP provide *error correction coding* (ECC) on all their embedded memory. This feature makes them especially useful in network gear destined for carriers that provision edge and core networks where reliability is critical. Powered from a 1.2V supply, the iCL is offered in a 560-pin EBGA package and consumes typically less than 2.5 watts.

The iAP is primarily used for address searching and, more specifically, for Ethernet MAC, *n*-tuple flows, and *virtual private networks* (VPNs) with tag lookup, or traditional *Internet Protocol version 4/6* (IPv4/v6) address lookups. It contains embedded memory, which can be filled with up to 256K table entries (producing the equivalent content of a 9Mb CAM) for IPv4 or 82K table entries for IPv6 addresses. The iAP is rated at 65 million lookups per second with deterministic result latency. No penalty is associated with the key size. It can perform more than two lookups per minimum-length packet at OC-192 speeds, and associated data fields can be modified on-the-fly by the on-chip *arithmetic logic unit* (ALU) simultaneously with any lookup operation.

In addition to the chip's *double cycle deselect* (DCD) *synchronous SRAM* (SSRAM) interface, its available *zero bus turnaround* (ZBT) SRAM interface enables it to connect without any glue logic to typical NPU chips. However, surprisingly, it cannot connect to the company's iPP. Therefore, Silicon Access provides *field-programmable gate array* (FPGA) code, which the company calls IZB. This allows the bridging between the iAP's ZBT bus and the iPP's *high-speed coprocessor channel* (HCC) interface. The latter is discussed in the next section. With the IZB code on an FPGA chip, a single HCC can be shared by up to four iAP chips, thereby allowing table sizes of up to 1M entries.

The iPP is clocked at 300 MHz, can process 30 Mpps, and can offer up to 115 Gbps bandwidth for connections with other look-aside or in-band coprocessors. The chip contains 4 clusters (called iAtom™ cores) of 8 packet engines, making a total of 32 programmable 8-way multithreaded engines that handle all the required packet modification as well as custom-written code. Therefore, these 32 packet engines provide a total of 256 concurrent threads of execution with a context switch of zero latency and an overall computing power of 9.6 billion operations per second. Of course, classical bit manipulation operations add flexibility to the tasks of adding, replacing, inserting, modifying, and deleting fields anywhere in a packet.

Silicon Access has created several hardware-assisted coprocessors that can parse and insert bit fields into packet headers or that can hash bit sequences, etc. A most interesting piece of assistance hardware inside the iPP chip is called the *Massively Parallel Branch Accelerator* (MPBX). This block of custom hardware increases the execution performance over traditional *reduced instruction set computer* (RISC) execution more than 100 times when code for complex conditional statements is run. The compiler simply detects the presence of these types of statements in the source code, and automatically reserves and schedules the use of the hardware-based MPBX unit. All packet buffering for the iPP is embedded on chip. Likewise, on-chip SRAM eliminates the need for external tables for protocol data and data-path state information. The iPP can contain up to 4K instructions. The company's own reference-design code is reported to only take up about half of this space, so plenty of room is available for custom coding. In addition to the advantages the on-chip TCAM offers, it can be accessed up to six times per packet.

The iPP has two *transmit* (Tx) and two *receive* (Rx) *System Packet Interface, 4.2* (SPI-4.2) interfaces. These are capable of 12.6 Gbps on each interface. The host interface is ensured over a standard 32-bit 33/66 MHz *Peripheral Computer Interconnect* (PCI) 2.2 bus. It also contains proprietary HCCs based on *low-voltage differential signaling* (LVDS), which are used to connect the iAC, iAP, and iCL chips with the iPP. Clocked at 400 MHz *double data rate* (DDR), an 8-bit HCC provides 6.4 Gbps of bandwidth for each direction. The iPP is available in a 1,170-pin HPBGA package and consumes about 12 watts.

As of this writing, the company has not yet disclosed details about the chipset's iTM. Consequently, current users are obliged to use the other members of the iFlow chipset in conjunction with a special *application-specific integrated circuit* (ASIC) that the customer must design to handle traffic management issues. The company has only alluded to the connectivity between the iTM and the switch fabric as being either SPI-4.2 or CSIX-L1. However, it seems that bandwidth throughput issues will occur with the CSIX-L1 if a fabric throughput of 25 Gbps is required (although this is not the case with the dual SPI-4.2 approach).

The iAC is a powerful platform that can handle up to 550 million operations per second. Its role is to assist the iPP by taking care of traffic policing and statistics gathering. It can match header values against policing contexts, and easily reject noncompliant packets. It is equally capable of handling color-blind and color-aware policing contexts. It contains 23.3Mb of memory that can be configured as 1.1 million 21-bit counters or 528K 42-bit counters. This means that the iAC can keep count of packets transmitted into a million parallel flows.

The ramifications are extremely important for service providers who bill their customers on a per-use basis. Competitive network processors must access statistics counters that are stored in external DRAM for the performance of billing operations. This usually implies the use of a read/modify/write sequence involving transfers of 42 and sometimes even 128 inefficient bits (if the memory interface is 64 bits wide) to update a 21-bit counter. The iAC handles this type of operation internally with a single command. Its horsepower allows the equivalent of roughly 20 counter operations per packet at a traffic throughput of 30 Mpps. The iAC comes in a 520-pin *ball grid array* (BGA) package and typically consumes 5 watts.

In a typical line-card application, such as the one shown in Figure 9.1, the packets arriving from the line interface are handed over to the iPP to initiate the required processing. The iPP extracts the desired search keys from the packet header (and in some cases, from the packet payload too). It engages the CAM of the iCL to look up the keys. Based on that, a route lookup is then executed on the iAP chip, which can yield per-hop or per-entry data in a single pass. The classification results are then handed over by the iPP to the iAC, which polices the packet and brings billing data structures up-to-date with the compilation of needed statistics.

Following the classification and policing work, the iPP sometimes modifies specific fields on the packet according to the application, such as creating encapsulations or updating bit fields in the header. An internal bit tag (flow identification number) is generated and attached in front of the packet for internal tracing by the traffic manager. The packet is then turned over to the iTM, which handles queuing and other typical traffic management functions.

In terms of development tools, the company provides a C-language compiler and a source-level debugger. Although the programming model keeps the individual packet engines away from the eyes of the software engineer as if a single engine was being programmed, the debugger provides the visualization of the status and progress of individual threads that are allocated over the multiple packet engines. Therefore, the programmer can inspect the interaction between threads.

Silicon Access also offers under a nice *graphical user interface* (GUI) a cycle-accurate simulator that covers all the chips of the set, including the IZB code, a packet generator, and a performance analyzer that monitors the packet engines and coprocessors that are embedded inside the iPP. During code execution, these are controlled by the packet engines. A time-accurate, but not cycle-accurate, model allows the emulation of the whole ingress and egress paths. This is obviously required to verify the performance of the entire chipset. Customers who use ASICs along with the company's chipset (as is the case with traffic management functionality) can add their own ASIC models to the suite and analyze/simulate the entire board design. The development environment has a powerful command-line capability that allows for scripting and the extension of the toolset. The company also offers several evaluation boards for many of these chips.

Last but not least, Silicon Access, like other NPU vendors, provides their customers with optimized-quality reference code for several networking applications and protocols. These include routing IPv4 and IPv6 traffic, *Multiprotocol Label Switching* (MPLS), DiffServ, bridging (layer 2 switching), IP tunneling, *virtual local area network* (VLAN) tagging per IEEE P802.3ac, and *Point-to-Point Protocol* (PPP) over *Synchronous Optical Network/Synchronous Digital Hierarchy* (SONET/SDH) per RFC 2615.

BAY MICROSYSTEMS' MONTEGO™ AND THE InP™ FAMILY

One of the most interesting architectural approaches in network processing is the *Internetworking Processing* (InP) family from Bay Microsystems (*www.baymicrosystems.com*). The first product of this family is the Montego network processor, which has been designed for the OC-192c realm. The designers had the following critical requirements in mind when developing this product: ultrahigh performance, scalability, service breadth and awareness, multiple-protocol intelligence, and ease of provisioning for its customers.

To properly focus the product design, the company correctly capitalized on the business importance of supporting the incumbent carriers. These carriers have massively invested in legacy circuit-switched technologies such as *time-division multiplexing* (TDM) voice, SONET, frame relay, and *Asynchronous Transfer Mode* (ATM). However, they also want to provision newer IP-based services such as IPv4/v6, MPLS-based VPNs, and DiffServ, as well as incorporate CoS- and QoS-based traffic management and billing capabilities. The company was fully cognizant of the magnitude of this task, unlike other vendors who simply embark on an IP-packet-centric product development spree. It knew that the work would require a combination of a powerful network processor and a sophisticated traffic manager in order to handle this new environment. It also understood that its architecture should be able to offer computational capabilities that allow the real-time management of millions of microflow counts and hundreds of thousands of queue counts in addition to the associated classification and billing requirements that these numbers entail.

Therefore, the designers took a fresh approach with the systems engineering and design of its first product—the Montego. They ensured a tight integration between the network processor and the traffic manager units within one and the same die. This resulted in a superchip that is impressively capable of providing five overriding types of functionality in a tightly integrated environment, which minimizes chip count. As a result, *printed circuit board* (PCB) real estate, power consumption, and cost are also minimized while offering 32 Gbps of switching capacity and a packet-processing speed that is rated at 31.25 Mpps. Its programming model provides direct access to the computational resources of the chip by enabling an application to be mapped onto the underlying engines that compose the architecture. More specifically, the model contains a multiphase dynamic classifier, a flexible transformation editor, a wire-speed capable *segmentation and reassembly* (SAR) unit (cells/packets), a robust queue manager, and, last but not least, a sophisticated traffic manager.

The chip provides native support for ATM, IPv4, *Packet over SONET* (POS), PPP, Ethernet, frame relay, MPLS, DiffServ, and IPv6. It can therefore easily be envisioned inside MPLS *label edge router* (LER) and *label-switched router* (LSR) switch or router systems. In fact, its AnyMapping™ programmable function allows the flexible internetworking mapping of any protocol to any protocol. The line-speed forwarding and bridging design arguably bridges the packet-processing gap between the legacy circuit-switched paradigm and connectionless world of IP. For example, the company's comprehensive MPLS support can simultaneously map multiple IPv4/6 microflows and ATM *virtual channels* (VCs) onto MPLS traffic streams at guaranteed data rates of 10 Gbps.

On top of all this, a whole series of programmable modification and editing functions is available, which can be engaged by the user to handle both standard and proprietary protocols. For instance, the Montego can seamlessly handle mapping, stripping, encapsulation, *cyclic redundancy check* (CRC), *Time to Live* (TTL), and even checksum operations.

We mentioned Montego's robust multiphase dynamic classifier. By directly interfacing to state-of-the-art TCAM lookup memories and in-band deep packet preclassifiers, this classification engine, which supports flexible packet parsing and key generation, has the impressive performance of 83 Msps. This can be expanded to 300 Msps.

On one hand, in terms of its channelization capability, the Montego chip provides support for the seamless mixed multimode operation of 64K virtual channels and up to 4,096 media ports operating across 16 physical channels. On the other hand, in terms of its traffic engineering, it allows hierarchical scheduling for QoS and CoS. This means that intrinsic support for class- and flow-based queuing, VPN-aware traffic isolation with guarantees, a variety of dequeuing algorithms, and even voice grade shaping are available. Policing with DiffServ occurs through the services of a *dual leaky bucket* (DLB) algorithm implementation, and congestion avoidance is implemented based on *Weighted Random Early Detect* (WRED), *Partial Packet Discard* (PPD), and *Early Packet Discard* (EPD). Both in-band and out-of-band versions of flow control are available. The programmable SAR facilities include *ATM Adaptation Layer Level 5* (AAL5) for ATM.

Multicast is natively supported for fabric, logical, or spatial modes. In terms of interfacing with a fabric and the rest of the word, Montego supports industry standard CSIX- and SPI-4-compliant interfaces. A 32-bit RISC *central processing unit* (CPU) running at 166 MHz assumes the executive supervisory role inside the Montego system and is capable of handling statistics up to 1 million counts per second.

With its native support for packets, cells, and frames and its seamless internetworking capabilities, the InP family is ideally suited to scale from requirements imposed on equipment designed for access networks all the way to carrier-class network gear designed and destined for deployment in long-haul carrier networks. As a result, the company targets its products toward designers of network equipment such as access concentrators for voice circuits, wireless base stations, xDSL gateways, multiservice switches and routers, cable head ends, and intelligent optical transport equipment (*dense wavelength division multiplexing* [DWDM] and SONET).

In order to achieve very high levels of performance while maintaining the maximum flexibility, Bay Microsystems has created a new technology that is optimized for the specific requirements of high-performance packet processing. The company calls this technology *Vertical Instruction Processing*™ (VIP) and *Vertical Data Processing*™ (VDP). The term *vertical processing* is used here to denote its principles. The basic idea is that sets of deterministic, programmable, and pipelined processor engines, which are optimized for specific packet-processing operations, are arranged in a data flow-through structure. As one can infer from Chapter 14, "Switch Fabrics", this flow-through structure is quite reminiscent of a shared-buffer switch complete with an ingress processor, shared output buffer memory, and an egress processor.

In addition to improving performance, the utilization of VIP and VDP technologies is in line with the school of thought that has consistently advocated structured *very large scale integration* (VLSI) design. Therefore, it allows for the undisputedly improved and structured integration of massive circuitry as opposed to other more traditional processor designs.

Unlike alternative architectures, the most distinguishing characteristic of the Montego architecture is the deterministic performance that it affords. The vertical-processing environment accomplishes this. Figure 9.2 shows how this principle is implemented. Imagine that data comes in from the lower-left side of the picture. By deploying the data on a dimension that is perpendicular to the actual data flow *input/output* (I/O), the Montego chip is applying a *multiple instructions single data* (MISD) model. A stream of packets is then processed by multiple high-performance, fixed-cycle pipes. Each pipe is composed of multiple engines (which are non-RISC-based in this case) that execute simultaneously,

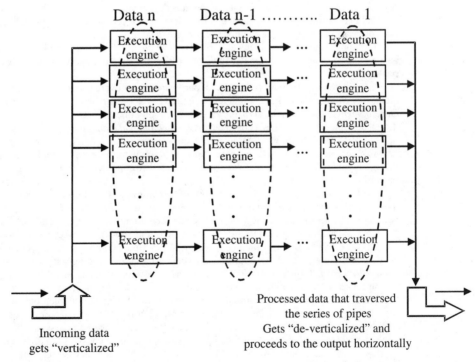

FIGURE 9.2   Vertical Instruction Processing (VIP) inside the Montego NPU. *(Source: Bay Microsystems)*

thereby eliminating the nondeterministic characteristics of sequentially programmed RISC core engines such as the ones found in other network processors. Each pipe executes a series of operations on the data stream. It then passes the data stream onto the next pipe in line. Each engine within a pipe is responsible for executing a particular network feature. By enabling (turning on) or disabling (turning off) the engine that is associated with a specific feature, that specific feature is applied on the processed packet or it is simply skipped.

In Figure 9.2, inside the first stage of pipes (shown as an oval) where the classification and policy instructions are executed, the data is subjected to engines that will parse and search, filter, and perform statistics. When the data is moved to the next stage of pipes, a traffic management set of instructions will take place. The data is then subjected to shaping and marking, and executing algorithms such as WRED and *weighted fair queuing* (WFQ).

Farther down the horizontal path, the data is treated to the forwarding and multicast-related instructions. Engines that handle pushing and popping, TTL, and checksums operate on the data, which is deverticalized at the end of the process and sent out to the next stop downstream in the switching system. We must note that the instruction memory is consulted on a per-flow basis for the next code steps to be executed. The Montego processor also preserves state-related information on a per-flow basis.

An interesting by-product of this architecture is that it can scale in both the horizontal and vertical dimensions. This translates into an ability to add more engines into a pipe in order to increase a pipe's capabilities and to increase the number of pipes in order to obtain an overall higher performance.

With this vertical-processing architecture, because all the associated network features are executing simultaneously and in parallel, it is completely irrelevant (from a performance measurement standpoint) whether an underlying packet requires and obtains the operations that correspond to features X or Y. This means that the performance remains deterministic, and the architecture is one of the pillars that help sustain this performance at the wire-speed levels. The other important pillar is the balance of performance from the memory subsystem design.

The Montego's core clock runs at 166 MHz. The chip, which is designed in a 0.18μ *complementary metal oxide semiconductor* (CMOS) process, is presented in a 1,600-pin BGA epoxy flip chip package.

To facilitate the parallel development of hardware and software inside a customer's network equipment, Bay Microsystems has created an integrated development environment called *Internetworking Development System* (IDS). IDS provides a cycle/pipeline-accurate C simulation and emulation design environment as well as a complete *original equipment manufacturer* (OEM) application development platform that the development engineer can replicate, modify, and/or scale to fit his or her network gear application. IDS is more than a development system for emulation, simulation, and debugging; it is a code-ready platform on which real-life applications can be made to run on real-life networks.

Besides facilitating the rapid convergence of hardware and software development, IDS can also be used to analyze performance and power, as shown in Figure 9.3. A series of traffic generators that can be random, protocol dependent, or even user defined complements the picture of the tools that are available inside this integrated tool suite. The base of the *Software Development Environment* (SDE) consists of a Java GUI, the company's NextWARE™ suite containing a comprehensive *application programming interface* (API), and industry-standard VxWorks, along with a *Transmission Control Protocol/Internet Protocol* (TCP/IP) stack, intermodule-communication software, systems administration server software, and the appropriate drivers. The development engineers can quickly apply, verify, and debug application examples on any desired traffic pattern or contemplated network service.

Bay Microsystems also offers several other protocol stacks as a series of options, such as IPv6, MPLS, and ATM. The IDS environment can be organized in various chassis (with one or eight line cards, respectively) of either 10 or 80 Gbps switch fabric. These chassis have different configurations such as 1×OC192c, Quad OC48c, 1×10 Gigabit Ethernet, and 16×1 Gigabit Ethernet, and support POS, ATM, and Ethernet interfaces. They also support a direct connection to third-party switch fabrics.

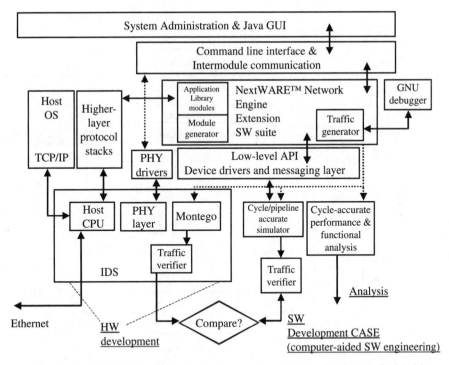

FIGURE 9.3 The parallel development of hardware and software using the IDS environment leads quicker to a converged design. (*Source: Bay Microsystems*)

COGNIGINE

A pioneering effort in the quest for architectural preeminence in the field of network processing is the Intelligent Network Processor™ by Cognigine (*www.cognigine.com*). The company calls its technology Variable Instruction Set Communications Architecture™ and VISC Architecture™ for short. It constitutes a scalable platform that is poised to handle traffic processing up to OC-768 levels of wire speed and beyond. It has intrinsic support for multiprotocol services such as Ethernet, PPP, IP, ATM, MPLS, TDM, and others; traffic management possibilities for up to 512K queues; and classification lookup capabilities for up to 1 million table entries in its product. The company is naturally targeting its products to metro, edge, core, and *point of presence* (POP) switches and routers, TCP termination systems, multiservice aggregation nodes, load-balancing server switches, and even *storage area networks* (SANs).

Figure 9.4 depicts this powerful multiprocessor platform. It is based on the integrated combination of five-stage pipelined 16 four-way multithreaded processors called *reconfigurable communications units* (RCUs) and a highly intelligent embedded switch fabric called an *RCU switch fabric* (RSF), which interconnects the RCUs. Figure 9.5 shows the five-stage pipeline that is located at the heart of the VISC Architecture.

The result of these combinations inside Cognigine technology's current implementation is a computational powerhouse of 38 billion operations per second, which can be executed in a single clock

RSF switch fabric interface

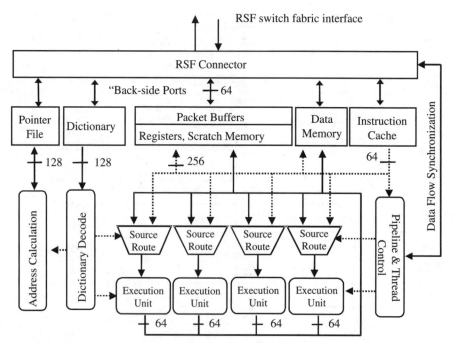

FIGURE 9.4 The architecture of each RCU inside the Cognigine network processor. (*Source: Cognigine*)

cycle. While other typical network processors will have a hard time even approaching that level of raw-speed performance, the Cognigine network processor provides a single-chip solution. It consolidates all classification for layers 2 through 7 as well as traffic management functions within one chip and can handle wire-speed fast-path packet processing at 10 Gbps. It has an internal overspeed of 40x, finally yielding a useful packet-processing performance of about 25 Mpps. The implementation of a full-duplex 10 Gbps data path requires only two Cognigine processors.

As each RCU is four-way multithreaded, it should not be surprising that each RCU has four 64-bit reconfigurable data paths and four 20-bit address paths. In addition to the fact that the hardware of each RCU provides support for operations such as timestamping and CRC, it also has the convenience of a 4K packet buffer and 2K of scratchpad memory inside each RCU. The RSF handles all communications from RCU to RCU or from RCU to peripheral units. Two programmable *Optical Internetworking Forum* (OIF) SPI-4.2 network interfaces provide external connectivity toward lines and/or external switch fabric *serialization and deserialization* (serdes), whereas the interface with a supervisory host CPU is handled over an industry standard PCI 2.2 bus.

The heart of each RCU contains an interesting concept called a *dictionary*, which decodes a VISC instruction (as soon as it is pulled out of an instruction cache) and decides which local computing resources need to be dispatched to execute the instruction based on its "meaning." This is a flexible way of reconfiguring complex tasks such as 8 operations of 32 bits each or 32 operations of 8 bits each, effectively using the maximum of locally available resources while minimizing the access to slower off-chip memory.

Figure 9.6 shows the beauty of the scalability that is obtained with the structured and extremely modular architecture that Cognigine has developed. Figure 9.6(a) shows the RSF—in other words, the crosspoint switch module that interconnects four RCUs. In Figure 9.6(b), multiple RSF modules

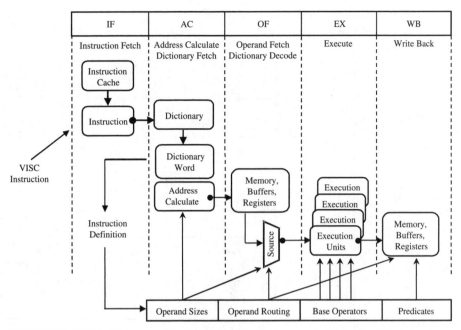

FIGURE 9.5 VISC Architecture pipeline in the Cognigine network processor. *(Source: Cognigine)*

combine with a series of RCUs to show how a much more powerful engine can be created to handle higher loads of traffic.

Designers of next-generation products usually have access to a series of options upon which they can capitalize, such as moving the previous silicon design deeper into submicron realms and consequently into smaller geometries, thereby taking advantage of the latest spectacular lithography progress. The silicon die savings in such a move can be extraordinary. A company can decide whether it wants to save costs and pass them to its customers with a smaller, faster, and less expensive product, or whether it prefers to use this advantage as a cushion (both geometrically on the silicon die and financially) that enables the designers to embed more (and previously unthinkable) functionality and therefore improve the integration and value of the new product.

However, the mere knowledge that the underlying architecture can be easily expanded is a tremendous advantage in the designer's mind. The designer is now confronted with a certain peace of mind that is rare in this industry. This is why Cognigine and industry analysts are so excited about the prospects of this technology in the OC-768 environment and beyond.

The optimization of the memory bandwidth in terms of balancing the memory read/write load and the cost and performance of memory access by intelligently managing that bandwidth in a hierarchical and distributed fashion is a very important task. Cognigine engineers have clearly done their homework in this regard. To start with, the NPU chip provides a first memory level of shared 2Mb of internal SRAM.

Most importantly, however, several memory controllers are integrated inside the chip. More specifically, four 64-bit DDR SDRAM controllers operate at 200 MHz for packet buffering. This means that the capability of 512MB of space for packet buffers is supported. A configurable SSRAM controller (2×64 bit and $4\times16/2\times32/1\times64$ bit) running under a 200 MHz clock provides access to classification memory and coprocessor interfacing. The NPU chip's memory-controller landscape includes a

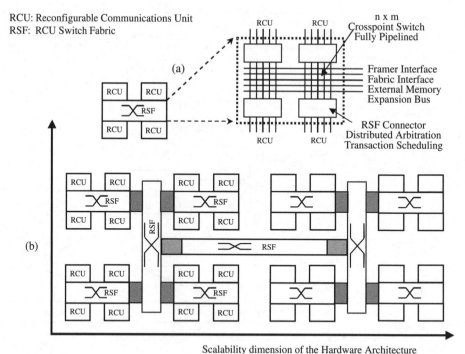

RCU: Reconfigurable Communications Unit
RSF: RCU Switch Fabric

FIGURE 9.6 The scalability of the Cognigine NPU architecture that is composed of structured combinations of RCU-RSF clusters. *(Source: Cognigine)*

programmable flash-memory controller for system boot operations. It is interesting to note that the SRAM peak bandwidth is 76 Gbps, whereas the peak DDR SDRAM bandwidth is 100 Gbps. The Cognigine network processor is available in a 1,517-pin HFC-BGA package.

The picture is completed with a GUI-based integrated development environment that offers a single-processor programmer's model; therefore, the software engineer does not have to worry about allocating tasks to specific engines. The development environment contains an application configuration tool as plenty of software components function on-chip, such as framing, parsing, traffic management, accounting modules, and so on. There is also naturally a C/C++ compiler, assembler, and debugger for code development; a clock-accurate software simulator; and a services library that facilitates the tackling of issues such as fabric access, parsing, traffic management, and so on.

EZchip TOPcore™

EZchip (*www.EZchip.com*) is an Israeli company that has very strong ties to IBM (EZchip's silicon foundry and strategic investor). It is poised to have a very significant impact on the evolution of the network-processing industry as it has developed extremely integrated products that eliminate multiple chips for the realization of complete switching cards. The company's current products include the NP-1 (a 10 Gbps seven-layer network processor), the NP-1c (a second-generation 10 Gbps network processor), and the QX-1 (a 10 Gbps traffic manager). It also provides the necessary software development infrastructure around these chips.

The company's NP-1 is a single-chip, full-duplex NPU with embedded search engines for 10 Gbps/OC-192 and 1 Gigabit Ethernet applications. The NP-1 chip provides fully programmable packet classification, modification, forwarding, queuing, and policing at wire speed. By using external DRAM only, the NP-1 requires no classification coprocessors, TCAMs, or even SRAMs. It provides full-fledged packet processing between layers 2 and 7 and classification.

The company has also integrated all search engines and eliminated the need for such external components. A series of proprietary and patented search algorithms ensure that the NP-1 can perform lookups in very large tables with over 1M entries at 10 Gbps throughput. The user does not have to worry about data or entry caching. Flexible, user-definable lookup table formats are inherently supported. Tables with variable-length keys and results can be included or wildcards can even be used.

It is particularly important to notice that the NP-1 processor seems able to reduce to about one fifth, the chip count, power dissipation, and cost of implementation of several networking solutions. This is feasible through the network processor's combination of embedded search engines and embedded DRAM, full-duplex 10 Gbps throughput, and integrated 10 Gigabit Ethernet and 1 Gigabit Ethernet MAC controllers. This all culminates to a situation that can have very serious ramifications not only for the company, but also for the evolution of this industry.

The company has based the design of its NP-2 processor on its TOPcore architecture, thereby scaling the original 10 Gbps NP-1 design to achieve 40 Gbps throughput. In fact, the NP-2 chip is implemented around the same *task optimized processing* (TOP) cores that EZchip used in the NP-1 design. As a result, software that has been developed for the NP-1 is portable and can be easily reused in higher-speed designs that are centered on the NP-2 network processor, thereby offering the customer a smooth migration path from 10 to 40 Gbps systems. Based on market input, EZchip is currently focusing on next-generation products based on its TOPcore architecture for 10 Gigabit Ethernet and multigigabit applications. The company has stated that development of its NP-2 product will take high priority when the market demand for 40 Gbps applications picks up.

The EZchip NP-1 network processor includes a PCI bus with the host CPU and a DDR interface with external SDRAM. On the fabric side, it includes a CSIX interface with the switch fabric itself (or to cascade multiple NP-1 chips and increase system capacity), or an XGMII interface with an integrated 10 Gigabit Ethernet MAC. On the line side, it includes an SPI-4.2 interface with an external OC-192 POS framer, another XGMII interface connecting with yet another 10 Gigabit Ethernet MAC, or a GMII/TBI interfacing with eight 1 Gigabit Ethernet MACs. This flexibility allows the NP-1 to function as a standalone box connecting a 10 Gigabit Ethernet port to another 10 Gigabit Ethernet port. It can also be configured as an aggregator of eight 1 Gigabit Ethernet ports onto one 10 Gigabit Ethernet port in addition to working in a more traditional PHY-to-NPU-to-serdes-to-fabric chain.

Obviously, systems designed around the EZchip NP-1 network processor can be programmed to deliver layer 2 functionality and MPLS switching, along with IPv4/IPv6 routing, packet tunneling, flow classification, QoS, and policing. In general, they can manipulate packet payloads with a large flexibility for numerous types of applications. As we mentioned in the beginning of this section, the NP-1 can handle up to layer 7 processing.

You may wonder which layer 7 functionality is required for a 10 Gbps processor. This seems more geared toward carrier-class applications. Parsing, classification, and modification capabilities are, of course, highly desirable in systems such as server load balancers or URL-based web switches. In general, the NP-1 enables advanced services that must rely on fine-grained flow classification, URL matching, and per-flow state updating. The beauty of layer 7 processing is that it can all be done by writing and executing software that runs on the network processor. The continuous content awareness of the NP-1 enables the programmer to code layers 2 to 4 switching and routing applications with granular flow classification. It also enables the programmer to handle layers 5 to 7 deep packet processing to address needs such as content switching, TCP offloading, security, and even traffic analysis.

The QX-1 is a single-chip traffic manager that can be used to extend the QoS features of the NP-1 when networks must be built with stringent requirements for advanced services provisioning. QX-1 can be used in either the ingress or egress path. It was designed to achieve optimal interoperability and performance when interfaced with the NP-1, whose companion chip it was designed to be all along.

A switching system that is built on a combination of the NP-1/QX-1 chips can provision QoS in accordance with the DiffServ model. In fact, QX-1 enables groupings of flows and queues to offer *per-hop behavioral* (PHB) QoS options. Features such as multiple queues that are flexibly mapped per destination port as well as a hierarchical scheduler are used for the implementation of all DiffServ services including *Expedited Forwarding* (EF), *Assured Forwarding* classes (AF1–AF4), and *Class Selector* (CS).

In a typical system that combines the NP-1 and the QX-1 chips, the general partitioning of the tasks between the two units is as follows. The NP-1 network processor executes classification over the seven layers, handles forwarding decisions, learns new information that must be kept in tables and updates all existing tables, handles policing (using *single-rate three-color marker* [srTCM]/*two-rate three-color marker* [trTCM] token bucket), performs per-flow statistics, and modifies packets when necessary. On the other side, the QX-1 handles all queuing, manages congestion, manages per-flow queuing, and is responsible for hierarchical scheduling.

When used in the egress path, QX-1 is the last device prior to transmitting the traffic to the *physical* (PHY) interfaces and enables the precision shaping of traffic directly to the network link(s). The QX-1 offers several types of interfaces that enable it to interconnect to the system switch fabric or line links. QX-1 offers a CSIX or SPI-4.2-based streaming interface when connecting to the switch fabric. It supports a 1×10 GbE, $1\times$OC-192, $4\times$OC-48, $16\times$OC-12, or 16×1 GbE channels when connecting to an external framer or Ethernet multiplexer through the integrated SPI-4.2 interface.

Instead of using the approach taken by some other network processors that integrate generic RISC processors, EZchip's TOPcore architecture consists of engines called *task optimized processors* (TOPs), which are typically 10 times faster than alternative RISC cores and are customized to perform a specific networking function at an optimal speed. Multiple instances of these fast and efficient processors are integrated inside the same die configured in a super-scalar architecture, which has been designed to optimize packet-processing tasks.

The following describes the four types of TOP engines:

- The *TOPparse* processors handle packet parsing. These processors can parse any type or format of frame or packet, regardless of whether it is encapsulated, and extract entire headers, tags, addresses, port numbers, protocols, bit patterns, keywords, and so on.

- The *TOPsearch* processors handle lookup and search operations by using the parsing results as keys for lookups in the tables maintained by the system for routing, policy, and classification.

- The *TOPresolve* processors take care of all forwarding and QoS decisions as well as updating state-related information and the tables themselves.

- The *TOPmodify* engines perform all required packet modifications by overwriting bit fields inside packets by inserting or adding bits, swapping bits, and/or rotating bit fields.

These four types of engines are cascaded in a four-stage parallel-pipelined fashion. As soon as one stage is done with its computing tasks, it passes the processed data onto the following stage downstream in the pipeline. The term *parallel pipeline* means that at each stage of the parse-resolve-search-modify pipeline, multiple TOPs engines perform identical functions. As a result, multiple packets are processed simultaneously at each stage. The multiple TOP processors at each stage execute the same code in principle, but they all have their own instruction memory. Therefore, they are able to preserve their independence and high efficiency while executing a series of tasks. A hardware scheduler transparently allocates incoming packets to available hardware resources at each stage of the pipeline while preserving synchronized frame pointers across the pipeline and ensuring that messages are passed on between the engines to coordinate processing. As a result, the programmer does not have to worry about addressing individual TOPs processors. It should be clear, however, that writing code for the NP-1 entails actually writing code for the four types of engines.

In terms of facilitating the development of software for its network processors, EZchip is offering several tools. To start with, EZdesign™ is a comprehensive suite of design and testing software tools for developers, which enables the rapid delivery to production of new designs based on the company's NP-1 network processor. EZdesign enables designers to create, verify, and implement NP-1 applications that must meet specific functionality and performance targets. EZdesign has the following components:

- A *microcode development environment*, which under a unified GUI allows the editing and debugging of code, including setting breakpoints, single-stepping program execution, and obtaining access to internal resources. Features of this environment include a code editor, a view of memory and register contents, performance charting, macro recording, and script execution. The microcode development environment can be used to develop and debug code that runs on both the NP-1 simulator and the actual NP-1 chip.

- A *simulator* that is able to provide cycle-accurate simulation of the NP-1 for code functionality testing and performance optimization.

- An *assembler* and *preprocessor* that generates optimized code for execution on the NP-1 network processor. The NPU assembly is interleaved with high-level macros. A C compiler is now available as well, although to create the most optimized code, the assembler is usually preferable.

- A *subroutine library* that contains the source code of many common networking tasks that the company provides with the intention of helping customers simplify and accelerate their code development.

- An *applications library* that contains reference code, which customers can consult or use to implement high-level applications when designing new networking platforms and services. EZchip offers reference code for applications such as layer 2 switching, MPLS, IP routing, *Network Address Translation* (NAT), and URL-based load balancing.

- A *frame generator*, which is essentially a GUI-based tool that guides the software engineer through the process of creating frames, layer by layer. It allows for the easy generation of frames of different types, protocols, and user-defined fields.

- A *structure generator*, which is another GUI-based tool that enables the definition of data structures used by EZchip's NP-1 network processor for forwarding and policy table lookups (such as hash and trees), their keys, and all associated result information.

Among the company's development tools, we should also mention EZdriver™, which is essentially a control processor API layer. This is a toolset designed to facilitate the development of software that is meant to be executed on computational resources of the control path CPU of NP-1-based systems. EZdriver contains a set of routines that execute on the control path CPU and provide an API for applications that run on the same control path CPU and need to interface with the NP-1. With EZdriver, software engineers working on control path development tasks can easily handle tasks such as configuring the NP-1 chip, loading the microcode, creating and maintaining NP-1 lookup structures, sending and receiving frames to and from the NP-1, and configuring and accessing the NP-1 statistics block.

EZdriver, in conjunction with the company's EZdesign tool, provides an extensive set of debugging capabilities by offering software-driven debugging features (such as breakpoints, single step, register, and memory access), which the code developer can activate on both the NP-1 simulator and the actual NP-1 chip.

To help expedite the development of a complete system based on its NP-1 and QX-1 chips, EZchip offers evaluation boards with a choice of 1 Gigabit Ethernet, 10 Gigabit Ethernet, or OC-192 POS interfaces. Their design enables two boards to interconnect in order to obtain a complete ingress and egress line-card path. In addition to this, multiple boards may be connected over an external switch fabric and backplane. In 2002, the company demonstrated interoperability with IBM's latest PowerPRS™ 64G switch fabric. All evaluation boards can be accessed through a standard PCI-bus connector that plugs into a standard single board computer on which control plane software can be developed. As a result, the interface is ensured with the network-processor chip (NP-1) on the evaluation board.

The company will be expected to expand its development tools if it wants to address the needs of customers of the NP-2 and encompass the 40 Gbps realm. Competition seems inevitable with the development of other 40 Gbps network processor and traffic management products from companies such as Xelerated. However, the slower economy of 2002 seems to have adversely affected the carrier investments for new equipment at the core level and has consequently kept the market emphasis on 10 Gbps and below.

In late 2002, EZchip introduced its second-generation network processor dubbed *NP-1c*. The intention was to better target a wide range of markets and, more specifically, systems that include multi 1 Gigabit Ethernet, OC-192, 4×OC-48, and even 16×OC-12 with a single chip. The NP-1c, which is manufactured by IBM Microelectronics, is pin compatible with the first-generation processor (NP-1); however, it has some striking differences. The NP-1c is built using IBM's cutting-edge Cu-11 semiconductor process, offering 0.11μ line widths and therefore extremely compact density. In addition to its other benefits, this process enabled NP-1c designers to double the processing power. It extended the headroom by 80 percent while reducing the cost of ownership directly by lowering the price by 30 percent for a full-duplex 10 Gbps processor and indirectly by lowering it by 80 percent when it comes to considering a switching card's chip count and power dissipation.

EZchip bases a lot of its arguments on the compelling case that IPv6 will be adopted more frequently in order to deal with the lack of IPv4 addresses, especially in the Far East, and to accommodate the wireless IP networks where an IP address is needed per telephone. Since the IPv6 addresses are 16 bytes as opposed to IPv4 addresses, which are 4 bytes long, it is clear that IPv6 routing and session tables will be approximately 4 times larger than with IPv4 routers. A significant advantage of NP-1/NP-1c-based routers is that no extra hardware is required to support such tables, whereas routers based on alternative network-processor technologies will probably need a significant number of extra chips.

To make the case more tangible financially, EZchip clarifies that a 10 Gigabit per second interface of an IPv6 router will need a single NP-1c processor and four DRAM chips, which is identical to what happens in an IPv4 router. The bit density of typical DRAM chips is approximately 30 times higher than similar capacity CAMs, whereas the power dissipation of the DRAM chip is roughly 280 times less than that of power-hungry CAMs. Even the cost per bit of a DRAM chip is almost 1,000 times less than the corresponding cost per bit of a CAM.

The total cost of the NP-1c solution for this example is estimated at $820 with 17 watt power dissipation. With other network processors, however, the same interface would have to be implemented based on the use of two network-processor chips and somewhere between 20 (especially for small routers) and 80 additional CAM and SRAM chips. These combinations total up to cost somewhere between $3,000 and $12,000, with 75 to 300 watts power dissipation. The NP-1c was scheduled to be sampled during the first quarter of 2003.

VITESSE IQ™ FAMILY OF NETWORK PROCESSORS

Vitesse (*www.vitesse.com*) has been a major player in the network-processing arena. It has gained even more presence since it acquired a startup called Sitera for its line of high-performance networking chips. The IQ2000 chip was its first important processor. The company now offers multiple NPUs and traffic managers, and it is also uniquely positioned to offer one-stop shopping for its customers. It provides essentially all the necessary components from optoelectronic transceivers, SONET and POS framers, and PHY and MAC chips all the way to powerful and scalable switch fabrics, backplane interconnects, and serdes chips.

The IQ2000 is positioned as an OC-48 processor capable of performing all the necessary operations for packet processing between layers 4 and 7. These operations include packet inspection, classification, filtering, encryption, modification, address translation, policy enforcement, traffic shaping, and multicast management.

The chip features four full-duplex 1.6 Gbps interfaces which combine to give up to 12.8 Gbps of aggregate bandwidth, properly designed to match the needs of four embedded 32-bit RISC processor cores that run at 200 MHz. The cores are inspired by the MIPS-I architecture, but their instruction set is not fully MIPS compatible. Vitesse provides all the required development tools. The IQ2000 can be configured in multichip schemes, enabling the company's customers to build and scale more powerful systems as needed. The IQ2000 is unusual among NPUs in the sense that it uses Rambus™-based RDRAM memory to store packet payloads. However, with only one RDRAM channel that provides a peak data transfer rate of 1.6 GBps, the IQ2000 does not perform as well as other competitive OC-48 network processors in memory bandwidth.

Vitesse is supporting development with a series of hardware evaluation/development boards/kits/platforms and software development tools, including layer 2 and layer 3 application reference code, software support libraries, compilers, and so on.

The latest member of the company's network processor family is the IQ2200. The IQ2200 is not only fully pin compatible with the IQ2000, but it also operates at twice the core frequency of the IQ2000 and therefore provides twice the packet-processing performance. In addition to providing OC-48 performance, another major characteristic of the IQ2200 is that it has an integrated *Common Switch Interface* (CSIX) interface that enables it to natively connect on Vitesse's GigaStream™ and TeraStream™ families of intelligent switch fabrics.

The IQ2200 is positioned by Vitesse as a powerful platform for the delivery of flexible and scalable applications in the areas of complex multiprotocol routing, address translation, classification, policy enforcement, filtering, traffic shaping and grooming, multicast, and so on.

Some of Vitesse components that allow a customized treatment of the QoS realm include RIO, RED, WRED, *weighted round robin* (WRR), and WFQ. Its scalable multiprotocol capabilities allow the easy deployment of added-value services such as MPLS, DiffServ, NAT, and *IP Security* (IPsec).

To address the needs of either high-density-port line cards or examples centered on small fabric designs, Vitesse also offers a switch-interconnect chip called FOCUS Connect. This chip allows for the easy connection of up to eight NPUs of the company's IQ2x00 family, but ASICs, FPGAs, or other FOCUS-enabled peripherals can be connected as well. Each FOCUS port is a point-to-point, high-performance 1.6 Gbps full-duplex link that is structured as eight channels that are clocked at 100 MHz. This means that multiple packets can be transferred at the same time. The chip supports 1,024 separate multicast distribution trees, 4 priority levels for data packets, and flexible clock modes for the easy integration of FPGAs. It is scalable to larger port densities by using multilevel stacking or grouping.

A single FOCUS Connect device can connect up to eight Vitesse IQ2000 NPUs with over 1 Gbps full-duplex bandwidth for each, or four IQ2000 NPUs with over 2 Gbps full-duplex bandwidth for each. The combination of two FOCUS Connect devices allows the rapid, glueless, and straightforward connection between eight IQ2000 NPUs with over 2 Gbps bandwidth for each one. In the latest IQ2200 network-processor chip, Vitesse has integrated the FOCUS interface. In fact, it supports either FOCUS16 or FOCUS32 (32-bit-wide transfers) links for higher bandwidth.

The company offers a series of advanced development tools in conjunction with evaluation and hardware development platforms from compilers all the way to models based on *Hardware Description Language* (HDL) for the FOCUS interconnect in order to facilitate and accelerate the overall system development.

WINTEGRA

Although we have not focused on the lower part of the performance spectrum where devices need access to 622 Mbps links or where at the worst case they must provide connectivity to 1 Gbps links, we will slightly deviate and mention a company that is setting some serious precedents in that arena. No one should be surprised if we start seeing the same trend in network processors that address the higher-speed links.

Wintegra (*www.wintegra.com*) is an interesting startup in this field. It is certainly not a coincidence that several major players in the industry such as Motorola and Marvell have participated in its funding. The company has introduced WinPath™, a family of single-chip solutions in the access network arena, based on a technology that is equally at ease with packetized traffic as well as with frames and voice *pulse code modulation* (PCM)/TDM channels. Wintegra has already announced important agreements and project breakthroughs in areas such as DSL, wireless base stations, or voice over network with major partners such as Texas Instruments in the *digital signal processing* (DSP) arena with whom they have created a full-fledged reference design. Rightfully so, it takes pride in multiple communication protocols that are implemented on board its chips. The respinning of silicon is not required by these protocols as they are implemented in RAM memory. The evolving list includes ATM AAL0, AAL2, and AAL5 SARing; ATM cell switching and AAL2 CPS switching; *ATM Circuit Emulation Service* (CES); *Inverse Multiplexing for ATM* (IMA); traffic management for ATM; IP and Ethernet *High-level Data Link Control* (HDLC); IP over ATM; IP over Ethernet; IP over PPP; IPv4 *longest prefix matching* (LPM) routing; IP classification; VLAN tagging and detagging; ATM to Ethernet interworking; and others. Every port can be immediately set up as an ATM, IP, or TDM port without any overhead or any hardware change.

WinPath provides a direct interface with any one of these PHY level standards: T1/E1, T3/E3, xDSL, OC-3 ATM, OC-12 POS, and 10/100 Ethernet. Gigabit Ethernet is supported through an external and proprietary POS. The *Universal Test and Operations PHY Interface for ATM Level 2* (UTOPIA 2) or POS interface is also meant to handle any external need of switch fabric interface. Multiple devices (in a one-master-many-slaves configuration) can be connected on the other UTOPIA interface, connecting up to 63 external DSPs for *voice over IP* (VoIP) applications (vocoding, compression, echo cancellation, and so on) or up to 6 octal DSL PHYs for DSL applications. Any of WinPath's various interfaces can be programmed by applications so they interwork with any other interface. For instance, one can have ML-PPP over the T1/E1 serial channels, interworking with POS running over the POS OC-12 interface, IP over 10/100 Ethernet, and IP over ATM AAL5 over a multi-PHY OC-3 configuration on the UTOPIA interface.

External memory is flash and *synchronous dynamic random access memory* (SDRAM). Both are 32/64 bits wide and three interfaces are available: one for host CPU interfacing needs and the other two for packet processing. Larger applications may need two chips: one for the ingress path and one for the egress path processing. As an added advantage, lower-end applications, where one WinPath chip can handle both, have only one SDRAM memory bank needed where both processing parameters and packet information can be stored, thereby further reducing the chip count and the cost of a solution.

The company has announced two major products so far. The first is called the WIN777. Since it embeds a 200 MHz 64-bit MIPS core CPU along with the rest of its packet-processing hardware, it can handle both control and data path functionality. The second product is called the WIN707. By the mere fact that it does not contain an embedded CPU core that could function as a control processor, it is meant to operate only in the fast data path, leaving all control path processing work to an external processor such as a PowerPC 750, which offers full bus compatibility.

One of the interesting abilities of the WinPath is the device's ability to balance dynamically multiple 200 MHz embedded processors with 200 MHz memory subsystems, thereby creating a very predictable performance environment that could otherwise be matched only by custom ASIC designs. This means that if an application requires more entries in the routing table than another, or if it needs access to more virtual channels than another, no degradation of performance will occur.

Last but not least, Wintegra takes pride in the fact that not one line of assembly code has been written for its chip. It touts its C compiler and integrated SDE as a key factor in accelerating the customer's time to market.

XELERATED PACKET DEVICES

While most vendors were struggling to stabilize their network-processing platform at OC-48 levels, to verify whether it is feasible to scale what they have at higher speeds, or to prove the actual scalability of their architectures to full duplex OC-192, some vendors were entertaining ambitions for higher-speed products. Others have presumably moved forward with the development of actual and concrete product plans. One of the major surprises in this industry was the sudden announcement in the summer of 2002 from a small Swedish startup called *Xelerated Packet Devices* (*www.xelerated .com*). It announced that it has not only designed, but is actually sampling an integrated network-processing chipset, which is the first one to be able to function at full wire speed in 40 Gbps[1] (OC-768) networks.

The chipset is based on an architecture that the company calls PISC™, which stands for *Packet Instruction Set Computing*. It is composed of two chips—the Xelerator™ NPU and the Xelerator™ traffic manager. They can be used either as a combination or as standalone units.

One of the development tools that the company provides is a cycle-accurate simulator, which is fed with files containing the executable code the programmer creates for forwarding plane application. The Xelerator chipset offers a single-threaded programming model to the programmer, who writes code as if he or she were faced with a single-image traditional sequential machine without the slightest need to know how parallelism will be involved in the actual code execution. The executable code is the result of the linking process, which occurs on the output of the assembler that generates compiled code by processing the PISC instructions (source code) that the programmer has to write. These PISC instructions perform the actual packet-processing operations (parsing, editing, encapsulating, modifying, and so on) and call on hardware resources such as engines, meters, counters, TCAM, and so on. The simulator is part of the GUI-based integrated development environment, which also contains a debugger and an integration support library along with ready developed code examples for several real-life applications such as IPv4, IPv6, MPLS, layer 4 packet filtering, and traffic conditioning.

The Xelerator network-processor units are available in three models, as shown in Table 9.1. Their packet-processing performance is always at wire speed and the deterministic processing delay of the chips offers very good jitter characteristics.

Conceptually, the internal structure of the Xelerator NPUs can be imagined as a large programmable pipeline fed from one side by one to four (depending on the model) Rx ports implementing the SPI-4.2 interface and fed from the other side based on the NPU model between one to four Tx ports implementing SPI-4.2. Four look-aside engines allow interfacing with external coprocessors, SRAM,

TABLE 9.1   Xelerator Network Processor Models

Chip Model	Number of 10 Gbps Ports	Packet-Processing Performance
X10s	1	25 Mpps
X10d	2	50 Mpps
X10q	4	100 Mpps

1. An interesting discussion on the advantages of using data flow architecture to process 40 Gbps traffic can be found in Gary Lidington's "Data Flow Architecture Must Match the Network to the Application," published by EE Times (May 9, 2003). The article can be found online at *www.commsdesign.com/story/OEG20030509S0035*.

and TCAM with the possibility of multiple accesses to each one of these resources per processed packet. The programmable pipeline has internal access to other hardware resources such as hash engines, classification hardware, counters, meters, and even an internal TCAM engine that manages the search process.

The programmable pipeline is the implementation of the company's PISC architecture. It is essentially a packet-editing chain that performs operations on packets as they traverse the pipeline from the Rx side to the Tx side. All memory access channels are equipped with integrated ECC for carrier-class reliability. In order to be able to consult memory at full-duplex wire speed, the company's traffic manager needs *reduced latency DRAM* (RLDRAM) that behaves like Rambus-based DRAM but with significantly lower latency.

The Xelerator traffic manager is available in two configurations—T10s and T10d. These are available with one or two 10 Gbps ports (either Rx or Tx), so they can work in simplex and duplex environments. Like the NPU, the structure of the traffic managers is based on Rx ports (one or two depending on the model) feeding the PISC programmable pipeline that takes care of classification and statistics counting. It now feeds an SAR module that outputs its work into a queue manager before the results go to the one or two Tx ports. The queue manager uses WRED and performs individual queue scheduling and shaping up to three levels. An embedded memory manager controls the interface to external *quad data rate* (QDR) SRAM and DRAM. A look-aside engine enables it to interface with an external coprocessor, SRAM, or TCAM again with the possibility of multiple accesses per processed packet. The queues are structured based on packets and different applications, and may require that the queues be combined upon specific structures. Such applications could be guaranteed-bandwidth VPNs or switch fabrics based on virtual output queuing.

In a full-duplex OC-768 environment on the ingress side of a line card, the OC-768 framer through the SPI-4.2 interface connects to the NPUs (Rx port), which connects to the traffic manager (through the Tx ports). The traffic manager then connects onto the switch fabric interface. The egress side is the exact opposite. The fabric interface is connected on a traffic management chip, which is cascaded with the egress path NPU, which connects via SPI-4.2 with the OC-768 framer. The originally implemented SPI-4.2 interface (which the company has promised to replace with SPI-5 when it becomes available) enables the convenient structuring of the I/O bandwidth as several OC-192. This allows a better utilization of the chipset's computational power.

OTHER APPROACHES

To describe the approach taken by large *network equipment vendors* (NEVs), we will use Cisco as an example of a company that has been very active developing its own internal designs of network processors. The Cisco PXF chip (better known in the industry as *Toaster*) has been reborn in three successive generations. Each one comes with 16 packet engines arranged in 4 parallel pipelines. It has been at the heart of several Cisco routers. A rough estimate of the computational power of a pair of PXF chips makes it approximately equivalent to an IBM NP4GS3.

Another approach that companies like Cisco take toward the evolution of the market and the rapidly advancing network-processing technology is the acquisition of a startup. Cisco recently acquired Navarro Networks, a secretive startup from Texas, which was led by industry-veteran management and was largely funded by Cisco.

SUMMARY

In this chapter, we discussed several promising architectures in the network-processor arena, coming predominantly but not exclusively from startup companies. We now have seen the trends toward integrating critical components inside the same die and the tendency to raise the performance bar toward higher wire speeds. A few players now offer unprecedented 40 Gbps processors and are probably a little ahead of the demand curve in the market. Stepping back for a moment, the field seems over-populated in the 2.5 Gbps arena with multiple vendors competing for design wins and market share. As this is by far the largest chunk of the market and as some of the players are true powerhouses, sooner or later some players will have to bow out of the race. They will either fail or be acquired by a larger vendor.

The jury is still out regarding the 10 Gbps market, which is definitely taking shape but in a very slow fashion. This is mainly due to the overall slow economy after the boom of the 1990s, something that is even further compounded by the significantly slower pace of investment from carriers who would like to upgrade their infrastructure but cannot afford to at this point.

CHAPTER 10

ALTERNATIVE APPROACHES TO NETWORK PROCESSING: NET ASICS AND DESIGNING WITH IP CORES

We have seen that network processing is a computational area that requires several resources to ensure good performance at wire speed while preserving flexibility of the network protocols and applications supported. In previous chapters, we saw how some of the most promising network-processor architectures address this problem. In order to complete our overview of the network-processing architectural landscape, we will turn our attention to a couple of different approaches toward achieving the same goal. More specifically, we will look at a special breed of microchips called *net application-specific integrated circuits* (Net ASICs). We will also look at specialized integrated solutions that some companies build around IP cores.

NET ASICS

Net ASIC is a generic name that has been adopted by the industry to denote a special type of network-processing integrated circuit that contains specialized assist hardware (sometimes referred to as *embedded coprocessors*) for most functions required in packet processing; however, there is one big difference—unlike *networking processing units* (NPUs), a Net ASIC is not programmable.

It could be argued that this lack of programmability is a mark of inflexibility, as users cannot change the behavior of the Net ASIC chip, depending on the application at hand. This is the price you pay for having the privilege of combining fast and deterministic performance (like the performance that these chips usually deliver) with most of the necessary packet-processing functions, which are already integrated into the same Net ASIC die. This combination, along with the associated trade-off, is somewhat appealing to many companies that are confronted with the dilemma of choosing between a more traditional network processor and designing a specialized ASIC for their project.

In order to understand the rationale behind the Net ASIC phenomenon, we must examine this dilemma. Looking at a contemplated ASIC, many companies that decide to use a Net ASIC lack the necessary design and engineering skills, lack the financial resources, or cannot afford the longer time to market that is associated with designing a complex fast-networking ASIC from scratch. This venture usually takes between 12 and 18 months.

On the other hand, companies that favor a Net ASIC seem to shun the idea of using a programmable network processor because of the amount of time it takes to develop software for packet processing over a vendor's proprietary development system. This task often must be implemented based on unusual instruction sets, unfamiliar languages, and the tools themselves. Application developers actually have to learn the underlying NPU architecture and how to activate its various parts. All this is deemed by such companies as time consuming. They would rather opt for a Net ASIC.

The word *deterministic* was used to describe the packet-processing performance of Net ASICs. This is not a coincidence. As a Net ASIC is completely hardwired, as long as the available integrated functions are exactly what a customer wants, the user has little to worry about regarding issues such as jitter or packet-processing latency when working at wire speeds, especially with time-sensitive applications such as slot-based *time division multiplexing* (TDM). Traditional NPU customers (such as ASIC designers) often struggle to fine-tune and balance multiple aspects of an entire design in order to maintain adequate levels of performance.

Implementing a complete solution around a design that is based on a Net ASIC requires some software development, but that development must occur along more traditional software-engineering directions. In fact, it entails writing control plane code that will run on a supervisory host *central processing unit* (CPU) and not in the packet-processing piece of fast silicon. The host is programmed with languages, tools, and methodologies that are familiar to anyone in the engineering field. Therefore, these companies are not confronted with the need to suddenly have their engineers climb up a new and steep learning curve. This further justifies the decision to use a Net ASIC instead of using an NPU or having to design a complex networking ASIC.

Traditional network processors and Net ASICs are in fierce competition. Given the global commercial and technological prowess of the main NPU vendors (IBM, Intel, and Motorola), it will not be surprising that some of the Net ASIC vendors will soon disappear. In fact, as of this writing, Entridia, a promising and well-funded startup from Southern California, which had actually been one of the pioneers of the Net ASIC concept, was forced to lay off its staff, close its doors, and sell its technology to Stratigos Networks. At the same time, Internet Machines (*www.internetmachines.com*) announced that it was suspending its Net ASIC offering.

These are just a few examples of the major shake-up and consolidation that this new industry will undergo before the fittest platforms, technologies, and vendors survive. These winners will then divide up the market in a pragmatic way. This usually happens in new industries right after the initial phase fades away and the associated excitement that attracts a shower of competing ideas, lots of entrepreneurial talent, and heavy investments usually in the form of venture capital disappears.

Table 10-1 compares two Net ASIC product families that are offered by two major vendors. The choice between these families is a direct function of the user's application at hand. One of these products has an edge in environments that combine *Asynchronous Transfer Mode* (ATM) and IP traffic, whereas the other is much easier to interface with Ethernet and Gigabit Ethernet realms where it is more likely that only IP traffic will be transmitted.

We will conclude our discussion about Net ASICs by highlighting a key industry fact: The tremendous programmability and flexibility of ordinary network-processor chip-based platforms in conjunction with free application code that network-processor chip vendors often offer to their customers place some dark clouds over the commercial viability of Net ASICs. Since the Net ASIC approach is questioned mostly for business reasons, it is not a surprise that as of this writing, major players in the industry have announced that they will suspend their development efforts on Net ASICs and concentrate their future development efforts on programmable network processors instead.

DESIGNING WITH IP CORES

Although IP-core-based network processing is not intended for mainstream users who are in search of solutions to the computational needs of their switching/routing project, we must discuss the approach taken by several companies to create state-of-the-art network-processing systems based on the use of intellectual property offered by third parties.

TABLE 10.1 A Comparison between Major Net ASIC Solutions (Source: ZettaCom and Marvell)

Feature	ZettaCom (*www.zettacom.com*)	Marvell (*www.marvell.com*)
Net ASIC	MSP-200 chip capable of full-duplex 10 Gbps	Prestera-MX two-chip offering: • 98MX20 for 1×10 Gigabit Ethernet • 98MX30 for 10×1 Gigabit Ethernet
Traffic manager chip companion	Yes, through the company's ZEN-QM two-chip (QMD-QMC)	• Congestion management (*Weighted Random Early Detect* [WRED]) • Traffic shaping and traffic scheduling only at egress
Integrated Ethernet *Media Access Control* (MAC)	No	Yes (easy connection with the company's *physical* [PHY] chips)
Packet over SONET (POS) and ATM suitability	Yes	Not easily; glue logic is needed for OC-192 framers.
Classification	Yes	Yes, at ingress only
Policing	Both on cells and packets	Yes, only on packets and only at ingress
Packet modifications	Yes, with support for ATM, IP, and *Multiprotocol Label Switching* (MPLS)	Yes, with support for IP and MPLS and only at ingress
Host interface	Generic bus that is 16 bits wide and works at 66 MHz	Standard *Peripheral Computer Interconnect* (PCI)
Search engine	External *content-addressable memory* (CAM) up to 1 million	No need for external engine
Types of memory needed	• Packets in *double data rate* (DDR) *synchronous dynamic random access memory* (SDRAM). • SRAM needed for the traffic manager • External CAM is needed for search engine implementation.	• DDR-SDRAM is absolutely needed, offering a cost advantage to the memory subsystem. • Other types of memory are optional.
Interface toward the line side	*System Packet Interface, 4.2* (SPI-4.2)	RGMII for Gigabit Ethernet and XGMII for 10 Gigabit Ethernet
Interface toward the fabric side	CSIX-L1 64 bits at 250 MHz	• Proprietary 15 Gbps and HSTL uplink bus • CSIX-L1 fabric adapter chip that also does ingress scheduling
Package	1,036-pin HPBGA	901-pin *ball grid array* (BGA)
Power consumption (max)	±10 watts	Not disclosed

The idea of using IP cores for the design of sophisticated integrated circuits is not a new phenomenon. In fact, it has become a widely practiced principle over the 1990s. The fundamental idea is as follows: Instead of designing a specific and usually very complex part of an integrated circuit, a designer licenses the use of a *core circuitry* from a competent and qualified third party. This core circuitry delivers the desired functionality, and has been designed, tested, and documented according to specific methodologies and industry-accepted criteria, thereby offering the possibility of connectivity and programmability as well as easy integration into a larger design, testability, documentation, and even scalability of performance. These characteristics can be combined with the inevitable accelerated time obtained by having to design less of the final product. The decision almost seems to favor using IP cores.

One of the advantages companies see in using licensing IP cores (or even internally generating their own cores) is that it promulgates a school of thought that believes in the merits of component reuse. Following the same evolution path that was taken by systems implemented on *printed circuit boards* (PCBs) in the late 1970s and early 1980s when *large-scale integration* (LSI) and *medium-scale integration* (MSI) components started replacing the discrete use of multiple transistors in the implementation of more sophisticated systems, designs of complete *systems-on-a-chip* (SOC) are now based on the structured use (and even reuse) of multiple cores that implement several functions.

Ample literature has been written on the subject of designing and verifying IP cores as well as on the methodologies involved in the reuse of hardware and software IP components. Interested readers should refer to several of the pertinent sources in the section "Suggested References" provided at the end of the chapter.

In the context of network processing, the IP core principle is applied to computational resources that facilitate, if not accelerate, the handling of specific tasks that are encountered in network processing. To be more specific, several companies offer IP cores that seem to be suited for network processing and/or for certain associated computational tasks. Again, the fundamental idea is that companies that must or prefer to design their own fast-processing networking silicon should take a close look at the cores offered and decide whether they should license one or more of these pieces of intellectual property.

The detailed mechanics of a cost-based make-or-buy decision obviously go beyond the scope of this book. However, based on their analysis, some companies may discover that it does not always make sense to license a specific IP core for their network-processing design. In some other cases, it might not make much sense either from a technological or economic standpoint. As these decisions are largely subjective and often based on personal preferences, they reflect previous experience or bias on behalf of members of the company's senior technical management. Other companies may just as likely make the exact opposite decision.

There are companies that offer for license IP cores for any function a person desires to license. IP cores can span the whole functionality spectrum from main CPUs and full-fledged *digital signal processing* (DSP) cores all the way to exotic cryptographic functions, and from simple communication-protocol converters to highly specialized functions such as MPEG4 video-compression modules. We do not intend to elaborate on those aspects. Our discussion is limited to IP core issues that are relevant to network processing.

The field of network processing consists of a few important IP-core contenders among several players. In this chapter, we will discuss the approach taken by MIPS Technologies Inc., ClearSpeed Technology, Tensilica, ARC Cores, and Improv Systems. Other vendors in this arena include established companies such as IBM Microelectronics (*www.chips.ibm.com*) and Motorola (*www.motorola. com)*, which license their respective families of PowerPC series of CPU cores, and companies such as ARM (*www.arm.com*), which outsource their IP know-how through a large team of licensee semiconductor vendors. We even look at companies such as Sun Microsystems (*www.sun.com/ microelectronics*), which offer a family of *Scalable Processor Architecture* (SPARC) CPUs and embedded Java processors.

As of this writing, Lexra (*www.lexra.com*) was considered a leading contender of network-processing IP, especially when compared with companies like Tensilica and ClearSpeed. A major lawsuit was brought against Lexra by MIPS for the alleged inappropriate use of MIPS's instruction set. This was finally settled, and Lexra had to formally license the MIPS instruction set. Part of the onerous agreement was that Lexra could not engage in IP licensing anymore. Instead, the company will have to design a full-fledged NPU chip that may be available later in 2003—that is, if the company survives the financial turmoil. Lexra technology is therefore not included in this chapter.

MIPS TECHNOLOGIES

MIPS (*www.mips.com*) is a startup that was originally set up to commercialize technology that was based on pioneering research that Professor John L. Hennessy and his associates had done in the early 1980s at Stanford University. MIPS was later bought by Silicon Graphics and then spun off again as an independent company. It is one of the few undisputed powerhouses in *reduced instruction set computer* (RISC) technology. In the last 15 years or so, it has managed to propel itself to one of the pre-eminent global positions in the embedded CPU market. Through the extremely wide acceptance of its technology platform, the company has created an impressive list of licensees and varied applications ranging from workstations to network routers and from digital cameras to laser printers. It has also helped create an entire industry of third-party software development tools, such as assemblers, compilers, debuggers, and simulators, that facilitate programming and enable applications to be smoothly ported from one system to another.

MIPS offers embedded, scalable 32- and 64-bit CPU platforms that are presented in the market as a base architecture or as a CPU core. Historically, MIPS CPUs have always been designed to handle general-purpose computing. As a result, they were never intended to become part of the unusual computational environment that ultrafast packet processing has become. This pushed the adoption of MIPS IP cores predominantly in control plane applications or in applications that were meant to be part of a supervisory computer system. These are applications where the classical development tools, methodologies, and programming models ensured that the MIPS approach would yield results. It is not surprising that MIPS IP cores were deficient when it came down to manipulating gigantic quantities of packets that needed sophisticated processing in real time and at wire speeds of several tens of gigabits per second.

In addition to the lack of powerful *input/output* (I/O) bus and speed capabilities, the following are the two most important reasons for this deficiency:

- The original MIPS CPU core instruction set did not offer provisions for such packet-processing functionality such as one-cycle bit-field extraction, swapping, insertion, modification, rotation, and so on. As a result, implementing them on a MIPS core meant that entire programs would have to be written. This is a painful experience in RISC assembly when referring to having to fine-tune the CPU's multistage pipeline—something that C compilers cannot do that well. These programs would have to be recalled numerous times from the main application as macros from an I/O or packet-processing library just to implement the necessary packet-processing functions.

  This proposition would entail many wasted cycles every time these programs were invoked. In fact, even if the direct cost (in the RISC programmer's time), the indirect cost (in the extra memory footprint of the embedded implementation), and the inconvenience of writing extra code for these packet-processing functions were discarded, and if the overall problem is considered purely from a performance standpoint, the idea is absolutely unacceptable when confronted with wire-speed processing requirements.

- More importantly, however, the MIPS RISC cores are unable to handle multithreading. Every packet being processed is associated with a computational context (thread) that is usually stored in temporary locations, which are usually on-chip registers. These contain parameters, return values, and lookup table pointers that associate packets with classification results, stack and heap pointers, timers, counters, and so on. A certain level of register sets is available inside the network-processing chip, but the main execution unit will often require that some overhead be spent before the hardware switches context from one thread to another. This implies a waste of clock cycles while the thread is being switched.

  Some network processors require the programmer to manually insert special instructions to switch the thread context at a specific point in time or under specific conditions. Others automatically switch the thread in one clock cycle even when a thread is simply waiting for data to be fetched from memory.

However, the inability to support multithreading is not just a MIPS problem. It is a general RISC and *complex instruction set computer* (CISC) problem. As a result, it plagues other IP core vendors such as Tensilica and ARC.

Recently, in order to play a more important role in the network-processing realm, MIPS has reworked its fundamental instruction set to create some extensions that allow a satisfactory addressing of the first one of these two problems. Although the company's RISC cores have not suddenly become specialized network-processor cores with the introduction of the extended MIPS instruction set, it offers an improvement in programming network-processing applications. Nevertheless, MIPS IP cores still cannot compete with any network-processor chip that we have discussed so far. NPUs have been designed to excel in data plane applications; therefore, MIPS technology remains largely a candidate for the embedded implementation of control plane processing.

MIPS was unprepared when it was confronted with the sudden arrival and the ringing market endorsement of configurable architectures and methodologies within the last three years such as the one Tensilica has evangelized. Many people who are not familiar with internals of computer architecture may be wondering what is so different between the two schools. For example, with the Tensilica approach, a quick comparison will show that in the MIPS extensibility and configurability scenario (at least as depicted in the MIPS presentation at the Embedded Processor Forum in 2002), a person must hack into the processor's pipeline by coding in *Register Transfer Language* (RTL) in order to make a new instruction work. That requirement alone lies well outside the skills territory of most experienced design engineers. Handling all issues pertaining to synchronization with the processor's pipeline is the customer's responsibility. As the customer must handle the new instruction decoding, this is scary for most people. As if this is not enough of a worry for those who may be contemplating the customization of a MIPS core to handle network-processing tasks, no discussion has taken place about any type of software support from the core vendor.

Even the company's latest M4K core, which has been touted as configurable and extensible, has significant functional issues when it comes down to these two dimensions of usefulness. It also has performance issues as it can only attain 200 to 250 MHz at best in a 0.13μm *complementary metal oxide semiconductor* (CMOS) technology. This compares poorly with Tensilica's numbers, which we discuss later in the chapter. More specifically, it has the following extensibility and configurability issues:

• The MIPS M4K core does not provide support for additional registers and additional register files.

• The configuration/extension capabilities are not automated but manual, requiring RTL coding and tool modification, which is tedious and also error prone.

• It does not offer real-time operating system support for extensions.

As we mentioned earlier, IP cores are only licensed by companies that can financially afford them and that will use them in their own design of integrated circuits. The licensing of a typical IP core CPU is usually negotiable, but it usually implies a licensing fee of a half to 1 million U.S. dollars, which must be paid in advance. It also entails a scaled structure of royalties usually based on a small percentage of the chip sales, which the licensee will realize over several years with the use of the technology. There are several variations on the same theme. A company usually licenses an IP core either for a single design use or for multiple design uses, but the fundamentals of the business model remain unchanged—it involves a significant license fee up front and royalties.

In the embedded network-processing arena, however, MIPS is not confronted just with NPU chip vendors. Some IP core companies compete squarely by the mere prowess of their IP technology, which has been designed modularly for scalability and performance at wire speeds. On one hand, IP from these companies seems to hold tremendous promise in the network-processing field, which would be considered good. On the other hand, the network-processing IP from these specialized companies has a very limited marketability as no other companies outside the small network-processing realm are susceptible of using it. This is definitely not as good for the future prosperity of such companies. This can be a major concern for large networking *original equipment manufacturers* (OEMs) in search of a long-term partner.

It should be obvious that small startups cannot often afford to use an IP core if they are not adequately funded. Of course, the counterargument is that if a company does not license technology from a third party, then it must develop it. Chances are that it will cost much more. Therefore, a company that is not funded for such an internal-development endeavor is simply not adequately funded. It probably is severely undercapitalized and consequently facing extinction.

This implies that the vast majority of potential customers for specialized high-performance, network-processing IP technology are large established *network equipment vendors* (NEVs) or extremely well-funded and staffed startups (a rarity these days), who for various technical or business reasons, are not satisfied with the available network-processor chip architectures and would rather contemplate designing their own fast networking ASIC's one way or the other. However, this is not a big market for an IP company. This fact raises the issue of the mere survival and future prosperity of companies that choose this avenue as their business model.

It is not a coincidence that once key NEVs are intrigued by a new technology, they often decide to invest in it by taking a minority equity position in some of their key suppliers to ensure their ongoing viability. In many cases, they simply decide to acquire them, thereby assuring themselves of the in-house unrestricted availability and access to the key technology and even to the design team that had created it in the first place.

CLEARSPEED TECHNOLOGY

On one side of the IP-licensing spectrum in network processing, we find a company with a unique and very powerful technology—ClearSpeed Technology (*www.clearspeed.com*) (previously known as PixelFusion). ClearSpeed is a leading vendor in the network-processing IP field. This young, but promising British company has introduced a modular and highly scalable architecture for realms well beyond OC-768 and 40 Gbps. In this section, we will take a closer look at the company's approach. The overall technology trade-offs should be compared to the context of alternative network-processing architectures we have seen so far.

The *multithreaded array processing* (MTAP) architecture rests at the heart of ClearSpeed's synthesizable platform. The MTAP architecture is available for licensing in either hard (synthesized against the technology library of a specific semiconductor foundry process) or soft IP (delivered in synthesizable RTL) form. It has been shown to scale to 40 Gbps and beyond. Figure 10.1 shows the principle of this architecture. Assume that the flow of information travels from left to right. The flow-through idea is immediately applicable in switching system designs, such as in line cards, as shown in Figure 10.3.

The MTAP idea combines and blends some of the traditional characteristics of *Single-Instruction Multiple Data* (SIMD), *Multiple-Instruction Multiple Data* (MIMD), RISC, and *very long instruction word* (VLIW) approaches in a clever hybrid solution. The result is a highly scalable, high-performance, low-power architecture that is very well suited for network processing.

An MTAP processor is able to contain an array of up to 2,048 *processing elements* (PEs). Each PE can execute several simple tasks in parallel and can therefore be roughly seen as the equivalent of a small VLIW engine. If the maximum number of PEs inside an MTAP sounds impressively large, it is. However, some basic characteristics of the PE structure enable the deep levels of integration that the company's IP can achieve when it is synthesized against various foundry technology libraries. More specifically, the data path of the PEs is 8 bits wide (as opposed to the typical case of 32- or 64-bit RISC cores). They only contain a small and efficient *arithmetic logic unit* (ALU), a register file, and local memory. If necessary, some of them can also offer special extension capabilities such as a hardware-based *multiplier-and-accumulator* (MAC) module used in DSP algorithm implementations. (This brings to mind the applicability of the technology in TDM-based voice applications such as voice coding and echo cancellation.)

Another significant characteristic of the PE structure facilitating large-scale integration is that PEs do not contain their own instruction fetch and decode units. Instead, the MTAP has a centralized

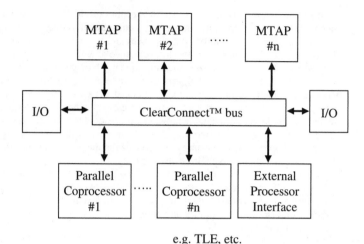

FIGURE 10.1 The architecture of the ClearSpeed IP technology. *(Source: ClearSpeed)*

control logic that according to the fundamental principle of SIMD architecture, fetches, decodes, and then issues one instruction, which is broadcast to all the PEs to execute on their own set of data. The MTAP processor assigns packets to the individual PEs. All the PEs inside the MTAP have to execute the same common instruction on their individual packets before the PEs are handed the following common instruction.

This approach has some positives and some negatives. On the positive side, the overall code is simpler as all PEs execute the same code. You do not have to worry about allocating code to the available computing resources and fine-tuning applications.

On the negative side, more resources seem to be wasted than with a more traditional network processor. This occurs especially when multiple protocols are executed at the same time. Some packets may require IPv4 processing, whereas others may require processing according to a different protocol, such as MPLS.

Code running sequentially on a classical network processor would first have to identify the type of protocol involved. It would then invoke the appropriate subroutines by conditional branching to handle it accordingly. However, in the approach taken by the ClearSpeed architecture, code is executed in parallel inside all the PEs of an MTAP processor and completely independently of what protocol is to be applied on the individual packets inside the PE. In this specific example, this means that both MPLS and IPv4 code will be executed in each PE, which wastes resources. However, you should not rush to conclusions for the following reasons:

- First, we will mention that at their presentation during the Embedded Processor Forum in June 2001, the company stated that their 400 MHz implementation, which was based on four MTAP cores that each contained 64 PEs, achieved 102.4 GIPS (102,400 MIPS). When combined in a die with 40 Gbps interfaces, for example, the ClearSpeed solution will still enjoy the astounding privilege of having 16 times as many MIPS per packet as the EZchip NP-1 network processor, even when EZchip NP-1 is only allowed to work in a 20 Gbps environment. This means that a lot of computational power can be "wasted" without even coming close to worrying about performance penalties.

- The individual PEs can nullify instructions that do not apply to their data context. Even more than that, they do not consume power while they are in that nullified state. This means for instance in the example just mentioned that the central instruction fetch/decode unit issues code pertaining to both

FIGURE 10.2   The *very large scale integration* (VLSI) layout of the basic building block for the PE array within the MTAP processor. *(Source: ClearSpeed)* Reprinted with permission.

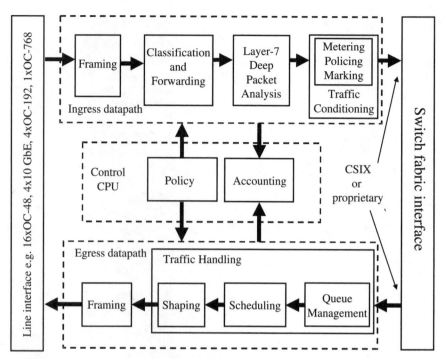

FIGURE 10.3 Architecture of a line card and based on ClearSpeed IP Technology. *(Source: ClearSpeed)*

IPv4 and MPLS and broadcasts this code to all PEs. However, if a specific PE is only dealing with, say, an MPLS packet, it will consistently nullify all instructions that it is handed that pertain to IPv4.

This approach seems extremely deterministic and efficient based on numerous simulations that the company has performed. For example, ClearSpeed has simulated a chip with four such MTAP processors performing simultaneous IPv4, IPv6, and MPLS protocol processing. It found that less than 30 percent of its available cycles was used for the actual packet processing. This discovery seems to justify the company's approach to solving the network-processing problem despite the fact that it obviously runs against the intuitive impression that this brute-force approach of throwing vast amounts of MIPS on the computational task at hand causes a waste of computational bandwidth.

- ClearSpeed claims that their deterministic software approach has particular benefits for network-processing software. If the worst-case performance guarantees are to be met, each path through "branchy" code must be proven to take no more cycles than the number available. Also, systemwide instruction fetch bandwidth must be guaranteed under all circumstances; otherwise, unnecessary packet drops may occur. In systems that have many units that can fetch instructions and that have branchy software following different paths on different cores, systemwide performance proof is next to impossible. A program on ClearSpeed's MTAP cores is essentially *straightline*, running the worst-case code on each core. Since every PE will now run that same code, instruction fetch bandwidth and instruction store are both massively reduced by more than an order of magnitude. This results in significant savings in power and area. Also, straightline code has predictable, deterministic performance, which provides obvious benefits to the user.

- Finally, ClearSpeed also claims that software can easily be written in a manner that minimizes the cost of running multiple code paths through every PE. For example, if code to process IPv4 and IPv6 packets is written separately, each code path takes about the same number of cycles to execute. Running them both on every PE would intuitively take twice as many cycles as running them on one code path. However, because much of the processing of *any* protocol is common to the processing of *all* protocols, a much more efficient code path can be written that performs both protocols simultaneously.

ClearSpeed claims that software that processes both protocols takes just 10 percent more cycles than software that processes a single protocol. The optimizations involved are all simple, common-sense transformations.

As mentioned previously, every PE must nullify specific instructions from the underlying common code that may not be applicable in its own context. This occurs through the following steps. Each PE has its own predicate stack. Instructions can be executed, such as conditionals, which push their result onto that stack. The current instruction will only affect this PE's state if all the bits in the predicate stack are true—in other words, all register and memory writes are gated by the OR-ing of the entire enable stack. This produces code that looks something like the following example, which is a parallel max function on 16-bit signed integers:

```
max:
if.gt      r_src1:p2s, r_src2:p2s    // 16-bit op, push results onto  enable
                                     // stack. 2 cycles
   mov     r_max:p2s, r_src2:p2s     // 16-bit op, only on those PEs where src2
                                     // > src1. 2 cycles
   otherwise                         // invert top bit of enable stack. 1 cycle
   mov     r_max:p2s, r_src1:p2s     // 16-bit op, only on those PEs where src1
                                     // >= src2. 2 cycles
endif                                // pop top bit from enable stack. 1 cycle
```

The **:p2s** suffix on the operands indicates they are poly (parallel) 2-byte signed. The code could consist of **:p1u** for poly 1-byte unsigned, **:p4s** for poly 4-byte signed, or **:m4u** for mono, or scalar, 4-byte unsigned. Mono variables are operated on in the MTAP's *thread sequence controller* (TSC), which is responsible for fetching and decoding instructions. As a result, it can actually execute real branches.

The sequence takes a total of eight cycles. Do not be misled by the **if.gt** instruction—it is not really a branch! It is simply the start of a new, nested level of predication. A PE's state will only be changed by the instructions in that basic block if all the conditions up to and including the most recent are true. Hence, by the time the **endif** is left, each PE has either written **src1** or **src2** to **max**, but not both. So this is just a straightline piece of code. It will always take eight cycles regardless of the conditions.

This technique is, of course, not new. It is quite common in several CPUs these days. Since branches are one of the biggest performance bottlenecks in modern microprocessors, many include predicated execution just like this for turning small branches into straightline code, which can then be executed in their wide, fast, multi-issue pipelines much more efficiently. ARM has had this for some time. It is also available in STMicro/Hitachi's SH5. The small difference here is that they have a stack of such enable bits and have multiple PEs using their different enable states to produce the effect of control flow but without the branches.

Incidentally, this could easily be microcoded into just one instruction—for example, **max**, which saves code space (4 bytes instead of 20). Also, some details of the architecture enable the **otherwise** to be performed at the same time as the first **mov** and the **endif** to be performed at the same time as the second **mov**. However, this can only occur when it is written in microcode. This means the microcode **max** will only take six cycles instead of eight. However, not everyone can or wants to code in microcode.

Compared to a small RISC core, a PE occupies on silicon about one-tenth of the area, offers about one-third of the computational horsepower of a RISC CPU, and consumes less than one-tenth of the

RISC's power. To give an idea of the PE capabilities, the company clarifies that a combination of 256 PEs clocked at 400 MHz offers the equivalent of 102,400 MIPS, 4MB of embedded memory, and 400 GBps of memory bandwidth. For those who like to think in terms of DSP MAC operations, they provide the capabilities of 100 billions of MACs/per second.

Figure 10.2 shows an image of a four-PE subpanel in a low-dielectric k constant, 0.13μm technology library from UMC with eight layers of copper metalization. The various logic blocks in the design use the lowest four metal layers, whereas the higher four blocks are used for overrouting—that is, for interblock stitching between the individual cores at the chip level. It is part of a good SOC design methodology. Its size is about 3 mm high by 0.5 mm wide. The outermost regular blocks are the PE memories, which are 4KB each in this example. The company uses them in their EV1 evaluation chip.

The next section down is the *programmed I/O* (PIO) logic. The PIO is a fairly complex, high-bandwidth *direct memory access* (DMA) engine per PE—hence the significant size. Below that section is another regular block. This represents the memory associated with the *stream I/O* (SIO), which is 128 bytes per PE for the EV1 chip. A smaller slice of logic appears before reaching the register files —one per PE, making up 64 1-byte registers. Each register file has five ports, so they end up being quite big. Finally, the main block of logic appears below the register file. This includes the ALU, the 8×8 to $16 + 48$ MAC, and the rest of the configuration.

It is extremely important to note that native hardware support exists for multithreading by the control unit, which is in charge of instruction fetching and decoding. The actual thread switching, which remains accessible under software control, can be triggered upon the occurrence of specific events, such as when an I/O operation has completed. I/O can be handled by two methods called SIO and PIO. The former is used for very-high-speed packet entry and acceptance directly into memory for subsequent processing. The latter is used when access is required to other coprocessors or memory. The number and type of these I/O channels can be configured by the user of the company's IP core architecture.

To facilitate the design of complicated SOCs based on the IP core architecture, ClearSpeed has developed an on-chip, high-speed, modular interconnect bus called ClearConnect™. It is a point-to-point link based on distributed arbitration. It is structured in segments that connect different SOC components to the bus. Each segment behaves like a local bus between the corresponding nodes. These links can be scalably structured with up to four lanes of bidirectional traffic where each lane provides up to 6.25 GBps of bandwidth for an aggregate bandwidth per link of 50 GBps between any attached nodes. The segmentation of the ClearConnect bus means that multiple transfers can take place simultaneously between unrelated nodes on the bus. In addition, ClearConnect uses standard *Virtual Component Interfaces* (VCIs) (as specified by the VSI Alliance) for the easy integration of third-party cores and other coprocessor or components on the same SOC design. ClearConnect is delivered in synthesizable RTL. It fits perfectly into any standard ASIC design flow and interfaces easily with place and route tools.

In addition to the embedded MTAP processors that share access on the ClearConnect embedded bus, the standard architecture that ClearSpeed proposes also provides for the potential presence of a series of parallel coprocessors (also known as *accelerators*) that can be either among those designed by the company or user or that can be licensed from a third party. ClearSpeed offers a series of IP cores that may be interesting to customers for the integration of a complete design. It offers among others accelerators for tree-search functions as well as for queue and state management.

However, the most prominent of these designs is a powerful *Table Lookup Engine* (TLE), which was designed for situations where lookup capabilities are needed for more than 300 million lookups per second. In a reference design, by embedding 24 lookup engines in the TLE and multiple banks of compiled SRAMs from third parties, ClearSpeed managed to attain an impressive performance of 350 million lookups per second while clocking at 400 MHz.

The TLE (which is further discussed in Chapter 12, "Search Engines") can be configured to work with internal SRAM or DRAM depending on the capabilities of the targeted semiconductor process. At the same time, support for external DDR SRAM or DRAM enables the creation of systems that match performance, table size, and key length requirements with actual budgeted design costs. As the design has been largely optimized for tree walking, multiple parallel *level compressed* (LC) trie search engines operating simultaneously provide results out of order because this increases the overall efficiency of the TLE.

ClearSpeed clarifies that its MTAP architecture will restore the order of operations automatically without any special buffering. The TLE can support tables with over 2 million entries at application wire speeds requiring 350 million lookups per second, and can use variable size keys from 32 to over 128 bits. It also has significant advantages as opposed to the traditional use of external CAM.

A *global semaphore unit*, which is usually unique in one SOC design, coordinates synchronization and communication between the multiple cores. Any major core has its own collection of private semaphores to which only it has access. The MTAPs have these semaphores to coordinate chores such as signaling when a memory transfer has finished.

The scalability of the technology stems from the fact that the architect-designer of a network-processing superchip using ClearSpeed IP can configure his or her design by judiciously playing with the following parameters in a five-dimensional space:

- The number of embedded MTAP processors in the chip.
- The number of PEs per MTAP.
- The amount of cache memory and instruction memory per MTAP.
- The number of lookup engines per TLE.
- The amount of table memory available per TLE.

In the implementation of a reference design of a classification engine, ClearSpeed has used 4 MTAP processors, which each have 64 packet processing engineers, a TLE embedding 24 lookup engines, and 1MB of embedded memory for the TLE. Such a device is capable of classification and forwarding in protocol environments such as IPv4/v6 and MPLS (*label-switched router* [LSR] and *label edge router* [LER]) sustaining a performance of more than 100 Mpps. If the reader consults a typical traffic-correspondence table such as the one shown in Appendix II, this translates to a simplex OC-768 link with 40-byte packets. The idea is that by replicating this device, a unit can be created that can condition the traffic by performing policing and metering, among other tasks.

Traffic management is a very important systems design issue, especially in realms of 40 Gbps and beyond. ClearSpeed presented a preliminary design of a programmable chip for multiple traffic management tasks and algorithms at the Network Processor Conference in October 2001. This traffic manager can work at either the ingress or the egress path. It can handle congestion avoidance and scheduling as well as run statistics in the background. All these algorithms run in software on the MTAP cores so simply altering the software may enable proprietary versions of the algorithms to be run.

The company has already proven the concept of its architecture by building an actual piece of silicon on which it integrated: a single MTAP core containing 1,536 PEs, 3MB of embedded DRAM, structures that provide 600 GBps of on-chip bandwidth, and computational power that amounts to 1.5 Teraops of integer performance and 3 Gigaflops (floating-point performance). All this was coupled with four Rambus™ channels that offered a bandwidth of 6.4 GBps in communications with off-chip devices.

ClearSpeed manufactured this proof-of-concept chip using a standard but now quite obsolete 0.25μm CMOS process from UMC and packaged the chip with roughly 1,000 pins. This is an impressive set of numbers, and it deserves the appropriate level of attention from the industry.

ClearSpeed is offering an elaborate *Software Development Kit* (SDK) for the development of complete applications. The SDK, which runs on standard platforms like Linux, Solaris, and Windows 2000, is comprised of the following:

- An ANSI-compatible optimizing C compiler along with a few extensions that allow the programming of the parallel features available in the MTAP architecture.
- An assembler.

- A linker along with a set of source- and object-code libraries, including standard functions and *application programming interfaces* (APIs).
- A debugger.
- A profiler.
- A microcode compiler.
- A full-fledged simulator and associated simulation tools for rapid prototyping, including models of the associated hardware IP cores.
- An *Applications Development Kit* (ADK), which contains a traffic generator, optimized libraries of key functions, reference implementations, and test code, thereby accelerating overall development time.
- A *Hardware Development Kit* (HDK), including tools that allow silicon configuration and design verification as well as operating system and drivers.

Helping promote the parallel development of hardware and software, ClearSpeed's integrated development environment enables users to first develop their code using the *Virtual Instruction Machine* (VIM). An application is initially debugged in terms of functionality before it can be compiled on the final underlying machine language. The SDK profiler helps identify what types of instructions are used most often and which parts of the programs actually consume the most resources, so that users can fine-tune their application by modifying the C-language source or by writing some inline assembly code, if necessary.

The Virtual Machine Simulator facilitates the improvement of application performance until the actual underlying hardware design, which evolves in parallel with the development of the software, arrives at a level of progress where the target instruction set has been finalized. ClearSpeed calls this the *Implementation-Specific Instruction Set* (ISIS). Once both the final instruction set and the application have been finalized, the application just needs to be recompiled against the target ISIS. The linked code is then executed on simulation models of the actual hardware, where performance measurements can be taken and instruction profiling can be performed. Finally, the application can be refined and fine-tuned before it is executed on the actual target hardware.

An interesting characteristic of the company's technology is that the user can create his or her own custom instructions. During development, the code compiler at configuration time reads the encoded instruction set from a special file, where the user has previously described the exact operations that each instruction is expected to perform, how these operations are to be done, and which computational resources from the system (ALU, registers, and so on) are involved. Through this straightforward process, the user can describe altogether new custom instructions, which should be expected to positively impact the performance of the contemplated application code. The compiler then will naturally choose the more appropriate instructions when generating code.

To further clarify the overall systems engineering context, we must point out that the generated code is microcoded. Understanding why this is so, is straightforward. PEs are CPUs that are 8 bits wide, but it may be that a new custom instruction revolves around a 16- or 32-bit operation. By microcoding everything in terms of available 8-bit operations, ClearSpeed allows the implementation of essentially anything. If you want to add two 32-bit numbers, depending on the exact addition algorithm's use of carry, you will need four 8-bit operations. As each native 8-bit operation is executed in one clock cycle, the number of cycles required to execute a custom microcoded instruction will depend on the actual operations involved. Our 32-bit addition example will take four cycles.

When an instruction is issued for execution, it is looked up in a special table that shows the steps of how to implement it in 8-bit PE operations. In this context, the lookup table is the actual microcode. For all practical purposes, one application may require different microcode than another. Therefore, microcode is loaded at run time from external memory (ideally at boot time) along with the actual application code to be run. In fact, the microcode space can be booted partially or completely, thereby affording an extra degree of flexibility around systems engineering.

We should briefly pause and compare ClearSpeed's approach of customizing the MTAP instruction set to the ones taken by Tensilica's configurable Xtensa™ CPU or even by ARC. The definition of a new instruction usually entails the (automatic) creation of a significant number of extra logic gates (increasing the size of the underlying hardware core), but allowing the design to stay more predictably close to the one-cycle execution rate objective of native instructions. These two latter examples are not in the same MIPS league as ClearSpeed's MTAP, which outperforms both of them by orders of magnitude. However, we are referring to customizing instruction sets in order to optimize application performance. We consider architecture/performance trade-offs involved in designing a system with various approaches.

We conclude our discussion on ClearSpeed's technology by saying that with all the computational power of its technology, it is not a coincidence that the company has pushed the emphasis of original applications on core networks that require 40 Gbps performance, but do not necessarily need intricate deep packet processing. As the technology can be scaled down rather easily, users will most likely come forward with designs that implement in-house-designed network-processing chips performing more elaborate tasks outside the core and at the edge level. Indeed, much of ClearSpeed's initial customer interest has been at lower line rates, from 2.5 Gbps to 10 Gbps, but with high levels of functionality—what the company affectionately calls the *high touch*. Other computationally heavy applications (from the network-processing arena) besides wire-speed classification/routing and *quality of service* (QoS)-based traffic management will also most likely emerge soon. We will examine some of these applications later in this book.

TENSILICA

The other side of the IP licensing spectrum, as applied to the network-processing realm, has a couple of promising IP companies. Tensilica (*www.tensilica.com*) apparently has the most significant technology proposition. Because the company has created a new paradigm of the design flow, we will discuss the actual look and feel of designing a configurable processor CPU with this technological approach.

Although other companies such as MIPS and ARM historically preceded Tensilica in the area of licensing RISC CPU IP cores, Tensilica along with Improv Systems can be considered pioneers of the idea of configurable processors. Although Improv Systems used the embedded VLIW approach with a tightly controlled toolset, Tensilica's current and prior products have worked on the RISC model while enabling customers to automatically generate their own customized tools. The company, however, recently unveiled *Flexible-Length Instruction Extension* (FLIX)—its new VLIW architecture, which was developed in partnership with a major semiconductor manufacturer. The new architecture can be configured to provide an optimal match to the application workload, thus making efficient use of all the processor's resources.

Returning to the origins of its configurable processor approach, Tensilica realized that in many designs users

- Actually need to be able to customize their CPU.
- Want to eliminate functionality that they do not need.
- Desire to change functionality (in many cases, altogether) to suit their own application needs.
- Want to add custom capabilities that would improve the performance of their CPU choice.
- Want to replace traditional hardware design functions (such as complex *finite-state machines* [FSMs], packet-processing functions, and *Transmission Control Protocol* [TCP] offload engines) with the flexibility of a software-programming model that only a programmable processor can provide. The company's recently patented technology is based on Xtensa, an extremely flexible CPU core, and a suite of associated tools that allow the generation of the configuration files that enable the company to generate customized development tools for its users.[1]

---

1. Tom R. Halfhill and Rich Belgard, "Tensilica Patents Raise Eyebrows: Legal Protection of Configurable-CPU Technology Could Frustrate Competitors," *Microprocessor Report* (December 9, 2002). This is also available online by subscription at *www.mdronline.com/mpr/h/2002/1209/164901.html*.

The basic Xtensa V (already in its fifth generation as of this writing) core is a fully configurable 32-bit RISC core that delivers above 420 MIPS, typically clocked at 350 MHz.[2] In the worst-case scenario, it is implemented in a 0.13μm line-width CMOS technology. It occupies only a small area (~0.3mm^2) in silicon real-estate—something extremely important when it is contemplated as part of a larger design. It can be ideally suited for low-power designs (<0.1 mW/MHz) when synthesized on typical 0.13μm CMOS technology libraries.

The Xtensa processor core is an implementation of a five-stage (or more) pipeline, as shown in Figure 10.5, which shows the involvement of different pieces of CPU hardware at each stage. More specifically, it shows the following:

- First, an instruction is fetched from the instruction cache.
- The instruction is then decoded and contents of needed registers are read.
- The ALU executes operations such as effective address generation and other operations as specified by the instruction opcode.
- Memory is then accessed for reference or a branch is taken.
- Results are written back into the register file.

The company's *processor generator* is an intuitive browser-like *graphical user interface* (GUI) tool that enables the user to enter the configuration details of the processor that he or she designs. We should clarify what we mean by "the generation of customized development tools" and show what an impressive feat this is. When the user has defined custom instructions or extensions (such as special multipliers, *cyclic redundancy checks* [CRCs], checksums, packet header checks, or DSP-needed blocks such as single or dual MACs) to add on to the licensed core technology, he or she securely submits to Tensilica through the company's web site the configuration files that the processor generator produces. Within an hour or so, the company's tools will generate a completed set of customized development tools that the user can download.

With the arrival of the company's fifth-generation technology in the fall of 2002, several important enhancements were made:

- With the intention to maximize the usable I/O bandwidth and to improve the communications between multiple embedded processors in an SOC, Tensilica enhanced the core processor's *Xtensa Local Memory Interface* (XLMI), which now allows multicycle devices to be attached with variable latency.
- A convenient incoming request feature for the Xtensa *Processor Interface* (PIF) now enables an Xtensa CPU to simultaneously execute instructions and handle read/writes to the processor's local data memory. This can be useful for some external functional modules in an SOC (such as DMA engines) that need to get in touch with a specific processor or, most importantly, for other tightly coupled processors. With configurable interface widths up to 128 bits, the Xtensa processor can deliver a peak I/O bandwidth of 45 Gbps.
- The addition of a processor ID register to the *instruction set architecture* (ISA) can identify each unique processor integrated on an SOC. This eases system software development when an overlay application must be broken down to pieces that need to be allocated to specific processors. It can

---

2. According to the company, this greater than 400 MIPS number is derived from a Dhrystone V2.1 benchmark. For the Dhrystone benchmark, with no in-line code or file-merging activities, Xtensa V achieves 1.2 MIPS/MHz. With optimizations (in-lining and file merging), Xtensa V has been reported to achieve an impressive 2 MIPS/MHz or over 700 MIPS if the core is clocked at 350 MHz. It should also be kept in mind that MIPS is not a good metric for a configurable processor for obvious reasons. For example, one single *Tensilica Instruction Extension* (TIE) instruction (using Tensilica's architectural extension definition language) can perform the equivalent of several instructions. One concludes that a single Xtensa TIE instruction produces work at a higher performance level than a single instruction in a standard 32-bit fixed RISC processor.

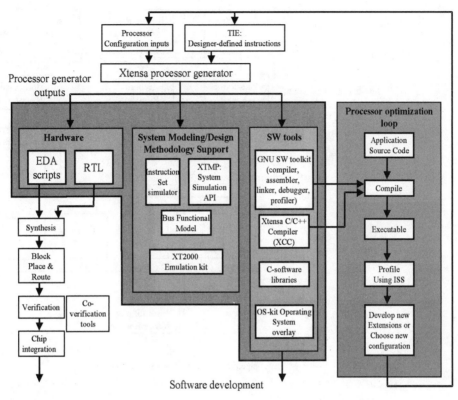

FIGURE 10.4   Parallel rapid development of hardware and software using the Tensilica approach of configurable processor cores. *(Source: Tensilica)*

also impact the possibilities of large-scale SOC integration for natively parallel applications that are based on multiple copies of the same configuration of the Xtensa processor, as each processor can be now uniquely identified while it communicates with other fellow processors.

• The company has also implemented designer-defined conditional load and store instructions. This has significant value in deep packet classification tasks, which are so often executed in network processing. When carefully used, it can result in programming that contains far fewer branch instructions. As a result, the executable code will have better performance.

Figure 10.4 summarizes this approach. The figure resembles a typical integrated circuit design flow except with two major differences: the underlying hardware and the instruction set of the embedded code can be changed in order to optimize performance, and the actual software development tools are automatically modified to reflect the latest changes, so they can match the development requirements and context perfectly.

The toolset is made up of the following:

• A standalone tailor-made GNU C/C++ compiler.

• An assembler/disassembler.

• A linker.

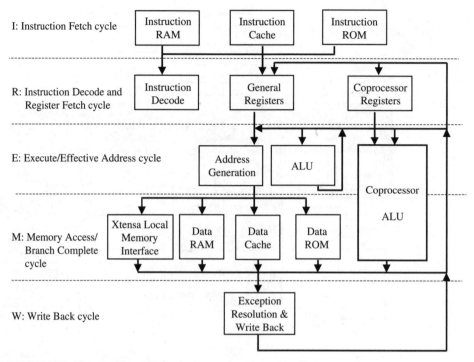

FIGURE 10.5 Five-stage Xtensa pipeline implementation. *(Source: Tensilica)*

- A debugger.
- A cycle-accurate instruction set simulator.
- An advanced code profiler as shown later in Figure 10.10 that allows the fine-comb scanning of the application at hand looking for oversolicited resources, potential conflicts, bottlenecks of perform-ance, and so on.

If the user's initial software analysis shows some areas of poor performance, especially in con-junction with the underlying architecture, some hardware resources (MAC, multipliers, registers, ALUs, comparators, and so on) or more specialized instructions may need to be added. If optional hardware additions must be made, the company allows the configuration of an instruction and/or data cache, a memory interface, interrupt control mechanisms, timers, and the size and count of registers. Most importantly, it allows the potential insertion of custom units that the company calls generically *designer-defined execution units*.

These execution units can be blocks such as a floating-point unit or even a full-fledged, very pow-erful, customized-width DSP engine that can even have multiple MACs for extremely fast DSP pro-cessing. In the case of instruction extensions, the configuration of the data path must be reiterated using the company's processor generator tool. Instruction set extensions (a feature that network-processing system designers using this platform seriously need to engage) are easily coded in what Tensilica calls *Tensilica Instruction Extension* (TIE) language. This is a Verilog-like language that describes the desired instruction mnemonics, operands, encoding, and semantics into what the company calls *TIE files*. TIE files serve as inputs to the processor generator tool.

This takes a matter of minutes if the user knows architecturally what he or she wants to achieve and not more than a few hours if the user must think through the architecture and experiment first. Once this done, the user uploads the newly produced configuration tools to Tensilica, who will generate a new set of development tools for the user.

The flexibility afforded toward configuring a CPU data path is revolutionizing the industry. It is no wonder that Xtensa cores have been chosen by several network-processing chip designers to be part of larger in-house created designs. These designers include companies such as Bay Microsystems, which uses the Xtensa core in the exception/control plane of the Montego™ *Internetworking Processor* (InP), and others such as Transwitch for its T3BwP (bandwidth processor), Onex for its Omni Service Processor, Trebia for its Storage Network Processor, Marvell for its NetGx coprocessor, and NEC[3] for its *Wideband Code Division Multiple Access* (W-CDMA) network infrastructure chip.

Depending on the exact function of the development tools, they not only produce executable code for and work directly with the new customized instruction set, but they also reflect the underlying design configuration resources, integration, and use that the user has stipulated. Tensilica's patented design database is an integrated repository for all pertinent information. It facilitates the parallel development of hardware and software. At the same time, the company has developed patented technology that allows the compression of code instructions in less than 32-bit words (decompressing them on-the-fly during operations), thereby optimizing the memory footprint of embedded implementations.

Up to now, we have described what happens in the software development process. For the hardware development process and depending on the actual hardware choices and performance constraints that are imposed (regarding power, speed, and size) on the design during the interactive processor generator session, the company's generator tool will also automatically generate the appropriate hardware tree of the newly configured core in synthesizable hardware description language (RTL). It also provides *Electronic Design Automation* (EDA) scripts for the subsequent synthesis step, the necessary verification suite, and a *bus-functional model* (BFM) to interface with the *instruction set simulator* (ISS) and other standard ASIC design tools for synthesis, functional, and timing verification. The processor generator GUI also provides an impressive set of dynamically changing colored bars that show in real time the impact and cross-influence between a user's architectural decisions and the underlying clock frequency (in MHz), the logic-gate count (number of gates), the silicon area (in mm^2), and the estimated core-power dissipation (in mW). If the architect knows what the power or space budget is for the corresponding system design resources, he or she can easily readapt his or her thoughts and ideas in a series of iterations that ultimately lead through balanced compromises and trade-offs to the satisfaction of the design requirements at hand.

The toolset is completed with a real-time operating system overlay that works with a hardware-abstraction layer on the custom-configured core processor data path. This layer also natively supports ATI's Nucleus PLUS™ or Tornado™ for VxWorks from WindRiver Systems. The company also offers a prototyping development system based on a board that uses either Altera *custom-programmable logic device* (CPLD) technology or Virtex II platform *field-programmable gate array* (FPGA) technology, which can be used for processor emulation and early software development for some types of applications. Customers configure the processor and download from Tensilica's servers generated tools for the emulated testing of the design on the CPLD.

Last but not least, it is worth mentioning that Tensilica and CoWare (*www.coware.com*) have been working very closely on a multiyear commitment, whereby the Xtensa V processors in configurations using multiple cores and peripherals along with multiple memory blocks are integrated into CoWare's

---

3. NEC engineers not only configured the Xtensa core, but they also designed 20 new powerful bit-handling instructions for ATM timer control and data queue manipulation in this ATM-centric communications chip by using the TIE language. ATM is used for the communications among base station nodes, radio network controllers, mobile services switching centers, and gateways to the *Public Switched Telephone Network* (PSTN).

N2C™ (the abbreviation of napkin-to-chip™) platform. CoWare enables C-based design, simulation, and analysis. Therefore, it facilitates parallel hardware and software design and coverification instead of the traditional hardware and software partitioning of the problem. Interested readers can obtain more information from each company.

Tensilica states in its product literature that in IP forwarding/routing, the addition of a few well-thought-out instruction extensions on its base instruction set and about 6,000 gates of extra logic on the fundamental core, which is usually a little more than about 100,000 gates, enables the achievement of around 12 times the performance of a typical 32-bit RISC equivalent. This is important, and it argues in favor of the company's technology as opposed the technology proposed by its few direct competitors. However, it does not allow the multigigabit handling of real-time traffic, which requires deep packet inspection, classification and modification in conjunction with traffic management, flow control, scheduling, and so on. It only allows this if a large number of multiple similar RISC engines are integrated.

We will discuss benchmarking network-processing applications later in the book. At this point, we will only mention some rudimentary benchmarking efforts coming from the *Embedded Multi-processor Benchmarking Consortium* (EEMBC) forum. This forum was originally created to objectively measure and rate standard CPUs. However, standard computing applications such as word processing, database querying, spreadsheet calculations, and graphics rendering have a completely different temporal statistics and spatial structure where caching works miracles. As a result, the traditional computing architectures and platforms, which have become the bulwark of mainstream computing, are simply not capable of handling the multiple facets of complex network-processing applications running on live packetized networks at wire speed.

At the same time, however, do not discard the fact the industry has been struggling conscientiously to address this need. Tests like the EEMBC benchmarks are a good first effort to solve the problem. They can also be found useful for evaluating and comparing the control plane. However, more work is needed to develop representative, universally accepted, and useful test suites.

The EEMBC Networking benchmark suite is based on applications that are drawn from the networking reality and that have significantly different characteristics than consumer or IT applications. They usually involve less arithmetic computation, generally show less low-level data parallelism, and frequently require rapid control flow decisions. The EEMBC Networking benchmark suite contains representative code for routing and analyzing packets. Figures 10.6 to 10.9 show some interesting results obtained by executing this code on multiple processors.

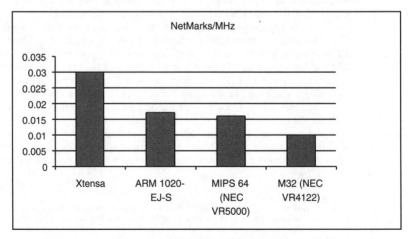

FIGURE 10.6 A comparison of EEMBC NetMarks/MHz of out-of-box scores for Xtensa 350 and several other architectures. *(Source: Tensilica)*

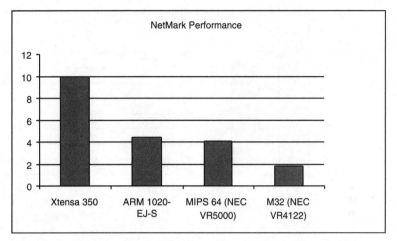

FIGURE 10.7   The comparative results of Figure 10-6 are only further exacerbated if the impact of the higher clock frequency now used in the Xtensa V pipeline is considered. The results shown here are in absolute terms. *(Source: Tensilica)*

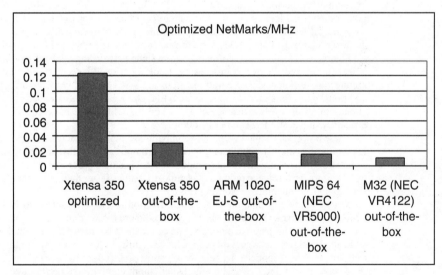

FIGURE 10.8   The same EEMBC benchmark shown in Figures 10-6 and 10-7 but with optimization of the Xtensa architecture for some networking applications. *(Source: Tensilica)*

More specifically, we compare EEMBC NetMarks/MHz of out-of-box scores for Xtensa and several other architectures, where the IDT 32334 (MIPS32) at 100 MHz has a performance reference of 1.0. Out-of-box means as shipped by the vendor and without any customer-performed architecture optimization. The results shown in Figure 10.6 indicate that Xtensa, even without any networking-specific extensions, consistently has twice the performance of some major alternative 64-bit RISC and

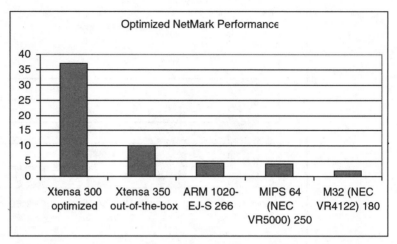

FIGURE 10.9 The performance increase that is obtained from properly configuring the Xtensa to suit the needs of the networking application at hand. *(Source: Tensilica)*

three times the performance of 32-bit RISC architectures.[4] Out-of-the-box testing is good because it gives a first good feeling about a basic architecture as well as the quality of the compiler. This performance difference is further magnified by the clock frequency advantages of the Xtensa pipeline, as shown in the absolute NetMark performance, which is shown in Figure 10.7.

Figure 10.8 shows the performance of the same networking applications, but this time it includes Xtensa optimized for packet processing. Looking at results per MHz provides a better idea of the architectural efficiency. These optimizations are small but highly effective, adding less than 14,000 additional gates (less than 0.2 mm² in area) to the processor. The extended Xtensa processor achieves about 7 times the cycle efficiency of a good 64-bit RISC processor core and more than 12 times the efficiency of a 32-bit RISC processor core.

These processors achieve generally comparable clock frequencies, though the NEC4122 (MIPS32) lags somewhat slightly behind, giving the overall optimized NetMark performance increase shown in Figure 10.9. The net result of these modifications is a new processor, which by its proper configuration attains a performance rating that is almost 10 times faster than other popular 64-bit RISC processors on high-throughput networking tasks.

More importantly than the exact quantification of any relevant performance improvement, the EEMBC benchmark results have been presented more for their qualitative conclusion. In other words, looking at these comparative numbers, one cannot help but notice the undeniable evidence that extensible and configurable processors can achieve significant improvements in throughput across a wide range of embedded applications, relative to good 32- and 64-bit RISC, DSP, and media processor cores.

Also keep in mind that results published about comparative performance between IP cores are based upon a simulated chip. This is because it would otherwise be prohibitively expensive for IP companies to design and build a custom chip just to compare their performance with an off-the-shelf processor. Also make sure that the appropriate clock frequencies, semiconductor process technology

---

4. In addition to checking out the details at the EEMBC web site at *www.eembc.org* for all the results that we discuss in this chapter and that have been independently certified by *EEMBC Certification Laboratories* (ECL), an interesting article was written by Michael Santarini called "Tensilica Aces Benchmarks, Actel Shoots the Moon," *EE Times* (September 16, 2002). It is also available online at *www.eedesign.com/story/OEG20020916S0023*.

library, and even power consumption factors are judged fairly. If they are not, a very erroneous set of conclusions can be reached. In other words, if core A implemented in 0.13μm CMOS library matches the performance of core B when it is implemented in an 0.18μm library, you cannot just brush the underlying silicon technology issue aside and state shamelessly that the two cores perform identically.

In a similar example with different parameters, if you compare a 1 GHz off-the-shelf processor X with a 200 MHz IP core Y and state that the former wins by a factor 5 in throughput, it may not be a

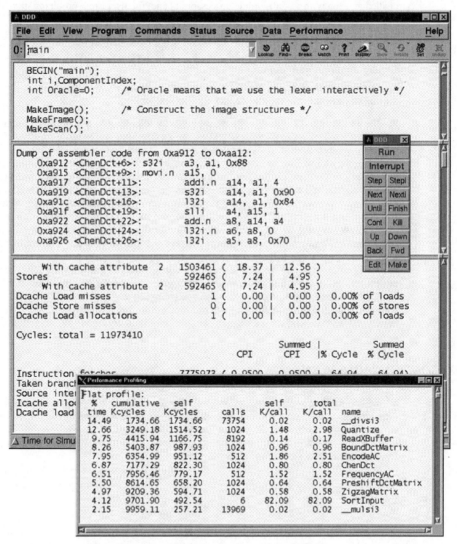

FIGURE 10.10 A performance analysis of custom-written code is done with Tensilica's profiler, which allows the detection of bottlenecks and the generation of statistics as to which subroutines, function calls, and operations occur during most of the time. This allows the definition of new instructions when and where needed, which will simply imply one more iteration in the cycle. *(Source: Tensilica)*

surprise. However, if Y does this at 17 times the power consumption of X, then the argument can be made that the winner has not really . . . won.

Let us look at a real-life example that further corroborates the context and argument. IBM Microelectronics is working with a nondisclosed (as of the writing of this chapter) NEV to integrate the nontrivial quantity of 174 Xtensa RISC cores into one chip.[5] In this case, IBM worked hard to trim down the individual core's gate count to around 92,000 per embedded processor in order to be able to fit the complete design in a die of 18 mm×18 mm using IBM's advanced thin-line lithography and copper-metal Cu-11 process, and to accommodate the staggering number of gates.

Returning to the technical considerations of a systems architect coping with a network-processing challenge, even with a configurable processor at hand, the list of real problems starts looking like the following:

- Deciding upon the memory structure of the overall system and figuring out which processor has access, when it has access, through which bus and mechanism, to which memory subsystem, and under which circumstances.

- Deciding how the processors communicate with each other and how they share access to resources using some scheme of arbitration and conflict resolution.

The list becomes elaborate. For now, we just want to give an idea of the task's magnitude. It should be rather obvious that this overall context creates an absolutely formidable computational "beast" that not many organizations really know how to formally tame—either from the hardware design side, or assuming they know how to logically partition the code for each processor (thereby allocating tasks at hand), from the mere challenge of tackling multiprocessor scheduling, coordinating execution and memory access, and even simply balancing the workload among the engines while respecting duplex multi-gigabit-per-second wire-speed I/O. This is where serious trade-offs will need to be considered by the systems architect and the true pros and cons of such a hyper-complex design become apparent.

Tensilica has publicly shared its noble vision of the computational future of its trademarked concept of Sea of Processors™. This concept portrays an SOC world to come where hyper-sophisticated integrated design tools will automatically map a customer's application code onto a large series of optimally configured and embedded processor cores, which in unison with each other will be able to perform the tasks as desired and thereby satisfy a system's application requirements.[6]

Although spectacular progress has been accomplished in computer architectures as well as in software and integrated-circuit design tools and methodologies during the last two decades, we are not there yet. However, you should retain a clear sense of industry trend from this short overview.

FLIX: CONFIGURABLE VLIW

We will conclude our short discussion of Tensilica's configurable processor technology as an interesting means to develop a specialized high-performance, network-processing ASIC or SOC by mentioning Tensilica's important recent introduction of the FLIX architecture, which embraces configurable VLIW principles.[7] For obvious reasons, we will examine the importance of this trend from a network-processing standpoint.

---

5. Anthony Cataldo, "Reconfigurable Processors Make Move into Big Time," *EE Times* (May 24, 2001). This is accessible *www.eetimes.com/story/OEG20010324S0001*. The same story is also mentioned in another article by the same author called "Comms Warm a Bit to Reconfigurable Processor," *EE Times* (March 23, 2001). This is accessible at *www.eetimes.com/story/OEG20010323S0071*.

6. See *www.tensilica.com/press/Tensilica_press_20011017_mprvision.html*.

7. The introduction was made on October 16, 2002 at the Microprocessor Forum conference.

Complex network-processing tasks, especially since they must be executed at wire speed, amount to extremely heavy computational loads that ordinary architectures cannot handle. With more data and applications to deal with per unit time, more work needs to be accomplished in the same short amount of time. The following are two fundamental ways of going about doing this:

- Increase the frequency clock when possible and force the hardware to complete more operations per second.
- Deploy a sense of parallelism into a design.

With dramatically shrinking lithography line widths and with IP core reuse methodologies proliferating by the need to meet shorter times to market, integrated systems become more complex by the massive piling up of multiple subsystems on one and the same SOC. This context makes the choice of increasing the frequency of the fundamental clock unacceptable as it drastically increases the chip's power consumption, which creates package choice (and therefore cost) and system cooling issues that may be difficult to confront given chassis-based power-consumption budgets and constraints. In order to cope with the increasing computational load, the network-processing architect has to match the need for parallel architectures. This has been corroborated by the creative approaches taken by the designers of many commercial off-the-shelf NPU chips.

Now parallelism in computer architecture does not just stand for one approach. For instance, a designer can deploy multiple cores inside an SOC and divide the work (when appropriate and feasible) to these resources. However, he or she will have to contend with managing access and resolving conflicts by some sense of arbitration that instead of resolving complexity, he or she simply shifts the design challenges from one hard issue to another equally difficult one.

Alternatively, a designer could consider engaging a wider data path on a CISC/RISC architecture platform and expect to accomplish more work per time unit. A 64-bit processor is expected (at least by some people) to perform more useful work than a 32-bit processor. However, this is not always true. Not all applications can benefit from longer word arithmetic or data transfers. A designer can also deploy superscalar architectures to tackle this design problem. However, if such an architecture is based on an extensible and configurable architecture like the Xtensa processor's, it will end up being a nightmare for the designer to manage all possible interdependency issues that can arise between custom instruction set extensions and the basic architecture itself. Tensilica architects have decided to follow a different path—the VLIW approach. Other companies such as Improv Systems whose approach has been marked by a history of distinctly less aggressive marketing toward configurability by customers are discussed in a later section in this chapter.

Tensilica's FLIX approach offers straightforward and easy configurability to the wide instruction community of custom ASIC designers who use embedded processors and who want to accomplish more work in a given amount of time. A new 64-bit long instruction format provides parallel access to multiple execution modules, which could be store-and-load units, ALUs, MACs, barrel shifters, and so on. By keeping the lowest 4 bits of all instruction words as the indication of the instruction length, FLIX allows the seamless mixing of 16-, or 24-, or 64-bit long instructions without a problem and with the possibility of aligning them at byte boundaries. It also guarantees the compatibility of preexistent Xtensa code with the new architecture. A designer can do the following tasks with such a flexible approach:

- Simplify the decoding of instructions based on a more rational instruction field allocation.
- Optimize the memory footprint especially if multiple streams of instruction sequences (threads) must be executed in parallel cycle by cycle in need of data from different areas of the addressable memory space.
- Save silicon space on a custom design by using a consolidated instruction sequencer.
- Take advantage of the possible and deterministic coordination between various on-chip modules.
- Adjust localized power management by software executing in real time.

The most significant advantage for a network-processing designer, however, is that this new architecture can simultaneously handle several instruction sequences (also known as *threads*) in parallel.

This has been one of the weakest spots as far as network processing is concerned in the original configurable architecture approach with which Tensilica started. The company is now addressing this issue.

A note of caution: It will be interesting to follow the arrival of the actual products and tools enabling the wide-scale acceptance of the configurable VLIW (FLIX approach). It will be especially interesting to see how code compatibility can be preserved between customer-extended Tensilica legacy instruction sets and the new parallelized technology.

ARC CORES

ARC (*www.arccores.com*) is a British IP core technology company that offers a configurable, extendable, and synthesizable 32-bit RISC architecture based on a CPU platform. called ARCtangent™. The heart of the ARCtangent technology is the A-5 32-bit RISC processor, which is based on a four-stage pipeline and implements the company's ARCompact™ orthogonal instruction set (meaning that all addressing modes and therefore all registers are accessible to all instructions). ARCompact combines a mixture of 16- and 32-bit instructions and intends to minimize instruction in the memory footprint. A core register file of thirty-two 32-bit registers can be doubled or extended with extra registers if desired.

The company's core technology is a little less configurable than Tensilica's, and its development tools do not exhibit the same possibility of customized generation based on the user-implemented extensions.[8] Nevertheless, the technology has been commercially accepted because of its simple and clear-cut approach and what appears to be extremely reasonable licensing terms.

In order to take advantage of its technological configurability, ARC offers a GUI-based configuration tool that enables a user to decide all the features and characteristics of the CPU. The user could decide to do things such as creating and adding extra instructions for specialized repetitive operations, customizing the cache configuration, or reconfiguring the interrupt-handling priorities and vectoring mechanisms. The user could also decide to use a Harvard-bus configuration (separate and parallel-running instruction and data buses accessing different memory banks for program code and data respectively) as opposed to a von Neumann structure that has one common shared bus for instructions and data. Once a processor is designed with customized extensions or options, the tool will generate the appropriate RTL code files.

The company also provides a series of ready extensions such as customized MAC instructions as well as an array of peripheral IP cores to help facilitate an SOC design. It also offers a complete series of development tools, high-level language compilers, simulators, and debuggers that facilitate and accelerate a systems design based on the company's technology.

The technology is very flexible, but for fast network-processing applications it suffers from the same generic weaknesses that simple RISC architectures exhibit across the board. In other words, a designer must do the following:

• Deploy a large number of multiple cores to share the load.

• Decide how to schedule work on each core.

• Sort out how to coordinate the cores on tasks that make part of a larger piece of work the chip must perform.

• Decide how to allow the multiple embedded cores to communicate among themselves and with a supervising host CPU.

---

8. See several pertinent articles published in the *Microprocessor Report*. A good example is Tom R. Halfhill, "Tensilica Xtensa V Hits 350 MHz" (September 16, 2002). This is available online to paid subscribers at *www.mdronline.com/mpr/h/2002/0916/163701.html*. This article discusses comparative results between Tensilica and ARC processor cores based on ECL-certified results of the EEMBC benchmark suites.

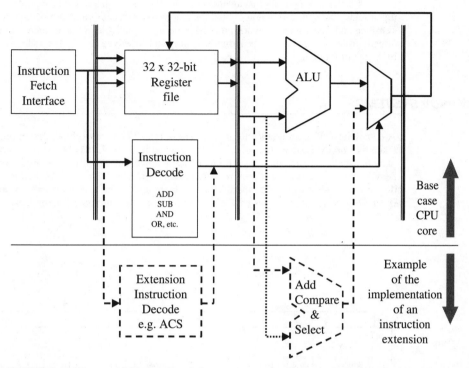

FIGURE 10.11 The example illustrated here is for a new instruction called here ACS, which adds some operands, compares some other entity with the obtained sum, and based on the comparison result selects the content of one among several registers. *(Source: ARC Cores)*

- Resolve a major resource-sharing problem that will be experienced by the cores, especially when it comes down to embedded and off-chip memory access, with scalable, real-time, and fair arbitration.

- Implement a convincing and (above all) functional scheme to address context-switching issues (multithreading with zero switch overhead) in an area where RISC has been traditionally incapable of addressing the problem efficiently.

- Last but not least (in order to compete with network processors), come up with a flexible and modular programming model that allows the efficient use of such a massive computational artillery in a transparent way, offering a single-image perception to software engineers, who do not need to worry about allocating software work to individual engines. The model should also eventually allow "hot" swaps or code upgrades in the field without requiring the chip to be redesigned every time just to accommodate new functionality.

Therefore, the assessment is that this type of technology can be used either in low-speed forwarding-plane designs for packet processing (*customer premises equipment* [CPE] or enterprise network equipment) or in multiprocessor designs where multiple embedded core processors are integrated into the same SOC. That inevitably brings along a whole series of systems architecture issues. However, in the network processing field, this type of configurable processor technology usually seems ideally suited for supervisory and control plane applications, where neither wire-speed performance nor complexity management of multiprocessor integration is required. Several interesting articles[9] and application notes about the engagement of configurable processor technology from Tensilica and ARC in the network-processing field are available. Some of them can be found either directly from the web sites of the individual companies involved or from the trade journals mentioned in the list of references.

IMPROV SYSTEMS

A completely different architectural approach to the IP core-based SOC design problem has been taken by Improv Systems (*www.improvsys.com*) and its Jazz™ VLIW CPU technology.[10] The company originally pursued the network-processing market, but recently it seems to have steered more heavily into applications that require extremely powerful scalable embedded DSP processing. This does not preclude the use of its approach in fast communications processors, which is why we discuss it here. Figure 10.12 illustrates parallel and scalable architecture based on the Jazz VLIW platform.

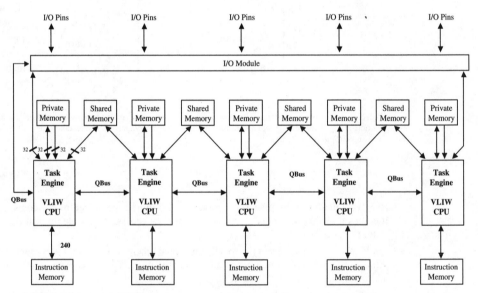

FIGURE 10.12 Parallel and scalable architecture based on Improv's powerful Jazz VLIW core platform. *(Source: Improv Systems)*

9. See, for instance, Loring Wirbel, "Onex Communications Corp.'s Omni Switching and Processor Architecture," which is available online at *www.eetimes.com/story/OEG20020721S0005*, and "Bay Microsystems Uses Xtensa Processor Architecture To Reach New Heights in 10G Integration and Packet Processing Performance," which is available online at *www.baymicrosystems.com/news/press_release_tensilica_07_29_02.html*.

10. A nice introduction to the Jazz architecture can be found in an article by Steve Leibson called "Jazz Joins VLIW Juggernaut," which appeared in the *Microprocessor Report* publication on March 27, 2000. It is available by subscription at *www.mdronline.com/mpr/h/2000/0327/141303.html*.

The Jazz platform enables the easy integration of multiple VLIW processors inside an SOC. Each one capable of executing between 8 to 12 operations per instruction. Each processor communicates with other processors through a proprietary on-chip fast Q-bus where control messages are exchanged. Data is passed on between processors by using shared banks of embedded memory. The advantage of this approach is that in a multiprocessor SOC, contention will never appear among processors for access to memory or a shared bus. The technology can be easily scaled to extraordinary computational capabilities. In addition to the hardware platform and architecture, the Improv approach deserves some serious attention for its advanced development tools and overall design flow approach.

Designers first describe the architecture that they have chosen into the company's interactive tool suite, either based on one of the company's several standard configurations or by embedding one or more *designer-defined computational units* (DDCUs) next to one or more Jazz VLIW CPU cores. The DDCUs can be essentially any piece of hardware logic that may be required to properly execute an application. The designers then use ordinary Java language along with a few extensions in some handy class libraries that the company has created as a notation tool to correctly describe behavior. The reason for this choice is that Java competency is much easier to find as a commonly available skill among software engineers than traditional hardware description languages, which are not so well mastered by the software community. Improv strongly believes that it is becoming more important than ever to control the complete SOC design cycle by software, as opposed to struggling with the integration of multiple and often incompatible or unverifiable IP cores. The role that software engineers play becomes more critical to the overall flow of work.

Solo, Improv's development environment compiler, reads the Java notation along with the description of the underlying architecture and generates the application image. It then maps this image code onto the configured multiprocessor hardware. As a result, a very complex application, which was developed with a single programming engine in mind, is automatically and seamlessly partitioned onto multiple processors, each working from its own private instruction memory. The SOC designer can simulate the complete solution using a cycle-accurate simulator and identify bottlenecks or decide on the necessary modifications in order to better balance loads or tasks or to change the architecture by adding more standard or optional customized hardware resources, when necessary. The final executable can also be emulated using standard FPGA-based boards. The results are both impressive and fast.

Improv has designed several multicore SOCs for and with its licensees. However, for our discussion, we will only mention one case where five embedded Jazz VLIW cores with their memory obtained sustained aggregate I/O throughputs close to 8 GBps on top of heavy-duty processing of packet-processing applications. This was achieved without pushing semiconductor die fabrication to boundaries of feasibility (meaning that more Jazz cores could be easily packed onto the same die if necessary).

This makes the technology a more than viable candidate in the network-processing field for custom-designed SOCs based on third-party IP cores.

SUMMARY

In this chapter, we looked at the idea of designing customized network-processing chips using IP cores obtained by multiple third-party sources. We discussed the cutting-edge performance offered by a leading supplier of network-processing-optimized IP technology as well as other mainstream configurable IP CPU cores—namely, those that are based on either RISC or VLIW approaches. Many of these approaches offer flexibility, but they may also decrease wire-speed performance. In other words, this flexibility comes at a serious price.

In addition to having to design the entire network-processing ASIC by themselves, which implies that an organization has the necessary design skills and money for in-house work based on this approach, IP-based network-processing design seems a viable approach for fast packet-processing ASICs if the design is based on the scalable and powerful ClearSpeed IP approach or on VLIW processors that are easily configurable.

If however, the configurable RISC or VLIW approach proposed by Tensilica and others who have tried to emulate its model is used, then this technology should be considered in lower-speed applications when dealing with the data forwarding plane, or as is more often the case control plane computational tasks or tasks where the daunting challenge of integrating multiple processors inside the same piece of silicon can be handled from the affordable silicon-die real estate and an architecture and systems engineering standpoint.

However, the latter case has a different result when dealing with programming and coordinating multiple embedded processors, scheduling and arbitrating their access to internal, scarce, and sometimes conflicting resources, while working under a real-time operating system and faced with traffic that is flying in and out of the chip at multiple-gigabits-per-second wire speeds. Classical RISC technology in that case, will be obliged to yield to more scalable and flexible architectures (off-the-shelf network processors) that can usually be procured and programmed more easily, more efficiently, and less expensively than custom ASICs.

SUGGESTED REFERENCES

Many good books are available on computer architecture. Interested readers can find valuable information in the following sources:

Gerrit L. Blaauw and Frederick P. Brooks, Jr., *Computer Architecture: Concepts and Evolution*, 2 volumes (Reading, Massachusetts: Addison-Wesley, 1997).

John L. Hennessy, David A. Patterson, and David Goldberg, *Computer Architecture: A Quantitative Approach* (San Francisco: Morgan Kaufmann Publishers, 2002).

Richard Y. Kain, *Advanced Computer Architecture: A Systems Design Approach* (Upper Saddle River, New Jersey: Prentice-Hall, 1995).

The following source is another book that provides a good discussion of the MIPS architecture and a complete software-based instruction simulator of the MIPS core along with many other relevant architecture-related references:

David A. Patterson and John L. Hennessy, *Computer Organization & Design: The Hardware/Software Interface*, 2nd edition (San Francisco: Morgan Kaufmann Publishers, 1998).

In terms of the reuse of IP cores in SOC designs and associated design and verification issues, the following sources are some good starting points for readers who may want to go deeper into the subject:

Peter J. Ashenden, Jean P. Mermet, and Ralf Seepold, eds., *System-on-Chip Methodologies & Design Languages* (Boston: Kluwer Academic Publishers, 2001).

Janick Bergeron, *Writing Testbenches—Functional Verification of HDL Models* (Boston: Kluwer Academic Publishers, 2000).

Henry Chang et al., *Surviving the SOC Revolution—A Guide to Platform-Based Design* (Boston: Kluwer Academic Publishers, 1999).

Alfred L. Crouch, *Design for Test for Digital IC's and Embedded Core Systems* (Upper Saddle River, New Jersey: Prentice-Hall, 1999).

Michael Keating and Pierre Bricaud, *Reuse Methodology for System-On-A-Chip Designs* (Boston: Kluwer Academic Publishers, 1998).

Thomas Kropf, *Introduction to Formal Hardware Verification* (New York: Springer-Verlag, 2000).

Rochit Rajsuman, *System-on-a-Chip: Design and Test* (Boston: Artech House, 2000).

Prakash Rashinkar, Peter Paterson, and Leena Singh, *System-on-a-Chip Verification—Methodology and Techniques* (Boston: Kluwer Academic Publishers, 2000).

Wayne Wolf, *Modern VLSI Design: System-on-Chip Design*, 3rd ed., (Upper Saddle River, New Jersey: Prentice-Hall, 2002).

Exploring some old concepts (and once considered heretic approaches) that now seem to come back to life with cutting-edge advantages that they offer, asynchronous interconnects inside an SOC allow the integration of multiple IP cores in unusual new designs using methods that are the complete opposite of today's best-design practices that have been taught at electrical engineering departments worldwide during the last 25 years and that have been systematically practiced in the industry so far. The following sources provides good coverage of this new school of thought:

John Bainbridge, *Asynchronous System-On-Chip Interconnect*, CPHC/BCS Distinguished Dissertations (New York: Springer-Verlag, 2002).

More information for this type of technology can be found from research done at Sun Microsystems at the web site *http://research.sun.com/features/tenyears/F3Async1JB.html.*

The following is a nice book that is focused on the issues surrounding integration of ARM RISC cores into larger designs, but it also discusses the general issues related with IP core integration:

Stephen B. Furber, *ARM System-on-a-Chip Architecture*, 2nd ed. (Reading, Massachusetts: Addison-Wesley, 2000).

The following are a couple of very good books on the fundamentals of ASIC design for readers who are new to this field:

Farzad Nekoogar, *Timing Verification of Application-Specific Integrated Circuits* (Upper Saddle River, New Jersey: Prentice-Hall, 1999).

Sung-Mo Kang and Yusuf Leblebici, *CMOS Digital Integrated Circuits Analysis & Design*, 2nd ed. (New York: McGraw-Hill, 1998).

Michael J.S. Smith, *Application-Specific Integrated Circuits* (Reading, Massachusetts: Addison-Wesley, 1997).

Jan M. Rabaey, *Digital Integrated Circuits: A Design Perspective* (Upper Saddle River, New Jersey: Prentice-Hall, 1995).

Neil H.E. Weste and Kamran Eshraghian, *Principles of CMOS VLSI Design*, 2nd ed. (Reading, Massachusetts: Addison-Wesley, 1994).

A nice source of study on the issues of parallel design (co-design) of hardware and software can be found in the following book:

Jørgen Staunstrup and Wayne Wolf, *Hardware/Software Co-Design: Principles and Practice* (Boston: Kluwer Academic Publishers, 1997).

An industry association that promotes open standards for the structured and disciplined use of intellectual property inside SOC designs is VSI Alliance (*www.vsi.org*). Their site contains some interesting links.

A good site with interesting information on IP components for reuse is *www.design-reuse.com.*

The following companies distribute trade publications that often discuss this technology in depth. These publications also include tutorials in new approaches:

EE Times (www.eet.com)

EE Design (www.eedesign.com/isd/issue)

Embedded (www.embedded.com)

Communications Design (www.commsdesign.com/csd/issue)

Integrated Communications Design (http://icd.pennnet.com/home.cfm)

EDN Access (www.e-insite.net/ednmag/index.asp?layout=webzine)

Several market research groups are also analyzing this market and consistently provide research material on multiple aspects of embedded processor technology.

The Microprocessor Report can be found at *www.mdronline.com.*

The Linley Group can be found at *www.linleygroup.com*. The Linley Group has also been the inventor and source of inception and industrywide launch of a more appropriate network-processing benchmark called LinleyBench™. Information can be found at their web site.

The Embedded Microprocessor Benchmark Consortium at *www.eembc.org* is an interesting association trying to standardize performance measurements between different embedded CPU architectures. It has gained significant industry acceptance and has developed some specific test suites to measure and analyze performance of a computing engine in multiple environments. As of this writing, the networking applications suite is probably extremely limited for historical reasons. Therefore, the consortium's work is more readily suitable for classical CPU rating. Meaningful network-processing benchmarks must include a realistic load of traffic as well as the need for multiple *class of service* (CoS)/QoS flows of processing to show performance that approximates real life. We can safely say that that the EEMBC benchmarks constitute a good path for the evaluation of CPUs that are intended for control plane applications.

A couple of important events in this industry include the Embedded Processor Forum (*www. mdronline.com/epf*) and the Communications Design Conference (*www.mdronline.com/mpr/h/ 2000/0327/141303.html*).

P · A · R · T · III

PERIPHERAL CHIPS SUPPORTING NETWORK PROCESSORS: STORAGE PROCESSORS, CLASSIFICATION PROCESSORS, SEARCH ENGINES, SWITCH FABRICS, AND TRAFFIC MANAGERS

CHAPTER 11
STORAGE NETWORK PROCESSORS (SNPS)

In this chapter, we discuss the influence of the evolution of storage networks on network processing and show how the storage network requirements create the demand for a breed of highly specialized processors that go beyond mainstream network processors. These *storage network processors* (SNPs) must be able to handle very high-speed data traffic while performing their tasks under much more stringent jitter and latency performance requirements than ordinary network processors. We discuss the various industry associations that are in the process of resolving the conflicts of interests among multiple technologies and vendors. We also review the approaches taken by a couple of major players in this emerging and specialized network-processing industry branch.

STORAGE NETWORK PROCESSING: THE CONTEXT

Originally, and to a large extent today still, the vast majority of storage devices used by computer systems were attached physically and directly onto the computer system they were supposed to serve. One would talk about *directly attached storage* (DAS) devices. Although this is a simple concept to grasp, it is obviously a limiting factor as a user must have access to the specific server on which the storage units are connected in order to access the stored data. DAS devices usually interface through standard interconnects such as the *Small Computer System Interface* (SCSI) bus. Its high data transfer rate, low latency, and reliability account for its wide-scale success in coupling computers with a plethora of storage devices.

Magnetic disks are the primary online storage medium. Tapes are considered more of a backup and archiving medium. Disk storage is usually found in one of two physical organizations: *just a bunch of disks* (JBOD) and *redundant array of independent (or inexpensive) disks* (RAID). On one hand, JBOD storage devices are usually individual, independent disks situated inside a cabinet and accessible individually by a server. They do not provide cache memory (disk buffering) for higher performance or an intelligent controller that allows operations such as data striping (replicating data on different disks) or parity checking for reliability. RAID storage devices, on the other hand, are controlled by such a controller (along with lots of memory) and provide functionality such as parity checking, data striping across drives, and even mirroring of critical data across multiple arrays for fault tolerance. Compared to JBOD, RAID provides larger storage capacity, enhanced availability, and significantly improved performance.

SCSI emerged as an 8-bit parallel bus in 1979. The *SCSI Architecture Model* (SAM-2), which is part of the *National Committee for Information Technology Standards* (NCITS) T10 standard, has created a layered model for SCSI implementation. The SCSI-3 command set converts the logical layer into a packet-based format, which can be transmitted over a network. As the protocol has evolved, we now have a serial SCSI as a layered, well-structured architecture of protocols that enables services to be requested from storage devices at a distance and over networks.

If a smaller chunk of data needs to be retrieved, an entire block that contains the desired pieces will still need to be read. Of course there are reasons as to why this is so, and along with the reasons there are obvious penalties in efficiency and speed. We will not elaborate on this subject. The references listed at the end of the chapter provide interested readers with more than ample documentation on any aspect of the storage technology and industry. In this discussion, we will ignore the small-capacity storage units that are found in small desktop computers such as PCs and workstations. For all practical purposes, these devices are connected directly on the PC or workstation bus and qualify as DAS devices.

An explosion in demand for storage capacity occurred as result of the exponential growth in *online transaction processing* (OLTP) during the 1990s, the need for flexible and wide-scale information access by employees and outside partners of a company or members of an organization, as well as the ever-increasing need to service requests for audio, video, and text/graphics files out of servers in several organizations. Organizations and enterprises must continue to revisit their approach to managing stored data in order to support several strategic organizational goals. Storage media must be easily accessible, reliably available around the clock, and scalable. With these characteristics, organizations can function with continuity around the clock and continuously improve their staff's efficiency and productivity (by granting them easy access to data they need, when they need it, and wherever they need it). Data must be so stored that it is straightforward to service, upgrade, and expand the organization's data storage infrastructure without disruption.

Two major technologies have appeared in the market over the last few years to address these trends:

- *Network attached storage* **(NAS)** EMC Corp., a leading industry player, has already qualified NAS as "suitable only for a small segment of the overall storage markets about 10 percent." This statement was made publicly by EMC Corp. at its annual stockholders meeting on May 9, 2001. In a NAS, host computer systems use a file access protocol such as the *Network File System* (NFS). They access directories and files on storage devices. NAS-attached devices require and/or provide complete files or directories to interested and qualified parties. Unlike DAS, they do not just require raw blocks of bit data. NAS is clearly more sophisticated and efficient than DAS.

- *Storage area network* **(SAN)** This technology is experiencing phenomenal growth, according to multiple research analysts. For instance, in a July 2001 report called "Reweaving SAN Fabrics: Worldwide Open Systems SAN Interconnect Fabric Forecast and Analysis, 2001–2005," IDC predicted that the SAN market will achieve an 80 percent *cumulative annual growth rate* (CAGR) by the end of 2004. SAN (which is reminiscent of a *local area network* [LAN]) is a generic name that describes a fully dedicated, reliable, and high-performance network that provides a direct connection between servers and storage devices. Figure 11.1 illustrates this principle. Storage devices are not coupled to specific servers.

Consequently, an entire organization can share resources since any computer system can be authorized to access any storage device directly over the SAN. This freedom of scalable configuration and management, in addition to the technology's flexibility and reliability, has attracted the industry's attention. It has been shown to ultimately lead to lower costs of ownership.

In this chapter, we will discuss a special breed of network processors intended to be used in SAN equipment. These network processors are already known in the industry by various names, such as storage coprocessors, storage processors, or SNPs. We will refer to them as SNPs to avoid confusion and capture their dual nature. More specifically, an SNP is a network processor that undisputedly spends time doing what all *network processing units* (NPUs) are supposed to do most of the time— while it churns data at wire speed, it should also perform deep packet inspection and classification/forwarding. At the same time, however, an SNP operates in the heart of a SAN instead of an ordinary high-speed switch or router. As a result, the SNP must meet some peculiar and stringent functional and performance requirements that are beyond the capability of a typical NPU.

The data storage area is extremely broad. It covers a very wide array of technologies from magnetic materials and laser optics all the way to fiber-optic transmissions and fast electronics, and from unbelievable aerodynamic designs of read/write heads over fast rotating disks to sophisticated

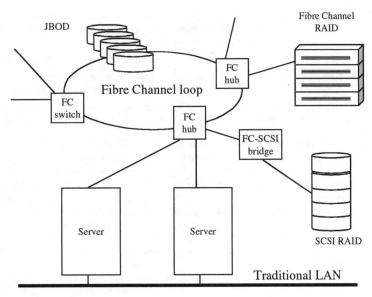

FIGURE 11.1   The principle of a SAN.

input/output (I/O) protocols and data management software. It is so technologically rich that we cannot discuss it in this chapter.

We will start our discussion by briefly mentioning fundamental concepts and technologies, but we will return quickly to the main thrust of the chapter—the SNP, why it is needed in the first place, and how it is different from other network processors. We will also provide some representative examples from industry leaders. Interested readers can obtain more information about these technologies in the sources listed at the end of this chapter.

SAN-ENABLING TECHNOLOGIES

For many reasons, Fibre Channel has been the de facto standard in SANs. At the same time, however, IT departments of large companies and organizations have undergone a revolution. The industry has identified the use of SAN technology as a key factor for the advancement of storage technology, especially if it can be deployed over Ethernet (with its two standardized and available fast varieties—1 Gigabit Ethernet and 10 Gigabit Ethernet). Unlike the current SAN technologies, an Ethernet-based SAN operates under a well-known *Transmission Control Protocol/Internet Protocol* (TCP/IP) infrastructure. This technology has several advantages:

- Leveraging of the vast current investment in network infrastructure.
- Consolidation of the same hardware and software tools, techniques, and methods in managing the storage network as part of a global enterprise or organization network at a significantly lower cost of ownership.
- Leveraging of established technical skills of an IT organization, such as TCP/IP.

- New arrivals in the IP storage era will not be accompanied by long learning curves for the people who must deploy the new technology within an organization as TCP/IP based tools and techniques are widely known and easily acquired.

- Established standards and protocols minimize the current SAN interoperability problems.

- IP-based SANs not only increase the management and support capabilities of centralized IT organizations, but they also enhance the usability of the storage resources within an organization.

- IP-based SANs can leverage the highly functional and widely accepted IP security technologies to provide an efficient, robust, and secure method to secure the transfer of data over the SAN as well.

In order for this evolution to occur smoothly, several things must happen. Current products are mostly based on Fibre Channel. Organizations will not just rip apart multimillion-dollar investments in order to accommodate the new trend, no matter how enticing it sounds. Therefore, a transition must occur that will require device compatibility. The industry is aware of this requirement. As a result, the first generation of IP storage products will have to function in a mode known as *protocol mediation*. Products that offer this capability will enable customers in a rather short term to connect their legacy Fibre Channel storage products through IP networks. The following wave of endpoint storage device products will support IP storage in native mode directly on an Ethernet medium (1 GbE or 10 GbE). We will discuss network-processing issues related to multiprotocol SANs later in this chapter.

During this rapid industry evolution and consolidation, the *Internet Engineering Task Force* (IETF) has been working on defining standards for IP storage to support the new storage network technology trends. These efforts include the following standards:

- iSCSI, which is a complete transport service for SCSI traffic
- FCIP, which tunnels Fibre Channel traffic through an IP network

FIBRE CHANNEL

The Fibre Channel is a standard from the NCITS T11:I/O Interface (X3.230-1994) effort of the T11 committee of the NCITS, which works on I/O interfaces. Fibre Channel defines a highly reliable, gigabit-plus-per-second class transport technology that allows servers, mainframes, workstations, switches, hubs, and storage devices to communicate using well-known SCSI and IP protocols based on multiple possible topologies. This combination of capabilities can tackle an organization's storage resource-sharing problem while still providing high performance, flexibility, reliability, availability, and scalability.

Fibre Channel is a network/channel standard that not only specifies the physical layer over copper or optical fiber, but also the control and transport layers. The specified fabric is self-managed, and different topologies such as point-to-point, arbitrated loops, and switched topologies are easily supported depending on the needs of a specific application. It offers connections over distances that can be up to 10 km (~6 miles) with speeds ranging from 266 Mbps to more than 4 Gbps. It allows multiple existing interface command sets such as IP, SCSI, IPI, HIPPI-FP, and audio/video. For example, SCSI is mapped onto a higher-layer protocol over the Fibre Channel stack. Fibre Channel topologies are sustained by switching fabric devices that closely resemble the switches found in more widely used packet networks.

Most current SANs are built around Fibre Channel infrastructures as they allow the efficient SCSI-based transfer of data over large distances. This is something that SCSI cannot do, but Fibre Channel does this very well.

Incidentally, the *University of New Hampshire Interoperability Lab* (UNH IOL) offers services for the certification of interoperability of Fibre Channel products from different equipment and SNP vendors.

Fibre Channel eliminates all scalability and bandwidth problems previously associated with the simple SCSI bus. It is important to note that current RAID storage devices ship with Fibre Channel loops directly integrated in their backplane for native support of Fibre Channel and for the modular capability of being hot swappable. *Hot swappable* means that one disk unit can be removed from the RAID array for service or replacement without affecting the availability of the overall RAID system.

Before we discuss systems that allow interoperability between the Fibre Channel and the IP world, we must mention some of the main technical characteristics of the Fibre Channel:

- It provides transmission reliability by offering the option of delivery confirmation. Alternatively, an implementation can completely bypass the Fibre Channel protocol stack to increase performance.
- It fully supports widely known mechanisms of network self-discovery, including relevant protocols such as *Address Resolution Protocol* (ARP) and *Reverse Address Resolution Protocol* (RARP). From a topology standpoint, it can accommodate dedicated bandwidth point-to-point circuits, shared bandwidth loop circuits, or scalable bandwidth switched circuits equally well.
- It offers extremely low-latency connections and connectionless service. The standard allows the automatic self-discovery of the specific Fibre Channel topology.
- It offers the flexibility of choosing between true connection service or fractional bandwidth and connection-oriented virtual circuits to guarantee the *quality of service* (QoS) for mission-critical operations such as backups.
- It can be instantaneously set up. This is done fast so the setup time is short enough to be measured in microseconds when a system uses the hardware-enhanced Fibre Channel protocol.
- It supports time-synchronous applications such as video, using fractional bandwidth virtual circuits. It provides efficient, high-bandwidth, and low-latency transfers using variable-length (0 to 2KB) frames.

It is important to realize the following characteristics in a Fibre Channel environment that contains a mix of both SCSI and IP:

- Native Fibre Channel storage devices as well as servers and workstations connect directly on the Fibre Channel.
- SCSI storage devices are connected onto the Fibre Channel by Fibre-Channel-to-SCSI bridges.
- The IP protocol is only used for server-to-server and client-to-server connections.
- Enterprise-wide Fibre Channel switches consolidate the various workgroups and departmental computing or storage environments in a hub-and-spoke approach to ultimately provide one scalable consolidated storage network that allows the sharing of storage across the whole organization.

According to the *Fibre Channel Industry Association* (FCIA), a Fibre Channel can routinely service critical database environments delivering a sustained bandwidth of over 200 MBps for large files while servicing thousands of simultaneous I/O requests. These numbers are important as they give us an idea of the magnitude of bridge-traffic load that can be expected in protocol mediation devices.

It is also interesting to see how the FCIA compares the Fibre Channel with alternative technologies. Table 11.1 provides a comparison that was compiled by FCIA. This table shows how the Fibre Channel technology stacks up against Gigabit Ethernet and *Asynchronous Transfer Mode* (ATM). We have deliberately placed question marks next to some parameters in order to call attention to their questionable importance. Part of the industry decided to pursue the investigation of establishing new combinations of SCSI-like techniques with TCP/IP over Ethernet (1 GbE and 10 GbE) networks to control the cost of ownership.

TABLE 11.1 A Comparison between Fibre Channel and Alternative Technologies[1] (Source: FCIA)

	Fibre Channel	**Gigabit Ethernet**	**ATM**
Technology application	Storage, network, video, and clusters	Network	Network and video
Topologies	Point-to-point, loop hub, and switched	Point-to-point hub and switched	Switched
Baud rate	1.06 Gbps and 2.12 Gbps	1.25 Gbps	622 Mbps
Scalability to higher data rates	4.24 Gbps	Not defined (?)	1.24 Gbps
Guaranteed delivery	Yes	No (?)	No
Congestion data loss	None	Yes (?)	Yes
Frame size	Variable and 0 to 2KB	Variable and 0 to 1.5KB	Fixed and 53 bytes
Flow control	Credit based	Rate based (?)	Rate based
Physical media	Copper and fiber	Copper and fiber	Copper and fiber
Protocols supported	Network, SCSI, and video	Network (?)	Network and video

A quick overview of the parameters shows that the arguments from the IP storage camp do have merits: TCP adds reliability of delivery. 10 Gigabit Ethernet is already two to four times faster today than what Fibre Channel will soon be. IP is a well-known protocol that people know how to configure, route, switch, manage, and even secure on an end-to-end basis. In the following section, we will turn our attention to this second major storage network technology—IP storage.

We do not advocate either one of these two technologies. This is a business decision every organization that envisions storage networks must make. It depends on how the potential deployment of each one of these two technologies maps onto the enterprise case or onto the users' organization case, business model and operational processes, budget and timing constraints, technical skills, manpower, expertise, the current computing and network infrastructure, the estimated position on the learning curve, and the disaster recovery and survivability constraints of the organization. We discuss these

---

1. The question marks shown on some parameters of this table are deliberately introduced as food for thought in order to question some of the arguments the FCIA has raised against the potential reliability and scalability of GbE networks carrying TCP/IP. GbE scales nicely to 10 GbE. If TCP/IP runs over it, the reliability of the sequenced delivery issue is well addressed. Of course, if TCP is not used over IP in order to improve performance, something like the *User Datagram Protocol* (UDP) would have to be used, which would validate the FCIA's reliability concern. However, when TCP is the transport protocol of choice, some storage applications may run out of steam when executed on hardware of limited computational horsepower. Retransmission latencies associated with TCP operations may also end up being prohibitively long in some cases, whereas the number of simultaneous TCP sessions with satisfactory performance may be constrained for a given hardware configuration. This situation, as expected, will improve decidedly if it takes place in a 10 GbE environment instead, but then the cost and complexity of the 10 Gbps adapter hardware to connect storage devices to such a network become significantly higher. In other words, the proverbial jury is still out; however, the IP storage camp makes a legitimate business case. Special GbE server and storage network systems using *TCP Offload Engine* (TOE) hardware (which is discussed later in this chapter) are enabled by high-performance iSCSI devices, but they will also require the support of potential future changes to the standards, which translates to the need to reprogram some of the key underlying component technologies. iSCSI enables native IP SANs to be built, thereby enabling SANs to be integrated into one organization-wide IP-based network infrastructure. This has distinct advantages over the expensive and complex alternatives that are encountered when creating and managing a Fibre Channel infrastructure. Large companies that have adopted the approach are using iSCSI to deploy IP SANs in departmental systems, whereas a similar trend can be found in some small- and medium-size organizations and businesses. iSCSI may end up coexisting in the data center with Fibre Channel, but it will most likely not displace Fibre Channel in the near term.

technologies in the context of SNPs that will be needed to service the rapidly growing network-processing context.

IP STORAGE

IP storage is a generic term widely used today in the industry to encompass a network-computing realm that is based on the combination of protocols, technologies, and products that enable the deployment of IP-based storage networks to execute and transport block I/O operations.

The last point is crucial for distinguishing the differences between IP storage and NAS-based storage networks. NAS devices operate based on a file transfer protocol such as NFS or *Common Internet File System* (CIFS). Consequently, all NAS I/O operations occur at the file level, not at the block level as with SAN technologies. When an I/O request is issued to a NAS device for a piece of information that lies inside some block of stored data, the NAS device in conjunction with the associated file system will resolve the request and extract it from the retrieved file to present it to the requester.

Many ingenuous organizations these days have come to a point where they combine both SAN and NAS technologies next to each other, thereby maximizing the value of their investment without ignoring technology advances or missing out on the financial benefits of embracing new technologies. SAN is used in performance-sensitive applications (such as transaction processing or data warehousing). NAS is used in more generic environments where common access to stored resources is important (such as engineering departments sharing access to design files).

IP storage technologies are divided into two categories: iSCSI and FCIP. IP storage is actively promoted by the *Storage Networking Industry Association* (SNIA) and, more specifically, by its Storage Forum. In the early fall of 2002, IP storage already managed to attract major attention from leading companies in the three major areas of products involved in storage systems:

- Designers and manufacturers of storage systems.
- Network equipment.
- *Host bus adapters* (HBAs).

We must first clarify some concepts.

Network Interface Card (NIC)

On one hand, if a NIC is used to interface with a LAN, it may seem that a NIC is all it takes to connect to a SAN. The industry has been using the term *NIC* as an equivalent to the term *adapter*; however, this can only be done inside an Ethernet realm. NIC cards are usually designed to transfer packetized file-level data among client devices such as PCs, servers, or storage devices. It is important to realize that NICs do not traditionally transfer block-level data. Such transfers are handled by a storage HBA, which could be a Fibre Channel HBA or a parallel SCSI HBA. In order for a NIC to process block-level data, the data needs to be encapsulated inside a TCP/IP packet before it can be transmitted over an IP network. By using iSCSI drivers that must be made available on a host or server, a NIC can be made to transmit packets of block-level data over an IP network. In that case, the server will handle the packetization process of the block-level data. It will obviously be responsible for the correct execution of all computational steps taken to process the TCP/IP protocol.

This entire computation-intensive scheme is extremely taxing on the server or host *central processing unit* (CPU). It can almost bring it down to its knees. This problem has been one of the main motivators behind the pursuit of powerful TCP termination engines. We discuss the functionality and requirements of TCP termination engines in a separate section. A TCP termination or offloading engine allows the completion of both TCP/IP processing and packetization on the HBA. Therefore, this SNP NIC, which is equipped with a *TCP Offload Engine* (TOE), operates like a storage network HBA rather than a traditional Ethernet NIC.

Storage HBAs

Unlike Ethernet NICs, storage HBAs are designed to transmit block-level data to and from storage applications. When an entire block is transferred from the software application to the adapter, the server or host CPU does not need to spend time trying to fragment the block into smaller frames for subsequent transmission. The HBA has the local intelligence to segment the block into frames. This process is usually handled by specialized *segmentation and reassembly* (SAR) chips, which are similar to the ones in ATM line cards. These chips are situated on the HBA.

iSCSI Adapters

A hybrid of the previous two categories is the class of iSCSI adapters. They combine the functionality of both categories (that of a NIC with that of a storage HBA). iSCSI adapters work with block-level data and perform the required segmentation and processing on the adapter card with the assistance of TCP/IP processing engines. The produced IP packets are then transmitted across the IP network. This allows the creation of full-fledged IP-based SANs without adversely affecting the host or server CPU.

STORAGE VIRTUALIZATION

Before we discuss some of the lower-level details, we must briefly mention the concept of *storage virtualization*, which is another state-of-the-art technology trend that is also contributing to the explosive growth of storage networking and depends on high-performance network storage processors. If we consult an industrial definition, such as the definition provided by Trebia Networks (*www. trebia.com*), storage virtualization is the "separation of the logical view of data storage from the actual underlying physical devices." If a storage infrastructure can arrive at that level of sophistication, then all physical storage is a shared pool of storage capabilities that can be used to service changing storage needs in an enterprise or organization. This can include, but is not limited to, online capacity expansion and reprovisioning. It is argued that the true potential of SAN can be maximized if IP storage is embraced by organizations and if equipment that supports storage virtualization is deployed.

Virtualization software is first used to collect data that may be originating from different types of storage devices. These devices can be SAN-attached devices, network-attached devices, or devices that are attached on a server. The virtualization software then consolidates all of this gathered data into a common pool that can be monitored, managed, supervised, and administered for broad use from a single console.

The term *storage virtualization* is widely used by several vendors, but each vendor approaches the issue differently: Some vendors implement virtualization on their own storage devices, whereas others provide storage virtualization on a variety of devices. However, currently, no single vendor provides across-the-board virtualization for any indiscriminate choice of storage device.

Storage virtualization can be implemented in three different ways:

• On a host CPU or server.

• On a storage array.

• On an appliance.

Another approach vendors take to attain the virtualization objective is to implement the following within one of these two categories:

- Symmetric storage virtualization (also known as *in-band storage virtualization*).
- Asymmetric storage virtualization (also known as *out-of-band storage virtualization*).

These names are derived from the fact that in the in-band approach, a device lies in the actual path of data that must be exchanged between a server and devices. It passes data and/or intelligence through to arrays that are attached to it. Conversely, in the out-of-band approach, data is passed between a server, switch, or router to the devices. The entire work is managed by the server or storage array.

If we take a closer look at the three fundamental platforms of implementing storage virtualization, we will notice some interesting characteristics:

- Server- or array-based virtualization was the first way to implement this technology. In this scheme, both the storage and the data-pooling intelligence reside on the server or array. Because this approach does not put any other devices on the path that the data must traverse, it scales better than network-based virtualization. When all virtualization work must be done on the server, no other devices on the network, such as Fibre Channel switches or other arrays, are affected. The downside of the approach is that this extra load on the server may cause server-based latency, which may be troublesome for specific applications.

- In network-based virtualization, the storage virtualization implementation usually depends on an in-band virtualization server (usually a Windows NT/2000 or Linux-based server) where all other network servers have to look for information about where their data actually resides. This performance requirement can be very exacting on the virtualization server. Typically, these implementations run on an Intel server, which some corporate IT people generically call an *NT box*. Despite caching attempts by some *original equipment manufacturer* (OEM) vendors to minimize server latency, its bus architectures are not designed for heavy loads like the ones handled by servers that are used by very large organizations to either manage data or I/O needs. The I/O structure of Intel-based servers is usually not optimized for system configurations that require the sophisticated capabilities of set mirroring, capacity on demand, snapshot backups, or data replication. Therefore, large organizations are typically very reluctant about engaging NT-based solutions in the heart of their enterprise-wide storage virtualization effort.

   The virtualization server remains one of the potential bottlenecks of this approach. It is often considered the Achilles heel of such implementations. Some storage system companies have announced their intention to come up with hybrid virtualization offerings that would sit out-of-band without affecting the flow of data back and forth between servers and storage devices. This would seem to scale nicely for large enterprises. However, it has one big downside—virtualization software must be put on each host CPU on the network.

- In-band appliance-oriented virtualization is a promising approach to manage and maintain. No code is needed on host servers, and all I/O requests and responses will have to first pass through the virtualization engine, which essentially requires nothing else in order to function.

An interesting trend in the industry is the move toward the *virtualization switch*. This is fast-switching network equipment that can provide storage virtualization at wire speeds without noticeable latency. A couple of companies with interesting technology are already active in this field, such as Maranti Networks, Pirus Networks, and so on.

Storage virtualization purports to offer organizations and enterprises an unprecedented value by ensuring access to their vast data without worrying about which system, what location, which format, or which operating system platform they are stored in. However, the very same abstraction layer that it provides from the actual underlying hardware may end up hurting the driving business application. This is because numerous commercially available data management applications take advantage of advanced hardware features to provide high value to their users by implementing sophisticated functions linked to the actual hardware, such as autoconfigure or autodiscover. By its mere dependence on this abstraction layer, storage virtualization may lose the capability to offer this type of sophistication unless the industry comes up with new ideas about how to handle this problem. However, for the moment, we cannot have our cake and eat it too.

iSCSI

The extraordinary advantage of iSCSI is that it provides access to and from block-level storage devices, such as disk arrays, single disks, tape drives, and libraries, directly over regular TCP/IP networks. Before the arrival of iSCSI, all TCP/IP-based access to networked storage in the form of NFS and CIFS servers occurred in the framework of NAS systems, which have always required TCP/IP host-to-host data transfers. The ramifications of this shortcoming cannot be overemphasized. Until the formulation and establishment of iSCSI, it was impossible for a TCP/IP computer to send data directly to a standalone disk array or tape drive that was also directly connected to a TCP/IP network.

iSCSI is a protocol that enables SCSI commands to be embedded inside TCP/IP session packets, which must be embedded into Ethernet frames for subsequent transmissions. To explain how it works, we will look at an example of configuration such as the one shown in Figure 11.2. The left side of the figure depicts a corporate traditional data-processing IP-based LAN on which some server (potentially among many) is connected. The same server is also connected on an IP-based SAN on which IP storage devices are directly connected. The IP SAN is composed of more than just servers and storage devices. Both these classes of systems (connected on the IP SAN) must have an embedded and specialized technology called an *iSCSI adapter*. iSCSI adapters can assume the role of an *iSCSI initiator* or *iSCSI target*. They can also be implemented in the physical form of either a full-fledged board or a sophisticated ASIC.

Now let us assume that one of the client devices (a workstation) (X) needs some specific file information from the server (where it believes the information is stored). It initiates a request over the LAN to the server for that piece of information. The server realizes through some indexing file directory that the information must be retrieved from a specific storage device on the SAN. It then issues specific SCSI commands for that device and passes the task to the iSCSI initiator. The iSCSI initiator will encapsulate these SCSI commands inside a TCP/IP packet(s) that will be embedded into Ethernet frames and sent to the storage device over a switched or routed SAN storage network. The iSCSI target device receives the Ethernet frame, strips it apart and recovers the TCP/IP content, decapsulates the packet, and obtains the SCSI commands needed to retrieve the required information. The process is reversed and the information is reassembled and reencapsulated into TCP/IP packet form. This information will be embedded into an Ethernet frame(s) and sent to the iSCSI initiator at the server, where it will be decapsulated and reencapsulated onto the IP LAN for subsequent transmission to the requesting client.

The iSCSI protocol stack is essentially an insertion of a few things right above the traditional layer 4, as shown in Figure 11.3. The main purpose of TCP, a layer 4 protocol that runs over IP, is to ensure the reliable transmission. The iSCSI layer, which is now at layer 5, runs right over TCP and ensures that the bit packaging of the underlying transmission becomes routable by serializing the inherently parallel SCSI structure. The native SCSI command set and SCSI bus protocol run over iSCSI (now in layer 6). The respective operating system layer and the final application software are located above that layer.

A whole array of industry players has adopted the IP storage realm and supports iSCSI with products that behave in a complementary fashion when deployed in a SAN:

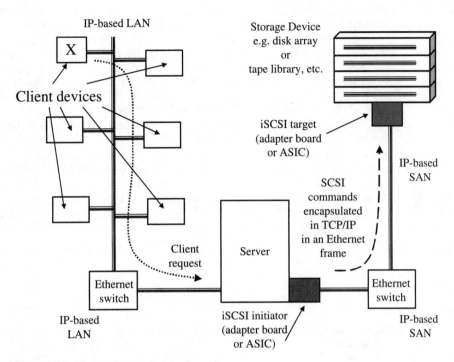

FIGURE 11.2 The iSCSI SAN principle of operation.

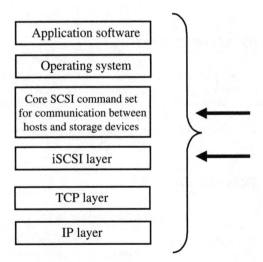

FIGURE 11.3 The iSCSI protocol stack.

- **iSCSI initiator manufacturers**   These include companies such as Adaptec, Alacritech, Emulex, Intel, HP, and Qlogic, which offer iSCSI storage NICs (also known as S-NICs) or HBAs. HBAs are used inside servers to enable the use of iSCSI for block-level access to storage systems.

- **iSCSI switch manufacturers**   iSCSI has been designed predominantly to enable end-to-end IP-based storage networking. This is done without requiring intermediate iSCSI-aware switches. As a result, several companies such as Cisco, HP, and IBM are working on or already offer multiprotocol storage networking switches, which enable bridging between iSCSI-based server devices and Fibre-Channel-based legacy environments, and/or provide storage virtualization capabilities.

- **iSCSI storage systems manufacturers**   These include companies such as Adaptec, IBM, and 3Ware, which offer native support for iSCSI in a new generation of storage devices.

- **FCIP switch manufacturers**   These include industry heavyweights such as Lucent and Cisco (working together with Brocade) as well as several startups such as Akara, LightSAND, Pirus, and SAN Valley. These companies are working on or already offer FCIP-to-iSCSI bridging products.

- *Network storage processor* **(NSP) manufacturers**   These include companies such as Platys (now acquired by Adaptec), Emulex, Silverback, and Trebia. These companies offer components (standalone as well as embedded and integrated) that enable low-latency TCP offload and IP storage protocol support for IP storage target and initiator products.

FCIP

FCIP is a tunneling protocol that allows Fibre Channel tunneling through the encapsulation of the Fibre Channel transfers inside IP packets, which can then be transmitted over a TCP/IP network and infrastructure. Through this method, users with Fibre Channel sites can connect them over the *metropolitan area network* (MAN)/*wide area network* (WAN), effectively expanding the scope and reach of their SAN.

FIBRE-CHANNEL-TO-iSCSI BRIDGING

The convergence of the Fibre Channel world with the IP network world requires a bridge so users can maximize the impact of their investment. The concept of *storage routers* is no longer foreign. Enterprises and organizations can use their TCP/IP infrastructure to make storage devices accessible by any system from anywhere in the corporate network, thereby optimizing the use of this strategic asset—the operational data of the enterprise or organization—to anyone who has the need to know.

TYPICAL APPLICATIONS FOR AN SNP

The following are some typical applications for an SNP:

- *Multiprotocol SAN switches* can use SNPs to handle protocol mediation. For example, consider a group of servers sitting on a 1 Gigabit Ethernet or 10 Gigabit Ethernet network. These servers access data transparently on an iSCSI RAID array situated on another 1 Gigabit Ethernet or 10 Gigabit

Ethernet network while targeting a legacy Fibre-Channel-based RAID array on a Fibre Channel network.

- *Tunneling* over the IP MAN/WAN Fibre Channel traffic between two or more geographically separated Fibre Channels is easily handled by an SNP, which sits on IP adapters of storage routers at the edge of the different Fibre Channel sites (in this example). It would then encapsulate Fibre Channel traffic and decapsulate it from TCP/IP packets that traverse the MAN/WAN.

- A logical variation of the previous example can be to use a *gateway* between a Fibre Channel network and an IP storage network (1 Gigabit Ethernet or 10 Gigabit Ethernet for that matter), which would definitely require the embedded services of an SNP.

- With the proliferation of IP-storage-based SANs, a plethora of SNPs will be required inside *storage network end systems*. More specifically, SNPs will be needed on target iSCSI adapters embedded in iSCSI-compatible RAID as well as in legacy Fibre Channel RAID arrays and more traditional server HBAs.

REQUIREMENTS FOR AN SNP

An SNP must meet the following requirements:

- Storage-related packets traveling over the SAN or the corporate IP network, which are encapsulated, and possibly multiple times, will often require *decapsulation* and *reassembly* at line speeds that can be 10 Gbps.

- A received packet must at least be submitted to *deep packet inspection* by the storage network-processing equipment.

- Correct, rapid, and deep packet inspection will lead to the appropriate decision regarding their appropriate *classification.*

- Classification is followed by *forwarding*.

- Reliability in transmission and *TCP offloading* are major issues. Offloading the heavy-duty processing of terminating the TCP protocol for multiple (possibly thousands) simultaneous active sessions is of paramount importance in order for the network equipment to function properly.

- Time-related performance must be optimized. *Jitter* and *latency* are exceptionally critical parameters.

- In some applications, the capability of running *multiple storage protocols* (iSCSI and Fibre Channel) at the same time is imperative.

- Connectivity to multiple different network physical layers is also very important; therefore, the SNP should be able to support 1 Gigabit Ethernet and 10 Gigabit Ethernet networks with embedded MAC circuitry as well as Fibre Channel and other networks, such as InfiniBand and/or others.

- The *level of solution integration* must be addressed, as a chip is preferable to a board because of its cost, power consumption, and reliability.

- *Ease of integration* into a final product should not be underestimated as this implies an advantage for the customer in terms of time to market.

- Last but not least, the *programmability* of the SNP is indispensable, as the OEM user must be able to program new functionality in order to upgrade or modify equipment to accommodate new protocols.

TCP TERMINATION ENGINES OR TCP-OFFLOAD ENGINES (TOES)

TCP/IP was developed 20+ years ago, at a time when its designers thought layer 3 and 4 protocols would only run on host CPUs with large computational resources and network bandwidth was at a premium. Many things have changed. The unprecedented proliferation of IP-based networks, the pervasiveness of embedded computing, and the explosive growth in demand for bandwidth by so many new generations of applications have completely reversed the computing landscape. TCP/IP is required now inside standalone devices such as a RAID system or a backup drive. The computing heart of such a system cannot be described as a host CPU in the traditional sense. Therefore, its computational capabilities are not up to par with a server. If a very high-speed throughput must be sustained and thousands of sessions with reliable transmission needs must be carried, the main processor must be offloaded from the protocol chores.

To be more specific, Adaptec (in its ANA-7711 TCP/IP Offload Adapter Datasheet [see *www.adaptec.com*]) estimates that a typical 1 GHz Pentium-type processor saturates about 70 percent of its capacity with TCP processing, if it must support a 1 Gigabit Ethernet link at line speed. Depending on the number of simultaneous TCP sessions, this performance constraint can degrade even more. If a security protocol such as *Secure Socket Layer* (SSL) that runs at layer 5 above TCP must also be supported, TCP must be terminated before actual SSL work can be performed.

Adaptec's estimates are much more liberal than the estimates in a 2002 Gartner Research Brief. The Gartner report describes the accepted rule of thumb as being that each bit per second of line capacity requires about 1 Hz of CPU horsepower to run TCP in software. In other words, a 1 Gbps link will completely consume a 1 GHz CPU, leaving no cycles for doing actual application work besides TCP. Even if we accept the Adaptec estimate, this obviously leaves very little headroom in the CPU for other meaningful application processing. Therefore, storage devices equipped with classical microprocessors cannot be expected to handle SNP chores sustained at multiple gigabits-per-second speeds.

This problem can be resolved in two ways: by replacing TCP with a more modern efficient and leaner-and-meaner protocol (an almost impossible prospect, which realistically should not be expected any time soon given the complete domination of the global market by TCP/IP) or by offloading TCP processing on an accelerator that will minimize the necessary intervention of the host CPU. Figure 11.4 illustrates the principle of TCP termination or TOE.

TCP offloading requires specific functionality. First, consider the sophisticated flow control and error recovery services that TCP offers. These require a significant amount of *protocol stack*, or protocol message processing, including the following:

- Copying TCP segments in and out of system buffers.
- Reassembling of IP datagrams that have been fragmented.
- Calculating TCP checksums across each data segment/packet.
- Processing acknowledgements on incoming and outgoing traffic.
- Detecting all packets that get lost or arrive out of order while trimming overlapping segments.
- Enabling/disabling retransmission timers and generating and processing retransmissions, if necessary.
- Updating congestion windows and slow start thresholds..
- Keeping all data transmission within the corresponding permissible windows.

TCP protocol-based sessions require access to memory where all the session-related data is stored. TCP, for example, requires easy access to station IDs and port numbers for each session. This data must be accessed every single time a session sends or transmits something. The reader must imagine the same process in a context where hundreds, if not thousands, of simultaneous and independent sessions are sustained. Implementing the TCP protocol traditionally meant that software engineers would

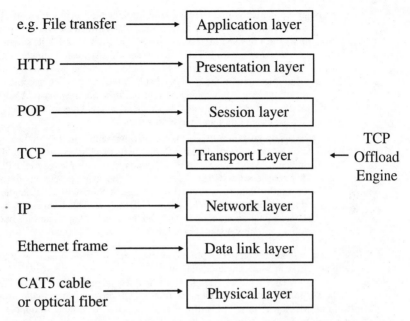

FIGURE 11.4 TCP termination engines assume the burden of running the TCP portion of the protocol offloading a significant weight from the host CPU.

create a lookup table where the session-related data for each active session is kept. However, this is good for simple software implementations on low-bandwidth devices such as a PC where a user will not have thousands of long-lived TCP sessions going on at the same time.

Compared to this realm, the SNP requirements are radically different. If the same traditional implementation approach is to be followed here, gigantic lookup tables will be required because multiple lookup operations occur per second for each session. Therefore, the multiplicative effect of traffic means that for thousands of sessions, hundreds, if not millions, of lookup operations per second must be performed. As a result, the following items of a TOE solution will be significant:

- The physical size
- Cost
- Speed
- Power consumption
- Reliability (as multiple components will fail more frequently than a single highly integrated one for different reasons, including heat emitted by other components)

If the TCP implementation does not radically change, an inefficient board-level product will have to be used. Of course, this does not mean that all boards are inefficient. The distinction is made more along the direction of traditional software-based protocol implementation versus integrated hardware-assisted acceleration in a chip. When coupled with a few ancillary and specialized chips, this can help provide cutting-edge performance at line rates.

The following are a few other important systems issues to consider when implementing an integrated TOE solution:

- In a board implementation, when the TOE implementation has to go off-chip to retrieve that session data, a severe time penalty occurs. This immediately translates into lower performance (latency and throughput) from the SNP device. An ASIC approach of the TOE solution, on the other hand, can combine *embedded memory*. This has a tremendously positive impact on the latency and throughput performance, along with the other parameters from the previous list. More importantly, to avoid making multiple visits to the memory bank to retrieve session-specific data, state-of-the-art TOE implementations utilize embedded *content-addressable memory* (CAM). The use of CAM enables a single memory lookup to retrieve all pertinent data that has been properly indexed. It is a key technology for the acceleration of TCP termination.

- A certain degree of *parallelism* is more than warranted in this application. Instead of having one single CPU sequentially deal with different parts of TCP processing for different sessions (no matter how powerful it is), it is much more efficient for multiple parallel CPUs to tackle TCP processing for different sessions. One CPU can handle checksum processing for one session, while another CPU can reorder data for yet another session. A third CPU can handle TCP flow control or be involved in the startup process for a new TCP session. This is an easy way to enhance performance in the TCP termination process. If you read the first 10 chapters of this book, you will see the applicability of several network-processor architectures in this field.

From a systems architect's point of view, plenty of room is available for subjective choices. We will now look at some implementations.

We will first provide an example from Adaptec, a leading provider of TOE technology. According to their estimates, an ASIC implementation of TOE provides sustained transmission rates of 900 Mbps to 1,000 Mbps with host CPU utilization of less than 20 percent, as opposed to board-based products delivering TOE functionality. Although the host CPU utilization remains at the same level (< 20 percent), these products can only sustain about 650 Mbps of traffic. In April 2002, Adaptec also announced that one of its products (ANA-7711 TCP/IP Offload Adapter) sustains 226 MBps full throughput of variable-packet-size traffic, including packets that are as small as 512 bits. This product, whose block structure is shown in Figure 11.5, is a board that contains the company's *storage*

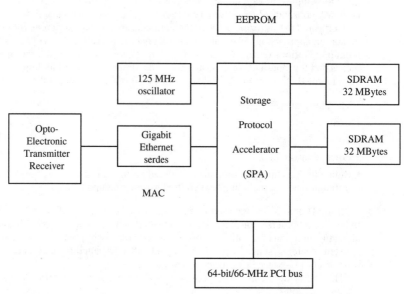

FIGURE 11.5 The block structure of Adaptec's ANA-771 TCP/IP Offload Adapter. *(Source: Adaptec)*

protocol accelerator (SPA), a highly optimized TOE, and a 1 Gigabit Ethernet MAC. This level of performance means that full-duplex (and almost saturated) Gigabit Ethernet traffic can be sustained in storage network processing using this level of TOE technology.

Adaptec has used several of these techniques in its ANA-7711 TCP/IP Offload Adapter to provide a cutting-edge solution. As shown in Figure 11.5, the idea is to connect a 1 Gigabit Ethernet SAN (or NAS) to the host *Peripheral Computer Interconnect* (PCI) bus. From an architectural standpoint, the *transmit* and *receive* paths are mapped onto different processors. The processing core of the adapter is a pipeline of network processors. Different processors work in parallel to handle different functions of the TCP/IP protocol stack. As a result, the technology scales easily to 10 Gbps.

The company provides drivers and *application programming interfaces* (APIs) and supports Linux and Windows environments so software can be upgraded to future platforms. *Synchronous dynamic random access memory* (SDRAM) is used for data and header storage, whereas the *electrically erasable programmable read-only memory* (EEPROM) bank is used to store systems code such as the serial bus interface and, more importantly, programmable MAC addresses. The Ethernet interface, which supports both copper and fiber, offers configurable TBI/GMII interfaces with full IEEE 802.3x flow control, IEEE 802.3z compliance, and *Remote Monitoring* (RMON)/*management information base* (MIB) support so traditional data-processing network management can also be expanded to encompass storage networks.

The TCP/IP engine in the Adaptec ANA-7711 can handle TCP segmentation and reassembly in the hardware; provides slow start, congestion, and sliding window; supports 1,000 TCP sessions; offers the capability of selective acknowledgements; and allows the choice of a configurable window size and *Maximum Transmission Unit* (MTU) size. It can handle all TCP encapsulation and segmentation requirements as well as TCP decapsulation and reassembly.

At the other side of the capability spectrum, a lower-speed approach is taken by Adescom (*www.adescom.com*), which offers a TOE on its IPAC E100 core that integrates TCP offloading with an Ethernet 10/100 Mbps MAC and supports up to 64K connections. This might seem like overkill as 64,000 TCP connections will rarely need to be supported on a Fast Ethernet link, but it is important to try to gauge an estimate of the number of realistic sessions expected at each end of the hardware offering spectrum.

In the case of offloading TCP or even the entire iSCSI work, such as Alacritech's *Session Layer Interface Card* (SLIC™) products (*www.alacritech.com*), performance constraints must be considered. Alacritech took an interesting approach: The whole protocol stack is collapsed and then statefully processed in an optimized fashion in order to decrease network latency and increase data throughput. The word *statefully* means that the silicon-based engine on the adapter (or as a coprocessor) simultaneously inspects and processes data structures that are traditionally handled sequentially and at different layers. Other TOE approaches offload TCP/IP, but process each layer of the protocol stack sequentially.

In addition to this processing acceleration (hence the name *accelerator* for the company's products), the SLIC approach does not keep multiple copies of data. It also distinguishes itself by two important architectural factors: it uses hardware *direct memory access* (DMA) to access memory buffers on the host system while transferring data to or from adapter memory, and it minimizes the interrupt load on the host. Traditional NIC cards using interrupt aggregation techniques have to interrupt the host CPU on which TCP/IP has traditionally been running with every packet or series of packets that require it. The SLIC approach is iSCSI sensitive so it interrupts the host CPU only at boundaries of iSCSI commands, just like an HBA would do. Alacritech server and storage accelerators based on the SLIC technology support the IEEE 802.3ad Link Aggregation protocol and Cisco Systems' Fast EtherChannel and Gigabit EtherChannel protocol for failover and link aggregation.

Contrary to the approach several network-processor vendors have been taking, efficient implementation of the solution to this problem is not just a question of taking the TCP protocol, breaking it arbitrarily or to one's best guess into component pieces, and deciding which ones will be implemented on the fast path (data path) and which ones on the slow path (control path). It is also a question of how to cleverly reorganize the protocol stack to ensure optimal processing and interfacing with other components and/or subsystems in the customer system's hardware, software, and firmware (drivers).

CASE STUDY 1: TREBIA NETWORKS' SAN PROTOCOL PROCESSOR (SPP)

Trebia's SNP architecture, which is shown macroscopically in Figure 11.6, enables the company's flexible approach to be applied in several SAN applications. The company's *SAN Protocol Processor* (SPP) is a high-performance and storage-network-specific network processor that offers the flexible support of emerging IP storage technologies, which often must be seamlessly connected with legacy Fibre-Channel-based storage network infrastructures. When we say that SNP protocols are supported, we mean that IP storage protocols such as iSCSI and FCIP run natively on the SPP and at line rates, for example.

An embedded, powerful, and feature-rich TOE completely removes the burden of handling the proper termination of multigigabits-per-second TCP flows from the proverbial shoulders of the host processor. As a result, plenty of processing horsepower is readily available for other critical storage network-processing tasks such as classification and security. Such a robust TOE approach is required for the efficient iSCSI termination in HBAs and endpoint devices.

The Trebia SPP architecture is optimized for storage I/O flows. More specifically, it can process pipelined storage commands and provides low-latency IP and FCP flow termination. It is also equipped with multiple reissue features (for both the IP and Fibre Channel realms) that are needed in order to support fundamental SAN capabilities such as SCSI termination, FCP-to-iSCSI mediation, and storage virtualization.

In terms of presentation, the SPP design is fully integrated in a *system-on-a-chip* (SOC). As shown in Figure 11.6, it packs the following items inside the same piece of silicon:

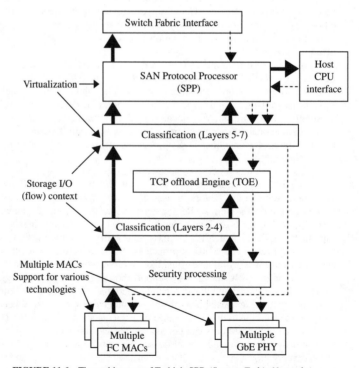

FIGURE 11.6   The architecture of Trebia's SPP. *(Source: Trebia Networks)*

- Multiple storage network interfaces.
- Two-tiered classification capabilities (below and above layer 4).
- The powerful TCP termination engine.
- The main SPP processor of the multiple storage network protocols.
- A module that handles security functionality.

In terms of scalability of performance, the Trebia SPP is capable of handling today's Gigabit Ethernet or Fibre Channel network infrastructure. However, more importantly, it is already capable of dealing with requirements for the next improvement in SAN throughput—the 10+ Gbps realm.

From a business standpoint, Trebia offers higher performance, higher integration, and lower cost per port than current solutions, which rely on a group of chips. At a minimum, these will include the following:

- An off-the-shelf TOE, which often performs nothing more than a checksum acceleration and automatic issuance of ACKs.[2]
- A previous-generation network processor, which usually does not have the stamina for the sustained classification workload at 10 Gbps full-duplex rates.
- The appropriate PHY/MAC and heavy-duty security coprocessing on a dedicated adapter board. A board is obviously bulkier, consumes much more power, costs commensurately more than a single chip, and is almost per definition less reliable than an integrated circuit.

The SPP approach from Trebia can be characterized as a next-generation design as it offers an improvement both in price/performance and the level of integration while allowing the flexibility through programmability to adapt the functionality to newer protocols and applications. This ensures that products do not become easily obsolete or field service and uprgradability is not compromised. These characteristics should allow the company's customers to accelerate their time to market for new affordable and high-performance SAN-related products. This technology should include the following products:

- SAN switches/routers/gateways.
- Legacy LAN internetworking systems (bridges/gateways), which are in need of storage networking interfaces in order to expand their marketability.
- Storage on LAN converged systems that offer both LAN and SAN solutions.
- Endpoint solutions such as servers/HBAs, storage systems, and even NAS devices.

CASE STUDY 2: SILVERBACK SYSTEMS iSNAP™ ARCHITECTURE

Silverback Systems (*www.silverbacksystems.com*) takes another relevant but distinct approach in the design of their *Storage Network Access Processor* (iSNAP) chip. The company embarked on the design of an SNP that has the following unique characteristics:

- Provides an integrated chip solution that minimizes component cost, solution cost by the mere impact on the chip count, and power consumption while maximizing reliability (as there are less components that can fail by heat radiation caused by other components, for example).

---

2. It is assumed that the reader is familiar with the detailed operation of the TCP protocol over IP, where in order to guarantee the reliability of the link, all received frames have to be systematically acknowledged by the transmission of ACK messages. Any good textbook on TCP/IP internals explains this topic in depth. See, for example, Douglas Comer, *Internetworking with TCP/IP Vol.1: Principles, Protocols, and Architecture*, 4th ed. (Upper Saddle River, New Jersey: Prentice-Hall, 2000), or Richard Stephens, *The Protocols (TCP/IP Illustrated Vol. 1)* (Reading, Massachusetts: Addison-Wesley, 1994).

- Fully terminates TCP in full-duplex Gigabit Ethernet, thereby offloading the host CPU.
- Natively supports multiple protocols such as iSCSI, NFS, and CIFS.
- Keeps *upper-layer protocols* (ULPs) fully aware of *protocol data unit* (PDU) content, placing incoming data directly into application buffers such as Fibre Channel.
- Fully maps the protocol onto the hardware, thereby achieving high throughput and low latency, which are two critical parameters that must be satisfied in order to be able to sustain wire-speed performance with the smallest possible I/O block sizes, as is the case with OLTP environments.

The company has combined several technological advances to achieve the objectives. For instance, a patent-pending technique of memory management eliminates the need to move the processed data around, which can significantly waste time. The management of multiple queues and two-tier classification allow a *class of service* (CoS) approach to flow management for ordinary networking data traffic and iSCSI traffic. Hardware-assist units execute fixed functions such as performing integrity checks and running traffic statistics. Therefore, the iSNAP chip is designed to avoid things such as interrupts, memory access bottlenecks, and context switching overhead. As a result, net performance is maximized. Current iSNAP implementations can handle up to 50,000 TCP connections. In future designs, this technology is poised to scale to 10 Gbps storage network performance.

Figure 11.7 illustrates the iSNAP hardware architecture. Incoming data are first placed in the SDRAM. The classification engine performs integrity checking. Based on its results, it generates an event in the appropriate queue. All TCP/IP and ULP events are dispatched by the Queue Manager to the processor nodes, which act directly on a packet's header data without having to move the packet around which other architectures would do. When the bit content is properly built and structured, the

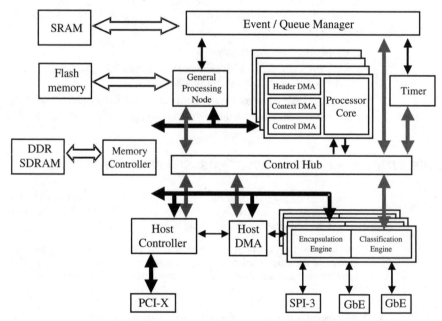

FIGURE 11.7 The hardware architecture of the iSNAP processor. *(Source: Silverback Systems)*

packet is transferred by DMA to the host. If the direction of the packet is outbound, the header information and the data will be forwarded separately to the encapsulation engine. The splitting of data from the header and the queuing of all TCP/IP and ULP events are key tasks in the elimination of access conflicts and bottlenecks, which ultimately maximizes the iSNAP chip's performance.

In a method that is reminiscent of Fibre Channel HBA designs characterized by optimal host CPU offloading and low latency, the iSNAP host interface is built so that PDU awareness allows the *direct data placement* (DDP) of host-bound incoming data into specific application buffers. This minimizes the number of interrupts issued to the host and improves the overall performance. Hardware performs the *cyclic redundancy check* (CRC) and checksum of all iSCSI PDU data. External memory is protected by *error correction coding* (ECC), whereas all internal memory is parity checked. Full-duplex Gigabit Ethernet ports are standard GMII interfaced, whereas the *System Packet Interface, 3* (SPI-3) interface is 2.5 Gbps (OC-48) and aimed at switch/router applications.

The iSNAP software architecture is equally versatile, as shown in Figure 11.8. It is modular and layered so that it provides the necessary flexibility that enables the future upgrading of equipment or modification to accommodate new requirements in protocols and functionality. Driver and firmware combinations enable services such as iSCSI and Link Layer. Link Layer's interface provides standard acceleration such as checksum offload or interrupt coalescing for TCP/IP stacks that may preexist in some systems. The firmware offers extensive management features. These are important factors for mission-critical data storage applications.

In addition, several services can be run simultaneously over the same port, such as iSCSI and NAS acceleration. The flexible approach enables a user to start out with just iSCSI and a native TCP/IP

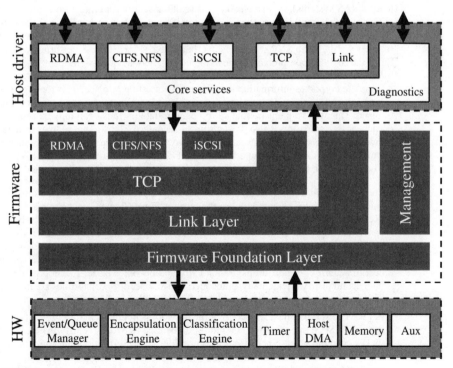

FIGURE 11.8   The software architecture of the iSNAP processor. *(Source: Silverback Systems)*

stack running over the Link Layer. Depending on the necessary functionality, the user can add other services such as TOE.

We do not intend to give a full-fledged description of each product. Interested readers can obtain this information directly from the companies. We introduced a couple of existing SNP solutions to show the direction of the storage industry and depict an important application area for high-performance network-processing technology. Without the appropriate SNPs, this evolution would be impossible.

SECURITY ISSUES IN STORAGE NETWORK PROCESSING

Security concerns are a high priority for SNP designers. Data that is in transit to and from a storage device must be protected at different levels from unauthorized attempts by third parties who may try to intercept parts of SNP traffic or from attacks that are likely to occur if a malicious insider such as a disgruntled employee or an outsider such as an attacker who, acting from outside, manages to hack his or her way inside the IP network. File-level security is influenced by how operating systems grant access to individual files. Storage devices, however, work with block-level I/O operations, which are a completely different story.

We discuss security processors and all the required functionality for confidentiality, authentication, integrity, nonrepudiation, and controlled access in Chapter 17, "Security Coprocessors." Here we only mention some critical issues that make the security-processing context relevant to storage network processing.

A judiciously chosen mix of cryptographic functions is needed inside all SAN devices (and in legacy NAS systems). A manageable and scalable security architecture is also critical. The phrase "judiciously chosen" is important because performance can be penalized if the computation-heavy cryptographic operations are not delegated to specialized hardware that can process data at line speeds. In other words, if the SNP security is overburdened, storage access performance will inevitably suffer. If the SNP is underprotected, the assumed risk may be quite significant.

The security function must be physically distributed at *endpoints* or in gateways. This implies that a complete corporate information security policy must be in place. This topic goes way beyond our discussion here, but session- and packet-level flexible authentication and access controls must also be in place. This will help ensure that stored data or data in transit can only be accessed by authorized parties. At the same time, it also prevents untrusted internal or external sources from launching attacks. In addition, the security infrastructure must provide flexible and interoperable data integrity mechanisms to safeguard against tampering or modifying data as it travels over the SNP.

Endpoints can be broken down to two constituents—the iSCSI initiators and the iSCSI targets:

- For the iSCSI initiator devices, network security capabilities within such products are an important value differentiator as compared to the current *Fibre Channel host bus adapters* (FC HBAs) and even Gigabit Ethernet NICs. In addition, built-in security capabilities within such devices also enable an improved cost of ownership for IP storage deployment. However, the economics of the proposition should not be underestimated. If the cost of security is distributed over a plethora of systems (both computers and storage devices) that are connected onto an infrastructure network, the cost of security per port drops dramatically and becomes easier for an organization/enterprise to budget and justify. Incidentally, embedded network security capabilities are mandatory in order for iSCSI initiators to fully comply with the IETF iSCSI specification.

- For the iSCSI target devices, securing the network is not just a matter of satisfying a critical part of the IETF iSCSI specification. Including embedded network security inside iSCSI target devices makes even more sense because it also provides some very important functional benefits. Having fine-grained embedded network security capabilities enables iSCSI switches, such as iSCSI/Fibre Channel bridges and iSCSI virtualization engines, to provide optimized support for several IP storage deployment scenarios. This can only be accomplished if an all-encompassing but flexible network security is in place. Last but not least, these network security capabilities enable the manufacturers of such products to compete effectively with the current Fibre Channel SAN infrastructure.

The IETF has decided that the security framework of iSCSI will be based on IPsec, a set of cryptographic technologies and specifications. IPsec can not only allow encryption and authentication at different levels of sophistication for all packets and participating endpoints, but it also brings along years of proven solidity in mission-critical circumstances. IP-savvy organizations usually have extensive skills in deploying IPsec schemes. Some leading storage system vendors, including the powerhouse EMC, have also proposed a similar security specification for the Fibre Channel environment to the IETF known as *FCsec*. In conjunction with IPsec, it shows which direction the industry is taking on both of these storage realms.

SECURE SNP TRENDS AND CONCERNS MOVING FORWARD

However, several important differences exist between classical IPsec, such as the IPsec encountered in firewalls, *virtual private networks* (VPNs), or typical modern routers, and IPsec as required and envisioned by secure storage networks. To start with, the performance requirements for multigigabit throughput and latency as well as the session traffic statistics of IP storage traffic require improved efficiency for the implementation of wire-speed IPsec, when compared to classical data-processing network traffic IPsec operations. The high percentage of large packets for IP storage networks as opposed to the classical data networks case should also be considered. Sessions between IP storage endpoints usually last longer than many traditional data-processing applications (for example, e-mail or *Hypertext Transfer Protocol* [HTTP]-based file transfer involved in web browsing).

Looking to the future and thinking along the paths of integration and consolidation, the question arises as to whether an SNP-optimized IPsec processor can be implemented next to a full-fledged TOE inside the same piece of silicon. Although the answer is not a flat-out no, the objective remains largely elusive if today's state-of-the-art *very large scale integration* (VLSI) design tools and semiconductor technologies are considered. This is because in addition to the massive cryptographic prowess that such a chip must have, it must also be able to correctly terminate several thousands of TCP/IP-based iSCSI and FCIP connections with low latency and at line speed. Each connection carries multigigabit-per-second traffic. It must also be superbly intelligent and flexible so that it can manage all of these sessions seamlessly and properly, ensuring compatibility with multiple systems and different software systems.

This seems quite a few years away from today's reality. Therefore, IPsec processing for IP storage networks in the short to medium term will most likely have to be implemented as a small daughter-board carrying a couple of complementary-function chips. This daughter-board would work as an adjunct coprocessor to the main SNP. With the evolution of semiconductor technology, the problem may be addressed in a more integrated way, especially if cryptographic advances allow the more efficient performance of several of the necessary operations in less time and using less silicon real estate.

The security industry is already taking some interesting initiatives to address these issues in a comprehensive manner. For instance, Hifn (a leading IPsec accelerator chip design house—see *www.hifn.com*) has formally teamed up with Trebia to formulate a next-generation security framework that addresses these concerns. NetOctave (another major IPsec-accelerator semiconductor company [see *www.netoctave.com*]) is pushing its flow-through architecture forward to efficiently tackle IPsec processing at endpoints in need of storage network processing at line speeds, as opposed to the traditional look-aside approach to IPsec computations. Tehuti Networks (*www.tehutinetworks.com*) is yet another promising startup. It combines the offloading of TCP termination with IPsec in the same silicon die, allowing gigabit-per-second-level performance and processing at wire speed. Chapter 17 discusses such approaches in more detail.

SUMMARY

In this chapter, we discussed the need for SNPs, which are used to ensure fast and adequate process-ing of data traffic transmitted on a whole new generation of SANs using IP technologies. We discussed the evolution of storage technology, which in its more recent forms, serves as the catalyst for the appearance of several of these chips and board-level products. We reviewed the requirements they are called to satisfy as well as the capabilities that they exhibit. We briefly reviewed a few cutting-edge technological approaches by leading companies at various levels of integration. To set up the discus-sion in Chapter 17, we provided an overview of communications and information security concerns that these SNP chips are required to handle on top of their expected network-processing operations.

SUGGESTED REFERENCES

Refer to the companies' web sites provided throughout the text to find more information about com-panies that design specialized board products and/or network-processing chips. Aristos Logic (*www.aristoslogic.com*) and Astute Networks (*www.astutenetsworks.com*) are two additional companies not mentioned in the text.

Hifn (*www.hifn.com*) and NetOctave (*www.netoctave.com*) are two security coprocessor design companies involved in SNP projects and plans.

References for more similar design houses, whose business emphasis might also eventually include the secure SNP arena, are listed at the end of Chapter 17.

Several white papers have been written covering all aspects of IP storage networks. These can be found on the web sites of companies such as Cisco (*www.cisco.com*) and Intel and (*www.intel.com*).

The FCIA (*www.fibrechannel.org*) maintains an extremely useful web site with tutorials, compar-isons with alternative technological approaches, white papers, and numerous links to pertinent sources of information, including global efforts of standardization.

The SNIA (*www.snia.org*) through its very comprehensive web site provides access to white papers, tutorials, publications, market research reports, an education center with articles and a list of numerous storage-related textbooks, links to multiple related industry-specific conferences and events, a certification program, an impressive glossary for storage technologies, and links to multiple industry resources, including their own recently launched *Storage Management Initiative* (SMI), which intends to develop the Bluefin specification. This creates the advanced object-based manage-ment technology that could lead to the manageable interoperability of multivendor SANs.

More information can also be found at the following web sites: UNH IOL (*www.iol.unh.edu/*) and IETF iSCSI FCIP *IP Storage Working Group* (IPSWG) (*www.ietf.org*).

A major industry-related trade show is the Storage Networking World. Information can be found at their web site at *www.storagenetworkingworld.com*.

Network World is a very interesting trade journal in this field and provides tutorials on new tech-nologies and business case presentations.

In addition to companies embedding their own TOE and TCP termination engine designs inside their SNP chips, several companies are working on a standalone TOE. Here are a few good examples.

Adaptec (*www.adaptec.com*), which acquired Platys (a major SNP startup) in 2001, maintains a large and extremely useful web site with numerous tutorials on storage technologies and TCP offload-ing, white papers, and links to other relevant sites of interest on the Internet.

Another company that provides helpful information is Emulex (*www.emulex.com*).

An excellent industry-specific report on SNPs that provides periodic updates of its content and presents the various products and companies in depth is available from the Linley Group (*www.linleygroup.com*).

Last but not least, several good books have been written on the subject of TCP/IP internals. The following are strongly recommended and contain a thorough review of the subject:

Douglas Comer, *Internetworking with TCP/IP Vol.1: Principles, Protocols, and Architecture*, 4th ed. (Upper Saddle River, New Jersey: Prentice-Hall, 2000).

Richard Stephens, *The Protocols (TCP/IP Illustrated Vol. 1)*, (Reading, Massachusetts: Addison-Wesley, 1994).

Both of these books have subsequent volumes in their series for those readers who want to see complete implementations of the protocols.

CHAPTER 12
SEARCH ENGINES

In this chapter, we discuss search engines. In the network-processing arena, they usually rely for their functionality on associative memory technology, which is also known as *content-addressable memory* (CAM). We discuss how CAM works in the context of search engines and review systems engineering issues as well as trade-offs. CAMs have pros and cons like any other technology. We then look at alternative approaches to the search problem that can provide higher performance than CAM-based search engines but are also more tuned for organizations that can afford them. This chapter provides background to the classification engines, which we describe in the following chapter.

THE PACKET CLASSIFICATION CONTEXT OF A SEARCH ENGINE

We will start by providing a sneak preview of the classification context. We do not intend to spoil the information provided in the next chapter, which discusses specialized classification engines, but we must clarify some basic concepts within the packet classification context. In fact, newcomers to this industry are often confused by the relationship between search engines and classification engines. The two engines will inevitably overlap since chip vendors in pursuit of product differentiation have confused matters. On the one hand, they have packed functionality that undisputedly adds value into their chips. On the other hand, the boundary between the two is blurred as one can find "search engines," "classification engines" and "search and classification engines."

A packet can be handed over to a network-processing system in two ways: either by its own host *central processing unit* (CPU) or, in the case of a switch/router, it can arrive at the *network processing unit* (NPU) as a member of a stream of unrelated or related packets. They may have been streaming by one of the line-card interfaces (following the switch/router's ingress path) or by the switch fabric interface (following the egress path). The NPU will have to conduct several operations with and/or on each one of these packets.

Classification is the very first task that needs to be performed on a packet arriving in a stream of other packets. However, in order to put things into the right context, we must clarify that a *classification engine* (also known sometimes in the industry as the *classifier*) receives as its input an aggregate stream of packets, which the majority of the time are rushing in at wire speed (which can easily reach 40 Gbps). By applying a set of application-specific sorting rules and policies continuously and indiscriminately to all packets (hence the term *classification*), it ends up compiling a series of new (parallel) packet streams (queues of packets) in its output. The packets of each individual output stream or queue (although potentially belonging to completely unrelated sessions, hosts, and/or users) will all share the same fate and short-term destiny. As a result of classification, the packet is *forwarded* to the appropriate output queue of the classification engine.

Given a specific application context, a few steps must be taken in order to correctly handle the classification and forwarding tasks for each individual packet that the network-processing subsystem of a high-speed switching/routing system receives. The NPU must consult a specialized memory bank, some sort of a knowledge base, a lookup table, an information base, or even a database where the appropriate rules are stored. These rules indicate how each arriving packet must be treated and processed prior to its being forwarded to the corresponding queue for subsequent processing.

For example, a *virtual private network* (VPN) box will at least have to look at the destination address field of each arriving packet and decide for every packet whether it must be treated securely. This decision is based on the policy tables that it was given at configuration time. If it must be treated securely, it will be steered to queue A, where secure packets are lined up. If it does not need to be treated securely, it will go to queue B. Once it arrives at queue B, it must decide whether it will be filtered and discarded or filtered and logged to a security management resource. Likewise, a *Multiprotocol Label Switching* (MPLS) switch looks at tags on incoming packets and decides which output port the packet should be steered to and whether some additional operations must be performed on the tag (such as label stripping and replacement). If additional operations must be performed, it also determines which operations are required based on forwarding policy tables.

This consultation of a lookup table or database based on rules and policies for the correct classification requires the use of a *search engine*. Search engines are mostly based on *associative memory*, which is also known as CAM.

CONTENT-ADDRESSABLE MEMORY (CAM)

During a read operation, traditional memory technologies receive as input the address location in the content of which one is interested. The memory produces the bit content of that address location as its output.

The principle of associative memory is based on the inverse mechanism of establishing a relationship between the input and a specific piece of information stored in the memory array. Therefore, it "associates" the input term with something already stored in its content in order to produce the output. In other words, the data—a search string of characters called the *search key*—is presented to the CAM. The CAM will produce an address if a match occurs with any of its content locations. The search key can be created in many ways from the several bit fields that calculate it. In the simplest form of looking up for instance the next-hop address from a routing table, the search key is the destination address itself. Assuming the search result is a hit and not a miss, it will then be used in network-processing designs as the index for access to yet another memory bank from where the system will retrieve the necessary data. This bank is known as the *associated memory*, which is usually an external *static random access memory* (SRAM).

The terms *Binary CAM* (BCAM) and *ternary CAM* (TCAM) are used in cases where the CAM stores 0s and 1s only (BCAM) or 0s, 1s, and "don't cares" (TCAM). Binary searches are still required for many lower-layer applications such as *Media Access Control* (MAC) table consultation or layer 2 security-related VPNs segregation. The latter is by far the most frequently used category higher on the protocol stack, as searches for *quality of service* (QoS)- and *class of service* (CoS)-inspired classification based on layers 3 and 4 must be performed with the use of wildcard characters. Therefore, the use of TCAM is predominant now in the industry.

In terms of available sizes, TCAMs come in 1Mb, 2Mb, 4.7Mb, 9.4Mb, and 18.8Mb chips. Like ordinary memory chips, they are measured in megabits. Unlike ordinary memory chips, the actual capacity of CAM chips is slightly higher than the corresponding powers of 2 found in traditional memory. This is because CAM entries are structured as multiples of 36 bits instead of 32 bits or even 8-bit bytes. Capacity figures are 4.5Mb instead of 4Mb, for example.

One advantage of CAMs is that they can deliver a lot of productive work per *input/output* (I/O) pin, especially compared to regular memory. This is because CAMs produce a result with fewer memory accesses compared to algorithmic approaches, which must use regular SRAM or *dynamic random access memory* (DRAM). Pins are a scarce resource on an NPU because more pins translate into larger NPU packages and the corresponding board size increases; therefore, the cost is rapidly escalating.

Newer NPU products from large and well-established companies such as Agere and even from a few of the more recent startups such as EZchip do not require CAMs or SRAM for lookups even when operating at 10 Gbps line rates. This means that after all has been said about them, CAMs do have some competition.

Pros and Cons

CAMs have the following powerful capabilities:

- They associate the input (comparand) with their complete database content within a single clock cycle. No other type of memory can accomplish this.
- They are configurable in multiple formats of width and depth of search that allows searches to be conducted in parallel.
- They enable multiple CAMs to be cascaded to dramatically increase the size of the lookup tables that they must store.
- They are able to learn what they don't know yet by updating specific entries into their table.
- They seem to have no competition at wire speeds above 2.5 Gbps.

On the other hand, CAMs are reproached for having the following disadvantages:

- They cost several hundreds of dollars per CAM even in large quantities.
- They occupy a relatively large footprint on a card.
- They consume excessive power.
- They suffer from several more generic systems engineering problems when dealing with issues such as painless interfacing with a network processor and updating table entries simultaneously while looking up requests. We discuss these issues and whether these four reproaches against CAMs have any objective merit later in the chapter.

CAM STRUCTURE

Detailed information for a specific CAM can be obviously found in a vendor's product literature (data sheets and application notes). In this section, we limit our discussion to the fundamental notions as applied to the network-processing realm.

The majority of CAMs are implemented in a two-port structure, as shown in Figure 12.1. The comparand bus is parallel (usually 72 bits wide) and bidirectional, because it is used for writing the search keys and for table updates (read/write). The results bus is obviously only an output. A command bus enables instructions to be loaded to the CAM so that it can configure the search operations according to the desired procedure.

CAMs are usually configurable in banks of various sizes, as shown in Figure 12.1. Some of these logical partitions can be set up to be ternary, whereas others can be binary. Parallel searches can be performed this way simultaneously at different parts of the table, thereby increasing the efficiency of the CAM design (which usually is pipelined for that purpose). For example, the Kawasaki[1] 9.4Mb CAM can be structured as 72 bits×128K, 144 bits×64K, 288 bit×32K, or even as 576 bit×16K.

---

1. Kawasaki LSI, "Preliminary Datasheet for 9.4Mb CAM."

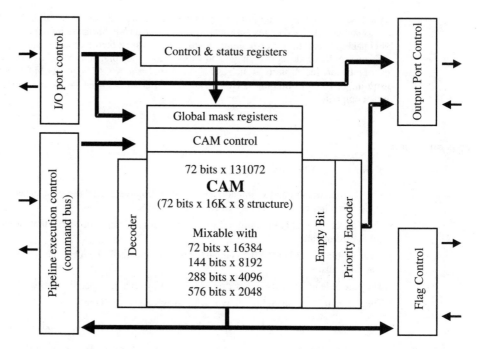

FIGURE 12.1 A typical block structure of a common CAM architecture — the 9.4Mb CAM (KE5BCCA9M) from Kawasaki LSI. *(Source: Kawasaki LSI)*

This specific CAM can be structured in eight banks. Any one of these banks can assume any of the four configurations we just described in the mixed-table example.

In order to retrieve the most pertinent information for the task at hand, the network processor (or custom-designed *application-specific integrated circuit* [ASIC]) issues commands to the CAM. The CAM then performs a search looking for an exact match or uses wildcard characters to extract relevant information. This is accomplished by two sets of mask registers inside the CAM. These mask registers are loaded with the specific bit template patterns against which the table memory content will be matched and the search and match operation will be executed accordingly. The two sets of registers are known as the *global mask registers*, which can remove specific bits from a comparison pattern, and a *mask register*, which is present in each location in the memory (in the case of TCAM). This combination together with the ternary encoding of data in the memory array allows prefixes of complete ranges of partial bit matches to be extracted. These are obviously critical capabilities for making classification and routing decisions involving functionality at layers 3 and 4.

The search result, depending on the CAM design, can be produced as a single output (for example, the result of the highest priority). In the case of multiple hits, it can be produced as a burst of successive results (for example, in order of priority) for subsequent processing by the system. The example shown from the Kawasaki LSI 9.4Mb CAM has an output port that is 24 bits wide. Other CAMs, especially smaller ones (1Mb/2Mb), have an output port that is a 32 bits wide.

Special flag and control signals available on a typical CAM usually show the status of the various banks of the array and denote the type of the search result (single or multiple hit). These signals also allow the cascading of multiple identical devices (at different levels of depth for different vendors) in a chain as a handy way to increase the size of the lookup tables, in many cases without incurring a

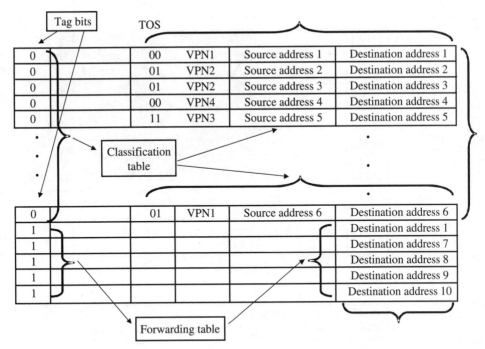

FIGURE 12.2 The concept of tag bits to improve the use of CAM. *(Source: NetLogic Microsystems)*

performance penalty in search time. For instance, our example of the Kawasaki LSI 9.4Mb CAM can be cascaded up to eight pieces with other CAMs without any glue logic and without any degradation in performance. This enables the systems designer to deal with a table that is 72 bits×512K. It is also cascadable (but with a degradation in performance) up to a maximum chain of 32 CAMs that together handle a very large lookup table of 72 bits×2M size.

When a CAM is initialized, some design-specific procedures need to be followed. These procedures depend on the vendor and the actual chip. One system may require that all bit positions in every possible table entry be reset to 0, whereas all bit positions in the mask registers may have to be set to 1 before the table to be loaded is written into the CAM. We say that "we write the table to be searched into the memory" by initializing the CAM. The term *learning* refers to updating specific table entries. The common industry phrase denoting a usual search operation is "writing search keys to the CAM." We must accept it even though it is an unfortunate misnomer since loading a comparand to initiate a search does not involve writing anything into memory.

Most CAMs use key (comparand) sizes that are 72 bits long. Some applications require wider keys that are 144, 288, and lately even 576 bits. In fact, many CAM designs can easily handle several of them in native hardware. It is also interesting to keep in mind that some applications still require shorter keys—namely, 36-bit keys. This can pose a performance problem to CAM devices that are designed to support 72-bit keys. However, these can be handled at the systems level in various ways. We discuss some impacts of these variations on the overall systems design later in this chapter.

CAMs are designed to run at different speeds and are typically clocked anywhere within the 66 to 133 MHz range. However, although a search is issued within one clock cycle, these frequencies only

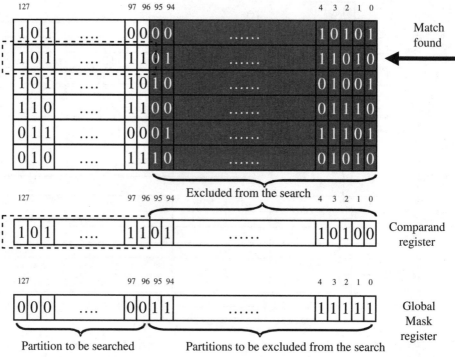

FIGURE 12.3 Partitioning a CAM array into multiple tables and the method of accessing them individually.

denote the *maximum search capability* of the CAM. This is because the *actual search performance* often depends on multiple additional factors, including the size of the key and the size of the lookup table, which may be so large that it requires multiple CAM chips to be cascaded. The speed of a CAM is denoted by *millions of searches per second* (Msps) or by *millions of lookups per second* (Mlps).

The *latency* in lookup table operations based on a CAM is another important measure of performance as the systems designer must know how much time his or her design must wait every time it issues a search and until the CAM yields an answer. In the case of CAM, latency is therefore used to measure the time between the moment when the search key has been presented to the CAM's input and the moment when the result has been produced by the CAM's output. This does not include the time needed to access the associated data SRAM by using the CAM output as an address index to retrieve the necessary data. Therefore, be careful about how the numbers are interpreted. One of the great characteristics about CAMs (as opposed to other types of hashed-index memory with which one might be tempted to build a content-addressable memory) is that latency is deterministic. However, this depends on the actual CAM design and clock frequency. Typically, latency can be two or three clock cycles long, but it can also be twice as much or more—namely, for cascaded CAM configurations.

In terms of lookup latency performance, it is also possible that vendor-published numbers may not necessarily be telling the truth. For example, a search engine's lookup latency numbers can be hidden in a system by cleverly adopting pipelining and multithreading functionality that is available inside the network processor. Turning the argument the other way, the most astounding (and most expensive) CAM component may not be necessarily needed in order to meet system search performance requirements. If the NPU architecture and software development toolset allow the tinkering of func-

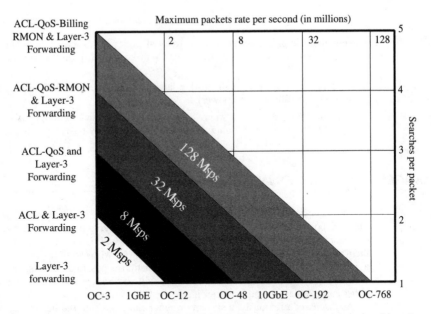

FIGURE 12.4 The interrelationship between typical applications, packet rate, wire speed, and the estimated need for search capabilities. *(Source: IDT)*

tionalities such as the creative and efficient allocation of computational load on multiple engines, stages on a pipeline, or threads running concurrently, then the systems designer suddenly has more freedom.

To get an idea of the technology evolution and where the search engine industry is headed, refer to Figure 12.4. In this figure, the interrelationship is shown between wire speeds, packet arrival rates, typical application loads involving a spectrum of cases spanning from simple *Internet Protocol* (IP) routing and *access control lists* (ACLs) all the way to consolidated environments with QoS provisioning, per-use billing, even network management using *Remote Monitoring* (RMON), and the commensurate number of searches needed per packet.

In order to meet carrier-class requirements for QoS and *service level agreements* (SLAs), the expected classification performance for OC-768 (40 Gbps) layers 4 to 7 applications will usually require a search performance of at least 125 Msps speeds and sophisticated classification-based forwarding that will be decided on rules applied to a set of up to eight fields. With millions of users and tens of millions of active sessions in a large metropolitan network, a router's classifier must be able to look up answers by searching through a database of two plus million rules.[2,3]

Given the current status of CAM technology, which is already pushing silicon to its limits (which keeps the costs high through semiconductor production yield), Figure 12.4 proves that in order for the trend to continue, CAM vendors will have to come up with new advanced techniques such as parallel lookup engines that will allow the concurrent execution of even more searches per second.

2. Jose Pereira, "Moving Classification and Forwarding to OC-768," a NetLogic Microsystems white paper.
3. IDT white paper, "Taking Packet Processing to the Next Level."

In summary, CAMs assist the systems designer when the following tasks are performed:

- Recognizing large bit patterns (they do a lot of work per trip across the I/O pins), where approaches using conventional memory typically need to make many trips as the bit pattern to be classified gets bigger.
- Handling tables that are small (storing large lookup tables in CAMs is prohibitively expensive in terms of both chip cost and power).
- An application environment where lookup latency is critical (although latency can often be hidden with a suitable use of pipelining and threads with memory-based approaches).

MANAGEMENT OF TABLES INSIDE A CAM

The direct cost of a CAM in dollars and the indirect cost of its use, such as power consumption and line-card board real estate, are probably the two main driving forces behind the creativity that designers must exhibit to optimize the functionality of CAMs. It is important to enter tables and maintain them properly while maximizing the usability of the CAM. We will look at some clever ways of managing the available space so that as much information as possible can be squeezed into the CAM real estate. As tables are periodically updated, we will also look at some issues resulting from relocating either entire tables or simple entries inside the CAM array.

As discussed previously, a system designer can maximize the occupancy of the array of useful entries by partitioning a CAM into segments. We will illustrate this point with an example from NetLogic.[4] More details on this approach and similar ideas can be found in product literature and application notes.

Imagine storing data into four tables that are IP addresses and are therefore 32 bits wide. If the CAM is 128 bits wide, unless the CAM array is partitioned into four tables, the storage capacity will be poorly used, as each entry will only store 32 bits. As a result, the rest of the bit positions that the CAM has available in each slot are wasted (128 available minus 32 used equals 96 wasted bits per slot). In this example, however, the four 32-bit-wide tables can be arranged next to each other. Every 128-bit slot is first split into four slices of 32 bits. These are numbered 3rd, 2nd, 1st, and 0th going from left to right. Each one of the four individual tables then occupies one of the four 32-bit slices of each 128-bit slot and runs the entire length of the CAM. If the CAM is, for example, a 1Mb array that was originally arranged as $8K \times 128$ tables, it can easily be structured as four $8K \times 32$ tables.

Figure 12.3 shows how to work with the global mask registers to access only one among these four tables in order to perform a search. Bits set to zero in the global mask register guide the search to the corresponding table. For the four tables (partitions), the global masks corresponding to the individual partitions would look like the following (in hexadecimal):

Mask 3: 00000000	FFFFFFFF	FFFFFFFF	FFFFFFFF
Mask 2: FFFFFFFF	00000000	FFFFFFFF	FFFFFFFF
Mask 1: FFFFFFFF	FFFFFFFF	00000000	FFFFFFFF
Mask 0: FFFFFFFF	FFFFFFFF	FFFFFFFF	00000000

In this example, only the specific 32-bit part of the comparand that is allowed by the global mask register is relevant. Consequently, searches can be conducted on any one of the four tightly packed tables.

A CAM can be partitioned in numerous ways. The specific design of each CAM chip enables the designer to use his or her imagination differently in each case. Judicious partitioning has been shown

---

4. NetLogic Microsystems Application Note, "Intradevice Configuration of Network Search Engine."

to enable the usability of almost close to 100 percent of an array, if the word sizes are chosen to be smaller than the default organization of the CAM.

Partitioning the CAM is an interesting way of enhancing its usability. However, a systems designer must determine whether entries are valid. This is accomplished by using a special bit at each word location that indicates whether the corresponding entry is valid data. This is similar to the remarks we made earlier about CAM initialization. When the system in turned on, the internal state of the CAM is automatically initialized to unknown values; therefore, the valid-entry bits are crucial for making some decisions.

We will examine a couple of interesting ideas as to how to further optimize the use of a CAM by loading tables more judiciously. For instance, NetLogic Microsystems has proposed the concept of *tag bits*. Tag bits enable searches to be performed on subsets of stored data. The idea is that a specific bit at each entry word is arbitrarily chosen to denote that this specific entry belongs to a defined subset (subtable) of the overall table. It is also tacitly assumed that the entry word is smaller than the organization of the CAM. In other words, if the entries are 128 bits long, in order to let one specific bit (say, the far-left one) among those be the tag bit, the table entry must obviously be shorter than 128 bits.

For instance, a systems designer may want to store two different tables that contain some common subdata inside the same CAM for economy. This could occur if a systems designer wants to limit the number of components on a specific line card and consequently wants to store two different tables inside the same CAM.

Figure 12.2 provides a simple example of this situation. A classification table (with a 32-bit source address, 32-bit destination address, 2-bit *type of service* [TOS], and 16-bit VPN number field in every entry for 82 bits total in this example) and a forwarding table based on a 32-bit destination address on every entry are stored in this example. Note that the destination address field appears in both tables. Unless there is a physical way of ensuring that the search operation is performed only against entries of the specific subtable that needs to be accessed, some unfortunate matches may be erroneously made when a miss should actually occur.

Tag bits easily solve this problem. One specific bit of each entry (let us say the far-left bit for convenience) is tacitly assigned to denote the corresponding subtable. If the tag bit is 0, it could mean that this entry belongs to the classification table. If the tag bit is 1, it could mean that the entry belongs to the forwarding table instead. During a specific search operation, the network processor loads the search key (comparand) into the CAM configuration register (or whatever this control register is exactly called in a specific CAM product) and the tag bit is set to the correct value corresponding to the subtable to be searched. This means, of course, that the software engineers in charge of developing that part of the NPU software must be careful to not set the corresponding bit of the global mask register in the CAM (by issuing the incorrect formatted command). This would completely inhibit the intended functionality of the tag bit.

Tag bits can also be used as the following:

- *Validity bits*, which are set to 1 for valid entry and to 0 for empty, which would allow the elimination of empty or inappropriate positions from participating in a search.
- *Skip bits*, which can be quite useful when multiple matches have been scored and they must be sequentially read out from a subtable of a larger table. The process starts with the skip bits for all entries cleared to 0. As soon as the highest priority match is read, the user sets the skip bit to 1 by performing a read/modify/write sequence and reinitiates the search, which will yield the next lower priority. The process can continue until all matches have been read.

SYSTEMS ENGINEERING ISSUES SURROUNDING THE USE OF CAMS

We will conclude our short overview of CAM-based search engines by discussing some issues that systems designers must consider in order to make the best decisions.

In many high-speed network-processing systems, several searches must occur simultaneously in order for the equipment to guarantee deep packet inspection and processing at wire speeds. Traditional classification applications need to look up a destination address to make a decision. With the current flow management implied by the differentiated services that carriers want to offer, which are based on stringent QoS and CoS requirements that are imposed onto the equipment designer, the packet classifier must be able to dive deep into the packet content and extract specific fields for subsequent processing. This means that the search engine that supports the classifier must be able to produce results within extremely short amounts of time. In many newer applications, several tables will need to be consulted at the same time.

For example, say that a MAC table, an IP table, a rules table, and a flow-management table must all be consulted in parallel, as shown in Figure 12.5. These tables will need to be loaded and maintained into four partitions of the same CAM, or four different CAMs (each with their own associated SRAM memory) will need to be searched in parallel.

What are the corresponding implications of these two approaches?

- The first solution is usually unacceptable as some tables are gigantic and others are small. In this case, some partitions may end up being too small to fit the larger tables, whereas the smaller table(s) may end up occupying more partitions than they should. This approach wastes expensive partitions that could be used more efficiently.

- The second solution is not negatively affected when larger tables are used, as they will each have their own CAM. However, it does suffer when smaller tables are used, as they don't easily justify an entire CAM of their own. The overall cost also increases significantly, because in addition to extra SRAM, some CAMs cost more than the network processor itself!

In Chapter 9, "Other NPU Architectures," we described an interesting approach that was taken by Silicon Access (with its integrated solution). With this approach, the associated SRAM is embedded inside its search engine chip. This definitely minimizes component count and power consumption.

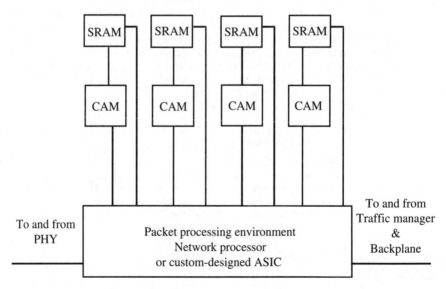

FIGURE 12.5 A typical approach based on a multiple-CAM arrangement for next-generation multitable parallel searches.

Another area of interest is when search keys are 36 bits long. As most current CAMs are designed with 72-bit search keys (comparands) in mind, some designs use two operations on 72-bit words to accomplish a search operation based on a 36-bit key. This is done by *soft* implementation. Although it gives the systems designer the convenience of both key sizes, it decreases performance for 36-bit search keys. Soft solutions for 36-bit search keys provide less than half of the maximum-rated search speed performance, precisely due to the problem just described. Some CAMs are hardwired to natively support both 36- and 72-bit search keys. These are fast in both modes and are easier to use if flexibility in the designs of various search keys must be maintained. However, because of the extra complexity in hardware, it should not come as a surprise that they are a little more expensive.

We have already discussed the performance rating of CAMs. An interesting case appears when the hardware limits the systems designer to a comparand bus that is 72 bits wide, but the actual application's search key is wider—for example, 144 bits. The systems architect has two choices:

- Use a *double data rate* (DDR) bus and load meaningful bits for the comparand at both the rising and dropping edge of the clock.
- Double the clock frequency of the bus that loads the comparands.

Now let's turn our attention to another important issue. We alluded to the fact that CAMs cannot be updated in a location while searching at the same location. Therefore, the systems designer must do some juggling. For example, all search requests can be steered to a backup CAM every time that an update operation must be performed on the primary CAM.

Some systems don't allow searches to go on while an update operation is being performed. It decreases the overall performance of the system. As a result, traffic will need to slow down and packets will need to be buffered up until the update operation is concluded and safe search operations can be allowed to resume. Some designs offer a third port that allows convenient table maintenance without inhibiting search operations. SiberCore CAMs are an excellent example; they are based on a nonintrusive interleaving technique and leave the search path unobstructed while external sources are engaged with the introduction of new table entries. Of course, this flexibility causes a significant increase in the CAM pin count, board real estate, and signals to route to the appropriate place. For more budget-conscious designs, two-port designs must be used where table maintenance can usually occur when a search is not occurring.

However, table maintenance is not just about introducing updates into the table. It may also involve relocating entries or even entire tables to different parts of the CAM because too much empty space may have been created between subtables following the continuous updating of entries. An example where this problem arises is in *Classless Interdomain Routing* (CIDR) (RFC 1519) routing, which was the *longest-prefix match* (LPM) algorithm, which is used in CIDR. The routes used in this scheme are described as a prefix and a prefix length. When a search is conducted (if the table has been properly structured and maintained), the location of the entry will produce the LPM.

If the table must be reshuffled because one segment is full, extra operations are required that eat up time. This is a critical factor when developing applications that must respect and sustain traffic arriving at wire speed. A read and write operation is used for every entry word that must be relocated. If the start addresses of entire blocks must be readjusted following such moves, the corresponding mask word must be reloaded each time—this also involves a read/write sequence. The software designer, who in this context is predominantly concerned about the search capabilities of his or her implementation, must take all these issues under consideration to ensure that production code remains robust under these circumstances. From a user's point of view, searches cannot be affected simply because tables must be reshuffled.

To further show the impact of efficient table lookups and good search-engine-based table management on the network behavior, consider the following scenario. In the previous paragraph, we mentioned that four operations are needed each time a table entry is moved to a new location. However, in cases like CIDR routing, the segments are created according to the prefix length and some empty slots are left in each segment to accommodate new entries. If a segment is suddenly filled up, the table must be taken offline to reshuffle the entries. This is an annoying situation.

The worst-case scenario is when all segments except one are full.[5] From that point, any new entries will require 31 move operations. Each move requires four commands (with one clock cycle per command) to the CAM—both reads and writes. This brings us to at least $4 \times 31 = 124$ clock cycles per move operation. This is conservative, because occasionally supervisory code must be executed to calculate the boundaries of entire segments. However, we will ignore this for now in order to make a point.

Approximately 3,000 route updates occur per second (if not 4,000 by the time this book is published) in a typical core/edge router. This means that $3,000 \times 124 = 372,000$ cycles per second must be spent on activities that update and maintain the table entries. If the packet-processing engine in the router is clocked at 100 MHz, the corresponding cycle time is 10 nanoseconds (10×10^{-9} seconds). This means that the 372,000 cycles spent on table updating and maintenance consume the following:

$$372,000 \; cycles \times 10 \; nanoseconds \; per \; cycle = 3,720,000 \; nanoseconds =$$
$$3.72 \times 10^6 \times 10^{-9} = 3.72 \times 10^{-3} = 3.72 \; milliseconds$$

In the case of OC-192 links, which are typically characterized by aggregate flows of around 20 to 30 *million packets per second* (Mpps), this means that 3.72 thousandths of 20,000,000 or 30,000,000 will be affected because table entries in the search engine of this simple example must be reshuffled. This is an increase from 74,400 packets per second to a staggering 111,600 packets per second. Therefore, at least 74,400 packets per second will not be classified properly. The router, which must struggle to sustain wire speed, will probably just discard them. This is a huge number of packets to lose!

Of course, if the NPU used in the heart of a switching/routing system like this can buffer some of the packets during a CAM update, they may not be entirely lost. However, this requires that the long-term average rate at which lookups can be done is greater than the rate at which lookups must be processed. This may not be possible for some applications and line speeds.

Now these same 74,400 discarded packets per second will cause their respective *Transmission Control Protocol* (TCP) sessions (assuming that they belong to typical TCP sessions) to time out because no acknowledgement response will be received at the source to confirm the safe arrival of these packets to their intended destinations. The TCP congestion algorithm specifies that if a TCP time-out occurs, the congestion window must be narrowed down to the size of a single packet.[6,7] This slows down the entire TCP session unbelievably. This is because as a consequence of the narrower congestion window, the transmitter at the source must wait for an acknowledgement from the receiver for every packet it sends before it transmits the following packet. This is a subtle, but rather spectacular, indication of how much the management of the search engine tables in the router affects numerous sessions.

REPROACHES AGAINST CAM-BASED SEARCH ENGINES

We mentioned earlier in this chapter that systems designers in general have a rather negative view of CAMs.[8,9] It is logical to ask how much this view is justified. To find the answer, we will go through some of the major reproaches that the switch/router industry has voiced against CAM technology and examine their merits.

5. NetLogic Microsystems white paper, "High-Performance Layer 3 Forwarding in CIDR."

6. V. Jacobson and M. Karels, "Congestion Avoidance and Control," ACM SIGCOMM (1988): 314—329.

7. M. Allman, V. Paxson, and W. Stevens, "TCP Congestion Control," RFC 2581 (obsoletes RFC 2001), April 1999.

8. Linley Gwennap, "Is It Time for CAMs?" EE Times (June 3, 2002). This is also available online at *www.eetimes.com/ story/OEG20020603S0011*.

9. SiberCore Technologies white paper, "Classification and Forwarding Co-Processors Come of Age."

CAMs have been accused for having "gargantuan" power consumption needs. Several industry players (both vendors and analysts) have played around with numbers like trial lawyers when comparing older generation CAMs with the more recent chips in order to show that the power consumption of CAM increases. Most of these comparisons do not help systems designers because comparing the power consumption between an older 2Mb CAM clocked at 66 MHz and capable of 66 Msps with a more recent 9Mb CAM that is clocked at 150 MHz and capable of 125 Msps does not make any sense in relative terms.

The issue is much more complicated as power consumption in a CAM is a combined result from multiple and unrelated factors such as the specific semiconductor manufacturing process, the number of searches per second the CAM is called to execute, and the storage density. All these factors come into play in a not-so-obvious set of ways. For example, the smaller the process geometry, the larger the storage capacity. This can cause a drop in the power supply and even an increase in the clock frequency. No wonder CAM vendors have been moving continuously to smaller line widths. 0.25μ processes were replaced by 0.18μ processes. Those were then replaced by 0.15μ processes, which are now being replaced by 0.13μ. The 90 nm realm is on the horizon for CAMs as it is already in use for other semiconductor products. Since power supplies are lower for each new smaller line-width process, CAMs that are built with 0.18μ processes exhibited almost 50 percent less power consumption than their 0.25μ predecessors for the same search rate and clock speed. A 30 percent further improvement occurred with the subsequent movement to 0.15μ.

Meanwhile, more megabits of stored information can be packed onto the same silicon die and more searches per second can be initiated thanks to the higher clock rates. Older products cannot scale to these higher expectations, making these comparisons inappropriate. In any event, the normalized power consumption trend has consequently been pointing downward, if power consumption is to be looked at as watts per megabit. This achievement must be credited to the CAM vendors who have worked hard to make their products more efficient.

From a systems architect's point of view, however, the real issue of power consumption is the absolute value in watts, not the relative value of watts per megabit. Even if vendors with advanced CAM technology can provide the welcome and spectacular performance of 0.95 watts per megabit in their chips, the search engine system's power consumption evil does not lie with the CAM. It lies with the evolving applications themselves, which require that larger tables be stored for lookup and classification based on their consultation and that this continues to happen at wire speed. This is what drives the quest for an increase in CAM size. Therefore, power consumption is a necessary by-product of the issue, or as the cardinal technology law stipulates, this is the price to pay for the luxury of more elaborate classification that continuously needs more powerful search capabilities of larger knowledge bases. The systems designer then has to tackle the following power consumption problem. When moving from one realm to another, and when such a move requires bigger CAMs such as using 18Mb CAMs instead of 9Mb CAMs, he or she must find twice as many watts in a usually very limited power budget.

Another dimension of the power consumption problem associated with CAMs is that consumption numbers as quoted in CAM vendors' web sites and white papers are not typically the worst-case ones. Therefore, designers are strongly advised to ask their CAM vendor early on in the component technology evaluation stage to confirm the worst-case numbers and what kind of offered load would generate the worst-case behavior.

The power consumption problem in CAMs is much wider than what might appear at first glance. In our view, tackling it involves much more than simply sticking the undesirable label on the CAMs.

Maintenance and table management is another area where the industry has been struggling with the ramifications of optimizing the usability of CAMs and minimizing the time to market with software, which can become extremely complicated and heavy at times. Some CAM products lack in this area, but others excel. For instance, the third port (*Synchronous Maintenance Interface* [SMI]) for SiberCore CAMs is an interesting way of having the control plane processor access the CAM out-of-band and modify the table boundaries without affecting the ongoing search processes. We also briefly mentioned the efforts of some vendors to provide sort-free CAMs, so the partial truth in this reproach quickly evaporates. Some leading CAM vendors will end up being successful, whereas others who do not innovate and keep up with the industry will lose ground and ultimately lose business.

The density and footprint of CAMs have also been called a major issue by several quarters, but this is an unfair statement. Only a few years ago, we had 1Mb CAMs. Now essentially all leading CAM suppliers propose their 18Mb models. The need to store large tables inside CAMs has traditionally been seen as a problem that is easily addressed by cascading multiple CAMs. For instance, a *Border Gateway Protocol* (BGP-4) routing table with 100,000 IPv4 routing entries takes two cascaded 2Mb CAMs from SiberCore or one 9Mb CAM instead. With 27 mm 296-pin and 27mm 336-pin PBGA packages for each one of the two chips, respectively, the real-estate savings become apparent. Likewise, a 1,000,000 entry IPv4 address table can be implemented in sixteen 2Mb CAMs or in four 9Mb CAMs from SiberCore.[10] The footprint savings are obvious, if the parallel need for larger tables is considered.

Inflexibility with table configurations is a very broad issue. Unfortunately, many current CAM products suffer in one way or another from this generic weakness. Some CAMs offer more flexibility than others, and the systems designer should verify what features each product offers and how they map to an application. Some systems need tables that are different sizes, but cannot afford the CAM structures that support such a solution. Others need the flexibility at initialization time and less at run time. The reproach has some validity; therefore, time will hopefully make it less pronounced. New research and development will undoubtedly continue to improve CAM products in this regard.

Most current systems designs are based on proprietary ASIC designs for the packet-processing engine, but it is expected that for flexibility and improved cost as well as time to market, this situation will be changing rapidly in the coming years with the engagement of more standard off-the-shelf network processors. One of the highest priorities for the industry is to gluelessly interface the search engine with the NPU that will handle the classification and forwarding. The first CAM designs have not reflected that fact for historical reasons. The wider acceptance of network processors will force the CAM vendors to optimize their interface mechanisms to accommodate at least the most widely used NPUs from established leading companies, such as AMCC, IBM, Intel, and Motorola.

Of course, the NP industry is still very young. Many players are still alive and active (although a couple have gone out of business as of this writing because of the financial rigors of the market). Until some inevitable industry consolidation occurs, vendors are entitled to their view of the world. Because some NPU vendors still claim that CAMs are not needed because they already provide embedded SRAM in their NPUs to store the associated date, the NPU-CAM interface problem is not even being discussed. However, the majority of established vendors do think differently. This is why announcements are constantly being made between CAM and NPU vendors about how they propose to tackle the problem.

One of the interesting efforts is the work that is being done at the *Network Processor Forum* (NPF) and, more specifically, at the Look Aside Task Group (organized under the Hardware Working Group). This organization strives to provide standardized mechanisms for interfacing between all types of coprocessors and network processors. This effort can have very important ramifications of the ability to natively connect CAMs (search engine coprocessors) on NPUs.

CAMs are accused of being expensive. In absolute terms, this is true, especially since some of the latest CAMs are almost more expensive than an NPU or a regular DRAM chip. The truth of the matter, however, is that CAMs are very sophisticated devices that are absolutely indispensable in most designs and cannot be considered as a commodity product. Their complexity and manufacturing entail yield issues precisely because they push the silicon technology process envelope to its quality limits. In our view, only time and consistently larger quantities (such as those experienced by other successful semiconductor products) will improve the situation. Hopefully, we will see an improvement in that front in the future.

---

10. SiberCore Technologies white paper, "Classification and Forwarding Co-Processors Come of Age."

GOING FORWARD

The industry has been actively looking at ways of dramatically enhancing the performance of CAM technology. As we mentioned earlier, CAM vendors have announced plans to create parallel architectures that will enable multiple simultaneous lookups as a clever means of accelerating systems-level throughput. Figure 12.4 shows the dire need for such techniques as we move to higher-value QoS functionality (involving classification based on multiple fields from layers 4 to 7) and higher wire-speed platforms.

One of the noteworthy new techniques that NetLogic has pioneered, which undoubtedly other vendors will want to emulate, is to produce *sort-free* CAMs. This technology purports to automatically manage the gaps between the table partitions so the user does not have to worry about reshuffling the entries. This will lead to streamlined performance and easier table-management-related development software.

Having the ability to restructure the various databases held inside the CAM as lookup tables that can be searched with various key lengths is important; therefore, it is currently one of the major areas of intensive academic and industrial research. For example, designers want to be able to partition the CAM array in 16 databases that can be searched each with a different key that can be anywhere from 36 bits all the way to 576 bits long.

Power management is a real issue for CAMs, and the next generation of CAMs that are denser, deeper, and faster will obviously have to have advanced power-management features in order to ensure their market viability. One of the areas of continuing development is trying to devise an improved, fast, and dependable means of powering down unused database tables. This way a highly structured and partitioned CAM will only keep the portion that contains the currently used tables active. This is expected to minimize power consumption.

Last but not least, CAM vendors are struggling to make their products easily usable by their customers. Users come from various camps. Some of them are proponents of the highest possible performance and are ASIC driven. Others are proponents of flexible network processors instead. CAM

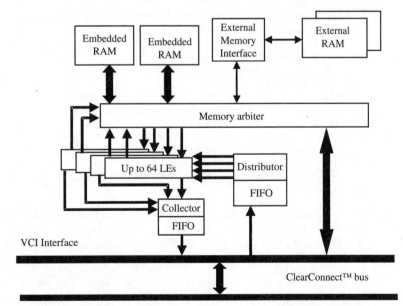

FIGURE 12.6   An example of an implementation of the ClearSpeed TLE using both embedded and external RAM. *(Source: ClearSpeed Technology, Ltd.)*

designers find themselves between a rock and a hard place as they try to please as many potential customers as possible. The proprietary interfaces to each CAM are rapidly becoming a context of the past. Consequently, more CAM vendors take pride in announcing that their latest designs allow glueless interface with most mainstream network processors such as AMCC and Intel.

ALTERNATIVE WAYS OF IMPLEMENTING A SEARCH ENGINE

A very elegant and highly scalable way of handling up to 1 billion searches per second without the use of CAM is the approach taken by ClearSpeed Technology in the implementation of their *Table Lookup Engine* (TLE) *intellectual property* (IP) core.

We have already discussed in Chapter 10, "Alternative Approaches to Network Processing: Net ASICs and Designing with IP Cores," ClearSpeed's approach to the implementation of a powerful network-processor platform that addresses the evolving wire speeds by deploying a massively parallel computation solution to the problem. The ClearSpeed TLE is an optional part of that solution. It can be used in application areas where more than 100 million searches are needed per second. Figure 12.6 shows the arrangement of the ClearSpeed TLE. It is based on a proprietary implementation of the *level compressed* (LC) trie algorithm with some improvements including the number of memory accesses needed per lookup.

Typically, each branch of the tree held in memory requires a memory access. Level compression causes branches to vary in depth; therefore, the algorithm causes lookups with a varying number of memory accesses. The table data is distributed over a series of memory banks, some of which may be SRAM, DRAM, embedded, and external. ClearSpeed provides the software tool that will reformat typical routing tables in a format that optimizes the use of the TLE. This software can run on the ClearSpeed NPU or on a typical control plane CPU. It has many interesting and powerful search features, such as programmable tree depth for table size versus performance trade-offs, LPM with and without false hits, multiple tables, and configurable search key size and results index.

The company's software also allocates the table entries to the various memory banks, so the most necessary ones will be accessible from embedded RAM. The memory banks are also accessible to any device that has access to the on-chip ClearConnect™ bus and can therefore be managed if necessary from another processor such as the control plane CPU. One of the advantages of this solution is that the table can be searched and managed simultaneously, minimizing the downtime for maintenance. The company also claims it has developed a strategy that allows the incremental replacement of a table by only requiring that 10 percent of the table be kept in memory as opposed to storing an extra copy of the entire table.

The *lookup elements* (LEs) are highly optimized state machines that can completely traverse the tree accessing any of the memory banks required for the search. Unlike other solutions that require deterministic latency, the ClearSpeed TLE will produce results of searches as soon as they are available. They can therefore arrive out of order without affecting the order of packet processing, which is designed to be in order for the engine to attain for maximum performance. Multiple tables are handled simultaneously, and atomic table updates can be done on-the-fly. The table structure is dictated by software and search keys can be larger than 32 bits. Both CIDR and structured wildcard searches are supported.

A series of LEs has access to embedded on-chip RAM (either SRAM or DRAM, depending on which one is available for the semiconductor process) where more critical entries should be stored and potentially off-chip memory, in cases where lookup tables are large. Current implementation of the architecture allows up to 64 LEs inside the TLE. The beauty of this approach is that each LE can provide the answer to the searches it launches whenever it obtains a result, which means that searches are fulfilled potentially out of order. This does not affect the macroscopic packet-processing order.

Typically, more LEs will be needed to perform more searches per second or depending on how long each lookup takes. This means that the systems architect will always need to make a trade-off at the system level depending on whether off-chip memory is used and whether this memory is expensive, small-capacity, and power-hungry SRAM or cheaper, high-capacity, and lower-power consuming DRAM. The main disadvantage in the external DRAM case is the latency involved, as searches will take more time to be performed by a specific LE. This situation can be resolved by increasing the number of LEs so the memory subsystem remains saturated and always busy. According to the company, if off-chip SRAM is used, then 24 LEs are enough to sustain 350 Mlps.

In an example implementation of a 40 Gbps 100 Mpps solution in need of search capabilities to support classification, the company has reported using a 260,000 entry table that has between 1.3MB and 4.8MB of memory for search key sizes between 32 and 128 bits. A 400 MHz implementation of the TLE shown in Figure 12.6 was used and there were 32 LEs. 2MB of embedded memory is enough to hold 330,000 trie entries. For the reported example, this was divided to two banks, which each had an access rate of 400 million accesses per second. External memory was structured in two banks that were accessible 200 million times per second. The size of the overall memory depends on the size of the search key. In a typical layer 3 classification application based on a source address, destination address, and TOS that requires a 72-bit search key, this ClearSpeed TLE performed 251 Mlps. A five-tuple lookup based on a 104-bit search key at layer 4 yielded a performance of 208 Mlps.

These numbers clearly command attention. The only downside to the approach is that it is only available as intellectual property (IP core) and not as a ready off-the-shelf component. Therefore, it can only be used by customers who intend to build their own integrated solution and who are consequently advised to take a close look at this approach as well.

SUMMARY

In this chapter, we examined search engines and predominantly those that are based on CAM for the support of critical classification and forwarding processors that we discuss in the next chapter. We reviewed how CAMs operate as well as how they are organized. We discussed several important systems engineering issues that need to be handled when trying to use CAM in larger designs of ever-increasing speed and functionality. We also looked at the trade-offs involved in using CAM technology and reviewed some of the rather serious image problems from which they seem to suffer—some justified and some unjustified. We identified major areas of current development for the improvement of CAM-based search engines and concluded the chapter by providing an overview of an alternative way of creating a fully integrated search engine.

SUGGESTED REFERENCES

The following lists some companies that design and build CAMs and search engines:
Cypress Semiconductor (*www.cypress.com*)

IDT (*www.idt.com*)

ISSI (*www.issi.com*)

Kawasaki LSI (*www.klsi.com*)

Micron (*www.micron.com*)

Mosaid (*www.mosaid.com/semiconductor*)

NetLogic Microsystems (*www.netlogicmicro.com*)

SiberCore Technologies (*www.sibercore.com*)

SiberCore Technologies white paper, "Packet Management Lookups in Modern Networks."

ClearSpeed Technology white paper, "The ClearSpeed™ Table Lookup Engine."

Data sheets and application notes are available directly from these companies. In most cases, however, this is an extremely competitive industry. As a result, the companies most probably will require that a nondisclosure agreement be signed prior to their release of the product literature to the requesting party.

CHAPTER 13
CLASSIFICATION PROCESSORS

In this chapter, we discuss the problem of *packet classification* in network processing. The fundamental notion of classifying the ever-changing dynamic outcome of various events into categories implies that consecutive events of this outcome must be distinguished from one another, some context-specific rules must be applied to these event outcomes, and some rational decisions must be taken based on specific criteria. If the dynamic event is extremely unpredictable, as is the case when packets arrive at a network node where a switching/routing device is located, then the shape and form of the outcome (in this case, the packet content) is hard to profile.

Consequently, several rules may be needed to cope with all the possible combinations of factors and parameters that lead to a classification decision. These rules must be stored somewhere in a lookup table or a rule database. They must be made available on an ongoing basis and be accessible at very high speeds in order to facilitate the process. This rule database must then be searched every time an event in need of classification arises. In network processing, the arrival of a new packet is an example of a dynamic event in need of classification. When network gear receives a packet among billions of other packets per second, it must decide within a very short time (a few nanoseconds) whether and how to forward each individual packet, how to process it, and so on. This constitutes the underlying realm of the classification problem. The combination of a rule database and decision engine has created the concept of *search and classification*.

In the previous chapter, we discussed search engines as one of the two legs required to successfully tackle search and classification in the context of network processing. In this chapter, we look at the second leg of the approach. We first define some terms and formulate the problem of classification. We then discuss various approaches and provide configurations for tackling it. We conclude the chapter with a case study of state-of-the-art classification processors to give a sense of the direction that the industry has taken.

TWO TYPES OF PACKET CLASSIFICATION

So what exactly do we mean by *classification*? The trade press and industry use some of the terms regarding classification in contradictory ways, causing some confusion. We must clarify that the terms *lookup* and *search* refer to the same thing in this book. The implementation of lookup and search was examined in the previous chapter.

For our purposes, it is assumed that a mechanism will always be provided that will perform one of two tasks when the classification process is initiated: It will either sieve through a database of rules or it will consult a lookup table (where the rules are stored) to provide a result that may be used for indexing into some other data structure, which will ultimately provide the answer to the classification problem. In network processing, a systems designer faces two general classification problems:

- Let us first look at what is commonly known as *lookup and classification*. This all-encompassing term, which is almost de facto used loosely to refer to one of three terms: *layer 2*, *layer 3*, or even *routing classification and forwarding*. It is mainly used in simple packet routing/switching contexts. The word *classification* has other connotations in network processing and is usually used as a synonym for *forwarding*. Classification consists of the identification of the correct output port/channel/interface of a piece of network gear (a switch/router) to which the packet must be forwarded. This decision is made when the packet's destination address matches the content of a lookup table, as shown in Figure 13.1. This lookup table is invariably known in this context as the *routing* or *forwarding table*. It summarizes routing information based on prefixes.

- The term *classification* or *deep packet classification* is used when a packet must be distinguished among several others (usually for purposes other than just routing or switching). This should not be done simply based on its destination address; it should also be based on several internal bit fields of variable length or format.

Notice that some interesting words have been used in this definition. These words were used on purpose so let's individually define them for this context:

- *Distinguished* means that implicitly different processing awaits each packet after it is singled out from a stream. These different types of processing correspond to *flows*.

- *Several* means that the deep packet classification process will usually occur based on the simultaneous application of more than one rule and criterion. This is unlike classification, which is only based on rules surrounding layers 2 and 3.

- *Internal* means that the bit fields, which the classifier seeks to retrieve in order to apply some of its classification rules and criteria on them, may be buried deeper inside the received packet than the traditional source and destination addresses, which are conveniently located in well-known positions in front of the payload. The bit fields needed sometimes carry a substantial amount of unrelated material that is piled up on top of them. This unrelated material could also envelope the fields of interest in tunneled or encapsulated applications such as some firewall or *virtual private network* (VPN) products. They also cannot be easily seen by most network equipment, which so far has only relied on using information from other previously easily accessible header fields.

- *Variable length or format* implies that the bit fields used in the rules applied to deep packet classification are not as straightforward as frozen 32-bit addresses, but they can represent ranges of values and can sometimes be of variable length (such as *Uniform Resource Locators* [URLs]).

Deep packet classification is required in cases where multiple rules must be applied to a combination of bit fields found directly inside a packet or calculated according to some quick procedure from the bit content of specific positions inside the packet. These bit fields reflect parameters pertaining to information that is specific to anywhere between layers 4 and 7. Therefore, it is not surprising that we encounter deep packet classification in advanced networks that provide *service level agreements* (SLAs) as well as *quality of service* (QoS) and *class of service* (CoS) guarantees. We will discuss these topics later when we cover issues such as *Integrated Services* (IntServ) and *Differentiated Services* (DiffServ). For example, deep packet classification could be based on the application of specific rules that require the simultaneous use of six bit fields. Those bit fields could potentially be the source and destination addresses (each 32 bits long), the source and destination port numbers (each 16 bits long), the *type of service* (TOS) field (8 bits long), and the type of protocol field (also 8 bits long).

Deep packet classification sorts out the incoming unstructured packet stream in a series of flows. The flow is an artificial concept that may bear no relationship with tangible reality. It is just a subset of a packet stream that satisfies specific sorting criteria that we impose arbitrarily to reflect what we expect the network service provider to manage. The flow assignment of packets distinguishes the actual service that a packet will receive. At an age where we talk more about QoS, SLAs, and billing per usage, it is imperative that equipment at the disposal of service providers be able to take a peek inside all the packets in order to ensure that the appropriate session-related information is extracted and fed into the SLA rules to ensure correct billing or to the QoS rules to ensure that the client obtains the service he or she expects.

The following are examples of different uses of packet classification:

- **To manage** *access control lists* **(ACLs) for specific applications**   For example, in a firewall, the systems manager may be required to block (filter) packets carrying e-mail from a certain *Internet service provider* (ISP) to all enterprise users.

- **To make forwarding and routing decisions based on policy**   For example, an enterprise with multiple connections to the outside world may decide to route all *voice over Internet Protocol* (VoIP) and IP telephony traffic over an *Asynchronous Transfer Mode* (ATM) network link of carrier A instead of letting it go through the IP network of ISP B.

- **To limit traffic based on rates**   A core network provider can stipulate to its equipment that no more than 75 Mbps traffic from a specific ISP can be accepted. They can also state that no more than 10 Mbps out of whatever comes in can be telnet sessions.

- **To impose traffic shaping**   For example, an intercarrier network where specific agreements are signed with regional carriers might apply a rule that stipulates that no more traffic above 75 Mbps from carrier X will be allowed into the core pipeline.

- **To provide accounting and billing per agreed levels of usage**   For example, some context could stipulate that all streaming video traffic destined to host X must be forwarded through the output interface Y, and accounting and billing must be performed on a per-packet basis according to some contract agreed to by the customer.

Obviously, the possibilities are endless. Let us now go back and look at each one of these two major categories of classification in a little more detail.

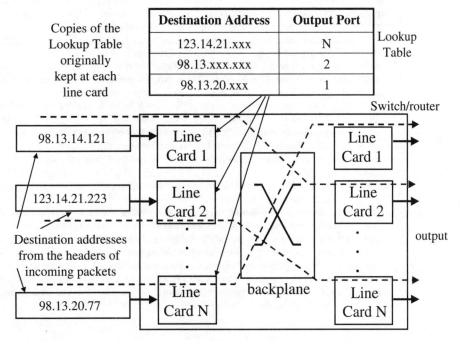

FIGURE 13.1   Lookup and forwarding based on tables kept in local memory at each line card.

LOOKUP AND FORWARDING

Figure 13.1 shows a typical example of a rudimentary classification system that implements a lookup and forwarding procedure. Originally, the implementation was based on flat lookup table copies that were kept in local memory on each line card. This scheme had obvious advantages since lookups did not saturate the main bus and did not interrupt the main *central processing unit* (CPU). Each card was responsible for the lookup operations it needed. This simple scheme is now a thing of the past. Routing tables are so large that it is almost impossible and extremely expensive to have several copies in the same router.

The most typical embodiment of a prefix-based address lookup algorithm, such as the *longest-prefix match* (LPM), seeks to obtain the longest prefix that best matches the search key. In this case, the search key is the packet destination address. The previous chapter offered some typical approaches for tackling this problem in the discussion on *ternary content-addressable memories* (TCAMs) and search engines. More information on the search part of this problem is available in the previous chapter. Here we will limit ourselves to the actual classification part of the problem.

On one hand, IP routers and switches in need of evolving classification capabilities must look up the output port identification in forwarding or routing tables to which they must forward an incoming packet. To make the routing lookup and forwarding decision, the longest prefix that matches the packet's destination address must be obtained among the entries in the stored database (table) of routes.

On the other hand, ATM and *Multiprotocol Label Switching* (MPLS) switches receive routing/switching-related information at the input upon arrival of a cell or packet that immediately correlates the port and the *Virtual Connection Identifier* (VCI) (or label). They directly proceed to the next step of the switching process. For example, with an MPLS switch, a previous router has already tagged every incoming packet the router receives. Therefore, the locally kept correspondence table will simply need to match the content of the tagged label with the router's output ports to find the correct one. If the label is 69451, for instance, it may indicate according to the label table that the packet must be forwarded to the output port 34, no questions asked. In the strict sense of the term, classification is therefore not a problem for this realm.

The original routing on the Internet was built on class-based addressing. For more details, consult the book *Routing in the Internet* by Christian Huitema.[1] Four major classes of 32-bit addresses were devised in what we now call *IP version 4* (IPv4). For convenience, the industry uses the dotted quad notation—for example, 128.23.34.5. This notation represents addresses as integers (distinct points) on the IP address line—in other words, a closed segment (including its two boundaries) between 0 and $2^{32}-1$. Therefore, addresses assumed the form K.L.M.N, which corresponds to the number K $\times$ $2^{24} + $ L $\times 2^{16} + $ M $\times 2^8 + $ N. The first possible address is 0.0.0.0 on the left side of the fictitious IP address axis segment, whereas the end address is 255.255.255.255 on the right side of the segment.

As shown in Figure 13.2, class A addresses occupy the space between 0.0.0.0 and 128.0.0.0, whereas class B addresses occupy the space between 128.0.0.0 and 191.255.255.255. Class C addresses are between 192.0.0.0 and 223.255.255.255, whereas class D addresses are between 224.0.0.0 and 239.255.255.255 for multicast. Class E addresses between 240.0.0.0 and 255.255.255.255 remain reserved.

The concept of the *network ID* (netid) was established to enable organizations to manage their own internal addresses. In class A addresses, the netid was between bits 1 and 7, whereas in class B addresses, the netid was between bits 2 and 15. In class C addresses, the netid was supposed to be situated between bits 3 through 23 in an address. However, the problem was that the address space available to IPv4 was quickly used up mainly because the netid boundaries were not flexible. Class A addresses would only be offered to large organizations that could justify hundreds of thousands of hosts. Class C was deemed extremely narrow, as many organizations needed addressing for more than

---

1. Christian Huitema, *Routing in the Internet*, 2nd ed. (Upper Saddle River, New Jersey: Prentice-Hall, 2000).

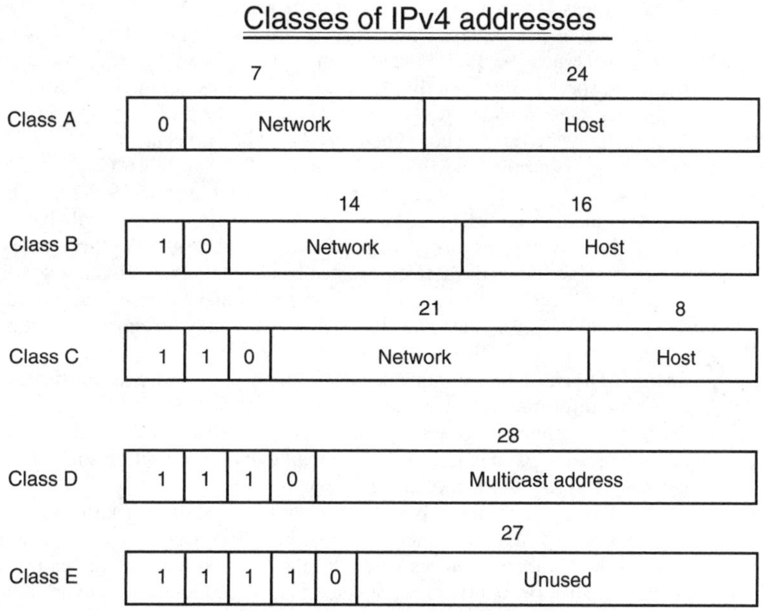

Classes of IPv4 addresses

FIGURE 13.2   The structure of the original address classes in IPv4.

the 256 hosts that the 8 bits allocated for that effect could allow. Since Class B addresses were the least evil of the remaining options for organizations seeking pools of addresses, the prospect that we would run out of useful addresses soon was real. To further compound the problem, the simultaneous explosive growth of the number of routes kept inside forwarding tables created an equally untenable problem.

In order to address these parallel problems, the *Internet Engineering Task Force* (IETF) worked on several issues. In addition to the appearance of more sophisticated routing protocols (such as the *Border Gateway Protocol* [BGP]), the following interesting developments occurred:

- *Classless Interdomain Routing* (CIDR) appeared and boundaries were eliminated between classes with the adoption of hierarchical addressing and routing.
- *Network Address Translation* (NAT) schemes were specified and implemented.
- Dynamically assigned addresses were proposed.
- IPv6 with addresses that were 128 bits long was invented.

As the industry has not yet fully moved over to IPv6, the status quo of modern networks forces us to determine what is of immediate interest for this discussion on lookup and forwarding (classification) among these four topics—that is certainly the CIDR approach. In order to cluster or aggregate addresses, a hierarchical notation had to be introduced. In response to this demand, CIDR launched the concept of a *prefix*. A prefix is a variable-length field that can be anywhere from 0 to 32 bits long. *Supernets* are implied through which a common access is subsequently provided to unrelated *subnets*. The principle seemed logical. From Boston, we first needed to go to Chicago and from then on we could go to Minneapolis, Omaha, or Sioux City. The latter three destinations are irrelevant to each other, but they all share something in common—we must first go to Chicago in order to reach any of them. This principle was going to be based on address prefixes. Route aggregations would then need to be implemented based on the use of the address prefix.

First, addresses were no longer going to be represented as ordinary 4-byte or 32-bit numbers. In the new decimal prefix-based notation, 11.0.0.0 becomes 11/8, 124.54.0.0 becomes 124.54/16, 193.19.3.0 becomes 193.19.3/24, and so on. Second, the routes needed to be represented in aggregate form to save space in the exploding routing tables. Consider the last example—193.19.3/24. If this is an aggregate route, it means that the first 24 bits are the prefix and the remaining 8 bits of any 32-bit address under its hierarchy should be treated as wildcard characters.

In our city example, if Minneapolis is 157.9.0/23, Sioux City is 157.9.2/24, and Omaha is 157.9.3/24, we could aggregate them in our routing table as 157.9.0/22 with a new 22-bit prefix that serves all of them. Another router away from us, say, in Chicago, would have to worry about the granularity of the network concerned with the eventual access to the three cities. From our perspective, whatever must go to one of these cities must be channeled toward the route that goes to Chicago. Our aggregation of the routes to one common city serves that purpose. The problem and the solution obviously scale up. We should also be able to aggregate our Chicago routes with routes that are destined for other cities, such as Cleveland.

In the CIDR case, the incoming packet header, which carries the destination address, does not carry any specific information that could help determine the length of the longest matching prefix. As a result, the search must be conducted not only in the space of all prefix lengths, but also in the space of all prefixes of a given length. From the mathematical algorithm standpoint, this makes it harder to satisfy the LPM requirement than a best match.

Performance is becoming a bigger issue with the size of the routing tables and the ever-increasing wire speeds. The BGP table contained 133,000 entries used by routers as of this writing. It was estimated that the number of entries would hit the 140,000 mark by the end of 2002. From extensively gathered statistics from Telstra,[2] it appears that some pieces of network equipment must be able to support between 5,000 to 10,000 route updates per second (based on new route advertisements from other routers) by early 2003 before becoming a bottleneck on the network's performance. Lightreading provides an example of a test enjoying wide-scale acceptance from the industry. In March 2001, Lightreading formulated a benchmark test for core router updates.[3] Systems designers usually want their designs to be rated as performing well in this benchmark, for instance at 8,000 updates per second. All this points to the same direction.

Actually, the industry estimates that limitations of the BGP will probably be the first reason why a bottleneck will appear. The bottleneck will most likely occur before TCAMs run out of updating steam. In any event, Figure 13.3 shows the tendency of explosion in the routing table content, which roughly tells us that we have been adding about 50,000 new prefixes every year.

In addition to the size of a default-free BGP table, other significant factors that will influence the need for classification performance include all IPv4 routing tables, the number of routing tables that must be consulted in some cases, and the need for all network equipment that also relies on other lookup entries to be able to support the following:

- Separate routing tables for each VPN.
- IPv6 routing tables, especially in the case of wireless networks as IPv6 may likely end up being driven by the fact that each third-generation wireless phone could have its own IP address.

---

2. The BGP table status along with numerous pertinent statistics as well as papers, links, IETF reports, and even tutorials on CIDR and routing can be found at *http://bgp.potaroo.net* or at Telstra's web site *www.telstra.net/ops/bgptable.html*. The drawing shown in Figure 13.3 is based on results produced by the BGP Analyzer, which is written and maintained by Telstra's Geoff Huston. Results before April 4, 1997 are interpreted from data provided by Erik-Jan Bos of SURFnet.

3. See "Internet Core Router Test," March 6, 2001, *www.lightreading.com/document.asp?doc_id54009*. In that test suite, one of the interesting issues was to test some routers in a problematic mode of operation called route flapping. This condition occurs in real life when numerous routes are withdrawn from the BGP table (which consequently changes state often) and then are re-advertised in a very rapid succession. This route-flapping test required the tested routers to learn and unlearn 200,000 routes over a period of 60 seconds while measuring forwarding performance. The so-called drop-dead requirement to run this test is less than 8,000 per second; however, it goes without saying that higher learning rates will improve how a piece of routing equipment scores. It could be argued that real-life routers usually enjoy a functionality called route dampening, which is a means of instructing the router to ignore the frequent changes of the BGP table if they occur often within a certain time interval. The router, however, must still be able to deal at least with one major change in BGP table state before the life belt of route dampening is activated.

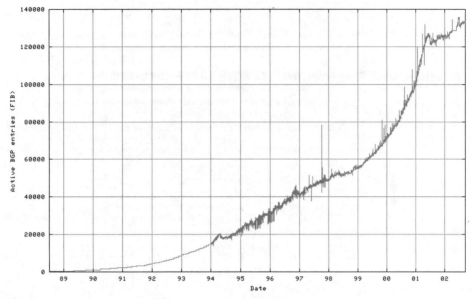

FIGURE 13.3   The explosion of the BGP table entries as of late 2002.[4]

- MPLS environments with multiple labels and tags.
- ATM networks.
- Classical Fast, 1 GbE, and 10 GbE Ethernet realms, which rely on extensive lookup information for their operation.

Some conservative system builders already require that the *network processing unit* (NPU) they choose must be able to economically support the equivalent of 2M lookup table entries in order to handle the short- to medium-term evolution of these needs.

From this context, we can imagine where this trend will lead us within the next 5 to 10 years when it comes down to compiling life-span requirements for network-processing and lookup circuitry capabilities inside new carrier-grade equipment for which ongoing investment will ultimately be contemplated.

This is why in the previous chapter we alluded to the importance of properly managing the content of TCAMs by reshuffling entries so that entries are arranged in the order of decreasing prefixes and that no empty space exists between entries. This actually leads to the fortunate situation where if properly ordered, the entries are structured in blocks, starting first with the 32-bit-long prefixes, then the block with the 31-bit-long prefixes, then the block with the 30-bit-long prefixes, and so on. The mere order of these blocks significantly accelerates the search time. By finding a match and knowing which prefix block was obtained, the length of the prefix can be determined, so half of the problem has already been tackled.

---

4. See *http://bgp.potaroo.net.*

The continuous arrangement of entries (deleting old entries, inserting new entries, and moving blocks of entries to accommodate dynamic updates) in such a way is a frequent problem and substantial research is being carried out.

ALGORITHMS FOR MANAGING LOOKUP TABLE UPDATES

It is foolish to think that every time a new entry must be suddenly inserted somewhere (at the right prefix block) into the TCAM-based lookup and forwarding table, we just start rolling all prefix blocks one by one (that is, shorter-prefix blocks that must be located in an orderly fashion below the new entry) downwards one memory slot at a time in order to open up one slot space for our new entry. The number of prefixes determines the amount of time it takes. Latency on an update operation becomes prohibitive for the intended wire-speed performance of the design. For example, if we have a 100 MHz TCAM-based device where the clock cycle is 10 nanoseconds, then this is the amount of time it takes to conclude one lookup operation. Now if there are 64,000 prefixes, as is usually the case, and these must all be relocated by one position to lower memory positions in order to allow a new clumsy entry, then such an update of the table according to this method will consume the following time:

$$10 \text{ nanoseconds} \times 64,000 = 640,000 \text{ nanoseconds} = 640 \ \mu\text{sec} \approx 0.6 \text{ milliseconds}$$

If this system is an application, for example, on an OC-192 link with small packets (40 bytes), then to achieve an equivalent rate of 31.25 *million packets per second* (Mpps), there would be a need for:

$$31.25 \times 10^6 \times 0.6 \times 10^{-3} = 18.75 \times 10^3 = 18,750 \text{ packets}$$

These packets would need to be buffered temporarily (multiplied by the size of the typical packet we chose, which yields 750K bytes of extremely fast and expensive buffer memory) while we wait for the table update to conclude.

If this process happens too often, then the buffer estimate is not even close to being enough, as new packets will continue piling up while others are still lined up in the buffer awaiting treatment. This will continue until we are forced to discard the new ones.

To avoid causing too much disruption, some designs intersperse a few empty positions between prefix blocks in case new entries must be inserted. This approach has many variations, and several interesting algorithms have been proposed to that effect. Their performance has been studied and analyzed in terms of the number of clock cycles they require for the insertion and deletion of an entry into a table. One of the fundamental constraints in the formulation of the efficient insertion/deletion problem is TCAM's *Prefix Length Ordering* (PLO) constraint. The PLO constraint in a TCAM has nothing to do with policies toward Middle East issues. It is just an easy-to-remember abbreviation. PLO is a mathematical problem formulation of how to best structure the CAM in such a way that prefix sorting remains invariant. The PLO concept is based on the observation that prefixes of the same length can be stored in any order inside their correct block.

Among the many algorithms proposed to tackle the reorganization of the TCAM-based lookup and forwarding table, the most preeminent are the PLO_OPT, CAO_OPT, and L-Solution algorithms. *L* is the number of possible prefix lengths; for IPv4 addresses, it is equal to 32.

These algorithms are discussed in detail and analyzed in the article "Fast Update on Ternary-CAMs for Packet Lookups and Classification" written by Devavrat Shah and Pankaj Gupta.[5]

---

5. Devavrat Shah and Pankaj Gupta, "Fast Updates on Ternary-CAMs for Packet Lookups and Classification," *IEEE Micro* 21, no. 1 (January–February 2001).

In order to empty one space inside the memory bank, the L-Solution algorithm takes no more than L memory shifts, whereas the PLO constraint applied to the same problem reduces that requirement by half to almost L/2 memory shifts per table-entry update.

CAO stands for *Chain Ancestor Ordering* constraint. This is a means of relaxing the exact storage constraints applied to some types of maximal-length chains. It has very important effects on the overall classification performance of the algorithm. We will just mention that the ratio of updating performance between both the worst-case and the typical (mean) number of memory writes for CAO_OPT, PLO_OPT, and L-Solution has been shown by Shah and Gupta to be roughly a proportion of 1 to 4 to 7, respectively. Some people have conjectured that CAO_OPT is an optimal algorithm, as it consistently allows updates to occur within one to two clock cycles. This is different from the L-Solution, which according to extensive simulations of the authors of that paper, seems to require seven clock cycles on average.

ALGORITHMS AND DATA STRUCTURES TO SUPPORT LOOKUP AND FORWARDING

We learned that routers parse the destination address of every incoming packet and then match it with some entry into a routing table in order to determine to which port or next-hop address the packet should be forwarded. This matching function is not trivial given the context of optimized real-time behavior in the router. The router cannot just find any match; it must be the best match possible.

For example, if an incoming packet has the destination address 154.32.4.57 and the routing table contains the following relevant entries—154.32.*, 154.32.4.*, and 154.32.4.57—only the last entry should be matched. This problem can be better understood if we consider that the routing table may contain hundreds of thousands of entries. One of the most favored algorithms for this kind of classification is the LPM algorithm, which is thoroughly discussed in *The Art of Computer Programming* by Donald E. Knuth.[6]

For example, a specific routing decision (layer 3) may have to be made for some packet based on a specific routing protocol. This protocol may use the LPM algorithm based on a trie, which we discuss in this section. With address prefix matching algorithms, the forwarding database (lookup table), which must be consulted, generally contains a *dictionary* of address prefixes. The algorithm finds the longest initial substring of the destination address that is included in the forwarding database. During a classification (lookup and forwarding) operation, the network processor (or classification processor, custom *application-specific integrated circuit* [ASIC], or other CPU for that matter) will have to traverse the trie looking for the LPM.

The *radix trie*, *Patricia trie*, and *leaf-pushed binary trees* are some common data tree structures used to produce the best match. Any good data structure book explains them in detail. Without going into details here, we will say that they are inverted trees. Each parent node represents a partial string of the packet addresses and each child node represents one possible single-component extension. The leaves represent the strings that are stored in the trie. Readers who have experience using tree-traversal algorithms in pattern-matching programs from the artificial intelligence realm or those who have experience with database tree parsing algorithms will immediately understand this principle.

In the example shown in Figure 13.4, where only some of the leaves are shown, an incoming packet with the destination address 154.32.4.57 will be routed to the correct interface after only four matching operations.

6. Donald E. Knuth, "Sorting and Searching," *The Art of Computer Programming*, vol. 3 (Reading, Massachusetts: Addison-Wesley, 1973).

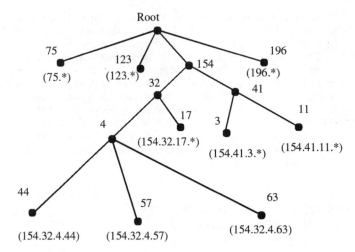

FIGURE 13.4 A trie is a data structure used in a router to efficiently produce the best matching route for the forwarding of the packet to the correct destination.[7]

Radix tries are simple to implement but they suffer from a potential performance disadvantage—namely, that a lookup may take O(m) (polynomial time depending on m, the number of bits in the prefix, which in the case of IPv4 is 32). In an interesting variation on performance-fine-tuning this trie-traversal algorithm documented by Radia Perlman, long nonbranching paths of the trie can be collapsed if they do not contain a dictionary entry. This enables the entire substring to be found with only one search in some cases.[7]

Another interesting approach, which is based on the idea that the search does not have to be initiated by chasing only one bit a time,[8] is to trade memory for search time.[9] As an example of 2-bit searches, the prefix 10 would have multiple child nodes instead of a 1 and a 0. It would therefore potentially expand to 1000, 1001, 1010, or 1011. Any address that matches the prefix 10 would simply have to match one of the elements of the set {1000, 1001, 1010, 1011}. The search is completed in two steps, as opposed to four. The principle can be expanded to 4, 8, 16, and higher bit-block searches that correspondingly increase the speed but also geometrically multiply the memory needed. Of course, such a trade-off between preprocessing time versus memory is not always feasible. Therefore, a systems designer must always be alert of ramifications of design choices within a certain context.

A couple of other clever approaches are to compress the expanded prefixes in the so-called Luleå scheme (from the Swedish city where the research team that invented it is based)[10] and to perform a binary search using special hashing functions on prefix lengths.[11] If we imagine that prefixes are expanded to lengths that are multiples of 16 bits, we will end up having 2^{16} child nodes, causing a very serious memory problem. If we suppose that only a small fraction of these 2^{16} possibilities are real prefixes, then most of the nodes will be the result of expanded smaller prefixes and lots of memory

7. Radia Perlman, *Interconnections: Bridges, Routers, Switches, and Internetworking Protocols*, 2nd ed. (Reading, Massachusetts: Addison-Wesley, 2000).

8. V. Srinivasan and G. Varghese, "Faster IP Lookups using Controlled Prefix Expansion," ACM Sigmetrics 1998, ACM Transactions on Computer Systems, March 1999.

9. Radia Perlman, *Interconnections: Bridges, Routers, Switches, and Internetworking Protocols*, 2nd ed. (Reading, Massachusetts: Addison-Wesley, 2000).

10. A. Brodnik, S. Carlsson, M. Degermark, and S. Pink, "Small Forwarding Tables for Fast Routing Lookups," *Sigcomm* (1997): 3–14.

11. M. Waldvogel, G. Varghese, J. Turner, and B. Plattner, "Scalable High-Speed IP Routing Lookups," *Sigcomm* (1997): 25–36.

will be wasted. The Luleå routing table scheme provides a handy method for compressing the data in such a case by minimizing both the size of the data structure and the number of memory accesses during operation.

This method is based on partitioning the data structure into three layers and a subsequent neat encoding based on bit vectors and masks. Some interesting issues arise in the binary search of prefix lengths—namely, what constitutes a good hash for every prefix length in storage and how to minimize the number of probes (trials). However, if it is properly designed, spectacular performance results can be attained. The Luleå algorithm data structure is a typical example of an algorithm that trades table-building (preprocessing) time (which in typical implementations takes close to 100 milliseconds) for lookup time. Of course, this is not a reproach. Routes are not supposed to change very often so this is a reasonable trade-off.

If the Luleå classification algorithm is implemented in software to run on a 1 GHz Pentium CPU, it can be expected to run roughly at a level around 10 *million lookups per second* (Mlps). Worst-case lookup has been seen to take about 100 clock cycles and, more typically, 50 clock cycles. To put things into perspective, a 1 GHz CPU has a clock cycle of 1 nanosecond (10^{-9} seconds). Assuming that it can force one instruction per cycle through its pipeline (something that is not always feasible in real-life applications), one lookup can be accomplished in around 100 cycles (100 nanoseconds = 0.1 μsec). Therefore, in one second, we can calculate that we can have $1/(0.1 \mu sec) = 10$ Mlps. This implies that the 1 GHz CPU is 100 percent dedicated to classification (an unrealistic proposition based on what we have seen elsewhere in this book). Even then, its performance should be compared with numbers attained by current TCAM products as discussed in the previous chapter. The question, however, seems to still remain as to whether this structure is desirable in IPv6 environments. A more detailed discussion of these techniques would require us to describe the corresponding data structures and discuss the efficient implementation of the algorithms, especially if they involve some clever architectural schemes like e.g. lookup tables, cache memory, etc. It is therefore beyond the scope of this book. For more information, consult the references at the end of the chapter.

Perlman has invented a parallel hardware-based approach that seems to outperform any trie-based approach.[12] It requires a lookup engine that contains several parallel registers (called *hash buckets*) that function in a multiple-stage operation. Each stage of the lookup operation compares a portion of the destination address against the content of all hash buckets. The winner is the register that contains the longest match. The winning register at the end of the operation points to a data structure, which will hold the bit patterns that the registers will need to compare at the next stage. This parallel hardware approach drastically expedites the search. It obviously costs more to implement in silicon, but it allows the rapid acceleration of the lookup tasks.

From a systems architect's point of view, algorithms used to implement lookup and forwarding operations are not surprisingly judged according to a short list of criteria that includes speed, instruction memory footprint, the necessary data memory to run efficiently, and the scalable capability of flexibly handling large tables that continuously evolve in size. In some cases, some algorithms trade lookup performance for the time needed to build the table. If this is an appropriate trade-off, then the systems designer is always looking for ways to decrease this preprocessing time.

DEEP PACKET CLASSIFICATION

We mentioned earlier that when classification must be performed based on the lookup on multiple fields, as is the case with deep packet classification, we create flows. We then assign packets to individual flows. The flow is a new concept that can be compared in granularity and conceptual importance with the datagram. As a result, routers (which from now on must become flow aware as opposed to being exclusively packet aware) treat different flows differently. However, when unrelated packets

---

12. Radia Perlman, *Interconnections: Bridges, Routers, Switches, and Internetworking Protocols*, 2nd ed. (Reading, Massachusetts: Addison-Wesley, 2000).

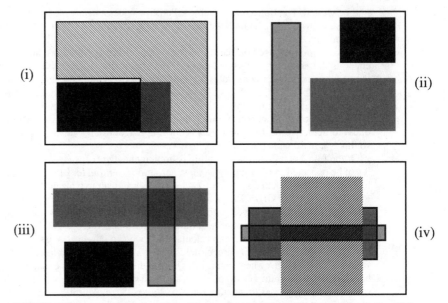

FIGURE 13.5 A geometrical visualization of a two-dimensional classification problem into multiple regions: (i) three regions, (ii) four regions, (iii) five regions, and (iv) six regions. The principle can be easily understood in three-dimensional and higher spaces.

are assigned to the same flow, they will all be treated in the same way. A flow is a virtual object that is created to better manage the performance of the network gear. As a result, it may not even have a concrete physical substance, yet it is important to monitor and process properly.

What does a flow look like? A flow can be a pair of source addresses and destination addresses, or it can take the form of an ordered list, such as the destination address, the type of protocol and so on. The list of possible flows is endless. So what do we mean by the phrase "process properly"? Every flow in the modern network realm, where QoS and guarantees are required, must be processed in a specific way to ensure compliance with an SLA. For example, all packets conforming to a specific flow may be required to be delivered with a maximum delay stipulated in the agreement that the carrier has signed with the customer. Or packets arriving from a specific router may need to be accounted for with comprehensive statistics for subsequent network management decisions. Or packets of a specific type with a specific destination host may need to be blocked for security reasons. And so forth.

The classification problem here is formulated as the process that a switch/router must perform in order to distinguish to which flow it must assign every new packet for subsequent processing. In order for this to happen, several fields inside the packet must usually be looked up and a decision based on several criteria must be made. This is the origin of the name *deep packet classification*, which is also known as *classification based on multiple fields*.

To better visualize the problem (in pursuit of a rigorous mathematical formulation), imagine two axes with the values of two fields along them. Multiple regions can exist based on the policy rules. For instance, Figure 13.5 shows a conceptual example of classification along two fields and the combinations of these fields may very well create three, four, five, or more regions. Each region is covered by one set of actions to be taken by the classifier. Every incoming packet in this case will have its two specific fields inspected. Based on the policy rules, it will be assigned into one of these regions for subsequent action.

The same principle can be applied in three or more dimensions, which ultimately leads to a set of complex multidimensional regions that correspond to specific actions to be taken following unequivocal classification. By applying methods of computational geometry to parameterize these regions and weighing the pros and cons of the computational load of the regions as the probabilities that a certain percentage of arriving packets are classified consistently into one or the other of these regions, we hope to identify ways of balancing the classification workload and optimizing the performance.

CLASSIFICATION BASED ON MULTIPLE FIELDS

Classification based on multiple fields is an extension of our discussion so far. A packet header carries multiple fields, such as layer 3 source and destination addresses, layer 4 port source and destinations, layer 4 type of protocol, TOS, and so on, all or some of which are important in the newer generations of network equipment. A classifier must apply elaborate and often numerous rules on these multiple fields in order to produce (according to the encoded policy rules) the appropriate action to be taken on each packet. This is a process that cannot be characterized as a best-efforts work. If a firewall is set up, the firewall is expected to be able to decide whether to accept or whether to filter each individual packet it encounters.

To rate algorithms in this area, we must look predominantly at the wire-speed performance of the algorithm and subsequently at the cost of implementation, which in this case implies the memory needed for the algorithm to work (the footprint for both instruction memory and working data memory). Other characteristics of interest are the setup time of the algorithm (also known as the preprocessing time) and the incremental update time when table content must change. The two characteristics are considered differently depending on whether the application is based on policy rule tables that change often. The data structure on which the classifier works sometimes needs to be modified. These modifications entail either inserting new entries or deleting old ones. This can either happen incrementally or it may require a radical reconstruction of the data structure (whether it is a table or a graph tree). The time needed for that process is called the *setup* or *preprocessing time*.

As these trade-offs can vary depending on the structure and the choice of algorithm, they must be considered because they spill over on the application itself. For instance, the policy tables on a gateway are not changed as often. When it happens, the network manager will usually carry them out manually. As this process is rarely done, it may be tolerable that a little more time is used than deemed acceptable for an organization. An ISP network edge router may be required to change its classification policies often based on dynamic conditions. This must be done with special tools without disrupting the network for long, which implies that the preprocessing time must be as short as possible.

Some approaches justify a naive approach based on a linear search of the rule space. In the simpler case of layer 3 lookups, this approach keeps a linked list of prefixes and its performance increases if that linked list is sorted by the prefix length. The concept is straightforward and exhibits acceptable memory requirements and update capabilities, but it can unfortunately sometimes yield a very long (and therefore unacceptable) classification time.

Some other interesting ideas have also been proposed. For instance, the concept of cross-producting introduced by V. Srinivasan exhibits powerful multidimensional classification time performance, but consumes an inordinately large amount of memory.[13] The idea of using a grid of tries, also proposed in the same work, performs nicely in classification time and provides the necessary storage for handling up to two dimensions of classification; however, it cannot be easily extended to problems in need of higher dimensions. Lakshman and Stiliadis discuss a bit-level parallelism that is suitable for hardware implementation and that can classify quickly along multiple dimensions;

---

13. V. Srinivasan, S. Suri, G. Varghese, and M. Waldvogel, "Fast and Scalable Layer 4 Switching," *Proceedings of ACM Sigcomm 1998* (September 1998): 203–214.

however, it suffers from large memory and bandwidth requirements.[14] Gupta and McKeown from Stanford University introduced the hierarchical intelligent cuttings approach, which is based on a heuristics-based partitioning of the problem space and a clever rearrangement of the decision tree in such a way that takes advantage of the classifier structure, thereby expediting classification performance while keeping the lid over the memory requirements.[15] Other applicable efforts are the *Area-based Quad-Tree* (AQT),[16] the *Fat-Inverted-Segment* (FIS) tree,[17] and the bitmap intersection classifier.[18] For more information, refer to the corresponding references as we will not be examining the internals of these algorithms. P. Gupta and N. McKeown provide an excellent tutorial of the most important classification algorithms in their paper "Algorithms for Packet Classification."[19] This paper can be a helpful starting point.

One of the most important algorithms that deserves to be mentioned separately is the *Recursive Flow Classification* (RFC), which has been proposed by Gupta and McKeown.[20] It looks at the classification problem as a mapping of S bits (composed from the concatenated content of all fields of interest) onto a series of T bits (the classification outcome). It is based on the observation that creating an ideal memory bank of 2^S entries with the corresponding class number stored in each entry slot and where only one memory access would be needed to produce the classification result is infeasible economically.

It then essentially breaks down the mapping task into three stages where it trades the number of memory accesses (therefore, the speed of execution) with the memory footprint. If we look at an example where S = 128 and say T = 10, instead of trying to map 2^{128} possible items onto 2^{10} ones, the RFC algorithm breaks this task down to three hierarchical mappings traversing four phases:

$$2^{128} \rightarrow 2^{64} \rightarrow 2^{32} \rightarrow 2^{10}$$

The basic idea is that the original sequence of S bits is partitioned into chunks, which are combined in pairs. The result is used to index into a set of tables that are usually precomputed. Combining results from previous stages and repeating the process, we advance down the sequence of mapping as just described until a final answer is obtained at the final phase. Figure 13.6 illustrates this principle. RFC has been shown to perform 31.25 Mpps classification using a three-stage pipeline with two 4Mb of *static random access memory* (SRAM) and four banks of 64Mb *synchronous dynamic random acess memory* (SDRAM) under a clock of 125 MHz. A fully dedicated 333 MHz Pentium running RFC in software achieves a little more than 1 Mpps as classification performance. RFC is considered based on simulations to be capable of handling classification based on a policy database of 15,000 rules in a 10 Gbps environment, if implemented in hardware. It is estimated that it can do the same for a wire speed of 2.5 Gbps, if implemented in software.[21]

14. T. V. Lakshman and D. Stiliadis, "High-Speed Policy-Based Packet Forwarding Using Efficient Multidimensional Range Matching," *Proceedings of ACM Sigcomm 1998* (September 1998): 191–202.

15. P. Gupta and N. McKeown, "Packet Classification Using Hierarchical Intelligent Cuttings," IEEE Micro 20, no. 1 (January—February 2000): 34–41. This was originally presented at Hot Interconnects VII in 1999.

16. M. M. Buddhikot, S. Suri, and M. Waldvogel, "Space Decomposition Techniques for Fast Layer 4 Switching," Protocols for High Speed Networks, 66, no. 6 (August 1999): 277–283.

17. A. Feldman and S. Muthukrishnan, "Tradeoffs for Packet Classification," Proceedings of Infocom 2000 3 (March 2000): 1193–1202.

18. T. V. Lakshman and D. Stiliadis, "High-Speed Policy-Based Packet Forwarding Using Efficient Multidimensional Range Matching," *Proceedings of ACM Sigcomm 1998* (September 1998): 191–202.

19. P. Gupta and N. McKeown, "Algorithms for Packet Classification," IEEE Network Special Issue 15, no. 2 (March/April 2001): 24–32.

20. P. Gupta and N. McKeown, "Packet Classification on Multiple Fields," *Proceedings of Sigcomm 1999*, Computer Communication Review 29, no. 4 (1999): 147–160.

21. P. Gupta and N. McKeown, "Algorithms for Packet Classification," IEEE Network Special Issue 15, no. 2 (March/April 2001): 24–32.

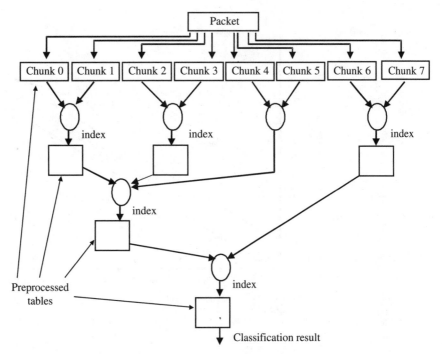

FIGURE 13.6 Flow through the RFC algorithm after a packet is split into chunks.[20]

IMPLEMENTATION

A classification engine is a state machine, and it is only natural that it can be implemented in hardware, software, or a combination of both. The predominant choice of venue is dictated by the wire speed it must sustain and by the size of the table of rules. Low-speed and small-number-of-rules contexts are easily amenable to a software implementation on a *reduced instruction set computer* (RISC) CPU. The more these two parameters start becoming sizable, the more one is confronted with performance choking—hence, it becomes critical to engage more powerful implementations. Later in this chapter, we look closer at the question of whether a TCAM or a classification processor is required. If a classification processor is required, it must be determined whether it should be an integrated one or a standalone coprocessor that offloads the classification task from the network processor.

Some products, especially search engines, require a search engine controller that supervises the operation of the search engine. This is rarely the network processor. In such cases, a separate processor must be included in the board design.

In an effort to differentiate their products, some vendors, such as NetLogic Microsystems, conveniently call their products *CFP processors*, which is short for *classification and forwarding processors*. These products usually combine a TCAM-based search engine implementing the LPM algorithm along with a programmable classification processor in one chip. The combination from a product-profile standpoint is a variation on a theme that fuses together what we discussed about search engines in the previous chapter with what we are discussing about classifiers here. To avoid sacrificing the generality of our treatment, we will not elaborate specifically on CFP processors.

Other vendors, such as Raqia Networks, FastChip, and Solidum, propose standalone classification coprocessors, whose applicability is much wider than just switching and routing table lookup, as

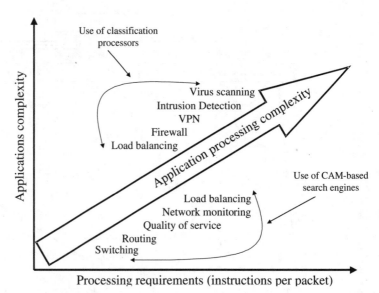

FIGURE 13.7 The complexity of applications requiring classification determines the implementation. *(Source: Raqia Networks)*

shown in Figure 13.7. Their model and applicability is broad and interesting; therefore, it is worthwhile that we review some of these approaches later in this chapter in a series of case studies.

The paper "Fast IP Packet Classification with Configurable Processor," by H. Michael Ji and Michael Carchia,[22] and the paper "Fast IP Routing Lookup with Configurable Processor and Compressed Routing Table," by H. Michael Ji and Ranga Srinivasan,[23] are two interesting works showing in great detail how to implement a classification engine on a configurable CPU by creating custom instructions and the associated modification in the processor data path. This is applicable to network processing contexts we discussed in Chapter 10.

CLASSIFICATION PROCESSORS OR CAMS?

The term *classification* is a broad concept that encompasses lookup tasks, which are handled by search or lookup engines. Lookup engines can be implemented either algorithmically or as CAMs. Therefore, a conceptual relationship exists between the two, as summarized in Figure 13.8.

A lookup engine, especially one that must operate at layer 7 where long and complex strings must be searched and matched, will first need to parse the incoming packet to extract the bit fields, which in the simplest case become the search key. In more complicated cases, the search key will need to be calculated from the extracted bit fields. Based on the key, the lookup process is executed, which yields the search result. A simple lookup or search engine, such as the ones used at layers 2, 3, and 4 for route-related next-hop address lookups, is not capable of both parsing strings and looking up matches

22. H. Michael Ji and Michael Carchia, "Fast IP Packet Classification with Configurable Processor," IEEE Global Telecommunications Conference, Globecomm 2001, San Antonio, Texas, November 25–29, 2001.
23. H. Michael Ji and Ranga Srinivasan, "Fast IP Routing Lookup with Configurable Processor and Compressed Routing Table," IEEE Global Telecommunications Conference, Globecomm 2001, San Antonio, Texas, November 25–29, 2001.

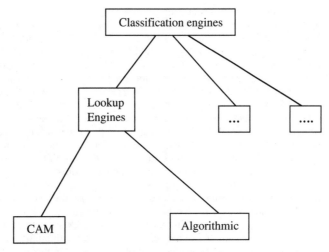

FIGURE 13.8   The conceptual relationship between classification processors and CAM-based lookup and search engines.

for multiple-field-based rules. In such a case, a lookup or search engine requires the help of an extra module that performs the parsing function. In order to build a full-fledged classification processor, a lookup engine must be augmented by a parsing engine.

This can be accomplished by two types of technologies: a combination of modules for the parsing and lookup functions, which would be performing these functions in two consecutive steps, or *programmable state machine* (PSM) techniques and methods to implement a classifier that can combine both steps simultaneously. A classifier usually produces two things at its output: a digest and a tag. The processor is programmed to parse the string of the packet and look for specific patterns. When a match is found, a numerical identifier (the tag) is calculated and produced at the output for that specific packet. The digest, on the other hand, is a user-defined extraction of the content of specific bit positions and/or bit fields, which are collectively put together in a new bit sequence called the *digest* in this context. For example, in a DiffServ application, a digest of up to 128 bits is then concatenated with a tag of up to 32 bits. As a result, 160 bits is produced at the output of the classifier.

Of course, the choice of algorithm depends on many factors. Various algorithms can be used, but the algorithms that we have encountered in route address and next-hop searches are completely inadequate here. This is why at higher layers, especially for layer-7-based search and classification problems, character-string-based search and match algorithms are used.

Several algorithms have been proposed and are being used in such systems. It may be surprising that much of the algorithm research in this field comes either from the multimedia areas of information retrieval or from computational biology teams, and more specifically, from researchers who have been mainly (but not exclusively) motivated by the need to match DNA sequences with large sequences of characters that may be repeated and/or structured in specific patterns. The sheer size of the DNA molecule, which has parts where pattern matching is needed for many aspects of genetic sequencing as well as protein- and drug-related research, has been a critical factor inciting work toward the improvement of the performance of such pattern search and match algorithms, if the researchers were ever to be able to get answers from experiments within a person's lifetime.

A couple of helpful references on the subject of this type of algorithm from the two different approaches we just mentioned include *Modern Information Retrieval* by Ricardo Baeza-Yates and Berthier Ribeieo-Neto[24] and *Algorithms on Strings, Trees, and Sequences: Computer Science and Computational Biology* by Dan Gusfield.[25] The latter is an especially good source of information from a tutorial standpoint. Some of the most significant algorithms currently used for layer 7 search and match classifiers such as the Boyer-Moore algorithm, its variation known as Horspool algorithm, the Sunday algorithm, the *Knuth-Morris-Pratt* (KMP) algorithm, the Aho-Corasick algorithm (an evolution of the KMP approach), the Needleman-Winsch algorithm, the Smith-Waterman algorithm, and others are discussed and analyzed in tremendous detail in this source. *Modern Information Retrieval* offers quantified graphs with performance comparison curves for several of these algorithms. We will not elaborate on the internal details of these algorithms here. Several research groups also maintain web sites where these algorithms are explained online through tutorials.[26,27] Their behavior can be visualized with custom-made simulation applets or tools.

Some string classification systems are based on a brute-force approach that tries all possible positions for a pattern before it can make a decision. This idea may work well with simple patterns when parsing a string, but it is not conducive to high performance for more complex tasks. Some other classification systems may be using much more complex and powerful algorithms such as the Boyer-Moore or even the Aho-Corasick, which work better with multiple string patterns. Boyer-Moore, for example, has the interesting, albeit bizarre, characteristic that it performs better when the pattern to be searched becomes longer.[28]

In addition, an algorithm such as Aho-Corasick may have to preprocess many patterns in which case it usually builds a state graph that can be used to simultaneously check multiple rules. This approach lends itself easily to hardware implementations using PSMs with the policies stored in commodity-priced *synchronous static random access memory* (SSRAM). From a performance standpoint, this definitely beats the CAM approach. The downside is the setup time of policy rules in the external memory.

Other types of algorithms may be more suitable for other aspects of layer 7 classification, such as suffix automaton or deterministic context-dependent parsers (as used in some advanced spam filters). Other types such as probabilistic *Hidden Markov-Model* (HMM)-based algorithms may also be used for something completely different such as making parsing decisions in a law-enforcement context on a subset of a collection of text-based intercepted files based on other pertinent but ancillary information, which may not necessarily be applicable to the entire collection of files under inspection. This subject is very broad and extremely dynamic, but it lies mostly outside the scope of this book.

Returning to our subject, however, CAMs do not require any setup time except for a few cycles needed to write the key. In cases with multiple matches, CAMs require the outputs to be prioritized, but this is usually taken care of by CAM vendors who integrate a priority encoder inside their chips. The reshuffling and reorganization of CAMs is a problem, but we have seen how state-of-the-art CAMs address that problem with superior organizational capabilities. The bigger problem for CAM is that when the search must occur based on a very long bit string—for example, thousands of bits long (as it can easily be the case with URL strings)—the search key must be segmented into manageable chunks. This drastically reduces the overall system performance as matches in different memory banks will not be attained within the same amount of time across the board. This dramatically increases the complexity of programming as results must be reassembled and structured before they can be delivered downstream in the processing chain. CAMs are more expensive and consume more power as opposed to SSRAMs used with standalone classification processors.

---

24. Ricardo Baeza-Yates and Berthier Ribeiro-Neto, *Modern Information Retrieval* (Reading, Massachusetts: Addison-Wesley, 1999).

25. Dan Gusfield, *Algorithms on Strings, Trees, and Sequences: Computer Science and Computational Biology* (New York: Cambridge University Press, 1997).

26. See *www.iti.fh-flensburg.de/lang/algorithmen/pattern/bm.htm*.

27. See *www-sr.informatik.uni-tuebingen.de/~buehler/AC/AC.html*.

28. See *www.cs.utexas.edu/users/moore/best-ideas/string-searching/*.

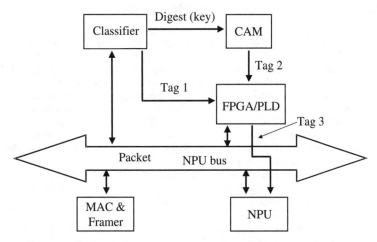

FIGURE 13.9   An interesting combination of a PSM-based classifier and a CAM for layer-7-based cookie detection and load balancing. (*Source: Solidum*)

So does it make sense to combine a CAM with a PSM-based classifier? The answer is yes in cases where the policy rules must be changed often and in a dynamic fashion. The CAM is then used to store the lookup keys instead of SRAM. Misha Nossik presents a very elegant implementation of a web load balancer based on intelligent switching decisions made based on real-time cookie detection and URL resource management.[29]

Figure 13.9 illustrates the scheme. A packet comes in through the *Media Access Control* (MAC)/framer and is immediately dispatched to the network processor. In the meantime, the classifier snoops the packet data, parses the packet content, and finds matches on-the-fly. For example, if the classifier in this application example detects the presence of a *Hypertext Transfer Protocol* (HTTP) GET request that a client issues to a server (with or without a cookie), it will extract the IP addresses and port numbers from the packet and produce the digest that serves as the search key when presented to the CAM. Part of Tag 1 is used as a learn mechanism by the CAM. When confronted with another subsequent packet from the same session, the CAM will be able to produce a match very fast if it sees the same session that produced Tag 1. The combination of consecutive tags (Tag 1 and Tag 2) can be used by the network processor to decide what to do with each individual packet. The drawing shows some configurable logic (*field-programmable gate array* [FPGA] or *programmable logic device* [PLD]) combining the two tags between the CAM and the bus. This is because depending on the choice of NPU bus and the precise CAM interface requirements, it may not be as straightforward for the systems designer to connect the two parts of the digest (the digest part and the command part) to the bus without any glue logic.

Another interesting question would be why CAMs and PSM-based classification processors are not integrated into one chip as a viable product. One answer is that as applications evolve in complexity, in order to guarantee the classification scalability of a systems solution, a systems designer must count on the physical independence of the two modules. Another answer is that because CAMs are still very expensive memory components, but whose prices are rapidly falling, it is sensible to allow the volume-based CAM price curve to evolve downwards as was the case with other memory products. This way, users can benefit the most of the advances of this technology as well.[30]

29. Misha Nossik, "The Relationship Between Classification Processors and CAMs," Solidum white paper.
30. Ibid.

INTEGRATED CLASSIFICATION OR STANDALONE?

The advantages of using a standalone coprocessor that is fully dedicated to classification, as opposed to programming the classification function as one more major task that runs on a multi-RISC-engine-based NPU, are forcefully and didactically explained and enumerated in Feliks Welfeld's white paper "The Case for a Standalone Classification Processor."[31] Welfeld argues that the advantages or disadvantages of the standalone classification processor must be measured in the speed and size of the code (as the *Network Processing Forum* [NPF] argues in its definition of a *network-processing element*) and uses an 8-RISC-core NPU running at 232 MHz on a 1 Gbps link with minimal (512 bits) 64-byte packets as an example.

According to this paper, which describes a DiffServ (RFC 2475) implementation in detail (assuming purely classification work with no packet modification or encapsulation), 20 percent of the NPU computational capacity goes to classification. With other tasks on the packet, this could easily become 50 percent, especially if variable-size headers are involved. A dedicated classification processor would immediately offload this function and therefore liberate these clock cycles from the network processor so that it can use them to do other useful work.

The same problem looked at from a code-size standpoint shows that a RISC implementation of the application requires about 3,000 instructions, where 400 are classification related. A classification processor produces the result as a 32-bit tag attached to the classified packet. Processing that tag by the network processor usually involves about 30 instructions. Therefore, the code savings in this heavy-duty-classification example is $400 - 30 = 370$ instructions. In a detailed example with meta-code showing the implementation for a simple content-inspection case such as the one needed for layer 7 web-based URL string processing, Welfeld shows that a multi-RISC engine must run at 1 GHz and be fully dedicated to classification just to produce the same results. A 100 MHz PSM-based classification processor can produce these results much more easily.

It is therefore safe to conclude based on what we have seen so far that depending on the complexity of the application, the wire speeds required, and the available computational resources, the standalone classification coprocessor is more than a viable solution that often cannot be brushed aside by network processor vendors who purport that their chip can handle everything.

CASE STUDY: RAQIA'S REGULAR EXPRESSION CLASSIFICATION COPROCESSOR

Raqia Networks (whose technology has now been acquired by SafeNet) has looked at the overall classification problem and realized that TCAM-based solutions can only handle applications where the number of classification-related instructions that need to be executed per packet is small and consequently the processing complexity of the classification needed is easily manageable. However, the higher we go on the protocol stack, the more complicated the classification becomes. When we are finally at layer 7 in load balancing for servers, where some complex tasks have to be performed at wire speed, as for instance things become very different: Typical examples would be:

• URL and cookie detection.

• On-the-fly switching of servers because some connections may be deemed to be (at least from an application standpoint) sticky and therefore unacceptable.

• Content that needs to be segregated from one server context to another (for example, streaming video must be delivered from one server while ordinary web pages must be delivered from another).

• High-speed *intrusion detection systems* (IDS).

---

31. Feliks Welfeld, "The Case for a Standalone Classification Processor," Solidum white paper.

(a)

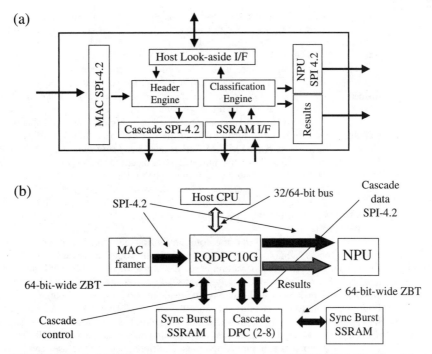

(b)

FIGURE 13.10 Raqia's 10 Gbps deep packet classification processor (RQDPC10G). (a) The internal structure. (b) How it fits into a network-processor-based system. *(Source: Raqia)*

- Virus scanning of packets.
- Firewall load balancing.
- QoS concerns (for example, some requests should be served by one server, whereas others should be channeled for service to yet another server and so forth).

The common-thread operation needed in these environments is a string search. This involves traversing both the header and the payload of packets (or strings) trying to achieve a match between regular expressions, including alternation, and based on ranges or wildcards, applying one or more policy rules or patterns.

Therefore, because these classification requirements often come from edge network equipment or equipment facilitating the correct behavior of powerful 10-GbE-based enterprise servers, a wire speed of 10 Gbps must be attainable. The systems architect must also be given the choice of using the classification processor in a look-aside or flow-through configuration depending on his or her system objectives. The classification system must be protocol aware, able to handle an entire header, and look at the payload of a packet in real time. Of course, it must be able to scale in numbers of rules that can be supported and the concurrency of rules needed at wire speed. In terms of programmability, it should be easy for software engineers to master.

Combining these requirements into their design specification, Raqia has come up with its impressive 10 Gbps deep packet classifier RQDPC10G (the name at first sounds a little kludgy and impossible to memorize, but there is logic behind it). Figure 13.10 illustrates the processor. The figure shows the internal structure of the chip and how it fits into an overall NPU-based system design.

The Raqia deep packet classification processor can sustain flow-through-mode classification at 10 Gbps. In the look-aside mode, it offers 4 Gbps performance. It is cascadable up to eight chips deep for the extension of the rule database at wire speeds, and it embeds *Packet over SONET* (POS)/ Ethernet framing for higher integration capabilities in a systems configuration. In fact, its MAC/framer can be configured as 1×OC-192, 4×OC-48, 16×OC-12, or 64×OC-3 interfaces. It can also be configured as 1×10 GbE, 10×1 GbE, or 96×10/100 Mbps Ethernets. In some applications, it can be used as a shared classification coprocessor where multiple subsystems turn for their classification needs. It can also operate as a standalone classification processor next to a high-speed network processor offloading the heavy-duty classification chores, enabling the NPU to benefit from plenty of saved capacity cycles that it can use for other types of work.

A 2MB on-chip rule memory can be accessed in a single- or dual-plane approach (for online and standby setups). In order to further extend the rule database, the RQDPC10G can easily interface to up to 128MB of external SSRAM (either *zero bus turnaround* [ZBT] or sync burst) through a 64-bit-wide bus. Such an approach, however, would degrade the performance by 50 percent to 5 Gbps maximum classification throughput, but only when all the rules are stored in external memory. The interfaces to the MAC/framer, network processor, and the one used for cascading multiple RQDPC10G chips are standard *System Packet Interface 4.2* (SPI-4.2). The results bus is 32 bits wide and SSRAM compatible, which is handy if subsequent indexing is required or more often for the NPU interface itself.

It had been announced that the company's first-generation product was being designed in a 0.18 µm process technology. Therefore, ample room would be available for relatively straightforward future improvement if the market requests it. It would be offered in a 550-pin *ball grid array* (BGA) that should consume approximately 5 watts. In fact, the ReGXP2G regular expression coprocessor, which is designed with a look-aside interface to the NPU, supports an aggregate data rate of 2.5 Gbps. It was taped out to manufacturing toward the end of 2002 and is available in a 648-FBGA package. It supports packets of up to 9KB and can handle up to 4 memory banks of 8MB of ZBT SSRAM, using up to 8,000 regular expressions per bank and up to 8 rule sets per bank. The Regular Expression Coprocessor supports rule matching across multiple packets. The user can be sure about its deterministic performance at wire speed regardless of the packet size, the protocol used, the complexity of the rules, or the actual number of rules.

Classification results are presented on the 32-bit-wide results bus. They can be optionally appended to the classified packet for subsequent processing by a network processor or traffic manager. Results can comprise the following:

- One packet offset for regular expressions (or many offsets if subexpressions are used)..
- The number of rule(s) matched (again one or many if subexpressions are used).
- A packet ID.

The classification results in a match that means one of the following:

- No rules were matched.
- Only one rule was matched.
- Multiple rules (or subexpressions) were matched.
- A rule was partially matched (cross-packet boundary assist). In this case, the current state is output as a result. It can be fed back into the classification processor as the next packet in this session through the look-aside interface, or it can be handed over through the cascade interface to another classification processor down the chain.

Unlike other approaches by other classification processor vendors, no special description language needs to be learned. Regular expressions and subexpressions are used to create the rules in ASCII files using the company's *graphical user interface* (GUI)-based tool or any usual text editor. No packet offset is required and the processor can match an unanchored rule anywhere in the packet payload. Raqia's RegEx compiler parses the input, verifies the syntax, and identifies errors. It then compiles the authored rules into special tables that will compose the classification processor database. Besides parsing and compiling the rules input, the RegEx compiler also reports memory usage, which is extremely important when building a complete system. It must then be decided whether external rule memory is needed, and, if so, how much of it. The compiler is written in C++ so it easily enables independency of the development platform. Its output (the rule tables) is then loaded into the classification processor, which processes the entire packet header and payload byte per byte for full wire rate classification at 10 Gbps. Interestingly, the RQDPC10G processor is fully protocol aware. As a result, it can perform *Transmission Control Protocol/IP* (TCP/IP) header checks and modifications on-the-fly.

Conscious of competition at the lower-end of the complexity spectrum (layers 2 to 4) with TCAM-based classification solution vendors, Raqia is justifiably concentrating its marketing attention on high-complexity applications such as the following:

• Virus scanning (for example, on incoming e-mail).

• Intrusion detection.

• Firewall-like filtering of packets (for example, in a gateway, some incoming packets—based on specific rules—will be accepted, whereas others will be denied entry to a corporate site).

In fact, the company's first product, the ReGXP2G Regular Expression Coprocessor is well targeted toward the IDS and web-switching realms, especially those that are based on HTTP 1.1. In the IDS case, every incoming packet must be parsed and analyzed byte per byte looking for attack signatures. This analysis must be performed in real time and is computationally intensive. Current IDS are implemented in software and usually monitor incoming traffic passively. Due to the heavy computation needed during their execution, they can only be selectively engaged periodically on some of the incoming packets. This leaves many holes continuously wide open for intrusion attacks. In some cases, they can only be engaged after an intrusion attack has been detected, which is essentially useless.

In the case of web switching involving HTTP 1.1, unlike HTTP 1.0, it not only supports multiple HTTP requests per TCP session, but it also requires that the URL parsing function inspect every incoming packet. At wire speed, this represents an unprecedented explosion in computational workload that current software-based implementations simply cannot handle. In the HTTP 1.0 case, the URL-based web switch in what has come to be known as *delayed binding* must first terminate the TCP session before it searches the HTTP request that contains the corresponding URL. This information is usually inside the second or third packet received by the web switch after the TCP session is established. Based on the URL thus found, the corresponding packets will be switched to the appropriate destination server. The technique is actually called *delayed binding* because the connection with the correct server is not established until after the HTTP request has been received and parsed. This can be easily handled in software without any significant performance degradation when only one HTTP request is encountered per TCP session; however, it is impossible to contemplate in software in a massively scaled-load case such as the ones allowed by HTTP 1.1.

SUMMARY

In this chapter, we discussed the problem of classification as a follow-up to the problem of search that we covered in the previous chapter. We defined two broad categories of classification: one that involves classical lookup and forwarding procedures, which are done based on search and match according to bits found in a single field, and one that is focused on deep packet classification, which implies that the search and match process has to occur based on lookup and match against bit patterns appearing in multiple fields simultaneously. The former is key for classical routing/switching applications. The latter is key for the new network world, where QoS, CoS, SLAs, and provisioning of differentiated services is the name of the game.

When comparing alternatives for lookups or many other functions, the following key measures of merit deserve a close scrutiny:

- Update rates (best, average, and worst case if they are not constant).
- Lookup performance on benchmarks of interest when doing updates at the advertised rate.
- Power consumption when running the previous benchmarks with background updates.
- Required board space and direct cost of purchase.
- Implications of interfacing with the NPU or other data path components on the previous measures of merit.

We discussed algorithms and data structures used in the implementation of techniques that provide these types of highly complex classification at astounding wire speeds. We concluded the chapter by discussing implementation issues and presenting a case study of a state-of-the-art classification processor product from Raqia.

SUGGESTED REFERENCES

The following white papers provide useful information on classification and forwarding:

IDT white paper, "Taking Packet Processing to the Next Level."

NetLogic Microsystems white paper, "High-Performance Layer-3 Forwarding in CIDR," *www.eetimes.com/story/ OEG20020603S0011.*

Jose Pereira, "Moving Classification and Forwarding to OC-768," a NetLogic Microsystems white paper.

SiberCore Technologies white paper, "Packet Management Lookups in Modern Networks," *www.sibercore.com/ pdf/scwp_00039_001.pdf.*

SiberCore Technologies white paper, "Classification and Forwarding Co-Processors Come of Age," *www. sibercore.com/pdf/wp_scwp002_1.pdf.*

The following companies design and build specialized classification and forwarding processors:

Cypress Semiconductor (*www.cypress.com*)

FastChip (*www.fast-chip.com*)

IDT (*www.idt.com*)

Mosaid (*www.mosaid.com/semiconductor*)

NetLogic Microsystems (*www.netlogicmicro.com*)

Raqia Networks (*www.raqia.com*)

SiberCore Technologies (*www.sibercore.com*)

Solidum (*www.solidum.com*)

Of course, other more traditional NPU vendors offer their own solution to the classification problem either with specialized silicon as member of a network-processor family or chipset or as an implementation that runs inside a network processor. The most important examples are Agere, AMCC, IBM, Intel, and Motorola. Refer to the corresponding chapters that cover each company's approach. Data sheets and application notes are available directly from all these companies.

Besides numerous tutorials on networking technology, *www.lightreading.com* offers an impressive array of documented testing of various types of network equipment. The associated discussion of both the methodology used each time and the results of each individual benchmark test can provide insight into what levels of performance are expected to typically qualify by the industry as acceptable.

The following are other important references on lookup and classification:

P. Gupta, S. Lin, and N. McKeown, "Routing Lookups in Hardware at Memory Access Speeds," *Infocom* 3 (1998): 1241–1248.

P. Gupta, B. Prabhakar, and S. Boyd, "Near-Optimal Routing Lookups with Bounded Worst Case Performance," *Infocom* 3 (March 2000): 1184–1192.

B. Lampson, V. Srinivasan, and G. Varghese, "IP Lookups Using Multiway and Multicolumn Search," *Infocom* 3 (1998): 1248–1256.

S. Nilsson and G. Karlsson, "Fast Address Lookup for Internet Routers," IFIP International Conference on Broadband Communications, Stuttgart, Germany, April 1–3, 1998.

V. Srinivasan, S. Suri, and G. Varghese, "Packet Classification Using Tuple Space Search," *Proceedings of ACM Sigcomm 1999* (September 1999): 135–146.

The *Tandem Repeat Occurrence Locator* (TROLL) algorithm is based on an Aho-Corasick variation. See *http://capb.dbi.udel.edu/main/slides/adalberto-4p/sld001.htm*. Another interesting site on the same subject is *http://finder.sourceforge.net/main.html*.

T. Woo, "A Modular Approach to Packet Classification: Algorithms and Results," *Infocom* (2000).

An interesting article discussing the way of connecting a search and classification engine to an NPU through the Look-Aside interface is "LA-1: Examining the Look-Aside Processor Interace," by Harmeet Bhugra, IDT May 20, 2003, CommsDesign.com, available online at *www.commsdesign.com/story/OEG20030520S008*. Its references include an interesting IDT paper comparing search-engine-based with algorithmic classifications.

CHAPTER 14
SWITCH FABRICS

In the first three chapters of the book, we took a glimpse at the unprecedented technology evolution that has occurred over the last 30 years. We discussed the reasons behind the development of network processors and saw typical operations that must be performed on packets transmitted in modern networks. Several critical factors have enabled the trend toward new sets of ever-increasing requirements for the performance and rapid deployment of sophisticated high-speed network equipment. These include the extraordinary progress in semiconductor technology, the spectacular advances in the development of distributed/embedded computing and applications and operating systems software, as well as the proliferation and pervasiveness of computer networks that will handle voice, audio, video, and data transparently, reliably, and at an optimal level of quality and cost (as perceived by the network users).

In the era of converged global networks, where from a transmission standpoint, almost no difference exists between voice or data, audio, or video, vendors must produce new products in record time while offering meaningful and different product features. These features not only provide a series of unique benefits, but they are also justified in an extremely competitive industry. Network processors promise to combine high performance with versatile functionality while providing ease of use, flexibility, and rapid development. They are a predominant enabling factor for this new phenomenon. In this book, we have been looking under the hood of these devices.

High-speed communications and network-processing equipment contain some clearly defined units of functionality. Once these are combined into the appropriate modules and system configurations by the manufacturer's engineering departments, they enable the creation of the broadband or high-speed switching/routing network gear.

In this chapter, we will look at some of the most successful and representative techniques for implementing the switching function. We will distinguish the underlying common denominators of the various technology modules and examine many trade-offs. We will then look at other functions required inside fast network switching/routing equipment. The next two chapters should hopefully provide a clear understanding of and better appreciation for the complexity and architectural trade-offs involved in the design and implementation of advanced switch fabrics and traffic managers—two of the most critical processing units with which network processors must be able to interface in order to carry out their complex mission in life.

So far, we have seen the architecture and internal structure of many different network processors. Hopefully, by discussing the internals of switch fabrics (and traffic managers in the following chapter), we can better appreciate the challenges of systems design, follow the internal technical description and/or justification of new or existing switch fabric products, better appreciate the differences in architectural decisions that the various vendor design teams have made as well as the reasons why the respective decisions were made either at the systems or component level, and better gauge the influence at the systems level of component capabilities or characteristics.

Hopefully, through this approach, the reader will not only acquire the basic knowledge of what the state-of-the-art technology is in this field, but he or she will also be able to independently perform a critical analysis and an objective evaluation of a platform or architecture comprising network processing, switch fabric, and traffic management and its suitability for a certain application and/or project.

THE DEFINITION OF SWITCH FABRIC

A *switch fabric* is a chip or a chipset that connects one or more among multiple inputs to one or more among multiple outputs based on some fundamental switching techniques and principles. For example, in a chassis-based switch/router, the switch fabric function is implemented on the *switch fabric card* or *switch card*, which is connected through the backplane to the multiple line cards. Switch cards serve as inputs and outputs to the rest of the network. Network processors and traffic managers usually are situated on the line cards. High-speed connection techniques such as *serialization/deserialization* (serdes) components are needed on both sides of a link over the backplane to ensure the efficient and fast connection between the line cards and switch fabric cards.

THE BASICS OF SWITCHING

Before we discuss the switch fabrics themselves, we must discuss some fundamental concepts and notions about switching. These will be introduced as system concepts. The actual implementation of switch fabrics is of secondary importance for our purposes as the switch fabric user needs to know its macroscopic behavior in a system that he or she designs, but not necessarily the transistor-level connectivity of such an extremely complex component. These system concepts can be applied in a switch at multiple levels of abstraction depending on the architectural intentions, design goals, and overall development context. For example, if we discuss concepts that we address later in this chapter, such as the scheduling of switching, arbitration of the crossbar, or backpressure, as a means of notifying upstream logic about congestion occurring downstream, it is instructive to note the similarities with the case that we would encounter when dealing with the issues surrounding the action of decelerating a vehicle.

An individual can decelerate a vehicle in several ways, including the following:

• Taking his or her foot off the gas pedal.
• Hitting the brake pedal.
• Pulling the handbrake.
• Shifting the gear-box down.
• Driving temporarily uphill without applying more gas.
• Ramming into the rear of another vehicle in the front that moves slower or crashing altogether onto a massive object, tree, or wall that does not move at all.

All these varied techniques undoubtedly achieve deceleration as a result, but from the vehicle designer and driver's points of view, they all tackle the problem in very different ways. Each approach involves trade-offs in efficiency and sometimes undeniable or undesirable side effects. A typical student driver does not need a degree in mechanical engineering to know how the depicted multiple scenarios function inside the decelerating vehicle. He or she only needs basic driver training (at the car's system level) and some common sense to decelerate the car.

The same thing occurs with switch fabrics. A switch user should have systems-level knowledge in order to understand the most important macro-issues. However, learning these basics does not need to be boring or frightening for the switch user.

Switching is an extremely active research area. As a result, the scope of this book is not intended to exhaustively cover the technology that has been invented in the field. However, we will discuss some basic notions of some of the most acceptable and widespread techniques along with several interesting and exotic methods in order to stimulate the imagination of readers who are new in this field. Several references are provided at the end of the chapter for a deeper study of the subject.

For our discussion, a switch is a network module with many inputs and outputs. A switch can provide a physical connection at any time between inputs and outputs and can transfer data that are available at any of its inputs to any of its outputs.

Network switches can be divided into two categories: *circuit switches* and *packet switches*. If we want to be extremely purist in our pronouncements, the packet switch category can be further divided into two subcategories: *packet switches* (also known as *routers* in everyday lingo, which connect the various Internet segments together), which can handle variable-size packets, and *ATM switches*, which are optimized for small, fixed-size packets (cells). A important variety of these switches is a hybrid of these two siblings known as the *multiservice router* (MSR). The MSR can service multiple ports with multiple protocols and technologies. In this study, we will mostly discuss packet switches and multiservice switches. Circuit switches are part of the traditional telephony or ATM network. These networks do not seem to be dominating the future networking landscape, although both will continue influencing it and playing a role.

Originally, it was thought that the traditional telephony and ATM networks were quickly becoming obsolete as we moved rapidly into the packet-switched networks. This has turned out not to be the case since the bad economy of recent years has dramatically forced carriers to massively scale back their investments for the introduction of new packet-switched technologies. This has injected more life into network equipment that can simultaneously switch both *Internet Protocol* (IP) packets and *time-division multiplexing* (TDM) and/or ATM traffic, where data traffic as well as digitized and compressed voice are carried next to each other, the latter using traditional approaches.

In previous chapters where we discussed search and classification, we learned that if the network is packet switched and therefore connectionless (something that is applicable for routers handling datagrams), it must have some intrinsic means to find the intended route that best services a desired destination. In these connectionless settings, the switch/router consults a table that maps destination routes to output ports. This dictates which output interface should be used at any moment in time. The information needed to conduct such a routing/switching decision process is based on destination information that a packet actually carries inside its headers; therefore, this approach is self-contained.

Otherwise, if the network is circuit switched and therefore connection oriented (as is the case with ATM switching systems), some switching-control mechanism must be available before switching can take place. This occurs in two consecutive phases—first, setting up a connection path and associating it with the data to be switched, and second, actually transferring the data.

Some types of switches/routers use routing tables to map inputs to outputs. One of the major tasks of a switching/routing system is to actually create, maintain, and update these tables. In addition to this function, however, switches/routers also perform several other generic functions. With the appropriate scheduling algorithms, for instance, switches/routers not only resolve contention at their outputs, but they also provide different levels of priority and therefore different levels of *quality of service* (QoS) to specific users or classes of users.

Some switches are also required to decide whether they will accept an incoming call (especially in the connection-oriented case). This is handled by the switch's ability to block calls. A packet switch usually does not experience this type of call blocking (also known as *admission control*). However, in more recent environments, the packet switch has the equivalent of a circuit-switch's admission control mechanism. It appears in the form of a *service level agreement* (SLA), which may be set up administratively and is strictly policed and enforced by the hardware. However, buffers may overflow at the input or output when more packets are stored than available buffers can preserve. Newly incoming packets will simply be discarded or dropped as no more space is available to buffer them. *Packet loss* is a highly undesirable situation that designers strive to minimize.

If we assume that all paths from inputs to outputs are simultaneously active, the switch has attained its theoretical capacity (or bandwidth). In practice, this is rarely the case, as only some of the possible paths are simultaneously active. Nevertheless, one of the switch designer's objectives is to maximize the available switching bandwidth for a specified budget and level of reliability.

A typical switching fabric can be visualized as a combination of input and output buffers, some port-mapping mechanism for packetized switches and the core, which is also known as the *crossbar*. When we dive into actual switch fabric chipsets later in this chapter, we will notice that these logical subcomponents are in a hierarchy of system complexity even inside commercial off-the-shelf switch fabric chipsets. Figure 14.1 shows a simple example of the basic concept. Depending on the crossbar state (which in this case can be encoded by a control bit of 1 or 0, denoting either one of the two possible states), every input can be connected to every output.

Arriving packets or frames[1] are stored into input buffers before the data are presented to the fabric's core. A mapping mechanism (a concept that is only applicable in packet switching) reads a packet's destination address or a VCI (in the case of ATM) from the packet's header. Then based on some lookup operation, it decides which is the appropriate output port. In many fabric cores, the term *packet* does not refer to an IP packet; it is a generic expression about a series of segmented frames that are created transparently by the switch fabric chipset. These frames have a different name with each switch fabric vendor—some vendors call them *frames*, whereas others call them *cells*. This process automatically segments an IP packet to optimize the internal switching over the crossbar. After the switching process has concluded, the fabric automatically reassembles the packet prior to releasing it downstream to the network processor or traffic manager. An individual time slot in TDM links that use circuit-switched systems always specifies which path is to be associated with the switching operation. Therefore, we can safely conclude that no port mapping is needed inside circuit switches.

The switch fabric core connects data from any ingress port to any egress port. It can be modeled and conceptualized as a simple *central processing unit* (CPU) that reads from one address (ingress port) and writes to another address (egress port). Recent switch fabrics are extremely sophisticated

Crossbar switch

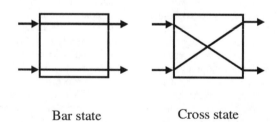

Bar state   Cross state

FIGURE 14.1   Rudimentary crossbar switch and its two possible states.

---

1. It is worth noting here that the *Network Processing Forum* (NPF) in the CSIX-L1 specification has defined the concept of the CFrame, which is a standardized shape and structure for the segmented packets while they traverse the fabric core in a protocol-independent format. It is characterized by a 6-byte header, a payload between 1 and 56 bytes (usually 64 bytes long), and a 2-byte trailer. The CFrame is supported by most commercially available switch fabrics. It is an interesting concept as it supports in-band flow control both per I/O port and per class per destination. On the other hand, the *NPF Streaming Interface* (NPF-SI) is not based on the CFrame and its flow control is structured as an out-of-band scheme instead.

multiprocessor units that can handle thousands of packet transfers simultaneously and over parallel paths using different techniques for the actual transfer between inputs and outputs. Traffic from input buffers is scheduled according to specific criteria and then transferred over the fabric core. Output buffers temporarily store data after the switching operation occurs and until any potential congestion at an egress port has been successfully resolved. At the egress ports, some sort of scheduler is often needed again to supervise buffering and arbitrate access to the line-card resources. Except for some fundamental definitions, we do not intend to expand on details over circuit switching, as network-processing technology is predominantly tuned to the packet-switching realm. All these concepts are covered in fundamental textbooks on the subject (see, for example, footnote 12).

A *time-slot interchanger* (TSI) is the basis of time-division switching. It reads a sequence of samples, which usually arrive in parallel from multiple ports, serializes them, and then reorders them into a new serialized sequence. If the input sequence is labeled 1, 2, 3, 4 or A-B-C-D as consecutive slots in a TDM system, then a TSI could rearrange them into new sequences as B-D-A-C or 2, 4, 1, 3 among multiple possible permutations of these labels. This essentially reorders the input sequence of samples that may be coming in from multiplexed lines. Therefore, it effectively behaves as a switch.

In space-division multiplexing, input samples follow a physically different path inside the switch core between input and output. The most rudimentary space-division switch is a crossbar such as the one shown in Figure 14.1. The principle can be extended geometrically. Signals are understood to be flowing horizontally into the switch from the ports shown on the left-hand side of the drawing, whereas ports on the right-hand side of the switch represent outputs from where signals exit the switch. At every moment and depending on which state the switch is in (among the two possible positions), we distinguish the bar state and the cross state.

Another basic concept is that of the *crosspoint switch*, which can be functionally visualized as horizontal bars (corresponding to inputs) physically running over a set of vertical bars (corresponding to outputs). Based on an implementation-specific scheduling mechanism, connections are established at the junction points, thereby effectively connecting a horizontal bar to one or more vertical bars. Figures 14.2 and 14.3 show the core principle of each connection junction. It works very much like the crossbar switch, but it is characterized by a slightly different topology and it can be in either the bar state or in the cross state. In the crosspoint switch, signals are propagated either in the vertical or horizontal direction. Input signals flow into the switch along the horizontal lines, whereas outputs are read out along the vertical bars. Active elements called *crosspoints* are placed between input and output lines. When a crosspoint is active, the signal flows from the corresponding input to the corresponding output. We can have the same or different number of inputs and outputs, which correspond to $N \times N$ or $N \times K$ configurations, respectively. The possibility of activating a different output forms the wider basis of space switching.

If the input lines to a crosspoint switch happen to be TDM multiplexed, then different signals (meaning the content of different slots) may or may not need to be switched to the same output line.

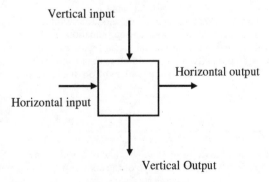

FIGURE 14.2 Crosspoint switch and signal flow configuration.

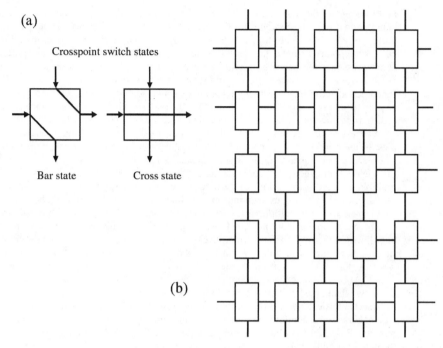

FIGURE 14.3 (a) The two possible states of a crosspoint switch and (b) a generalized crossbar switch with multiple inputs and outputs.

A scheduler is therefore needed in the switching system. It will be responsible for switching on the appropriate crosspoints at the appropriate time slots, so that the same input can be reliably connected to different outputs at different moments. The architect must then decide where and how to implement such a scheduler.

Despite its simplicity and functional appeal, an $N \times N$ crosspoint switch has some obvious shortcomings. More specifically, it requires N^2 elements, which can be prohibitively expensive for large switches. In integrated form, however, inside modern switch fabrics, state-of-the-art microelectronic design and semiconductor manufacturing processes allow the integration of hundreds of millions of transistors, so this is not the obstacle per se. Ancillary circuitry must also be embedded inside the same die to handle buffering, scheduling, and arbitration and often to provide serdes functionality to connect with the rest of the world. The combination of all these integrated blocks shows why switch fabrics appear in the form of complete chipsets.

Most importantly, a potential failure at one crosspoint junction ensures that the specific output to which it connects is effectively isolated from the corresponding input.

BLOCKING

Output blocking occurs in a switch when two packets are simultaneously destined for the same egress port. Obviously, in this case, one of the conflicting packets must be buffered (or blocked) in order to allow the other to exit the switch. No switch can avoid the output-blocking problem. A technique that can be used on TDM switches based on crosspoint architectures in order to avoid this undesirable eventuality is to use flavors of *time-space* (TS) switching, which essentially means using a TSI on each input and/or output line before and/or after the crosspoint module. One talks then about *time-space-time switching* (TST).

We say that *internal blocking* occurs when two or more packets from different inputs that are destined for different outputs run into an internal bottleneck and consequently one of them must be temporarily buffered. For example, the simple crossbar is *internally nonblocking* in the sense that no sample is blocked in the switch while waiting for an output line to become available. A *nonblocking* switch is one, which is immune to situations of internal blocking.

In general, a switch is *rearrangeably nonblocking* (RNB) if any new connection from a free input link to a free output link can be physically made, but it may require the switch to be reconfigured. A switch is *strictly nonblocking* (SNB) if a new connection from a free input link to a free output link can be made at any time without rearranging the internal configuration of the switch. A *fully interconnected switch* has N inputs going to N different switches, which each corresponds to one of the N outputs. A fully interconnected switch is necessarily a nonblocking one.

We talk about *head of line* (HOL) blocking when in some designs at one of the input stages of a switch a specific packet is denied transmission across the switch fabric core and it therefore blocks all other packets following it at the same input. This causes a backup in the queue that is entering the switch through that same port and stage.

BASIC SWITCHING ELEMENTS

We discussed how the two fundamental functions of a switch can be physically separated from each other. We will now show how this functionality can be implemented for a 2×2 switching element, which is often referred to in the industry as a β-element (pronounced "beta element"). In the case of larger switches (which involve more inputs and outputs than shown in this example), synthesizing a network that is composed of similar β-elements easily creates this functionality. Figure 14.4 shows the structure of a β-element.

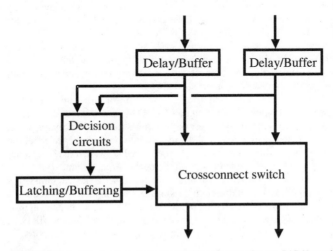

FIGURE 14.4   The structure of a basic switching element. *(Source: RAD Networks)*

The two input signals arrive at the element shown here from the top. Each must undergo a certain delay stage in order to synchronize the packet/cell contents with the decision that will be made based on the decision logic block. The decision logic block is operating on the packet header based on some application-specific sets of procedures. The shown latch is required to temporarily hold the result of that decision for the duration of the packet or, in the case of ATM, the duration of the cell, which is defined as 53 bytes long. The first 5 bytes contain header information such as source and destination addresses, whereas the remaining 48 bytes are the payload.

The crossconnect block is a dual multiplexer, which can be set in either a bar or a cross state. This means that each input can be routed to either one of the outputs, while the remaining input is invariably routed to the remaining output.

It should be clear that the complexity of the decision-making process will ultimately determine the complexity of the decision-making circuits. This is the key component of the β-element design. In some cases, the activity bit of an ATM cell will determine the crossconnect element's state. In other cases, the decision process on the destination of the cells is based on an elaborate inspection by a *finite-state machine* (FSM) of virtual circuit numbers inside the cell headers.

This may or may not involve a lookup of appropriate tables, an action that further complicates the design of the β-element. However, the design remains so simple and straightforward that it can be easily integrated in large numbers in very regularly repetitive circuit structures inside a *very large scale integration* (VLSI) chip. In the case of packets, the delay stages are simply buffers. The header of an incoming packet is inspected and, according to some routing rule, the packet is directed to one or both outputs. For example, if a specific 1-bit header is 0, then the routing rule states that the packet will need to be forwarded to the upper output, whereas the packet will need to be routed to the lower output of the switching element when the same 1-bit header of the packet is 1. If packets that arrive from both inputs must compete for the same output, only one will be forwarded to the output whereas the others will need to be temporarily latched (if storage space is still available). If this is not done, packet loss may occur.

An interesting variance of the basic design shown in Figure 14.4 is that provisions of buffering traffic may appear at the input, at the output, or at both of the switching element. In addition to buffering inside the switching cores, this more generic approach of buffering at the input or at the output of a switch offers flexibility and various possibilities of configuration to the designer who strives to achieve an overall design objective. Both approaches have pros and cons.

In previous chapters when we discussed the search and classification processes in relation to packet routing, we discussed the mechanisms through which input packets will be ultimately routed to the appropriate output ports by a switch/router. However, a packet switch must be able to perform an equally important role. The resolution of any potential output-port contention must be carried out correctly, which involves some sort of buffering.

One design approach is to physically separate routing decisions from buffering management in order to avoid contention at the output. In that case, a switch consists of an internal fabric (crossbar) routing network that can connect any input to any output and a set of buffers either at the input or output ports of this fabric or internal routing network.

Datagram has been used as a generic term for either an IP packet or ATM cell. The difference between a datagram switch/router and an ATM switch in our discussion is that the former can switch packets of variable length, whereas the latter's load will always be of fixed length. In ATM switches, the port-mapping function will decide what to do with a specific cell based on the VCI that it contains. However, in packet switches/routers, the hardware must first parse the entire destination address before it can decide what to do with the packet or datagram.

In addition, packet size also affects performance in a switching network to a certain extent. Chapter 13 discusses this addressing, parsing, and routing decision context in more detail. Packets at the input can now be broken by the fabric into smaller frames that are appropriately formatted and tagged and then routed over the fabric core at the output of which they will be stripped of their tags and automatically reassembled in orderly queues at the intended output ports.

GENERIC TYPES OF SWITCHING PLATFORMS

Three generic types of switching/routing platforms are available in the market today:

- Multilayer *Synchronous Optical Network/Synchronous Digital Hierarchy* (SONET/SDH) switches are network devices that can support TDM-based voice traffic in addition to IP and ATM traffic. IP and ATM switching are handled at a logical layer below the TDM switching unit (crossconnect). Incoming traffic is dispatched to the fabric according to the type of switching needed. For example, IP packets are switched and then the traffic is statistically multiplexed with the ordinary TDM load before it is forwarded to the crossconnect unit. The big problem with the SONET/SDH switch is that it must be designed for a certain mix of traffic. If the actual load mix varies from the designer's target values, the switching efficiency suffers dramatically. This is because the TDM channels must be able to provide fixed bandwidth to the customer. Unless the load is available, this bandwidth is not utilized well, even if many of the TDM slots are assigned statistically to IP or ATM traffic.

- The hybrid multiservice switch is a networking device that implements parallel processing paths for different types of traffic. This implies that the entire structure expected in network switching/routing devices, such as switch fabrics, line cards, and so on, must be replicated and their design must be adapted to the types of traffic the switch is built to support. It is obvious that the number of these cards directly affects the load factor of each type of traffic that the switch can support. Therefore, although the problem manifests itself with the same symptoms as with the SONET/SDH switches (namely, the lack of switching efficiency under various load conditions), the cause is essentially the same—the traffic mix can be varied and often unpredictable.

- Recently, some startup companies have emerged that are developing several types of advanced routers. By replacing the multiple switch fabrics with one common underlying switching architecture, these companies purport to offer native support for multiservice networks regardless of whether the traffic is TDM voice, frame relay, IP packets, or ATM cells.

THE EVOLUTION OF THE MULTISERVICE ROUTER/SWITCH

The packet switch/router originally started as a simple bus-based computer with multiple *input/output* (I/O) adapters (also known as *line cards*) providing the correct physical and data link interface to the various networks that the packet switch connected. The main CPU either polled the input line cards periodically for any input activity update or it was interrupted by a line card upon the arrival of a packet, in which case the line card would buffer-deposit the incoming packet into the switch's main memory for subsequent processing by the main CPU of the switch. The switch/router CPU would read the packet's destination and would consult the routing table in order to determine to which output buffer this packet should be forwarded. This was the same CPU that arbitrated any access to output lines at all times.

Most low-end packet routers and even simple Ethernet bridges were implemented according to this model. The shortcomings of this approach were the lung capacity of the main CPU, the extent and performance of the main memory, and the performance of the I/O bus or the line cards themselves. Different implementation technologies always suffer from bottlenecks that may be caused by different architectural components. In other words, the designer can use a fast CPU to solve the performance problem, but if the I/O bus is not fast, a problem still arises.

Subsequent I/O bus designs brought faster transfer speeds than were previously available. This allowed the faster CPU processors to be utilized more efficiently, but given the ever-increasing demand for traffic, a designer then had to struggle with the performance limits of the line cards themselves, so some sort of decentralized intelligence had to be adopted and implemented. Instead of sending packets back and forth between the routing CPU and the line cards, packets would now be temporarily stored at the line card upon arrival and only some information from their header would need to be escalated for study and action. A centrally located CPU would receive only that information (as opposed to all the packets in their entirety) and decide upon the forwarding fate of each packet. The forwarding instructions would then be communicated back to the requesting line card, which would possess the local intelligence to forward the packet from its local buffers directly to the correct destination. Designers realized that there was (i) work that needed to be performed *on* the packet content and (ii) work that needed to be done *regarding* the packet and its destination. The principle of *slow* and *fast* path processing planes was born.

Line cards connect to the switch fabric over a backplane that carries data traffic to and from cards as well as control and arbitration signaling. The control function can be implemented on the switch fabric card, or it can be distributed on the line cards. In some cases, it can also be implemented on a separate control card.

In order to improve this original packet switch/router architecture, designers started to come up with evolutionary thinking. When the bus became the bottleneck despite the decreased bus traffic, the concept of the switch fabric was developed. This component would switch traffic from input ports to output ports much faster than could ever be imagined at that time.

However, when the I/O bus was not the problem anymore, the line cards became the problem. Line cards soon evolved from their original limited status. Instead of being dumb I/O adapters, they found themselves locally possessing some network-processing intelligence, which could enable each line card to decide a packet's destination and output port without the intervention of the main switch/router CPU. In order for this to happen at wire speed, designers created fast *application-specific integrated circuits* (ASICs), which would operate directly on arriving packets. Splitting the work into the slow path and fast path processing (in other words, a slow path CPU had to be placed on the line cards) enabled designers to minimize the routing dependency from a centralized CPU.

These tasks were soon to be expedited through the local caching of routing tables in fast memory that designers made available on each line card. This meant that the macroscopic port-mapping function of the generic switch architecture was now effectively distributed among the line cards. Switching and routing were no longer central CPU functions. The main CPU of a switch/router would now act more as an overseeing controller to which exceptional tasks are handed over. For instance, when no route could be found locally on a line card, the main CPU would handle a request for routing such an exceptional packet and would also install a new route entry in the lookup table that line cards would use where the problem originated.

In the case of ATM switching, the line-card network-processing intelligence made sure that it would find a match between the cell's VCI and an entry in the locally cached routing table. This match was necessary because at call setup time, the supervising controller created that specific entry. In the case of non-ATM packet switching, the route entry may or may not exist inside the cache. If it was found, then the line card logic would switch it accordingly by sending the packet to the appropriate line card that was associated with the intended output. If it was not found, then the task would be handed over to the supervising CPU of the switch/router. The line cards in such a switch architecture would share access to (and communicate with each other through) a common bus or logical ring. This would be used to directly forward packets from line card to line card and could obviously be the potential source of a bottleneck in the performance of such an architecture.

In ATM switches, in order to decrease the cost of the line cards, which because of the local routing intelligence and cache memory would necessarily carry a significant cost, an interesting approach was to create a special centralized port-mapping card, which would be shared by all line cards. On this port-mapping card, the architect would provide the routing table, which would be updated by the switch controller at call setup time. All incoming packets were to be forwarded by their ingress line card to the port-mapping card. The port-mapping card would read the header and consult the routing table. It would then forward the packet to the appropriate egress line card. The danger of local cache misses was eliminated as all setup calls would have a table entry in this central location. However, in this case, the probability of contention on the bus/ring doubled, as all packets would have to traverse it twice—once when they went from the ingress card to the port-mapping card and once when they went from the port-mapping card to the switch/router's egress port. An interesting, but obsolete, variation of this approach was to use two unidirectional buses in order to effectively minimize this probability; however, this also increased the complexity and cost.

The ever-increasing capacity requirements from the exploding demand due to new IT applications and required Internet connectivity pushed the switch/router designers to confront the challenges of further improving this generic design. The fundamental breakthrough was the decision to replace this common internal bus or ring that was becoming the bottleneck with a switch fabric. The switch fabric can be conceptually visualized as an intelligent network that provides multiple parallel connectivity paths from any ingress port to any egress port. Figure 14.5 shows the concept of the evolved multiservice switch.

To increase efficiency and maximize the available switching capacity or bandwidth, incoming packets would be fractured into meaningful smaller pieces of information called *frames* or *cells*. A shared control CPU in the switch fabric would tag these frames or cells performing the work of the original port-mapping function with the correct destination port number. The frames/cells would then be injected into the switch fabric. The switch fabric would be responsible for their automatic routing to the appropriate egress port through a sequence of actions from the various switching elements inside the fabric. At the egress port, the fabric would strip those frames/cells from their internally created tags and headers/trailers needed to traverse the core. They would then be automatically reassembled into their original parent form by combining their multiple segmented offspring frames/cells. Different types of switch fabrics are available. Some can only handle fixed-size packets (such as ATM cells), whereas others can handle variable-size packets.

In order to fine-tune and optimize the performance of the evolving switch fabric architectures, a major issue was deciding where buffering would need to be introduced. Three obvious options are buffers introduced at the input, the output, or both. Because controlling or avoiding the potential blocking conditions altogether remains one of the primary objectives of switch fabric designers, many techniques and combinations of methods have been used to address this problem and this is discussed in-depth in a good fundamental switching texbook like S. Keshav's book (see footnote 12).

A major technique is to use buffers inside the switch fabric core, for example, at junction points. As a result, conflicting packets can be judiciously delayed where and when it makes sense. In the so-called shared memory switch fabric approach, which is favored by IBM, packet memory appears inside the switch fabric where incoming packets are written and from where outgoing packets are read.

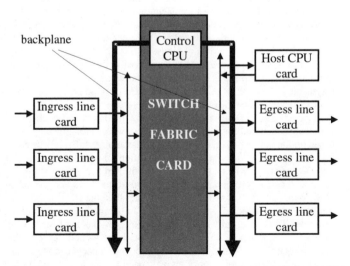

FIGURE 14.5 The conceptual architecture of a typical switch-fabric-based MSR/multiservice switch. Ingress and egress functions reside physically on every single line card.

For some applications or, more specifically, in cases where congestion appears, this embedded memory may not turn out to be sufficient. The systems designer must then introduce buffering on the ingress card as well. Whether this is feasible or desirable will largely be dictated by the answer to a separate question — namely, whether in this case network processors that are used on the line card have enough embedded static random access memory (SRAM) for such buffering.

IBM fabrics contain a 4K-packets buffer (as shared memory) for congestion. In addition, IBM provides a 2K virtual output queue (VOQ) per line card in the card interface chip of the fabric chipset. Therefore, a 32-port system would have 4K 1 2K332 5 68K packets to handle congestion, since without congestion, this memory is free. Fabric vendors such as IBM then expect some VOQ capabilities from the network processing unit (NPU) on the line cards, handling things at the flow level (as the fabric is completely unaware of flows). Their success in design wins and factory orders indicates that they are doing the right thing as far as the market is concerned.

This is an unrelated issue for the specific topic we are covering here, but it is important one for the overall architecture of a switching system. The intention is to highlight overall systems-related issues. In some designs, the scheduler will not allow packets to enter the switch fabric core at all before it ensures that a link to the desired output is actually available. This is not a magic solution, however, as planning more internal activity than there is switching bandwidth (a sad situation known as over-commitment of the switch fabric) will inevitably lead to packet loss. Therefore, the burden rests squarely on the shoulders of the scheduler to ensure that incoming packets can be admitted.

Some systems use in-band or out-of-band backpressure (see for example [a] Katevenis, M., D. Serpanos, and E. Spyridakis. "Switching Fabrics with Internal Backpressure Using the ATLAS I Single-Chip ATM Switch." *Proceedings of the IEEE GLOBECOM'97 Conference* Phoenix, AZ [November 1997]: 242–246. It is available online at *http://archvlsi.ics.forth.gr/atlasI/*, in Postscript [230 KB], or gzip'ed Postscript [53KB] and [b] Motorola text at *http://e-www.motorola.com/collatoral/SNDF2002_N302.pdf*). This technique minimizes the amount of necessary buffers especially in multistage switches. This is because whenever an output is blocked, special in-band or out-of-band backpressure signals communicate this condition to previous upstream stages and prohibit them from forwarding more packets downstream. In-band backpressure removes useful bandwidth, but it simplifies the silicon design. Out-of-band backpressure mechanisms make the design more complex, but do not waste switching bandwidth when applied. Once it is activated, backpressure is not a remedy against the abundance of incoming traffic. As a result, it will simply cause input buffers to be filled up instead, which sometimes may lead to packet loss at the inputs. In order to put things into perspective, notice that when transient congestion occurs and as a result transient backpressure is being experienced, the concept of buffering comes to the rescue of the systems designer so the system can ride out this undesirable state. The inter-relationship between these seemingly unrelated and independent factors should be obvious.

An interesting approach is the use of credit table mechanisms extensively, for example, by IBM (an interesting example is also discussed in Kung, H. T., T. Blackwell, and A. Chapman. "Credit-Based Flow Control for ATM Networks: Credit Update Protocol, Adaptive Credit Allocation, and Statistical Multiplexing." *Proceedings of the ACM SIGCOMM '94 Conference*, London, UK [August–September 1994]: 101–114). In this scheme, the fabric issues the input sources a number of credits, which are accounted for by the system. The sources are allowed to access specific output port buffers based on the number of credits they have accumulated. Each time contested resources are used, credits are withdrawn from the requesting source's account. Some radical action is taken in terms of assigned priorities when a source runs out of credits before it starts collecting new credits again.

In order to accommodate the arbitration of conflicts and the scheduling of transfers, the switch fabric core can be designed to be much faster than the I/O capabilities of the chipset. The fabric core links that traverse the crossbar from inputs to outputs are much faster than the inputs themselves. For example, if two internally competing packets are destined for the same egress port, they will both manage to be transported fast enough through the switch fabric to their destination port where they can be sequenced for output before a new incoming packet shows up at the ingress. Most current commercially available fabrics offer an overspeed factor that lies between 60 and 100 percent (these are also sometimes referred to in decimal form such as a factor of 1.6 or 2, respectively).

Some designers partition parts of the overall fabric function, the scheduling, the arbitration, and especially the necessary queuing buffers at ingress and/or at egress ports in a distributed fashion such as on line cards. Alternatively, these functions can be kept centrally on the switch card. However, trade-offs are involved, such as switching performance versus power dissipation versus cost (the complexity of the switch fabric card and the backplane) versus the overall scalability. These trade-offs affect the overall performance.

Some systems offer link redundancy through multiple fabrics (and we will discuss this later in this chapter). The use of multiple switch fabrics (in a load-sharing mode) that operate in redundancy connecting inputs to outputs is another effective way of reducing collisions inside the switch fabric itself. However, it is obvious that in this case an appropriate review of the output buffering requirements becomes imperative, as several incoming packets now may have traversed the switch fabric and will compete potentially for the same output interface.

Network equipment vendor (NEV) companies, when confronted with the maze of all these performance issues, were forced to design their own in-house switch fabrics in ASIC form. In addition to the effort, cost, and pressure on engineering teams, the economic reality of business life imposed unusually harsh life expectancies for switch fabrics. Enterprises typically expect their equipment to last anywhere between three and eight years. Carriers are a different market with exotic needs that include an expected life span of 5 to 20 years in addition to their already tough reliability, availability, and in-service upgradability requirements. The complexity of the task, the cost (if and when the skilled design team and infrastructure was in place, otherwise the whole issue could not even be envisioned as a project), and the risk of failing to deliver a working product that meets the customer requirements within budget and on time carried a massive penalty when judged against the ever-present need to excel in terms of time to market. Established and startup companies started coming up with off-the-shelf switch fabric chips and chipsets that would offer high performance and eliminate the risk of failure. As a result, a trend appeared and network equipment companies started moving from expensive, risky, and cumbersome do-it-yourself switch fabric ASIC designs to off-the-shelf commercial chipsets.

With the arrival of these high-performance commercial off-the-shelf switch fabrics, the predominant remaining problems include deciding where buffers must be placed in the architecture, how to schedule access to these buffers, and how to manage the switching bandwidth more efficiently. Unlike the fabric architects, users of these fabrics may not have to worry about them as they can pretty much choose the corresponding switch fabric architecture that matches their system requirements.

BACKPLANE DESCRIPTION

Figure 14.5 shows the typically modular structure of a modern MSR. Line cards for the various supported network types are connected to a switch fabric card (or cards) through a set of *high-speed serial* (HSS) links. A supervising/monitoring control processor is also provided. This processor sometimes appears on the switch fabric card or as a separate control processor card from where clock signals are distributed, ensuring the synchronization of operations, multiple clock domains, and so on.

The typical line card can be conceived in three cascaded stages. First, regardless of what the medium or *physical* (PHY) layer is (such as Ethernet, Gigabit Ethernet, SONET, and so on), the network side contains the corresponding PHY chip(s) implementing the PHY interface layer, the *Medium Access Control* (MAC) chip(s), and the framing functions. The network processor is right behind this layer of logic and takes care of the packet/frame processing functions that we described earlier. The NPU is then connected through a standard interconnection or through an *original equipment manufacturer* (OEM) proprietary scheme with a block of queuing logic that handles the interface with the switch fabric. Typical interfaces between the NPU and the switch fabric interface logic include CSIX, *Universal Test and Operations PHY Interface for ATM* (UTOPIA), *System Packet Interface 4* (SPI-4), *Network Processing Forum Streaming Interface* (NPF-SI), and so on.

The switch fabric card obviously contains the switching unit (also known as the *crossconnect*, which can be physically present in the form of a switch fabric chip or a set of complementary switch fabric chips). It also contains a control unit, which handles all scheduling and arbitration tasks. Sometimes the logical splitting of these two components places the scheduling control unit on the line card communicating with the crossconnect over the backplane. Scheduling is important in order to coordinate the traffic flow to and from the line cards as well as to and from the various stages of the switch fabric.

A control CPU handles several tasks from error logging, activity monitoring, and configuration all the way to keeping track of operation parameters such as power supply voltage variation, temperature, and so on. The switching unit provides connection paths from all its inputs to all its outputs. Through the switched backplane, every port on a line card can communicate with any port on the same card or on any other line card. Serdes stages ensure the appropriate modification of how information is presented to and from specific submodules (serial or parallel). For instance, the backplane is always handling information in serial form, whereas framers on the line cards need to be able to look at information as deserialized and properly aligned words. The backplane is structurally and electrically a key component of the overall switch fabric. As a result, it is usually of proprietary design and completely incompatible with other vendors' switch fabric chips. Despite this fact, they all essentially share a similar layout, which is physically limited to about 1,000 *printed circuit board* (PCB) traces given the very high speeds it must meet without signal reflection and interference. They are also manufactured most often out of the same so-called FR4 material that is used for advanced performance PCBs.

Backplane operation can be either *synchronous*, which means a central clock signal must be distributed along with data signals, or *asynchronous*, which requires very precise (typically to 100 parts-per-million levels of precision) clock generation on all system cards. In asynchronous operations, to make up for the potential data drifts that will inevitably occur due to the statistical lack of synchronicity between the multiple clock domains, the switch fabric will periodically inject an idle cycle in order to reset the local clock phase, which otherwise will disallow the decoding of the so-called 8b/10b-encoded data and inhibit operation. Buffering is typically foreseen in the backplane interface design next to the serdes transceivers. This ensures that data undergo a series of well-calculated delay stages whenever crossing into new clock domains. In fact, this first-in first-out (FIFO) path allows the deskewing of bits and the consequent alignment of frames. The fabric control CPU keeps track of many of these parameters and intervenes accordingly in the fine-tuning of the configuration at initialization time besides handling typical transmission-related issues such as error correction, cyclic redundancy coding, and so on.

The majority of switch fabric manufacturers provide high-speed serial links from line cards to their switch fabric chipset through their backplane. These links can be designed to be either synchronous or asynchronous and the distribution of a global synchronizing clock signal from the control processor board can become a very serious issue for the system designer. The links typically perform at 2.5 Gbps, although manufacturers sometimes quote the speed at 3.125 Gbps. A trend to design 5 Gbps links in the next-generation backplanes also seems to be appearing. Current backplane high-speed serial links are implemented with fast switching bus drivers designed around various high-speed, low-impedance, circuit technologies such as *low-voltage differential signaling* (LVDS),[2] *current mode logic* (CML),[3] and *differential-mode positive emitter coupled logic* (DM-PECL).[4] We will not expand on these circuit techniques here. See the references provided at the end of this chapter for more information.

---

2. For instance, Agilent Inc. provides a nice series of white papers with a solid introduction to the theory and advantages of LVDS in signal integrity contexts (serdes). See *www.measurement.tm.agilent.com/insight/2000_v5_i2/insight_v5i2_article01.shtml* and *www.measurement.tm.agilent.com/insight/2000_v5_i3/insight _v5i3_article05.shtml*. Vitesse also offers several helpful white papers and application notes explaining this technology in detail. See www.vitesse.com.

3. Any good advanced microelectronics textbook provides ample description of CML technologies. Some examples are provided in Chapter 3, "Packet Processing." An interesting presentation can be found at *www.bol.ucla.edu/~ive/ee215b.htm*.

4. Good textbook sources of analog integrated circuit design also cover this subject in detail. However, an application note from Philips explains the basics of both PECL and LVDS. See *www.philipslogic.com/support/appnotes/an253.pdf*.

We mentioned the presence of serdes modules that serialize data for transfer over the backplane and deserialize it for the line cards. Many backplane products are currently encoding the high-speed serial links using a coding scheme originally developed by IBM, which is known as 8b/10b coding. The 8b/10b coding technique is very robust for error resilience and link reliability. As a result, it is also used in the basis of the Fibre Channel used in *storage area networks* (SANs) and other media such as the ubiquitous switched Ethernet. However, it suffers from a significant degree of transmission inefficiency since only 80 percent of the available bandwidth is utilized for actual data transfer. The rest is eaten up by the redundancy of the code. This weakness has pushed many backplane manufacturers to design their own proprietary coding methods to improve the use of their backplane's bandwidth.

The design of high-speed backplanes remains a tricky engagement and consequently manufacturers usually go to extreme lengths to acquire, compile, and present to their customers simulation runs that can confirm and prove signal integrity and actual data delivery results. They also provide ample application engineering help in order to help their customers with the development of high-performance products, which are necessarily dependent on precise PCB trace lengths, correct placement and routing, and so on.

THE SCALABILITY OF SWITCH FABRICS

The concept of scalability in switch fabrics can affect multiple aspects of the design. The problem of scaling is usually three dimensional based on the following:

- Varying the number of ports.
- Varying the speed of ports.
- Varying the type of ports (such as protocols).

Note that the different types of traffic impose different buffering and scheduling constraints on the data traffic produced by line cards for the switch fabric. More specifically, the following is expected in a multiservice switch/MSR:

- ATM traffic requires deep buffering and extensive scheduling capabilities from the switch fabric.
- TDM traffic needs tight jitter and delay control from the switch fabric, but not necessarily buffering or scheduling.
- Ethernet-type traffic is required to be able to count on the switch fabric's ability to provide deep buffering and some rudimentary levels of scheduling.

The available capacity (or bandwidth) of the switch fabric that is given by the technology and architecture upon which it has been designed is of paramount importance as it decides the context within which the switch fabric can operate and function properly and efficiently. It is unfortunately not an infinite number, as this would make designing a complete multiservice switch/MSR a straightforward process.

As we have learned in a typical implementation by industry leaders in earlier chapters, in order to enforce and monitor/manage the flow of traffic according to the industry trend toward SLAs that rely on QoS and *class of service* (CoS) approaches and where users are billed based on usage, a traffic manager must typically be positioned at the egress paths of the switch fabric to manage and shape the traffic. The traffic manager, which we revisit as a generic concept in the next chapter, either by itself or in conjunction with a network processor manages the outgoing traffic and hands it over judiciously to the MAC/PHY stage of line cards to be injected into the actual network. However, this model of using a traffic manager only at the egress path is not necessarily the only way.

THE REDUNDANCY OF SWITCH FABRICS

Bordering the scalability problem and entering into the territory of reliability and/or availability, we briefly discuss how to go about adding capacity and/or redundancy in a switch-fabric-based system. Several approaches can be taken: At the switch fabric design level, more ports per chip could potentially be added or the speed of the ports can be increased. However, at one level of hierarchy above, the switch fabric must be replaced with a more powerful one of higher capacity or multiple fabrics must be added in parallel.

Redundant switch fabrics are introduced by adding more switch fabric components in parallel on the same or different switch fabric cards in order to increase the reliability and resilience of a switching system by increasing the availability of spare links onto which the system can fold back in case of malfunction. Figure 14.6 illustrates this principle. Appropriate design techniques can ensure that in case of malfunction, the system will automatically switch over to the backup switch fabric card with little or no traffic loss. The context may ensure what carriers call a *graceful degradation* of service and the cards may be *hot swappable*—that is, replaceable or serviceable while the switch is operating. This in-service serviceability is tremendously valuable to carriers who cannot afford to have the entire chassis powered down in order to search for and solve a problem at a specific card.

Three types of redundancy are available for switch fabrics: passive, active, and load-sharing-based redundancy.

Figure 14.7 illustrates the idea behind *passive redundancy*. For example, a configuration may have N active switch fabrics (in the drawing $N = 2$) and one extra fabric that behaves as the backup fabric (shown in gray). In such a configuration, the switching system architecture offers *N:1 redundancy*. If N backup fabrics are operating besides N active fabrics, the configuration is said to be *1:1 redundant*. A 1:1 redundancy will for $N>1$ be more expensive than N:1 redundancy. Therefore, the issue of man-

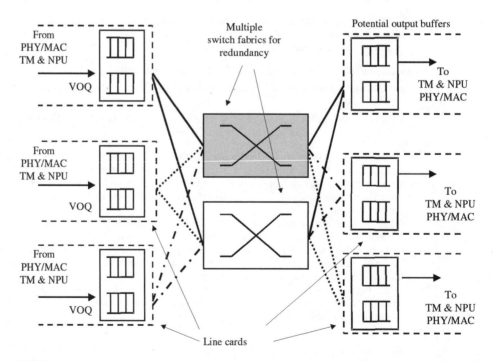

FIGURE 14.6 Multiple switch fabrics operating side by side increase the capacity and/or reliability/availability of a switching system.

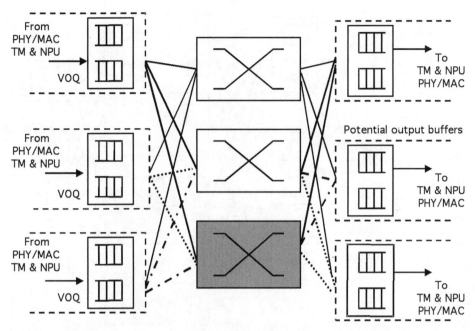

FIGURE 14.7   The principle of passive redundancy of switch fabrics configured as *N*:1 redundant. If there are as many backup fabrics as active ones, it has 1:1 redundancy.

aging resilience and availability becomes an application-specific issue that must be addressed in the framework of customer expectations and projected budgets.

Figure 14.6 depicts a configuration of *active redundancy*. In this example, two parallel sets of switch fabrics carry the same traffic in duplicate. If during operation any of these two sets of fabrics runs into a problem or fault, the switching system will simply lean back on the redundant fabric and no traffic or bandwidth performance loss will occur. This approach offers *1+1 redundancy*.

Figure 14.8 illustrates *load-sharing redundancy*. This example contains multiple switch fabric cards and they all carry traffic as in active redundancy. Enough fabrics are available to handle the normal traffic, but some fabrics in the combination also offer a resort in case of trouble. When a fault occurs, the problem card is isolated and traffic switches over automatically to switch fabrics where capacity is still available. Following a faulty condition, the new state of the switching system may reflect a certainly lower aggregate switching capacity, but operation continues unimpeded.

This is the principle of graceful degradation, which is of paramount importance on critical links, such as those owned and managed by carriers. If the total number of available switch fabrics in the starting configuration is twice what is needed for the anticipated traffic, this is the ideal case of carrier-grade QoS because no performance degradation will be observed. If not, then some delays may occur on some links and some traffic loss may occur, but the switch continues to function and connectivity is not lost. Depending on the arrangement the systems architect has created, such an approach has *N*+N, *N*+1, or *N*−*1 redundancy*.

ROUTING/SWITCHING SYSTEMS CONSIDERATIONS

Switch fabrics as systems modules are structured into two large conceptual subparts that need continuous access to the high-speed serial backplane: the *queuing manager*, or scheduling control unit, and the *crossconnect* switching unit.

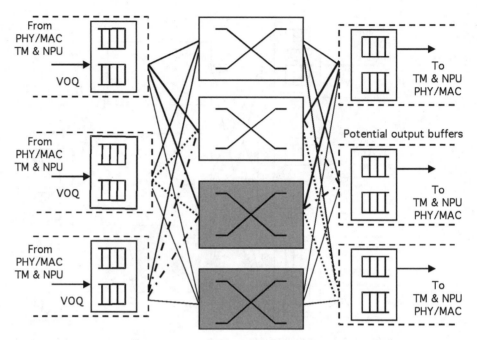

FIGURE 14.8 Redundancy based on load sharing between two sets of parallel active switch fabrics.

In most current vendor architectures, this partitioning reflects a physical distinction between two actual chips—one on the line card (the queuing manager) and the other on the switch fabric card (the switch fabric chip). Serdes controllers for input and output can exist in both of these chips. They also sometimes appear in the implementation as separate serdes chips inserted in the bidirectional data path between the queuing manager and the switch fabric chips. In some rare cases, the queuing manager chip is physically located on the central switch fabric card in which case high-speed transceivers on the line cards will ensure communication with the serdes controllers on the switch fabric card.

The queuing manager implemented on each one of the line cards, as the front end of connection to a network, disassembles and reassembles the incoming packets to and from cells, frames, or packets. However, this is done with the tacit understanding that these cells are just small units of internally switchable traffic that the fabric can handle and manage properly and that these cells have absolutely nothing to do with the well-known ATM cells, either in nature and structure or in size. The queues of cells are then built up, which will need to be mapped onto and scheduled for delivery to the corresponding output ports that may very well lie on the same or on another line card.

As instruments of advanced routing schemes that implement some sophisticated packet forwarding, switch fabrics can handle different classes of traffic simultaneously. Consequently, multiple class-based queues (*virtual queues*, also known by various other names in the industry or in pieces of product literature) are being set up for scheduling at the input ports on the line cards. Some traffic classes will inevitably receive a higher priority than others based on the choice of one or more among multiple algorithms. The switch fabric calculates the desirable route and order of events that must occur on each packet. It then switches the corresponding cells to the correct output port. Virtual queues issue *requests* to the scheduler notifying it that they have received fresh traffic (from the packet disassembly unit) for switching. Based on its rotating priority decisions, the scheduling unit of the fabric issues back *grants* to the queues, so everything is coordinated.

Any potential congestion or contention will be managed in the following ways:

- In a different way depending on whether it is input congestion (contention for the switch fabric) or whether it is output congestion (more than one packets competing for the same output port at the same time).

- By an array of arbitration units that can be implemented by vendors on the line cards in a system-distributed but coordinated fashion, or on the switch fabric card itself as a powerful and centralized unit.

Special provisions enable the correct handling of multicast packets. Unicast packets are switched from an input port to an output port and represent the bulk of the switch fabric's typical workload. Different architectures handle multicast packets in different ways. Some switch fabrics will switch the original packet only once over the crossconnect. They will then replicate it to the corresponding output port buffers without adversely affecting the crossconnect switching bandwidth. However, other architectural approaches copy the input packet and then switch it over the crossconnect in multiple instances in order to feed the appropriate output port buffers with the output of a single thread through the crossconnect.

However, the switch architect has to deal with a series of trade-offs regarding the overall efficiency, switching bandwidth usage, line-card logic complexity translating into development cost and time, and so on. We will not be expanding on them here. For more information, consult the recommended reading at the end of the chapter.

As some processing overhead needs to be expended in order for the fabric to create, affix, and remove internal special tags (facilitating the internal routing inside the fabric) to and from these internally created cells (after disassembly of the incoming packet), to enable the assembly and disassembly (internal to the switch fabric) of outgoing and incoming packets, and to properly handle and manage multicast traffic packets over the crossconnect, some extra switching bandwidth is required. A fabric's *overspeed* is the ratio of the actual data throughput over the aggregate user-port bandwidth. In this definition, the backplane encoding is not taken under consideration. For example, if a switch fabric is using a backplane and crossconnect structure that switches data with a speed of 20 Gbps and if traffic is received from a port that operates at a wire speed of 10 Gbps, then this switch fabric has an overspeed of 100 percent or an overspeed factor of 2. Most switch fabrics offer an overspeed that goes well above 60 percent. Some interesting theoretical modeling work has been done to show the advantages of 100 percent overspeed toward approaching an ideal switch design.[5]

Regarding the performance of a switch fabric, the industry has not yet agreed upon standard traffic patterns. Consequently, the measurements of throughput, delay, variations in delay (*jitter*), and the probability of loss are not easy to define consistently and compare systematically and objectively among different architectural approaches. Jitter is especially a concern in the voice-over-TDM realm or in the case of *voice over IP* (VoIP) transmitted over an ATM backbone. This is because even a slight misalignment of the jittered digitized bitstream chunks will likely compromise the integrity of time slots created for it on a trunk connection, thereby leading to potential loss of end-to-end bits, which implies poor link quality issues and consequently customer satisfaction problems. Last but not least, it is important to note that the NPF is in the process of standardizing switch fabric benchmarks.

SWITCH FABRIC ARCHITECTURES

The term *switch fabric* has been used liberally so far. This term has no formal and universally accepted definition; however, in this book, this term refers to a network composed of interconnected switching elements that handles the reliable transfer of packets from its inputs to its outputs. Several helpful ref-

5. Balaji Prabhakar and Nick McKeown, "On the Speedup Required for Combined Input and Output Queued Switching," *Computer Systems Technical Report* CSL-TR-97-738 (November 1997).

erences on switch fabrics are available such as the book *Broadband Packet Switching Technologies* by H. Jonathan Chao, Cheuk H. Lam, and Eiji Oki,[6] the article "Fast Packet Technology for Future Switches" by J. J. Deegan, G. W. R. Luderer, and A. K. Vaidya,[7] the article "Survey of Switching Technologies in High-Speed Networks and Their Performance" by Y. Oie et al.,[8] and the article "Fast Packet Switch Architectures for Broadband Integrated Services Networks" by F. A. Tobagi, T. Kwok, and F. Chiussi.[9] We will not cover the multiple possibilities here. However, we will look at the most common architectures.

A straightforward switch fabric is the crossbar switch. We have mentioned that the crossbar switch also requires a scheduling module, which at each moment in time will tell the fabric which inputs it must connect to which outputs. If the connections are functioning at a *constant bit rate*, the schedule can be computed in advance so the fabric simply executes the results and sequencing of operations as dictated by the scheduling algorithm. If the connections are supporting *variable bit rate* links, however, the schedule will have to be calculated by the fabric on-the-fly. This already points at significant intelligence that must be present in the fabric circuitry block.

If two incoming packets from different inputs compete for the same output at the same time, output blocking will occur. The problem is addressed either by running the switch fabric much faster than the inputs, which is usually hard and very expensive to accomplish, or by placing buffers inside the crossbar switch fabric at each junction. The latter approach is also characterized by the presence of an arbitration module that must be incorporated at each egress port and that decides at each moment which one among multiple buffers should be allowed to output a packet (or more) for routing to that specific egress line.

Despite its simple concept and structural elegance, the fundamental problem with the crossbar-based switch fabric is that it unfortunately does not scale well with the number of inputs and outputs. Other interesting approaches have been researched, such as the *broadcast* switches or *multicast* switches, where the desired egress port number is tagged onto each packet and then it is sent simultaneously to all egress ports by the switch fabric. Egress ports will buffer in their own output queues only those packets with a tag that matches their own port number. The advantage of such an approach is that as scheduling is only needed at the outputs and not over the entire switch fabric realm, this is an overall less complex system design despite the extra logic needed at each output to compare the port identifier with packet tags.

Some other rather creative techniques are based on methods such as call splitting, where the traffic load and the number of active input lines will influence the decision of whether one or more copies of the same cell (frame/packet) will be switched in a specific time slot.[10] The overall conclusion is that these are generally complex approaches and some other more intelligent means must be devised in order to create larger complex switch designs. One of these approaches is to use the Banyan switching network and the possibilities of augmenting it with other techniques.

In the following sections, we will look at buffered switch fabrics and, more specifically, at the two main variants—namely, when the buffering occurs at the input or output. We will then look at some of the most important architectural approaches—namely, those that are based on shared memory, the buffered crossbar, and the arbitrated crossbar. The arbitrated crossbar is available based on a centralized or distributed arrangement.

---

6. H. Jonathan Chao, Cheuk H. Lam, and Eiji Oki, *Broadband Packet Switching Technologies: A Practical Guide to ATM Switches and IP Routers* (New York: Wiley-Interscience, 2001).

7. J. J. Deegan, G. W. R. Luderer, and A. K. Vaidya, "Fast Packet Technology for Future Switches," *AT&T Technical Journal* (March–April 1989): 36–50.

8. Y. Oie, T. Suda, M. Murata, D. Kolson, and H. Miyahara, "Survey of Switching Techniques in High-Speed Networks and Their Performance," *Proceedings of IEEE INFOCOM '90* (June 1990): 1242–1251.

9. F. A. Tobagi, T. Kwok, and F. Chiussi, "Architecture, Performance, and Implementation of a Tandem Banyan Fast Packet Switch," *IEEE Journal on Selected Areas in Communication* 9 (October 1991): 1173–1193.

10. X. Chen and J. F. Hayes, "Call Scheduling in Multicast Packet Switching," *Proceedings IEEE ICC '92* (1992): 895–899.

Input-Buffered and Output-Buffered Switches

We refer to *input-buffered* (or *input-queued*) *switches* when packets are buffered before the actual switching occurs. In Figure 14.5, these buffers are assumed to be implemented on the line cards. This type of switch fabric suffers from HOL blocking because if any internal- or output-blocking situation arises, the single blocked packet at the head of the corresponding input queue will block all of the packets that are following right behind it. In input-buffered switches, packets can only enter the fabric core under conditions that depend on the actual design. An arbitration module must be used in many cases to decide which one among the conflicting packets will be scheduled to traverse the fabric core at each time and from which input buffer.

We say "in many cases" because *self-routing fabrics* do not need an arbitration scheme. Self-routing is a very important property that some switch fabrics possess in their architectural design, such as the crossbar shown in Figure 14.3(b), where the switching element at each junction can switch to the appropriate output state based on the content of the specific cell that it receives at its input. In such a case, the routing of a specific cell to the output does not require any extra knowledge from the switching element about the destination or fate of other cells traversing the core. Therefore, the overall switching decision function is conveniently distributed over the entire fabric core.

The most common algorithms used in the arbitration context are *dual round robin matching* (DRRM), *iterative round robin with matching* (IRRM), *iterative round robin with SLIP* (iSLIP), FIRM, *Parallel Iterative Matching* (PIM), and *round robin greedy scheduling* (RRGS). The arbitration context usually involves a handshake protocol starting with a request by the input queue for access to the shared resource (a fabric path in this case) and a grant issued by the arbiter followed by an accept acknowledgement from the input queue. Several sources discuss various arbitration algorithms in depth.[11–13] We will not discuss these algorithms. Although the arbiter must be able to scan all the inputs periodically and make a decision fast without causing a bottleneck, the fabric does not need to run faster than the input lines in such a configuration.

We talk about *output-buffered* (or *output-queued*) *switches* whenever the packets are queued at the corresponding output stages after the packets have been actually switched over the fabric core. The HOL blocking phenomenon is eliminated in this case because any potential output congestion has already been transferred from the fabric core to the output buffers. In general, output buffering requires more storage space per output line than input buffering. The speed with which the write operation of switched packets can be accomplished on the output buffers is the critical factor. For these two reasons, it must be implemented in a faster circuit design. On both situations, however, this means that output buffering is a more expensive proposition. That is the price one pays for avoiding the HOL problem.

An interesting and highly desirable side effect of output buffering is that per definition, it provides the systems designer with the possibility of a very fine granularity of traffic management. Decisions can be applied to the buffered packets that are available at the outputs, as the packets have already been switched past the fabric core and therefore are ready for dispatching downstream essentially at any time.

The HOL problem can be overcome by setting up queues at each input that are destined for each output. This scheme is called *Virtual Output Queues* (VOQs). This means, in the case of N outputs, that each input port maintains locally N queues where it buffers packets destined for each of the N different outputs. This can be translated from an implementation standpoint as saying that the input buffers of a line card are divided into N logical queues corresponding to each one of the N outputs the switch must be able to service. If more than one VOQ is available, some contention will occur for simultaneous access to the shared medium (such as the fabric core's bandwidth); therefore, arbitra-

---

11. Refer to Note 6.
12. Srinivasan Keshav, *An Engineering Approach to Computer Networking: ATM Networks and the Telephone Network* (Reading, Massachusetts: Addison-Wesley, 1997).
13. N. McKeown, V. Anantharam, and J. Walrand, "Achieving 100% Throughput in an Input-Queued Switch," *Proceedings of IEEE INFOCOM '96* (March 1996).

tion and scheduling capabilities need to be considered. At each time step, an arbiter will choose at most one packet from each input port, so that all selected packets can ultimately make it through the switch fabric to their intended outputs without any conflict (for the same output port).

This theme has many variations as switch fabric vendors strive to differentiate their products. For example, IBM whose switch fabrics we discuss later in this chapter, are based on a shared memory architecture with VOQs. This enables some low-priority traffic to actually make it from time to time through the fabric, no matter what higher-priority traffic is being mostly switched.

In addition to the basic function of arbitration that we mentioned a little earlier, another important function inside the switch fabric is the scheduler. *Scheduling* decides from which port data must leave for the next step on their intended path. It is equally applicable to input queues for deciding which ingress port's data will be allowed to enter the fabric core and to output ports for deciding which egress port's data will be transferred to the traffic manager. Typical algorithms for scheduling are *strict priority* (SP), *round robin* (RR), *weighted round robin* (WRR), and *weighted fair queuing* (WFQ). Scheduling is also responsible for deciding (when necessary) whether one or more (and most importantly which) packets must be dropped altogether due to lack of capacity. An ongoing debate among some switch fabric vendors exists regarding the problems that scheduling entails, especially if multiple levels of hierarchy are involved in shared resources.[14] This issue seems to be more of applicability in the context of arbitrated crossbars.

We said a little earlier that the switching bandwidth of real-life switches is limited. This finite capacity imposes constraints into what one can realistically expect to accomplish in an entire switching system design. Chances are that besides the implementation of VOQs at the inputs (which is irrelevant to the possibility of having or not having queues formed also in output buffers), a traffic manager must also be placed at the ingress path next to the traditional NPU in addition to the traffic manager/NPU pair we have already come to expect at the egress path of the switch fabric. The reason why a traffic manager/NPU combination may be needed at the input of the switch fabric as well is that in order to maximize the usability of the available switching capacity, egress ports must not be starved by inefficient fabric core scheduling or by poor VOQ management while priority traffic does not get delayed at the input. In such an arrangement, as shown in Figures 14.9(a) and 14.9(b), switch fabrics appear commercially as a chipset of two components—one as the actual core of the switch fabric and the other as the combination of input-based VOQs and the potentially existing output queues. Provisions are available to handle the serialization and deserialization of traffic in both directions over the backplane using serdes devices.

The QoS requirements of today's switches are of paramount importance in the minds of the switch fabric designers. More specifically, the logical segregation of traffic in classes, flows, and queues occurs. Each port has multiple queues and multiple classes of traffic are treated at different levels of priorities. Queues are usually structured on a per-flow basis, and flow control is handled by the switch fabric, which works in conjunction with the traffic manager.

---

14. The following is an interesting article summarizing the debate: Craig Matsumoto, "Startups Look to Supersize the Switch Fabric," *EE Times* (June 3, 2002). This article is available at *www.eetimes.com/issue/fp/OEG20020603S0026*. Make sure you read it critically, however. This is not because the arbitration logic is distributed so that the arbitrated crossbars suffer, as implied in the article. If the arbitration algorithm cannot keep up with the requests of all the input queuing managers, the result will be crossbar throughput degradation, leading to poor output link utilization. If collisions occur in the core, the solution is the fabric-overspeed idea (invented years ago by Clos). Furthermore, it does not sound intuitively correct that a centralized scheduler would be more efficient than a distributed one. To our knowledge, PowerX was the only fabric vendor who had tried it, and it is now out of business. Also although the statements in the article about iSLIP are correct, namely, that guaranteed bandwidth cannot be provided by it in a large system, especially as it requires tedious synchronization with all cards, be extremely circumspect about the thought of integrating the crossbar and queuing managers into one component, as this would completely negate the scalability approach of the architecture. This is a classic case where the economic need to integrate contradicts the modular future extensibility of a solution that is based on the arbitrated crossbar.

Buffered Crossbar

In the buffered crossbar switch architecture, the same principle is followed as shown in Figure 14.6. However, buffers will be placed on all three major stages, that is both input and output buffers as well as fabric buffers, right on the switch fabric card itself. With this approach, which is favored for instance by Vitesse switch fabrics, queues are naturally formed at each one of these 3 stages. At the output of every stage, some serious and coordinated scheduling must occur using RR or WRR or other scheduling algorithms.

Arbitrated Crossbar

This approach is also based on the basic principle of Figure 14.6; however, as discussed earlier, an arbitration scheme is implemented this time between the input queues and switch fabric on one side, and between the switch fabric and the output queues on the other side, if the specific fabric architecture requires output queues. VOQs are implemented in the arbitrated crossbar. They engage in a similar request/grant/accept handshake scenario with the arbiter that supervises access to the fabric core.

The arbitrated crossbar has essentially three variations: one where the scheduling and arbitration are based on a centralized arrangement, another one that revolves around a concept of distributed switching, and a configuration like the one we showed in Figure 14.9, where a traffic manager is needed not only on the egress paths, but also on the ingress paths.

By centralized arrangement, we mean a configuration where all switching decisions are handled by logic that is integrated on the switch fabric card itself. All input and potentially output queues are on the centralized switch card. Line cards are connected over the backplane through high-speed serdes

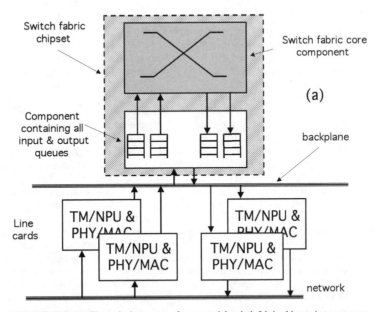

FIGURE 14.9(a)   The typical structure of commercial switch fabric chipsets in two components: one housing the fabric core and the other handling the input and output queues in conjunction with serdes devices that ensure the high-speed interface with the traffic manager and network processor.

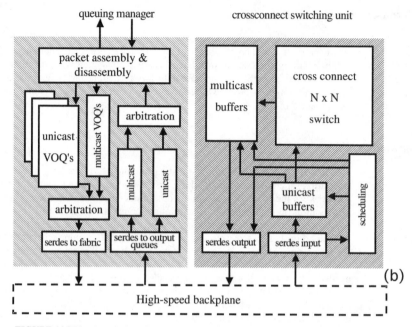

FIGURE 14.9(b) A typical partitioning between the two switch fabric components.

links. In a chassis-based system implementation, where typically two to four switch fabric cards are to be expected, this approach is translated into a straightforward backplane where paths do not need to cross. This type of switch fabric can be made redundant using the passive approach discussed earlier. Agere, whose switch fabrics we discuss later in this chapter, is a good example of a company using such an architecture.

By distributing the switching function, we obtain a configuration where the VOQs are implemented on line cards and where the backplane will necessarily include more complicated crossed paths. If each output port has its own VOQ allocated at the inputs by the queue manager (to avoid the HOL problem) and on top of this each CoS is also allocated its own VOQ, then for i inputs and k classes, a queue manager must be able to coordinate and manage ik queues. All of these sources will need to have access to the fabric core. With backplane, traces are physically limited to around 1,000 and the physical complexity of the backplane becomes an issue, along with the power consumption and cost that comes with the increased number of needed components to support the high-speed links.

In the distributed-switching case, redundancy can be implemented as either active or passive and may or may not be based on the load-sharing principle. In fact, the in-service scalability of this approach is enabled by the load-sharing redundancy principle, which enhances the advantages of this concept beyond its impact on simple reliability and availability. The architecture can scale very well because in a single stage the scheduling problem remains manageable. Scalability also looks good if different schedulers can be allowed to handle independent crossbar data paths, thereby affording a creative systems designer a wider dimension. From a performance standpoint, the configuration can offer low latency (because no queuing occurs in the data path crossing the fabric core). However, it may suffer from the need to reorder traffic because the presence of a shared medium (the fabric core) necessitates a contention resolution that may affect the sequential ordering of internal transfers. Multicast traffic is not natural for this approach so extra design creativity is required to adequately handle this type of data with the arbitrated crossbar when it is implemented using the distributed switching.

An arbitrated crossbar can be combined with traffic management and packet buffering also at the ingress ports, whereas the egress ports may or may not be equipped with queuing capabilities. This configuration allows better systems design and performance fine-tuning by enabling a close relationship between the switch fabric and traffic management, which enables the QoS to be scaled in a very fine granular and sophisticated way. This is also discussed in Chapter 15, "Traffic Managers."

Shared Memory Switches

In a shared memory switch, as shown in Figure 14.10, the inputs feed a multiplexer, which sequentially stores the (now serialized) incoming packets in a common memory bank that lies in the heart of the switch fabric. This embedded memory is shared by both inputs and outputs for the storage of frames/packets and for the management of queues. A control module is responsible for the extraction of stored cells/frames from the shared memory. By their recombination, it reassembles the original packets into an output stream, which is then subjected to the opposite operation from the one that occurred at the input. In other words, it is demultiplexed and its parts are routed into the corresponding intended output ports. The control module can often be easily reloaded to reflect new executable code that manages different QoS requirements.

The size of the shared memory can be shown to directly influence the possibility and rate of cell loss, so this is a rather convenient situation for the designer.[15] Also the memory technology used for the implementation of the actual shared memory is one of the fundamental constraints of the performance of such a switch fabric by the mere fact that all operations of the fabric will invariably revolve around writing to and reading from the shared memory. Consequently, memory access times are important. This can be a problem if the chosen memory technology does not allow data to be written into the shared memory N times faster than the rate at which data arrive at the N inputs.

In terms of scalability, as the Figure 14.10 shows, multiple switch fabric cards can be used, which combined with the *striping* (IBM calls it *byte slicing*) of frames/cells (similar to the striping concept discussed in the context of data storage in Chapter 11, "Storage Network Processors," where we reviewed storage network processors) across multiple fabric cards allows an easy performance scaling. The advantages of this technology include the in-service scalability, cost efficiency (as no external memory is required), and relatively easy implementation without a complex centralized scheduler like those needed in the input- or output-buffered approaches.

Some vendors claim this approach has a serious drawback. This has caused many designers to worry each time they are asked by a vendor to stripe data. If striping is implemented across multiple serdes lanes, then a designer needs to worry about what happens when even one serdes in the set fails. In some design implementations, this problem will prevent the other serdes lanes in the set from carrying any traffic. This will force the entire system to switch over to the backup fabric plane. This latter combination of possibilities is extremely worrisome for a carrier-grade-reliability systems architect as the mere possibility exists that two serdes lanes on different fabric planes fail. This will inevitably cause the system to go down.

The counterargument is that it all comes down to trade-offs. IBM, for instance, is an example of a major vendor using shared memory architecture in its switch fabrics along with striping. As we will see later in this chapter, their approach includes some interesting features such as native multicast and unicast support and in-band flow control. More specifically, regarding the striping of traffic, IBM's philosophy has been that once a link fails, the entire port must switch over to the backup fabric. IBM believes it is much more important to offer a graceful degradation of service. When talking about a switching system with several ports at 10 Gbps, for example, degrading a single port to 7.5 Gbps can cause congestion on the other ports and have an impact on some port performance that was not supposed to be affected. IBM's striping (byte slicing) enables its fabrics to operate in *full multicast* mode,

---

15. Refer to Note 6.

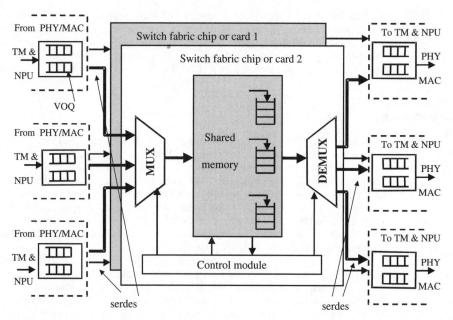

FIGURE 14.10   Shared memory switch fabric.

whereas other implementations (such as sending a full packet per serial link) will cause a bandwidth degradation by a factor of 2 (in the case of multicast packets).

MULTISTAGE SWITCHES

We discussed earlier that in the case of a crossbar network, a potential failure at one junction might catastrophically affect the entire switch. One solution for that problem is to design a multistage switch, where the architect configures a manageable matrix of interconnected crossbar switching modules, potentially of different but compatible sizes, arranged into two dimensions. The idea is that with such a matrix of switching modules, any input can be connected with any output through more than one switching path.

From a different angle, this switching-module-matrix concept is simply carrying the fabric scalability and redundancy principles that we discussed earlier into a link scalability and redundancy context, which this time is inside the fabric itself. If one path fails due to a junction failure inside one of the switching modules involved in the matrix, other alternative paths are possible and they will be traversing a different module sequence. We will now discuss some typical multistage switches before we take a look at a couple of state-of-the-art commercial switch fabric platforms.

Banyan-Based Switches

Let us now turn our attention to the highly significant Banyan switch and its augmented variants to address some of its shortcomings. A Banyan class network can be produced by merging multiple binary tree topologies, such as the one shown in Figure 14.11(a) where each stage routes traffic to one of its outputs according to whether the incoming bit value at the input is 1 or 0, allowing access to one of the two outputs each time a 1 goes to the top output while a 0 goes to the bottom output.

Based on this network, an entire class of configurations and multistage switches can be built. Figure 14.11(b) shows an example. Banyan networks have the property that exactly one path exists from any input to any output. By the mere structure of their binary-topology constituent trees, Banyan networks are self-routing. This means that an incoming serial bit sequence at any input with content that corresponds, for example, to a cell header, a frame header, or a packet header, will automatically be routed to the corresponding output. The switch does this by reading and applying one bit at a time extracting bits from the header and doing so for each stage.

For example, an arriving input sequence as the header of a frame containing the bits 010 in the arrangement of Figure 14.11(b) will mean that in the first stage, the first 0 (read from the left) will be used to route traffic to the bottom output lead. Then the second bit of the sequence, here a 1, will be used to further route the path to the top output lead of the second stage (wherever this stage might be in the web network). Finally, the 0 will be used in the last stage to route the connection to the bottom lead of the last stage. The switching path has thus been produced between the input and output based on the address header of the frame. These networks are discussed in more detail in the fundamental switching literature listed at the end of the chapter.

Batcher-Banyan Switches

Banyan switches have earned the reputation of being efficient, but they can suffer from internal blocking, which requires buffering to avoid possible traffic loss by discarding packets. This situation can be avoided if no idle input exists between any two active inputs and if the output addresses of cells/frames are sorted in either ascending or descending order. Once a special network presorts the frames/cells, they can safely enter the Banyan-based fabric core. Such a presorting-based topology is called the *Batcher-Banyan switch*. The sorting network is based on the *Batcher* architecture, which combines multiple simpler merge networks (all the way down to simple 2×2 modules) that sort their

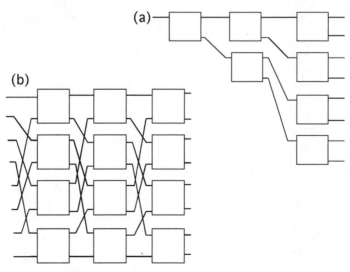

FIGURE 14.11 (a) Binary tree topology (b) Banyan class network.

inputs according to a certain order, such as ascending or descending. More details of the Batcher behavior are too specific for this discussion. For more information, refer to Batcher's original paper.[16]

Several possibilities are available for innovative architectures and plenty of references have been provided at the end of the chapter for more information on this topic. For example, multiple Banyan networks can be used in a tandem configuration. Packets without collisions are routed by the first Banyan to the output and are untagged as usual. Misrouted packets from the first Banyan are tagged as such. They are then introduced into a second Banyan where they are given another chance to make it to the desired output. This process continues (potentially with multiple Banyans in tandem) until they are dropped altogether or they have made it safely to the intended output. This specific approach is called *Tandem Banyan Switching Fabric* (TBSF) and the principle of deviating a misrouted packet to a new Banyan network is known as *deflection routing*.[17] It is reported that a 32×32 switch with uniform random traffic, based on a tandem configuration of 9 Banyan stages, decreases the probability of packet loss down to a respectable 10^{-9}.

OTHER EXAMPLES

The subject of switching is vast and this chapter cannot possibly cover the field. See the references for many other examples of creative architectures and systems designs. We will conclude this discussion by mentioning some representative cases.

The knockout switch originally proposed in "The Knockout Switch" by Y. S. Yeh, M. G. Hluchyj, and A. S. Acampora[18] and discussed in depth in *Broadband Packet Switching Technologies* by H. Jonathan Chao, Cheuk H. Lam, and Eiji Oki[19] is an interesting approach to the problem of reducing the cost of output-buffered switches. The idea behind the knockout switch is that from N possible inputs, the majority of the time, only a subset among the N (say, L of them) packets coming from different inputs need to arrive at the output simultaneously. Therefore, the output queue must run L times faster than the inputs (where $L<N$) as opposed to N times faster, which is the case with regular output-buffered switches. The key problem in the design of a knockout switch is to ensure that the packet losses are fairly distributed among the incoming virtual circuits.

The abacus switch is a multicast scalable architecture with input and output buffering.[20,21] It was designed to avoid the packet loss problem associated with the lack of routing links in the switch fabric of knockout switches. Although such a switch can be designed to achieve a satisfactory probability of packet/cell loss of 10^{-10}, for example, the fundamental assumptions for such calculations are that the traffic from different ports is uncorrelated and that the traffic load is uniformly distributed to all output ports. In many Internet-based web services, this is simply unrealistic. A popular web site may be faced with heavy traffic destined toward it at the same time, creating what we know as a *hot spot* situation where packet/cell loss probability becomes dramatically higher.

The Sunshine switch provides multiple paths from its inputs to all its outputs.[22] In order to accomplish this, it uses several parallel Banyan networks, which are fed by a Batcher sorting network. In a

16. K. E. Batcher, "Sorting Networks and Their Application," *Proceedings of Spring Joint Computing Conference, AFIPS* (1968): 307–314.

17. F. A. Tobagi and T. Kwok, "The Tandem Banyan Switching Fabric: A Simple High-Performance Fast Packet Switch," *Proceedings of IEEE Infocom '91* (1991): 1245–1253.

18. Y. S. Yeh, M. G. Hluchyj, and A. S. Acampora, "The Knockout Switch: A Simple, Modular Architecture for High-Performance Packet Switching," *IEEE Journal on Selected Areas in Communication* 5 (October 1987): 1426–1435.

19. Refer to Note 6.

20. H. J. Chao, B. S. Choe, J. S. Park, and N. Uzun, "Design and Implementation of Abacus Switch: A Scalable Multicast ATM Switch," *IEEE Journal on Selected Areas in Communication* 15, no. 5 (June 1997): 830–843.

21. H. J. Chao and J. S. Park, "Architecture Design of a Large-Capacity Abacus ATM Switch," *Proceedings of IEEE GLOBECOM* (November 1998).

22. J. N. Giacopelli, W. D. Sincoskie, and M. Littlewood, "Sunshine: A High-Performance Self-Routing Broadband Packet Switch Architecture," *Proceedings of the International Switching Symposium* (June 1990).

Sunshine switch of an $N \times N$ configuration with k parallel Banyan networks, k parallel paths are available for each N output. The idea is that if more than k packets/cells need access to an output during a time slot, then some of these excess cells (say, L) will be recirculated through a shared queue on a feedback loop of L parallel recirculating paths. They will be resubmitted to the switch at L specially dedicated input ports among the N input ports. In order to synchronize the recirculated packets with the ones that freshly arrive into the switch, an appropriate delay block is inserted into the recirculation feedback loop paths. In terms of its operation, a Batcher network sorts the arriving packets based on their destination address and priority. Then a trap network resolves output port contention by selecting the k highest priority cells for each output port destination address. Finally, the k parallel Banyan networks ensure that each output port can receive up to k cells in each time slot. If more than k cells are competing for the output, the excessive ones will need to be recirculated.

Another interesting approach is the *helix switch* design.[23] A hub-and-spoke topology is used in the helix switch. This speeds up the switching time by minimizing the distances over which signals must propagate. The helix switch works by sending packets forward by one segment of the ring with every time step. When an input packet reaches the correct output port, it joins the end of the local output queue, and a new input coming from this same card/port takes the slot, which was just vacated by the exiting packet on the inner ring of the hub. The helix switch is very fast because it uses very short interconnects between all I/O ports and because an open slot (such as the slot vacated by the packet of our example) is immediately used again to move the packets.

A COUPLE OF COMMERCIAL EXAMPLES

We will now turn our attention to some world-class switch fabric platforms from a couple of solid vendors to show the concrete form that the concepts we discussed take in a commercial product setting.

IBM PowerPRS™ Switch Fabrics

So why do we include a section on IBM's switch fabrics? The truth is that each one of the major vendors on whose architectures we elaborate has taken a different approach to tackle the overall systems issue. In the NPU design trade-offs, IBM's architecture has a very unusual approach as compared to vendors who propose their own traffic manager chip. Likewise, in the case of switch fabrics, IBM has its own separate philosophy based on the use of a powerful, shared memory switch fabric architecture. As IBM's offerings deeply influence some choices for an NPU user, the IBM switch fabric realm definitely requires some coverage at this point.

At the end of 2001, IBM started delivering its fourth-generation switch fabric chips called PRS-Q64G (PRS stands for *packet routing switch*). The original form of the technology called PowerPRS was used for internal work at IBM in the early 1990s. The second-generation switch fabric chip technology materialized into a commercial product that was dubbed the PRS-28.4 chipset. This continues to be a very important revenue generator for IBM today as it supports up to 16 OC-48 ports. The third-generation chipset was named PowerPRS-64G. It scaled the connectivity and performance to 32 OC-48 ports. The PRS-Q64G switch fabric provides support for 32 OC-192 or 128 OC-48 ports with a scalable aggregate throughput up to 512 Gbps. This effectively quadruples the performance of the third-generation product.

---

23. B. Patel et al., "The Helix Switch: A Single Chip Cell Switch Design," *Computer Networks and ISDN Systems* 28 (1996): 1791–1807.

IBM switch fabrics of the PowerPRS architecture implement several of the concepts we have already studied such as VOQ and shared memory, flow control, priority levels, credit tables and scheduling, multicast, and redundancy. Refer to "PowerPRS™" by René Glaise[24] and "A Combined Input and Output-Queued Packet-Switched System Based on PRIZMA Switch-on-a-Chip Technology" by Cyriel Minkenberg and Ton Engbersen[25] for more information on the switch fabric internals. These features offer multiterabit growth capability within a common architecture while supporting redundant switch fabrics and enabling manufacturers to design scalable, compact, and nonblocking switches ranging from 8 OC-48c/2.5 Gbps ports to 64 OC-192c/10 Gbps ports. The 64G and Q64G switch fabrics are available either as single, highly integrated silicon components or as board-level products.

The Q64G has been designed with a speed-up factor (overspeed) of 60 percent, sustaining a throughput of 16 Gbps per port (on 10 Gbps connection links). The extra headroom is required in order to take care of the overhead needed to disassemble, label, strip, and reassemble the cells that are internally created following the fragmentation of incoming packets.[26] The Q64G is a combination of output queuing with a shared memory. The advantages of both these techniques were discussed earlier in this chapter when we discussed switching internals. With output queuing, no internal blocking occurs and the delay is minimal. With shared memory, the best overall buffering usage can be achieved, and as it is shared between inputs and outputs, it is cost effective. Scheduling in the PowerPRS family is completely distributed and it is one of the keys for attaining high performance with IBM switch fabrics. Traditional crossbar, memoryless switches implement a centralized sophisticated scheduler that supervises the input and output queues by applying an overloaded and usually complicated algorithm. This type of scheduler needs to collect a lot of information from all input adapters to reconfigure the crossbar during every single packet cycle. In addition, in multicast cases, traditional crossbars will create multiple copies of the packet to forward to output queues.

With PowerPRS, the switch fabric output port queues keep pointers to the one and only copy of a packet in memory (in fact as many pointers as needed especially in the case of multicast). Based on their load, output queues decide on their own whether and when a packet residing in memory will be sent out from their port. Obviously, a packet is not erased from memory until all the pertinent queues have been served.

Conversely, all input queues can decide on their own without any higher-logic intervention whether an incoming packet can be allowed to be read into the fabric's memory, unless backpressure signals from the switch fabric core have signaled to all input line cards that they must refrain from doing so. The Q64G has room to internally store up to 4,000 packets entering the switch fabric through thirty-two 16 Gbps input ports. That is a rate of 1 packet every 32 nanoseconds potentially on each port. As packet fetching may be happening simultaneously at any (maybe all) of the 32 output ports, up to one memory operation (read or write) must be able to be performed within half a nanosecond if a full traffic load is to be sustained on every input and output port.

This has led to the implementation of a four-port spatial/time shared memory implementation, as shown in Figure 14.12. Two independent read and two independent write ports are used so that two writes and two reads can always occur simultaneously. The rest of the sharing is done on a time-sharing basis by 16 over a packet cycle (which is 32 nanoseconds, with typical switch packets consisting of 64 bytes, although the PRS can be configured to handle up to 80-byte packets equally as well).

24. René J. Glaise, "PowerPRS™: A Single-Stage, Reconfiguration-less, Packet Routing Terabit Class Switch Fabric with Distributed Scheduling," IBM Networking Technology Development, Microelectronics Division, 06610 La Gaude, France, 2002. This is also available online at *www.ibm.com/chips/techlib/techlib.nsf/techdocs/AEE3588F9F253EE587256B680060F11B/$file /PowerPRS_Tech_Overview.pdf.*

25. Cyriel Minkenberg and Ton Engbersen (IBM Research, Zurich Research Laboratory), "A Combined Input and Output-Queued Packet-Switched System Based on PRIZMA Switch-on-a-Chip Technology," *IEEE Communications Magazine* (December 2000): 70–76.

26. René J. Glaise, "PowerPRS™: A Single-Stage, Reconfiguration-less, Packet Routing Terabit Class Switch Fabric with Distributed Scheduling," IBM Networking Technology Development, Microelectronics Division, 06610 La Gaude, France, 2002. This is also available online at *www.ibm.com/chips/techlib/techlib.nsf/techdocs/AEE3588F9F253EE587256B680060F11B/$file/ PowerPRS_Tech_Overview.pdf.*

(a)

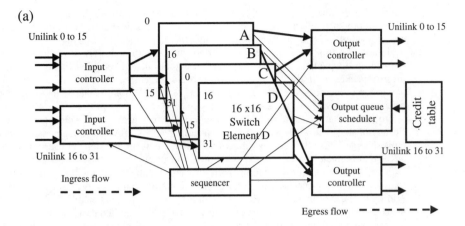

(b)

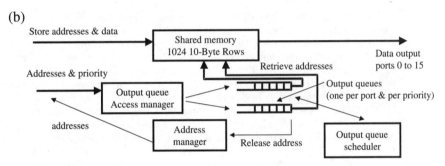

FIGURE 14.12 (a) The block structure of IBM PRS Q64G fabric switch and (b) the internal architecture of each 16×6 switch element. *(Source: IBM)*

In order to ensure that the shared memory never overflows and that the core of the switch fabric is lossless, rigorous flow control has been implemented. The fabric core itself does not handle flow control, which is instead implemented with the help of the IBM C192 *Common Switch Interface* (CSI) chip. This companion chip handles a pair of I/O ports known as the CSI. The C192 CSI chip is instrumental in avoiding HOL blocking at any of the input ports.

We discussed the HOL-blocking problem elsewhere in this chapter. The situation can be summarized as follows. In order to prevent an overloaded output queue from overflowing when the shared memory is already quite full, the switch fabric core decides that all input ports must be notified that no packet (destined for that specific output queue) be allowed into the core memory anymore until new instructions to the opposite effect are issued. The input card normally buffers incoming packets. The first packet in this input queue that happens to be destined for that clogged output queue is now blocked at the input and cannot enter the core memory. However, behind it on the same input port, other inoffensive packets may be present that are destined for other potentially idle output queues. Due to the HOL-blocking effect, these packets are not allowed into the switch and are thus unduly penalized. The following section discusses how IBM has addressed the problem with the C192 in the context of the PowerPRS architecture.

The CSI implements an input-queuing scheme based on VOQ. All incoming packets are sorted per destination port at each input port (that is, at each CSI ingress point). This means that if an output line card is experiencing excessive traffic load and potential congestion, input traffic destined for another output port destination can be chosen for forwarding instead with the appropriate feedback signal. The result of such an arrangement is that no packet will be blocked at the input just because another one in front of it is waiting to be admitted into the fabric's shared memory.

In addition to sorting per destination port, the output queues are also sorted per priority levels (tagged on the packet headers) and when an output queue is enabled for forwarding, the highest-priority packets will always be admitted into the shared memory. Likewise, even after a packet's admission to memory, the priority-sorted pointers at the output queues will ensure that the highest-priority packets will exit the shared memory first.

To ensure that the four available priority levels, which are always applied preemptively onto traffic, do not abuse the weak position of low-priority packets, IBM has implemented an elaborate credit table scheme. This allows the buildup of sufficient credit (in one step per packet cycle and up to 256 steps) that ultimately overrides the preemptive and strict priority-based scheduling.

This method ensures deterministically that less-fortunate links will at some point get a chance to transmit some of their traffic, despite the potentially simultaneous presence of consistently high-priority traffic. For more information regarding the technique that enables a situation where no links are starved because of other traffic priorities, along with a detailed description of the feedback mechanisms that communicate to the VOQs any output congestion through backpressure, refer to Glaise's article mentioned previously.[27]

The C192 has a throughput of 16 Gbps, which operates just like the fabric itself at 60 percent over-speed. Therefore, it can be configured as one OC-192 port, one 10 Gigabit Ethernet, or four OC-48 subports. This means that one C192 Queue Manager chip can handle up to eight Q64G chips arranged in each one of two different banks per OC-192 port. This configuration example, as shown in Figure 14.13, can be multiplied up to 32 times depending on how many OC-192 links must be supported. One of these Q64G banks is usually the primary fabric, whereas the other one is used as redundancy support for protection in case of failure or field service.

The CSIX-L1 interface at the other side of the C192 ensures an adaptable link operating with a clock of 125 MHz or 250 MHz and a 32-, 64-, or 128-bit-wide interface (breakable in up to 4 subports). Communication with a network processor on top of the regular inbound flow-control modes

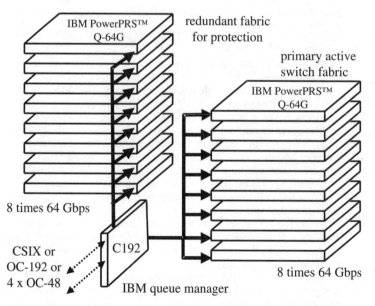

FIGURE 14.13 Two redundant IBM PowerPRS Q64G switch fabric banks (with eight fabric chips each) are used with one C192 Queue Manager per OC-192 port. *(Source: IBM)*

defined by the CSIX-L1 specifications can also be handled via an Xon/Xoff out-of-band protocol and support of CSIX CFrames. IBM therefore supports natively both of these techniques. More information about CSIX is provided in the Appendix III.

The 64G fabric interfaces directly with the NP4GS3 network processor as we saw via the *Data Aligned Synchronous Links* (DASLs). Other line cards that have been implemented around different NPUs or protocol processors can be interfaced with the switch fabric through two options: Take the approach of the standard UTOPIA-3 interface and link through an IBM PRS UTOPIA-to-DASL Serial Interface Converter chip into the switch fabric's DASL ports, or use the C192 Queue Manager with its CSIX interface on one side toward the NPU. Figure 14.14 shows the inner structure of the C192. This C192 will convert the traffic to and from an HSS link that supports the *Switch Core Interface Chip* (SCIC) chip (discussed in the following paragraph) and which in turn interfaces with the switch fabric. Figure 14.15 shows these configurations.

It is worthwhile to note that the standard CSIX-L1 interface of the C192 ensures that non-IBM network processors can be used with the IBM switch fabric chips. The potential of this interoperability is already apparent with EZchip's NP-1 network processor.

IBM also provides as we said, a new chip called Switch Core Interface Chip (SCIC). This chip allows the portability of line cards that were designed for the previous PRS product (the 64G), so that they can now be upgraded to take advantage of IBM's new 2.5 Gbps serdes used with the Q64G. As shown in Figure 14.16, the SCIC interfaces via a DASL link on one side with the older 64G switch fabric and on the other side with the C192 Queue Manager via an HSS link (in fact, a group of eight 2.5 Gbps links). The diagram in Figure 14.16 is straightforward. Data coming in on the ingress flow

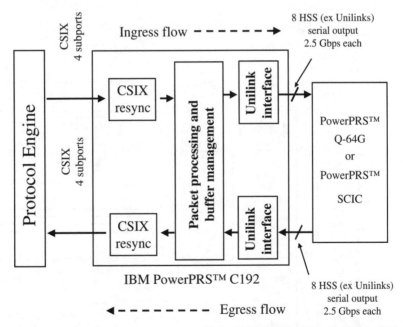

FIGURE 14.14   The inner structure of the IBM PowerPRS C192 Queue Manager. *(Source: IBM)*

---

27. Ibid.

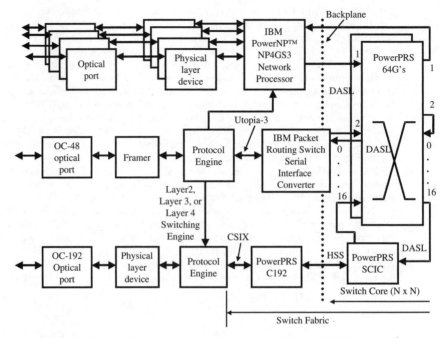

FIGURE 14.15 The connectivity options of network processors with 64G switch fabrics. *(Source: IBM)*

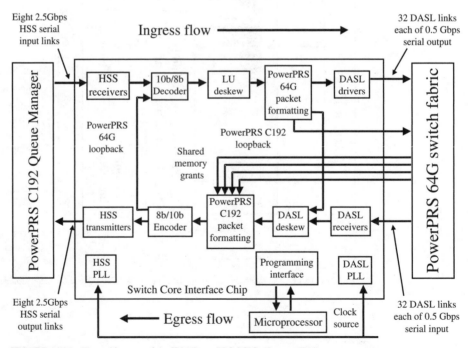

FIGURE 14.16 The architecture of the IBM PowerPRS SCIC. *(Source: IBM)*

from the C192 Queue Manager chip through the HSS links (each one of them being of 2.5 Gbps) is received by the SCIC. It is decoded, and then the *logical units* (LUs) (an IBM term used for the cell fragments of 16 to 20 bytes at a time from an initial packet, which was created for compatibility with the 64G), are realigned to avoid the potential skew that may occur with packets of differing delays while they traverse the backplane. After the appropriate formatting of the packets, the data are ejected onto the DASL, which will now take the data bits into the switch fabric.

On the egress flow, the exact opposite occurs; data exiting the switch fabric through the DASL links are received by the SCIC, deskewed, and prepared for the appropriate formatting. Then they are 8b/10b encoded (because both the C192 and the SCIC use Fibre Channel Standard 8b/10b codes) and transmitted on one of the HSS links toward the C192 Queue Manager for onward transmission.

The SCIC upgrades the older 64G switch fabric system backplane from 500 Mbps DASL to the 2.5 Gbps HSS high-speed serdes (previously known as Unilink). This IBM-proprietary link is a serdes serial link that IBM introduced for the latest and future announcements in the PowerPRS family. The 64G switch fabric can be directly connected to the NP4GS3 network processor through the latter's DASL interfaces over an IBM proprietary DASL backplane. The SCIC also allows an NEV to use a queue manager from IBM. The C192 is the first chip that IBM has introduced in this function.

The aggregate full-duplex throughput of the Q64G, which supports direct HSS interfaces (ex-Unilink), is 64 Gbps. For higher bandwidth needs, multiple devices can be cascaded in master-slave configurations. A cell is then sliced and distributed across multiple HSS interfaces to the different switch fabric chips. Each subfabric switch operates on a slice of the original cell. The master is responsible for all scheduling, synchronization and sequencing of operations, and control of the entire process. The slave switches just provide the rest of the necessary data path. Figure 14.17 provides a sample configuration for 256 Gbps aggregate throughput.

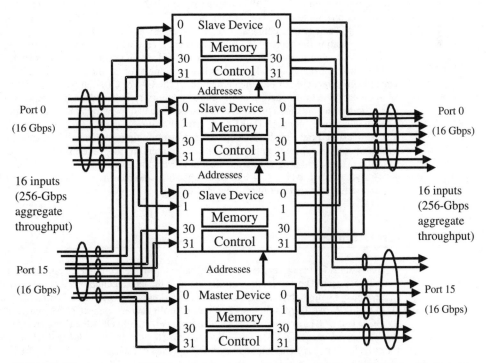

FIGURE 14.17 An aggregate 256 Gbps throughput configuration using multiple PowerPRS-Q64G switch fabric chips arranged in a master-slave scheme. *(Source: IBM)*

The Q64G is housed in a 624-ball IBM *ball grid array* (HyperBGA™) package. Its maximum power consumption is 19 watts (when configured as master in a multiple-chip configuration). The slaves consume about 20 percent less. An older 64G fabric chip just like the NP4GS3 network processor is housed in a 1,088-pin *ceramic column grid array* (CCGA) and consumes 22 watts. For comparison, we will mention that the SCIC is housed in a 399-ball BGA package consuming 5.2 watts and the C192 Queue Manager is housed in an 840-ball BGA with a consumption of 13 watts.

In September 2002, IBM introduced a new chipset comprising the PowerPRS 64Gu and the C48 CSI. The PowerPRS 64Gu has the same function as the PRS-64G, but IBM has now replaced the DASL (low-speed serial link of 500 Mbps) with the IBM HSS link (2.5 Gbps). This change simplifies the layout and wiring of boards significantly. The 64Gu offers up to 80 Gbps of switching capability and port speeds of up to 10 Gbps/OC-192c. It supports both frame-based (through *segmentation and reassembly* [SAR]) and cell-based traffic. In a protocol-agnostic nonblocking capacity, it offers an aggregate throughput of 64 Gbps and 128 Gbps when configured in a 40 Gbps or 80 Gbps switching system, respectively. The afforded overspeed as compared to aggregate user bandwidth is obvious.

A few comments regarding SARing are needed here. IBM considers the SAR function traffic manager related, not switch fabric related. Consequently, any SARing must be handled by the traffic manager regardless of whether this function is embedded inside the network-processor chip itself or whether a separate standalone traffic manager chip is available. The 64Gu is a fixed packet length fabric, whose packet length can be adjusted. This is why IBM calls its switch fabric protocol agnostic.

The 64Gu can be configured as 16 input and 16 output ports with 40 Gbps of aggregate user bandwidth if engaged as a single chip. It can also be organized in a set of two chips, thereby offering 80 Gbps of aggregate user bandwidth along with 32 input and 32 output ports. Each port can have up to four GbE or one OC-48c, or each group of four ports can have one 10 GbE or one OC-192c.

The functionality of the C48 is similar to the C192, but used for a single OC48/2.5 Gbps CSIX-L1 connection (single port). The C48 connects to the 64Gu by two HSS links. It is interesting to note that the C192 can also be connected to the PowerPRS 64Gu, so IBM's customers can have 10 Gbps pipes (eliminating the need for the SCIC, as was the case with the PRS-64G). To support redundancy over two switching planes and to enable either load sharing or packet-lossless switchover in case of maintenance or trouble, the 64Gu offers native support for an alternative/redundant switching path, which is implemented through the help of either the C48 or C192 chips.

As mentioned in the section about shared memory switches, the 64Gu implements data striping (byte slicing) and opts for a graceful degradation instead of penalizing links in multicast traffic cases. The 64Gu generally continues IBM's powerful platform for multicasting combined with QoS, where a packet is stored once in shared memory and then multiple copies will be transmitted to the corresponding outputs, where the memory indexes (pointers) reside. In terms of QoS, it offers four levels of traffic priority and programmable threshold levels for flow control.

The 64Gu is housed in a 624-ball CBGA package and consumes between 13 and 19 watts, the latter representing 100 percent of traffic capacity.

AGERE SWITCH FABRICS

Agere Systems is one of very few vendors that can provide a complete solution from fiber to fabric. Therefore, it would be unfair not to briefly mention the latest technology that the company has to offer in the area of switch fabrics. Agere Systems has a presence in the TDM fabric realm as well as in the ATM and IP network areas. In the TDM realm, it offers crossconnect TDM switches for high-capacity OC-3 and OC-12 systems, access concentrators handling OC-48 capacities, T1/E1 multiplexors, *digital loop carriers* (DLCs), and cellular infrastructure. In the ATM and IP network areas, until recently, it offered switching fabrics with throughputs of 5 Gbps expandable up to 25 Gbps for ATM, TDM, and IP.

More importantly, however, Agere has recently introduced a protocol-independent switching fabric—the PI40 family of chips, which can scale from a single-chip 40 Gbps fabric to a multichip 2.5 *terabits per second* (Tbps) system, as shown in Table 14.1. The PI40 supports fabric port rates from OC-12c to OC-768c and connects without any glue logic to the company's 10 Gbps network processors. It consists of three types of devices:

- The PI40X, a full-duplex switch fabric that has up to 40 Gbps throughput and can handle issues such as aggregation and concentration as well as queuing and scheduling
- The PI40C, a full-duplex 160 Gbps crossconnect device for crossbar arbitration and switching
- The PI40SAX series of 20/40 Gbps protocol independent stand-alone switching fabrics

The PI40X works in two modes: ingress and egress. In the ingress mode, the PI40X multiplexes incoming traffic to the crossbar switching stage. Traffic is buffered and queued separately per egress port taking a worst-case approach. The queuing structure is a good way to provide a sense of traffic isolation and fairness. A QoS-capable schedule handles all the queues, and traffic cells are forwarded to the crossbar planes over multiple 2.5 Gbps serdes links. For large systems, bandwidth can be dynamically allocated by the formation of link groups. Embedded memory is also present and, more specifically, 8K cells handle unicast traffic and 1K cells handle multicast traffic.

The crossbar itself is a 64×64 switching matrix with each link being 2.5 Gbps. Lossless self-routing functionality is provided by the internal QoS-aware arbitration logic. The crossbar can be configured as 1 off 64×64 matrix, 2 off 32×32 matrices, 4 off 16×16 matrices, 8 off 8×8 matrices, 16 off 4×4 matrices, or even 32 off 2×2 matrices.

In the egress mode, the PI40X demultiplexes the traffic that just arrived from the crossbar switching stage over multiple 2.5 Gbps links, keeping in mind that link groups provide a dynamic bandwidth allocation. Traffic is then buffered and queued. As in the ingress mode, not only is traffic isolation and fairness present, but a QoS-capable scheduler serves all the queues. The PI40X's embedded memory provides room and capabilities for unicast and multicast cells similar to those found at the ingress mode.

The PI40 family offers a modular, scalable switch fabric design because it can be configured easily as a single-chip 40 Gbps switch or as a multistage system in a larger switch configuration up to 1,024 fabric ports. One of the nice characteristics of the family is that the ultimate chip count needed for a multistage solution increases linearly with the desired capacity growth. Switching is protocol independent and can be programmed to switch cell payloads of variable size (64, 72, or 80 octets). All switching cell headers contain information that is exclusively used by the switch fabric. A user's protocol-related data are encapsulated deep inside the payload of the fabric cells.

TABLE 14.1 Configuration Examples of Switch Babrics based on Agere's PI40 Family

	Number of Ports				Number of Chips		
Capacity	OC-768c	OC-192c	OC-48c	OC-12c	PI40X	PI40C	Total
40 Gbps	—	4	16	64	*1	—	1
80 Gbps	2	8	32	128	4	1	5
160 Gbps	4	16	64	256	8	2	10
320 Gbps	8	32	128	512	16	4	20
640 Gbps	16	64	256	1,024	32	8	40
1.25 Tbps	32	128	512	—	64	16	80
2.5 Tbps	64	256	1024	—	128	32	160

* Requires the single chips PI40SAX

In the PI40 switch fabric family of chips, which implements an arbitrated crossbar design, a sophisticated queuing structure ensures that no HOL blocking occurs and that native support is available for multicast traffic. The availability of four different buffer management classes and two different schedulers, in conjunction with end-to-end flow control and self-routing capabilities that decouple the arbitration complexity from the system capacity issue, are considered among this fabric's strong points. In addition, advanced traffic management and traffic isolation capabilities enable the provisioning of bandwidth through SLAs and the preservation of QoS in realms such as VoIP and *virtual private networks* (VPNs).

The PI40 family chips contain embedded high-speed serdes and *clock-data recovery* (CDR) circuits. This combination enables the tight integration of a systems design with high capacity and high performance.

In systems where line cards support different wire speeds, the PI40X can be configured as an ingress or egress switch in the fabric, with the PI40C providing the crossbar and arbitration functions. Figure 14.18 shows a 160 Gbps system built from two PI40C devices and eight PI40X chips. The first stage of PI40Xs aggregates the incoming user traffic and connects the traffic to all the second-stage devices. The second stage is essentially the crossconnector, which handles the actual switching function.

For example, the PI40SAX can be connected directly to line cards, as shown in Figure 14.19, in a 32×2.5 Gbps I/O system that can be configured as a 32×32 switch operating at 2.5 Gbps. This enables the creation of a shared memory, nonblocking switch for up to 32 user ports. It can support 4 OC-192c ports, 16 OC-48c ports, 32 OC-12c/GbE ports, and other combinations. The afforded overspeed (speed-up factor) is 2:1. This means that for user traffic with a throughput of 40 Gbps, the fabric has an actual throughput capability of 80 Gbps. This leaves plenty of headroom for internal cell segmentation and reassembly as applied to variable-length packet size even under the worst traffic conditions.

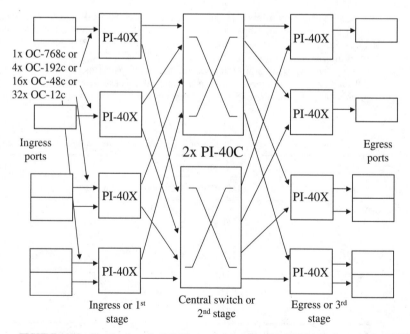

FIGURE 14.18 The structure of a 160 Gbps switch based on the PI40 fabric. It uses eight PI40X and two PI40C chips. *(Source: Agere)*

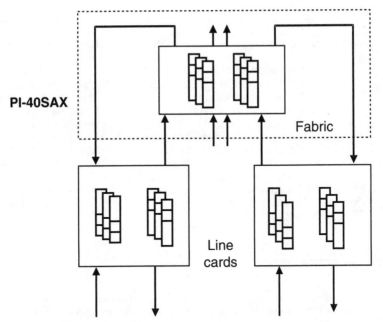

PI-40SAX

Fabric

Line
cards

FIGURE 14.19   A standalone switching system using the PI-40SAX fabric chip. *(Source: Agere)*

The PI40 switch fabric offers a flexible and programmable queuing structure based on multiple routing queues that can be associated with a single fabric port in order to ensure significantly improved QoS capabilities at that specific fabric port. Each routing queue has two scheduling subqueues—one for the real-time traffic (such as voice over TDM traffic or streaming video) and one for the non-real-time traffic. QoS scheduling is handled on an SP basis for delay-sensitive traffic with a bandwidth/delay guarantee for real-time traffic that may not be as sensitive to delay, on a granularity basis of 478 cells per second, or even on the basis of scheduling any excess bandwidth for so-called best-efforts traffic.

Fairness is provided by selective backpressure and the entire system can be configured as a centralized, distributed, stackable, or single-chip fabric. The distributed fabric configuration is done with each PI40C crosspoint operating independently from each other. If one fails or is removed, the operations of any other PI40C in the system are not affected. In its maximum configuration, up to 64 first-stage and 64 third-stage PI40X devices can be connected to 40 PI40C devices to create a 2.56 Tbps switch fabric capacity. By replacing the PI40C devices with a higher-capacity crosspoint element, this configuration can achieve a total throughput of 10 Tbps.

As part of the company's *Integrated Development Environment* (IDE), a performance simulator is available for the PI40 switch fabric. It is fully integrated with a powerful traffic generator (that can be made to generate traffic according to specific characteristics) and allows a *graphical user interface* (GUI)-based co-simulation with Agere's network processors and traffic managers for a full-fledged configuration evaluation and performance analysis. For instance, a designer can specify the number of switch fabric devices in a configuration, the number of ports per device, and the number of serdes links per device. Backpressure thresholds can be decided on a per-device, per-port, or per-queue basis. This may require constraining the latency, especially in a bid/grant approach. The simulator will provide at the output the minimum, maximum, and average values for all per-traffic flow delays, all per-queue occupancies and delays, all per-port and per-traffic class delays, and all per-device delays. It also calculates the throughputs for all of these categories.

SUMMARY

The market research organization InStat/MDR reports that the switch fabric market will experience a *compound annual growth rate* (CAGR) for shipments of 141 percent from 2002 through to 2006. Although early opportunities are in router and switch markets, by 2006, significant opportunities will be available in at least 13 different target markets, leading to the phenomenal growth, according to the same firm. In this chapter, we provided an overview of the established techniques for fast switching and, more importantly, how this is being accomplished through the use of switch fabrics. We reviewed fundamental concepts and identified the underlying trade-offs involved in many of the approaches that designers have taken. As two representative examples of different state-of-the-art switch fabric architectures, we discussed the IBM PowerPRS family of switch fabrics along with the associated IBM chips that support the actual fabrics in building a powerful switch/router and Agere's latest PI40 switch fabric chipset that can connect seamlessly with the company's network processors and that can scale up to 2.54 Tbps throughputs. A very extensive list of technical literature references has also been provided for subsequent study for those who may be more interested in different aspects of this extremely wide subject.

SUGGESTED REFERENCES

Several interesting publications cover all aspects of the switching realm. Consult these publications to delve deeper into any of the aspects of fast packet switching. The technology is rapidly evolving in so many directions that are outside the scope of this book that we cannot address all these changes, much less cover them in depth. To stay abreast of the fast packet switching technology evolution, the main sources of information in the area must be continuously monitored. In our opinion, the following publications are especially important for a full view of the areas that are most important in the evolution of fast packet processing. For the full bibliographical information for each of these sources, please refer to the reference section at the end of this chapter.

F.A. Tobagi provides a general overview of the field (including some good coverage of switch fabric topologies) in his article "Fast Packet Switch Architectures for Broadband Integrated Services Networks." It serves as a helpful tutorial, albeit slightly outdated by now since it was published 13 years ago. A nice overview of pertinent network traffic characterization is the article "MPEG-4 and H.263 Video Traces for Network Performance Evaluation" by F. Fitzek and M. Reisslein. The same principles are discussed in the article "On End-to-End Architecture for Transporting MPEG-4 Video Over the Internet" by D. Wu et al. In this article, the transport of compressed video traffic over the Internet is analyzed within a context of low bit rate and varying network conditions.

Scheduling of access to a crossbar fabric can be done either online or offline. For offline scheduling, consult T. Rodeheffer and J. Saxe's "Smooth Scheduling in a Cell-Based Switching Network," where QoS guarantees are addressed. Other good sources on the same subject are M. Bonuccelli's "Incremental Time-Slot Assignment in SS/TDMA Satellite Systems," I. Gopal's "Minimizing Packet Waiting Time in a Multibeam Satellite System," and T. Inukai's "An Efficient SS/TDMA Time Slot Assignment Algorithm." The latter reference is still very applicable although it is quite old. For online scheduling, the article "Two-Dimensional Round-Robin Schedulers for Packet Switches with Multiple Input Queues" by R. LaMaire and D. Serpanos discusses a two-dimensional round-robin approach, whereas some of the work from Nick McKeown's Stanford team on iSLIP and Tiny Tera can be found in "Designing and Implementing a Fast Crossbar Scheduler," "Tiny Tera: A Packet Switch Core," and "The iSLIP Scheduling Algorithm for Input-Queued Switches."

D. Serpanos and P. Antoniadis' article "FIRM" and Anthony Kam and Kai-Yeung Siu's article "Linear-Complexity Algorithms for QoS Support in Input Queued Switches with No Speedup" discuss scheduling when confronted with multiple input queues, the latter with QoS guarantees.

Regarding a sensitive speed-up requirement issue that pops up in implementations when emulating output queuing, the article "Analysis of a Packet Switch with Memories Running Slower Than the

Line Rate" by S. Iyer provides the context background. Articles written by P. Krishna et al., Chuang et al., and S. Iyer and N. McKeown elaborate on several approaches to tackle it.

For coverage of switch fabric topologies, take a look at the paper "The Theory of Connecting Networks and Their Complexity" by M. Marcus. In the book *Gigabit Networking*, C. Partridge covers the Batcher-Banyan networks. In their book *High Performance Communication Networks*, J. Walrand and P. Varaiya provide among other things an especially nice discussion of modular switches and performance. V. Benes' "Optimal Rearrangeable Multistage Connecting Networks" is also a classic. Articles written by C. L. Wu and T. Y. Feng, and C. Leiserson cover several interesting topologies.

The article "Nonblocking Copy Networks for Multicast Packet Switching" by T. Lee discusses a nonblocking copy scheme for multicast switches. F. M. Chiussi and F. A. Tobagi, and D. Khotimsky cover resequencing protocols and issues. B. Suter et al., M. Arpaci and J. Copeland, and S. Floyd and V. Jacobson have written about buffer management issues when a switch must control the flow by dropping packets. Backpressure is introduced in the article "New Directions in Communications (Or Which Way to the Information Age?)" by J. Turner. This work has been expanded to discuss multicast in J. Turner's article "Design of a Broadcast Packet Switching Network."

In their article "Deadlock-Free Message Routing in Multiprocessor Interconnection Networks," W. Dally and C. Seitz first talk about the so-called wormhole routing approach, where variable-length packets are segmented into *flits*. W. Dally continues this discussion in his article "Virtual-Channel Flow Control," which examines virtual channel for backpressure as a means to reduce the HOL blocking. Li-Shiuan Peh joins W. Dally to expand on the idea in the article "Flit-Reservation Flow Control."

An interesting combination of a buffered crossbar, WFQ scheduling, and backpressure is presented in "Implementing Distributed Packet Fair Queueing in a Scalable Switch Architecture" by D. Stephens and Hui Zhang. Atlas-I is an interesting single-chip ATM switch project where a credit-based multilane backpressure protocol has been implemented as a significant improvement against classic virtual channel wormhole backpressure. M. Katevenis and G. Kornaros et al. have written some reports regarding the Atlas-I project. Credit-based flow control is discussed in the article "Reliable and Efficient Hop-by-Hop Flow Control" by C. Ozveren, R. Simcoe, and G. Varghese. The articles "Fast Switching and Fair Control of Congested Flow in Broad-Band Networks" by M. Katevenis and the article "Credit-Based Flow Control for ATM Networks" by H. T. Kung, T. Blackwell, and A. Chapman are also informative. Interesting information can also be found at the web site of the Quantum-Flow Control Alliance at *www.qfc.org*.

Some interesting aspects of implementing WRR scheduling are discussed in the articles "Service Disciplines for Guaranteed Performance Service in Packet-Switching Networks" by Hui Zhang, "Implementing Scheduling Algorithms in High-Speed Networks" by D. Stephens, J. Bennett, and Hui Zhang, and "Pipelined Heap (Priority Queue) Management for Advanced Scheduling in High Speed Networks" by A. Ioannou and M. Katevenis. Issues regarding fast packet classification and route table lookup are covered nicely in the articles "IP Address Lookup in Hardware for High-Speed Routing" by Andreas Moestedt and Peter Sjodin, "Packet Classification on Multiple Fields" by P. Gupta and N. McKeown, "Fast Updating Algorithms for Ternary CAMs" by D. Shah and P. Gupta, and "High-Speed Policy-based Packet Forwarding Using Efficient Multi-Dimensional Range Matching" by T. Lakshman and D. Stiliadis. The March–April 2001 issue of *IEEE Micro* also contains several articles on the same subject. Finally, the articles "WDM Optical Communication Networks" by B. Mukherjee and "Advances in Photonic Packet Switching" by S. Yao, B. Mukherjee, and S. Dixit are good updated sources on optical switching applicable on the backbone of the *wide area network* (WAN).

An excellent (usually biannual) report on the switch fabric industry is published by the Linley Group at *www.linleygroup.com/npu/fabrics.html*.

Companies offering commercial off-the-shelf switch fabric chipsets include the following:

Agere Systems (*www.agere.com*)

AMCC (*www.amcc.com*)

Dune Networks (*www.dunenetworks.com*)

Erlang Technology (*www.erlangtech.com*)

IBM (*www.chips.ibm.com*)

Internet Machines (*www.internetmachines.com*)

MindSpeed (*www.mindspeed.com*)

PetaSwitch (*www.peta-switch.com*)

Sandburst (*www.sandburst.com*)

Tau Networks (*www.taunetworks.com*)

TeraChip (*www.terachip.com*)

TeraCross (*www.teracross.com*)

Vitesse (*www.vitesse.com*)

Zagros Networks (*www.zagrosnetworks.com*)

ZettaCom (*www.zettacom.com*)

White papers and application notes discussing each company's technology approach are available from these companies directly.

For readers who want to get a flavor of backplane technology, two introductory tutorials on backplane-related design issues and signal integrity on multigigabit per sec links can be found at the IEC web site at *www.iec.org/online/tutorials/design_backplane/* and *www.iec.org/online/tutorials/signal_integrity/*.

A couple of interesting articles on issues related to serdes influencing the backplane performance are: "Proper Serdes Selection Solves Serial Backplane Design Woes" by PMC-Sierra's Rachelle Trent. The article appeared in *Communications Design* magazine on November 21, 2002. It is also available online at *www.commsedesign.com/story/OEG20021121S0005*.

"Design 5-Gbits/s serdes: Steering throug a Road Filled with Potholes", by TriCN's Ronald Nikel, *Communications Systems Design*, March 2, 2003. Available on line at: *http://www.commsdesign.com/story/OEG20030327S0010*.

Also an interesting article on issues involving 10Gbps interfacing of switch fabrics is: "10 Gbits/s Switch Fabric Interface Shootout", by Tau Networks' Phil Brown, Communications Systems Design, January 3, 2003, available on line at: *http://commsdesign.com/story/OEG20030103S0058*.

Another interesting article discussing how the maturing switch fabric technology has far-reaching effects beyond the network-processing industry and how it is bound to affect main-street *system-on-a-chip* (SOC) and bus-based ASIC designs is "Switch-based Interconnects Solve Comms Designer's Woes" by Warren Miller. The article appeared in *Communications Design* magazine on November 13, 2002. It is also available online at *www.commsdesign.com/story/OEG20021113S0014*.

The High Speed Backplane Initiative (HSBI) formed in late 2002 is an interesting industry effort to advance serial link technology in this critical area of communications infrastructure to higher speeds. A very informative article in two parts (and strongly recommended reading) has been written by John D'Ambrosia from Tyco Electronics, Ryan Carlson from Velio Communications and Bill Woodruff from BitBlitz. It is titled "Analyzing the Challenges Facing HSBI: Part1:", Communications Design Magazine on May 29, 2003. The two parts are also available online respectively at: *www.commsdesign.com/design_corner/OEG20030528S0036* and *www.commsdesign.com/design_corner/OEG20030529S0004*.

REFERENCES

Ahmadi, H., and W. E. Denzel. "A Survey of Modern High-Performance Switching Techniques." *IEEE Journal on Selected Area in Communication* 7, no. 9 (September 1989): 1091–1103.

Anderson, T. E., J. B. Saxe, and C. P. Thacker. "High-Speed Switch Scheduling for Local Area Networks." *ACM Transactions on Computer Systems* 11, no. 4 (November 1993): 319–352.

Arpaci, M., and J. Copeland. "Buffer Management for Shared-Memory ATM Switches." *IEEE Communications Surveys* 3, no. 1 (first quarter 2000).

Bellamy, J. *Digital Telephony*. 3rd ed. New York: John Wiley, 2000.

Benes, V. "Optimal Rearrangeable Multistage Connecting Networks." *Bell Systems Technical Journal* 43, no. 7 (July 1964): 1641–1656.

Black, Uyless. *ATM: Foundations for Broadband Networks*. Upper Saddle River, New Jersey: Prentice-Hall, 1995.

Bonuccelli, M., I. Gopal, and C. Wong. "Incremental Time-Slot Assignment in SS/TDMA Satellite Systems." *IEEE Transactions on Communications* 39, no. 7 (July 1991): 1147–1156.

Chao, H. J., and B. S. Choe. "Design and Analysis of a Large-Scale Multicast Output-Buffered ATM Switch." *IEEE/ACM Transactions in Networking* 3, no. 2 (April 1995): 126–138.

Chiussi, F. M., and F. A. Tobagi. "A Hybrid Shared-Memory/Space-Division Architecture for Large Fast Packet Switches." *Proceedings of SUPERCOMM ICC '92* (June 1992).

Chiussi, F., D. Khotimsky, and S. Krishnan. "Generalized Inverse Multiplexing for Switched ATM Connections." *Proceedings on IEEE GLOBECOM Conference*, Sydney, Australia (November 1998): 3134–3140; it is available online through Bell Labs Data Networking Systems Research in .ps format.

Chuang, S., A. Goel, N. McKeown, and B. Prabhakar. "Matching Output Queueing with a Combined Input Output Queued Switch." *IEEE Journal of Selected Areas in Communications* 17, no. 6 (June 1999): 1030–1039, Stanford CSL-TR-98-758. It is available online *http://elib.stanford.edu*.

Cidon, I. et al. "Real-Time Packet Switching: A Performance Analysis." *IEEE Journal on Selected Areas in Communication* 6, no. 9 (December 1988): 1576, Stanford CSL-TR-98-7581586.

Clos, C. "A Study of Non-Blocking Switching Networks." *Bell System Technical Journal* 32, no. 3 (March 1953): 406–424.

Dally, W. "Virtual-Channel Flow Control." *IEEE Transactions on Parallel and Distributed Systems* 3, no. 2 (March 1992): 194–205.

Dally, W., and C. Seitz. "Deadlock-Free Message Routing in Multiprocessor Interconnection Networks." *IEEE Transactions on Computers* 36, no. 5 (May 1987): 547–553.

Degermark, M., A. Brodnik, S. Carlsson, and S. Pink. "Small Forwarding Tables for Fast Routing Lookups." *Sigcomm '97* (1997).

Demers, A., S. Keshav, and S. Shenker. "Design and Analysis of a Fair Queuing Algorithm." *Proceedings of ACM SIGCOMM '89* (September 1989).

Dias, D. M., and J. R. Jump. "Analysis and Simulation of Buffered Delta Networks." *IEEE Transactions on Computers* 30, no. 10 (October 1981): 273–282.

Ferrari, D., and D. Verma. "A Scheme for Real-Time Channel Establishment in Wide-Area Networks." *IEEE Journal on Selected Areas in Communication* 8, no. 3 (April 1990): 368–379.

Fitzek, F., and M. Reisslein. "MPEG-4 and H.263 Video Traces for Network Performance Evaluation." *IEEE Network Magazine* 15, no. 6 (November–December 2001): 40–54.

Floyd, S., and V. Jacobson. "Random Early Detection Gateways for Congestion Avoidance." *IEEE/ACM Transactions on Networking* 1, no. 4 (August 1993): 397–413.

Fore Systems Inc. "ATM Switching Architecture." A white paper, November 1993. The original company has been acquired in the meantime by Marconi plc.

Fraser, A. G. "Towards a Universal Data Transport System." *IEEE Journal on Selected Areas in Communication* 1, no. 5 (November 1983): 803–816.

Goke, L. R., and G. J. Lipowski. "Banyan Networks for Partitioning Multiprocessor Systems." *Proceedings on 1st Annual International Symposium on Computer Architecture* (December 1973): 21–28.

Golestani, S. Jamaloddin. "A Self-Clocked Fair Queuing Scheme for Broadband Applications." *Proceedings of IEEE INFOCOM '94* (June 1994): 636–646.

Gopal, I., D. Coppersmith, and C. Wong. "Minimizing Packet Waiting Time in a Multibeam Satellite System." *IEEE Transactions on Communications* 30 (1982): 305–316.

Gopal, I. S., I. Cidon, and H. Meleis. "PARIS: An Approach to Integrated Private Networks." *Proceedings of ICC '87* (June 1987): 764–773.

Goyal, P., H. Vin, and H. Chen. "Start-Time Fair Queuing: A Scheduling Algorithm for Integrated Services Packet Switching Networks." *Proceedings of ACM SIGCOMM '96* (August 1996).

Gupta, P., and N. McKeown. "Designing and Implementing a Fast Crossbar Scheduler." *IEEE Micro* (January–February 1999): 20–28.

———. "Packet Classification on Multiple Fields." *ACM SIGCOMM '99 Conference*, Harvard University (September 1999): 147–160. It is available online at *http://tiny-tera.stanford.edu/~nickm/papers/Sigcomm99.pdf*.

Hluchyj, M. G., and M. J. Karol. "Queuing in Space-Division Packet Switching." *Proceedings of IEEE INFOCOM '88* (March 1988): 334–343.

Huang, A. K., and S. Knauer. "Starlite: A Wideband Digital Switch." *Proceedings of GLOBECOM '84* (December 1984): 121–125.

Hui, J., and E. Arthurs. "A Broadband Packet Switch for Integrated Transport." *IEEE Journal on Selected Areas in Communication (JSAC)* 5, no. 8 (October 1987): 1264–1273.

Inukai, T. "An Efficient SS/TDMA Time Slot Assignment Algorithm." *IEEE Transactions on Communications* 27 (October 1979): 1449–1455.

Ioannou, A., and M. Katevenis. "Pipelined Heap (Priority Queue) Management for Advanced Scheduling in High Speed Networks." *Proceedings on IEEE International Conference on Communications (ICC 2001)* Helsinki, Finland (June 2001): 2043–2047. It is available at *http://archvlsi.ics.forth.gr/muqpro/heapMgt.html*.

Iyer, S., A. Awadallah, and N. McKeown. "Analysis of a Packet Switch with Memories Running Slower Than the Line Rate." *IEEE INFOCOM Conference* Tel-Aviv, Israel (March 2000). It is available online at *http://tiny-tera.stanford.edu/~nickm/papers/* in PDF.

Iyer, S., and N. McKeown. "Making Parallel Packet Switches Practical." *IEEE INFOCOM Conference* Alaska (March 2001). It is available online at *http://tiny-tera.stanford.edu/~nickm/papers/* in PDF or Postscript.

Jacob, A. R. "A Survey of fast Packet Switches." *ACM SIGCOMM Computer Communication Review* 20, no. 1 (January 1990): 54–64.

Kam, Anthony, and Kai-Yeung Siu. "Linear-Complexity Algorithms for QoS Support in Input Queued Switches with No Speedup." *IEEE Journal of Selected Areas in Communications* 17, no. 6 (June 1999): 1040–1056.

Kanakia, H. "Datapath Switch." *AT&T Bell Laboratories Internal Technical Memorandum* (1994).

Karol, M., M. Hluchyj, and S. Morgan. "Input Versus Output Queuing on a Space-Division Packet Switch." *IEEE Transactions on Communications* 35, no. 12 (December 1987): 1347–1356.

Katevenis, M., D. Serpanos, and E. Spyridakis. "Switching Fabrics with Internal Backpressure Using the ATLAS I Single-Chip ATM Switch." *Proceedings of the IEEE GLOBECOM'97 Conference* Phoenix, AZ (November 1997): 242–246. It is available online at *http://archvlsi.ics.forth.gr/atlasI/*, in Postscript (230 KB), or gzip'ed Postscript (53KB).

Katevenis, M., D. Serpanos, and E. Spyridakis. "Credit-Flow-Controlled ATM for MP Interconnection: The ATLAS I Single-Chip ATM Switch." *Proceedings of HPCA-4* (4th IEEE International Symposium on High-Performance Computer Architecture) Las Vegas, NV (February 1998): 47–56. It is available online in Postscript at *http://archvlsi.ics.forth.gr/atlasI/atlasI_hpca98.ps* (230KB) or in gzip'ed Postscript at *http://archvlsi.ics.forth.gr/atlasI/atlasI_hpca98.ps.gz* (58KB).

Katevenis, M. "Fast Switching and Fair Control of Congested Flow in Broad-Band Networks." *IEEE Journal on Selected Areas in Communications* 5, no. 8 (October 1987): 1315–1326.

Keshav, S., and R. Sharma, "Issues and Trends in Router Design." *www.cs.cornell.edu/skeshav/papers/routertrends.pdf*.

Khotimsky, D. "A Packet Resequencing Protocol for Fault-Tolerant Multipath Transmission with Non-uniform Traffic Splitting." *Proceedings IEEE GLOBECOM Conference* Rio de Janeiro, Brazil (December 1999): 1283–1289. It is available online at Bell Labs Data Networking Systems Research at *www.bell-labs.com/org/113480/* in *.ps format*.

Knuth, Donald E. "Sorting and Searching." In *The Art of Computer Programming*. vol. 3. Reading, Massachusetts: Addison-Wesley, 1973.

Kornaros, G., D. Pnevmatikatos, P. Vatsolaki, G. Kalokerinos, C. Xanthaki, D. Mavroidis, D. Serpanos, and M. Katevenis. "Implementation of ATLAS I: A Single-Chip ATM Switch with Backpressure." *Proceedings IEEE Hot Interconnects 6 Symposium* Stanford, CA (August 13–15 1998): 85–96. It is available online at *http://archvlsi.ics.forth.gr/atlasI/hoti98/*.

Krishna, P., N. Patel, A. Charny, and R. Simcoe. "On the Speedup Required for Work-Conserving Crossbar Switches." *IEEE Journal of Selected Areas in Communications* 17, no. 6 (June 1999): 1057–1066.

Kung, H. T., T. Blackwell, and A. Chapman. "Credit-Based Flow Control for ATM Networks: Credit Update Protocol, Adaptive Credit Allocation, and Statistical Multiplexing." *Proceedings of the ACM SIGCOMM '94 Conference*, London, UK (August–September 1994): 101–114.

Kyas, Othmar. *ATM Networks*. 2nd ed. Boston: International Thompson Computer Press, 1995.

Lakshman T., and D. Stiliadis. "High-Speed Policy-Based Packet Forwarding Using Efficient Multi-Dimensional Range Matching." *ACM SIGCOMM '98 Conference* Vancouver, BC, Canada (September 1998): 203–214.

LaMaire, R., and D. Serpanos. "Two-Dimensional Round-Robin Schedulers for Packet Switches with Multiple Input Queues." *IEEE/ACM Transactions on Networking* 2, no. 5 (October 1994): 471–482.

Lee, T. "Nonblocking Copy Networks for Multicast Packet Switching." *IEEE Journal of Selected Areas in Communications* 6, no. 9 (December 1988): 1455–1467.

Leiserson, C. "Fat-Trees: Universal Networks for Hardware-Efficient Supercomputing." *IEEE Transactions on Computers* 34, no. 10 (October 1985): 892–901.

Marcus, M. G. "The Design and Analysis of a New Type of Time-Division Switching System." M.Sc. Thesis, MIT (June 1969).

Marcus, M. "The Theory of Connecting Networks and Their Complexity: A Review." *Proceedings of the IEEE* 65, no. 9 (September 1977): 1263–1271.

McKeown, N. "The iSLIP Scheduling Algorithm for Input-Queued Switches." *IEEE/ACM Transactions on Networking* 7, no. 2 (April 1999): 188–201. It is available online at *http://tiny-tera.stanford.edu/~nickm/papers/ToN_April_99.pdf.*

McKeown, N. "Scheduling Algorithms for Input-Queued Cell Switches." Ph.D. Thesis, University of California at Berkeley (May 1995).

McKeown, N., M. Izzard, A. Mekkittikul, W. Ellersick, and M. Horowitz. "Tiny Tera: a Packet Switch Core." *IEEE Micro* (January–February 1997): 26–33.

Moestedt, Andreas, and Peter Sjodin. "IP Address Lookup in Hardware for High-Speed Routing." *Proceedings on IEEE Hot Interconnects 6 Symposium* Stanford, CA (August 1998): 31–39. It is available online at *www.sics.se/~am/HotI.ps.*

Mukherjee, B. "WDM Optical Communication Networks: Progress and Challenges." *IEEE Journal of Selected Areas in Communications* 18, no. 10 (October 2000): 1810–1824.

Oki, E., and N. Yamanaka. "A High-Speed Tandem-Crosspoint ATM Switch Architecture with Input and Output Buffers." *IEICE Transactions on Communications* E81-B, no. 2 (1998): 215–223.

_____. "Scalable Crosspoint Buffering ATM Switch Architecture Using Distributed Arbitration Scheme." *Proceedings on IEEE ATM '97 Workshop* (1997): 28–35.

Oki, E., N. Yamanaka, and S. Yasukawa. "Tandem-Crosspoint ATM Switch Architecture and Its Cost-effective Expansion." *IEEE BSS '97* (1997): 45–51.

Oki, E., N. Yamanaka, and M. Nabeshima. "Scalable-Distributed-Arbitration ATM Switch Supporting Multiple QoS Classes." *Proceedings on IEEE ATM '99 Workshop* (1999).

_____. "Performance of Scalable-Distributed Arbitration ATM Switch Supporting Multiple QoS Classes." *IEICE Transactions on Communication* E83-B, no. 2 (2000): 204–213.

Onvural, Raif O. *Asynchronous Transfer Mode Networks: Performance Issues*. Boston: Artech House, 1995.

Ozveren, C., R. Simcoe, and G. Varghese. "Reliable and Efficient Hop-by-Hop Flow Control." *IEEE Journal on Selected Areas in Communications* 13, no. 4 (May 1995): 642–650.

Partridge, C. *Gigabit Networking*. Reading, Massachusetts: Addison-Wesley, 1994.

Pattavina, A., and G. Bruzzi. "Analysis of Input and Output Queuing for Nonblocking ATM Switches." *IEEE/ACM Transactions on Networking* 1, no. 3 (June 1998): 314–328.

Peh, Li-Shiuan, and W. Dally. "Flit-Reservation Flow Control." *Proceedings of International Symposium on High-Performance Computing Architectures (HPCA)* (2000).

Perlman, Radia. *Interconnections: Bridges, Routers, Switches, and Internetworking Protocols*. 2nd ed. Reading, Massachusetts: Addison-Wesley, 2000.

Quantum Flow Control Alliance. "Quantum Flow Control: A Cell-Relay Protocol Supporting an Available Bit Rate Service." Version 2.0, July 1995, *www.qfc.org.*

Rabaey, Jan M. *Digital Integrated Circuits: A Design Perspective*. Upper Saddle River, New Jersey: Prentice-Hall, 1996.

Rodeheffer, T., and J. Saxe. "Smooth Scheduling in a Cell-Based Switching Network." *DEC SRC Research Report #150* (February 1998). It is available online at *http://gatekeeper.dec.com/pub/DEC/SRC/research-reports/abstracts/src-rr-150.html.*

Serpanos, D., and P. Antoniadis. "FIRM: A Class of Distributed Scheduling Algorithms for High-Speed ATM Switches with Multiple Input Queues." *IEEE Infocom 2000 Conference* Tel Aviv, Israel (March 2000).

Shah, D., and P. Gupta. "Fast Updating Algorithms for Ternary CAM's." *IEEE Micro* (January–February 2001): 36–47.

Shreedhar, M., and G. Varghese. "Efficient Fair Queuing Using Deficit Round Robin." *Proceedings of ACM SIG-COMM '95* (September 1995).

Srinivasan, V., and G. Varghese. "Faster IP Lookups Using Controlled Prefix Expansion." ACM Sigmetrics '98, *ACM Transactions on Computer Systems* (March 1999).

Stallings, William. *ISDN and Broadband ISDN with Frame Relay and ATM.* 3rd ed. Upper Saddle River, New Jersey: Prentice-Hall, 1995.

Stephens, D., and Hui Zhang. "Implementing Distributed Packet Fair Queueing in a Scalable Switch Architecture." *IEEE INFOCOM '98 Conference.* It is available online at *www-2.cs.cmu.edu/People/hzhang/publications.html* in ps.gz or in PDF.

Stephens, D., J. Bennett, and Hui Zhang. "Implementing Scheduling Algorithms in High-Speed Networks." *IEEE Journal of Selected Areas in Communications* 17, no. 6 (June 1999): 1145–1158. It is available online as shown in the previous reference.

Suter, B., T. Lakshman, D. Stiliadis, and A. Choudhury. "Buffer Management Schemes for Supporting TCP in Gigabit Routers with Per-Flow Queueing." *IEEE Journal of Selected Areas in Communications* 17, no. 6 (June 1999): 1159–1169.

Suzuki, H., H. Nagano, T. Suzuki, T. Takeuchi, and S. Iwasaki. "Output-Buffer Switch Architecture for Asynchronous Transfer Mode." *Proceedings IEEE ICC '89* (1989): 99–103.

Tanenbaum, Andrew S. *Computer Networks.* 3rd ed. Upper Saddle River, New Jersey: Prentice-Hall, 1996.

Tobagi, F. A. "Fast Packet Switch Architectures for Broadband Integrated Services Networks." *Proceedings of IEEE* 78, no. 1 (January 1990): 133–167.

Turner, J. S., and L. F. Wyatt. "A Packet Network Architecture for Integrated Services." *Proceedings of GLOBE-COM '83* (November 1983) 2.1.1–2.1.6.

Turner, J. "New Directions in Communications (or Which Way to the Information Age?)" *IEEE Communications Magazine* 25, no. 10 (October 1986): 8–15.

_____. "Design of an Integrated Services Packet Network." *IEEE Journal of Selected Areas in Communication* 4, no. 8 (November 1986): 1373–1380.

_____. "Design of a Broadcast Packet Switching Network." *IEEE Transactions on Communications* 36, no. 6 (June 1988): 734–743.

Verma, D., H. Zhang, and D. Ferrari. "Guaranteeing Delay Jitter Bounds in Packet Switching Networks." *Proceedings of Tricomm '91* (April 1991).

Waldvogel, M., G. Varghese, J. Turner, and B. Plattner. "Scalable High Speed IP Routing Lookups." *Sigcomm '97* (1997).

Walrand, J., and P. Varaiya. *High Performance Communication Networks.* 2nd ed. San Francisco: Morgan Kaufmann Publishers, 1999.

Wu, C. L., and T. Y. Feng. "On a Class of Multistage Interconnection Networks." *IEEE Transactions on Computers* 29, no. 8 (August 1980): 694–702.

Wu, D., Y. T. Hou, W. Zhu, H. J. Lee, T. Chang, Y. Q. Zhang, and H. Jonathan Chao. "On End-to-End Architecture for Transporting MPEG-4 Video Over the Internet." *IEEE Transactions on Circuits and Systems for Video Technology* 10, no. 6 (September 2000): 923–941.

Yao, S., B. Mukherjee, and S. Dixit. "Advances in Photonic Packet Switching: An Overview." *IEEE Communications Magazine* 38, no. 2 (February 2000): 84–94.

Yeh, C. *Applied Photonics.* New York: Academic Press, 1994.

Zhang, L. "A New Architecture for Packet Switching Network Protocols." *Technical Report MIT/LCS/TR-455,* Laboratory for Computer Science, Massachusetts Institute of Technology, Cambridge, MA, August 1989.

_____. "Virtual Clock: A New Traffic Control Algorithm for Packet Switching Networks." *Proceedings of ACM SIGCOMM '90* (September 1990).

Zhang, Hui. "Service Disciplines for Guaranteed Performance Service in Packet-Switching Networks." *Proceedings of the IEEE* 83, no. 10 (October 1995). It is available online at *www-2.cs.cmu.edu/People/hzhang/publications.html* in ps.gz or in PDF.

CHAPTER 15
TRAFFIC MANAGERS

After having discussed several commercial platforms, systems architectures, and network processors, we have examined most of the necessary components a systems designer requires in order to put together a working switching/routing system. These include network processors, classification and forwarding processors, search engines, and switch fabrics. In this chapter, we turn our attention to the last important piece of this architectural jigsaw puzzle—traffic managers.

THE DEFINITION AND PURPOSE OF A TRAFFIC MANAGER

In some networking realms, not all transmitted packets can be treated in the same way for profitability and efficiency reasons. As a result, different *classes of service* (CoS) must implement differentiated levels of *quality of service* (QoS) as expected for each packet that traverses the network. This QoS framework reflects the potential critical, important, urgent, lucrative, or undesirable nature of specific packets. *Traffic management* is a wide conceptual area of network processing that deals with how the underlying flow of real-time traffic, which can be composed of a continuous, massive, and time-varying collection of disparate and largely unrelated session packets from a staggering multitude of applications, must be treated in order to implement, monitor, and enforce specific QoS requirements of the network.

TRAFFIC MANAGERS AS STANDALONE CHIPS

As we have seen in the first 10 chapters of the book, most powerful network processors already contain certain embedded capabilities that enable them to perform in some cases adequate and in other cases rudimentary traffic management operations on the traffic stream that they monitor and process. However, in order to perform high-speed, fine-granularity traffic management on information streams, highly specialized standalone chips called *traffic managers* are required. Traffic managers have several advantages:

• Traffic managers are off-the-shelf commercially available components. This means their cost is tremendously lower than if a designer designs a traffic manager from scratch. On many occasions, in principle, traffic managers can be mixed and matched with network processors or switch fabrics from different vendors. The best of each breed can be chosen to suit the design requirements. In reality, however, this practice may create all kinds of unexpected difficulties and suboptimal system behavior. In addition, the wheel does not have to be reinvented in terms of devising, coding, simulating, testing, validating, and optimizing algorithms that handle the various logical tasks associated with traffic management.

- Traffic managers interface with network processors in a more or less standardized way; therefore, the time to market is accelerated for a systems architect since time is not wasted on unnecessary issues.

- By offloading the traffic management function from the network processor, finer traffic management granularity is enabled through the use of a specialized traffic manager chip, whereas significant amounts of silicon real estate remain available to the network-processor implementation for the materialization of an even higher-performance architecture for live packet classification and forwarding.

Standalone chips are now required inside switching/routing equipment to enable service providers and carriers to provide QoS based on *service level agreements* (SLAs) to their major customers who insist on (and, more importantly, are ready to pay a premium for) having specific response times, bandwidth guarantees, and assured levels of availability for their network applications.

In addition to the QoS-driven functionality that caters to the carrier customers' needs, traffic managers can also allow carriers to optimize their own operational finances by ensuring an optimal allocation of the available bandwidth as well as a graceful degradation of network performance in case of oversubscription (an undesirable situation, where more demand exists for bandwidth resources than is actually provided).

Like security coprocessors, which we discuss in Chapter 17, "Security Coprocessors," traffic managers are proposed in either one of two major architectures: *look aside* and *flow through*. In the look-aside approach, the traffic manager shares some functionality with the network processor. More specifically, they can share the same packet buffer memory and even the same buffer management function. In the flow-through approach, all the required modules out of which the traffic manager consists are laid out in a pipeline-like fashion and treat the incoming information in a sequential way, like in an old-fashioned assembly line. Each TM submodule receives packets from one side, processes them accordingly, and outputs them on the other side to the following TM submodule.

The requirements for traffic managers depend on the nature of the network in which they are called to function, as shown in Table 15.1. This means that not all traffic managers are able to offer the same level of help in every single type of envisioned network. In fact, some may not even be able to function in all possible settings. Consequently, this is an area where vendors have tried to differentiate themselves by capitalizing on the types of networks where they already have an advantage. This is particularly true for vendors who also propose network processors and/or switch fabric chipsets, as they try to promote one-stop shopping for their customers. Conversely, customers often find themselves confronted with the dilemma of choosing a superior traffic manager from another vendor or making their systems life easier by simply opting for the traffic manager of the same chip family with their choice of *network processing unit* (NPU) or switch fabric platform.

Before we look at how traffic managers work, however, we first need to introduce some basic notions about their macroscopic functionality. We will then put things into context by stepping back one moment and looking at the QoS realm within which the traffic manager is called to perform its intensive, complex, and noble tasks.

FUNDAMENTAL CONCEPTS IN TRAFFIC MANAGEMENT

Modern multiservice networks where *Internet Protocol* (IP) has ended up the de facto network protocol create a situation that is only going to be exacerbated in the future. Numerous applications that share the common network medium are sometimes radically different in nature and consequently have completely different performance requirements (as dictated by the expectations of users). Because fundamental network resources are finite, an optimal use of these resources is mandated, not only for the sake of efficiency, but also (and more importantly) for the sake of profitability.

TABLE 15.1 Different Traffic Manager Requirements for Different Applications

Applications	Capabilities	Number of Queues	Profiles	Support Needed
Carrier-class metro and core routers	Fast forwarding and simple *behavior aggregate* (BA)-based classification	1,000–4,000 aggregated flows	Small number of *Random Early Detection* (RED) profiles	RED
Multiservice switching	Fast forwarding, and sophisticated classification and policing/shaping	Large number	Large number of RED profiles	Fine granularity with *Weighted RED* (WRED), *Early Packet Discard* (EPD), and *Partial Packet Discard* (PPD), both per-class and per-flow behaviors, embedded *Segmentation and Reassembly* (SAR) for an *Asynchronous Transfer Mode* (ATM) interface of packet-based networks, and support for frame-relay switching
Boundary routers at the *network access point* (NAP)	Fast forwarding, and sophisticated classification and policing/shaping	Large number	Large number of RED profiles	Fine granularity with WRED, EPD, and PPD, both per-class and per-flow behavior, embedded SARing for an ATM interface of packet-based networks and end capabilities for frame-relay and ATM
Ethernet and access	Less fast forwarding and limited classification	1,000	Small number	Usually WRED, two to three priority levels

Until recently and to a large extent still, networks have been treating transient traffic on a best-efforts basis. Unfortunately, this is not good anymore. In addition to e-mail and file transfers, remote login, or typical web browsing, numerous telephone conversations or videoconferencing may also be occurring, or multicast multimedia material may be downloaded on a pay-per-use basis for entertainment purposes.

All these unrelated applications must share the common network, and it would be unfair to treat them all in the same way. In addition to fairness, the concept of profitability is important as service providers and carriers must manage their single most important resource—bandwidth (or throughput, which can be translated to *used capacity* and is a metric that will tell the carrier how much more revenue could be generated if traffic were managed in a different way). Bandwidth must be allocated differently among paying customers at different times of the day in order to maximize revenue and customer satisfaction. From this fundamental premise, the concepts of QoS and guaranteed-level service were born.

The idea is not new. For more than 20 years now, *type of service* (TOS) bits have been available as part of the definition of the classic IP packet. The basic notion was already in place. This has only facilitated the wide-scale adoption of the QoS vision. However, as we will see, many issues still need to be agreed upon and implemented.

In any discussion on QoS, keep in mind that QoS seeks to specify and control five fundamental network variables:

- **Bandwidth or throughput**   For example, a critical telemetry application may require a higher priority for the transmission of the data it carries than an *Hypertext Transfer Protocol* (HTTP) session browsing the Internet.

- **Latency**   This is another common way of referring to end-to-end delay as experienced by traffic originating from point A in the network and destined to travel to point B. Voice communication is a good example. Once voice latency goes beyond 150 ms from mouth to ear, it is perceptible by the average human operator and becomes distracting. If the delay becomes longer than a certain value, it may even negate a network application based on voice transmission. Videoconferencing is a notorious example where sound may not be synchronized with the video images of a participant opening and closing his or her mouth. This can become very irritating to customers and thereby prohibit further business opportunities.

- **Jitter**   This is essentially the variation in delay between two consecutive packets. A typical example would be the interfacing of an IP realm with *Synchronous Optical Network/Synchronous Digital Hierarchy* (SONET/SDH) or *time-division multiplexing* (TDM) networks, where the slightest jitter may cause a loss of read or written slots in time and therefore cause application errors.

- **Packet loss**   This is immaterial in some applications, whereas it is undesirable in others. Sometimes it is simply not an option.

- **Link availability**.

Because link availability is an issue that transcends the network layer or data link protocol and spans concerns as low as the physical layer itself, we will not spend much time on it. The network-processing computational systems cannot do much to ensure it. It remains undoubtedly in the list of obligations that a carrier may have when serving a corporate customer. However, this is not a problem that can be addressed by the mechanisms and techniques we will be discussing that purport to deliver QoS along the other four dimensions of the puzzle.

Now that the fundamental idea is clear, how would we go about implementing a realm that enables its realization? Several tools of the trade are available.

First, *policies* should be compiled about specific applications that are expected to run on a network that tries to implement QoS. Based on the collection of such disparate applications, the specific variables that QoS seeks to specify and control are mapped onto each application. For example, jitter and latency requirements for *voice over IP* (VoIP) telephony are not the same as *File Transfer Protocol* (FTP) file transfers, and e-mail is not as high a priority as network management queries based on the *Simple Network Management Protocol* (SNMP) from the management console.

Likewise, a long-distance carrier may have purchased a certain bandwidth of traffic on a backbone provider's network. This approach means that all traffic originating from the carrier must be monitored so that a certain level of bandwidth can be guaranteed by the backbone. This implies the need for continuous monitoring on behalf of the network of the quality it delivers. In the previous example, if the intended performance drops below the preagreed levels, then *bandwidth provisioning* has to be revisited and perhaps modified on-the-fly.

In addition to the vast amount of pertinent statistics of which modern network equipment keeps track, robust QoS-oriented *metering* processes and techniques need to be implemented. Metering helps detect specific situations and counts the instances and frequency of their occurrence. This way network management can monitor in real time whether they fit the intended profile of traffic. If they do not fit the profile, then special algorithms are used that will handle the *shaping* of traffic so it conforms to the desired profile. A special *shaper module* inside the traffic manager chips is responsible for doing this. It usually has its own external or internal memory subsystem within which it keeps and works with *rate tables*.

In realms where thousands of different applications may be simultaneously sharing the network, exhaustively compiling policies can be impossible. Two approaches can be taken:

- One approach is to not do anything, capitalize on cheaply available bandwidth and sharply dropping prices of gigabit routers, and hope that the backbone throughput will always be more than enough when spikes of demand appear. The problem with this approach is that no one can accurately forecast traffic loads in packet networks and major customers insist on preferential treatment for some of their key applications. This can only be achieved through SLAs, which can only be implemented when traffic is segregated into classes. SLAs stipulate the number of classes possible and the amount of traffic (bandwidth) allowed per class. SLAs can be *static* and therefore be negotiated at some predetermined frequency (monthly, yearly, and so on), or they can be *dynamic*, which requires a signaling protocol such as the *Resource Reservation Protocol* (RSVP) for the application to request more resources from the network in quasi-real time.

- The diametrically opposite approach is to first define a limited number of specific *classes of traffic* and then assign individual applications to a class. All applications in one class of traffic receive the same treatment from the network.

As the classes benefit from different treatment standards we can also talk about *differential QoS*. Carriers employ SLAs, which stipulate service level guarantees and consequently the levels of the network performance that the customer requires, expects, and pays for. This way carriers can also actually guarantee the minimum assured level of performance. Traffic managers facilitate this process through a judicious use of traffic *policing*. More specifically, flow tables are stored in a memory subsystem with which the *policer module* of traffic managers works.

In terms of classes of service, how many classes are good? This is an open and ongoing debate, but the widespread consensus in the industry is that four classes of traffic seem sufficient. Some *network equipment vendors* (NEVs) offer the possibility of more, but this seems superfluous since the industry still cannot agree on what levels of granularity make sense for two levels of markets that are positioned downstream from NEVs—that is, the carriers and the carriers' customers (enterprises and end-user organizations).

Network applications may at some point require (and perhaps be formally entitled to) better treatment for their packets. For example, consider the *Transmission Control Protocol* (TCP). TCP requires the reliable delivery of its packets to their respective destination. It not only requests acknowledgements of reception, but based on feedback that it obtains on congestion conditions from the network, readapts the specifics of its parameters to optimize its behavior, and so on. In some cases, special *signaling* techniques must be implemented that are ideally understood across the wider network and that allow the dynamic reallocation of portions of shared resources. This is similar to what happens in bandwidth provisioning.

Sometimes this may not be possible or even desirable. In these cases, an acceptable QoS architecture may require that individual packets be marked with specific bit fields set appropriately, such as the *Differentiated Services* (DS) field of the IP header. This would enable packets to be immediately visible when they traverse the network and, more specifically, when they enter and exit its individual systems at network nodes, such as routers, switches, and hubs. It would make it straightforward in principle to ensure that each packet received the treatment to which it is entitled.

In a multiservice switching or routing system, the hardware can use such flagging bits to detect the presence of specific types of packets either at an ingress port or at an egress port and act accordingly. If packets are not handled with the appropriate level of attention or with the necessary sense of urgency at an ingress port, input buffers may overflow due to traffic congestion. This could lead to packet loss. If the same traffic congestion occurs at an egress port, then until the state of congestion is resolved, the latency and jitter figures may suffer unduly.

This is somewhat similar to the revolution that *Multiprotocol Label Switching* (MPLS) has brought onto some IP networks. To form this revolution, MPLS borrowed the basic tagging idea from ATM with its *available bit rate* (ABR), *constant bit rate* (CBR), and *variable bit rate* (VBR) classes of cells. In this case, special labels are tagged onto packets. Based on that marking, the network knows exactly how to treat these packets. The whole field of *traffic engineering* has almost acquired a new meaning following this revolution. As a result, new lucrative services such as the *virtual private network* (VPN) have become possible.

Series of techniques and algorithms have been proposed, analyzed, documented, and implemented in order to materialize various methods that help create the framework of such traffic manipulation. Among them, we will mention methods of *queuing*, which allow the dynamic change in priority that some packets expect, as well as methods of deciding when, how, and which packets must be *dropped* (discarded). The traffic manager chips handle this through a *scheduler* module. Like other traffic manager modules, the scheduler module has its own memory subsystem where it keeps queues.

The following are a couple of other functional modules inside traffic manager chips:

• The *active queue control* module, which uses its own memory subsystem where it keeps the so-called *Random Early Detection* (RED) tables.

• A *statistics* module, which often works in conjunction with a host or control plane processor, which is connected with the traffic manager in the majority of the cases over a *Peripheral Computer Interconnect* (PCI) bus.

• For the case of multicast traffic, the embedded SAR module, which works with its own multicast tables, which are stored in its own memory subsystem.

As we have learned in previous chapters, the memory technologies for the various subsystems do vary between chip vendors. Different memory technologies address different needs. We summarize the choices of memory technologies and the rationale behind their respective use in Chapter 16, "Systems Engineering Issues." We will therefore not expand on this issue here.

Incoming packets arriving from a line interface (*physical* [PHY] or framer) are classified either by embedded functionality offered by the network processor or by an external classification coprocessor. A special descriptor (or tag) is generated, which is attached to the packet which is then forwarded for subsequent buffering or processing. In the case of look-aside architecture, the traffic manager will receive the traffic descriptors and only then will packets be fetched from the buffering memory. In the flow-through approach, the packets flow through the traffic manager. Deep packet analysis or further classification may be required. Another specialized coprocessor usually accomplishes this. Security coprocessors may also be used in some contexts. Packets may have to be buffered by the traffic manager either at ingress or at egress (with respect to the switch fabric) and sometimes at both.

Not all traffic managers support all of these capabilities. Traffic managers may contain the interface with the switch fabrics. Standards interfaces are usually chosen, such as *System Packet Interface 4* (SPI-4) and *Network Processing Forum* (NPF) CSIX, to connect the traffic manager to the switch fabric chipset component that lies on the line cards. This will then handle the interface with the crossbar component on the switch fabric card.

We will conclude this section on QoS by saying that the field remains largely in a desired state. Although much progress has been made in understanding the technical problems associated with the modules that compose it, this is just the tip of the iceberg. It will arguably take much more than just mechanisms and tools of the trade such as algorithms and signaling protocols. A great deal of standardization work is still needed. This is obviously the mission of industry bodies such as the *Internet Engineering Task Force* (IETF). However, several issues are still unresolved regarding how to interpret the available bit fields of mechanisms consistently across the industry in order to create meaningful differentiated services or determine how to deal with QoS when crossing the boundaries of autonomous systems or network domains. This may sound mundane, but it is actually one of the thorniest problems in this field.

QoS-ORIENTED PROTOCOLS

One of the key notions in the subsequent discussions is the term *flow*. This is a loosely defined concept, but for our purposes, we will clarify that any quantity of traffic consisting of a series of packets that can be viewed or treated in some common way may be characterized as *flow*. More specifically, this implies that traffic packets with the same source address, port number, protocol ID, destination address, and destination port number should be treated as one flow. Consequently, we can use the term to describe packets belonging to a specific session, originating from a specific network node, or fitting a specific profile. The term is generic, but serves an important objective when it comes down to finer-granularity traffic management.

Packets belonging to multiple and unrelated sessions are usually aggregated in multiple flows. Each flow will then be further logically assigned a priority level (in conjunction with queuing structures) and then mapped onto classes. Each class is mapped onto one of many ports. Some architectures even go so far as to make a point of splitting ports into a two-layer hierarchy of virtual ports and physical ports. Each class maps into one of multiple virtual ports and each virtual port can then map to one of multiple physical ports. The idea of hierarchical traffic-management granularity should now be a bit clearer. This example can be summarized as follows:

$$Flows \rightarrow Classes \rightarrow Virtual\ ports \rightarrow Physical\ ports$$

RSVP

The RSVP (RFC 2205) was one of the first protocols that attempted to provide some sense of rudimentary QoS or guaranteed bandwidth. It is currently used on IP networks. It works by requesting from downstream network equipment that network resources and capacity be reserved in advance—in other words, before a specific flow of traffic is sent out into the network. Its mechanisms are simple compared to the requirements of the complex realm of QoS. It may or may not receive a positive signal. The signaling works in the following way.

The originator of traffic (transmitter) A sends a PATH message to receiver B that describes the characteristics and needs of the upcoming traffic. Every intermediate router will forward this PATH message to its downstream peer on the next hop all the way to the receiver. If receiver B accepts the PATH message, it will issue a RESV message back to the transmitter that is requesting network resources for the flow. The RESV message now travels backwards on the same link hopping from one router to the other and hopefully making it eventually back to transmitter A. If a router cannot honor the request, then an error message is sent to the receiver and the signaling handshake is ended without setting up a connection between A and B. If every stage has honored the request, the appropriate buffering and bandwidth resources are set up at each router to accommodate the oncoming flow.

IntServ

The *Integrated Services* (IntServ) protocol (RFC 1633) is structured upon a framework that uses four fundamental components:

- A *signaling protocol*, which requests that specific network resources be reserved downstream in anticipation of transmission.
- An *admission control* scheme that will decide whether the requested resources will be reserved as requested.

- A multifield-based *classifier*.
- A *scheduler* that will handle a packet as needed in order for it to obtain the necessary QoS.

For signaling purposes, IntServ uses RSVP (RFC 2210) extensively, As mentioned previously, this requires that flow-related state information also be kept in intermediate nodes, not just in end systems. In conjunction with the following three classes of traffic that IntServ defines, it essentially behaves as a connection-oriented protocol that is similar to ATM switching. Packet flows are assigned to one of three classes defined in IntServ:

- *Guaranteed service*, which is used for applications that need assurances for no packet loss and controlled/specified latency bounds.
- *Predictive service* (also known as *controlled-load service*), which is used for applications that require probabilistic delay bounds, a realm where the presence of either light or heavy traffic on the network will not affect this class of traffic.
- *Best-efforts service*.

In order for this approach to function, specific resources need to be reserved from the network in anticipation of the needs of individual packet streams or flows. Consequently, routers must retain flow-related state information. The guaranteed service requires that all routers be capable of IntServ, which did not materialize as a vision. The predictive service, however, could work in principle if it was configured in conjunction with RSVP as its signaling protocol and if it was installed at all perceived bottleneck nodes of the network. They could run the controlled-load level of service with RSVP messages tunneled through other parts of the network, but this situation turns out to be difficult to manage.

Running IntServ imposes large computational (in *central processing unit* [CPU] cycles) and buffering loads (which are in proportion to the actual number of flows) to routers all over the network. This is one of its weak points. Since it is preoccupied by the flow's fate at the packet level, it necessarily splits traffic logically into numerous flows. However, without aggregating them into classes before launching the traffic into the network, the processing load is obviously commensurate with IntServ sessions. Therefore, it does not scale well into large core networks. Its undeniable capability to guarantee network performance and QoS levels by merely applying similar mechanisms to the ones used by circuit-switching networks when they decide, for example, on call admission has earned it friends for its abilities, and foes for delegating the advantages of packet switching to the proverbial second seat.

DiffServ

A different approach is taken by RFC 2475, which defines the *Differentiated Services* (DiffServ) protocol. DiffServ is computationally a much more realistic approach than IntServ. Based on its wide-scale acceptance by vendors, it seems to represent a promising and major attempt to implement an acceptable base platform upon which elaborate QoS can run.

DiffServ uses the previously defined TOS bits. Upon them, it builds a series of aggregate flows. We discussed multifield classification and the five-tuple lookup algorithm in previous chapters. The result of the classification process is that each packet is assigned to one of many flows. This is done by deriving a 6-bit value, which the DiffServ specification calls the *DiffServ Code Point* (DSCP). The DSCP is written into the DS field of the packet header. More specifically, it is the TOS octet in the IPv4 header. In the IPv6 header, it is the traffic class octet that composes the so-called *Differentiated Services* (DS) field where traffic classes are encoded. At origination, a packet's DS field can be marked by a customer for whatever service is desired. It can also be left to the first DS-compliant router (a leaf router in the routing tree) to decide how to modify the DS field based on the multifield classification of the packet, which we covered in length in Chapter 12, "Search Engines," and Chapter 13, "Classification Coprocessors." The scheme requires that all ordinary network-application flows will be consolidated into one of these aggregate flows.

The idea has merits from several angles. The number of bits in the DS field is very small—hence, the number of classes is not astronomical. Therefore, the classification that happens in the core of the network must occur based on *behavior aggregates* (BAs) and not based on individual flows as would be expected on a campus or at the edge. Different levels of service will be assigned to these aggregates before they enter the network. This is the scalability advantage that DiffServ offers. A packet header also contains the DSCP field, which identifies the aggregates. The DSCP can also be passed through a network junction (node) and be handed over to a subsequent router/switch. As a result, the DiffServ protocol is quite flexible in the sense that it does not require that all possible nodes between a source device and a destination device be DiffServ enabled.

This represents a serious advantage of DiffServ over IntServ. In order to run the protocol, all computational requirements on packet flows are handled by the local network equipment *before* the traffic is injected into the network. As a result, downstream routers will not be required to find out by extensive real-time analysis on a per-packet basis what treatment should be applied to each packet. Instead, routers use a series of *per-hop behaviors* (PHBs) to determine how to handle a specific flow. The PHB is the externally observable behavior of a packet in a DS-compliant router. The IETF has defined three major PHBs, which are in effect specific packet forwarding treatments:

- *Assured Forwarding* **(AF) (RFC 2597)**   This PHB has four classes of traffic, which each have three possible levels of drop-precedence.
- *Expedited Forwarding* **(EF) (RFC 2598)**   Also known by some as *virtual leased line service*, it provides capabilities of low latency, low jitter, no packet loss, and an assured level of guaranteed bandwidth.
- *Best Efforts Forwarding*.

Note that DiffServ only defines the DS field and a series of PHBs. These do not represent multiple levels of network services per se. The individual carrier or service provider is responsible for creating different levels of service and for deciding which combination of DS bits and PHBs reflect which service and how they are mapped onto each other through tools that network equipment manufacturers develop and put at the carriers' disposal.

It should be emphasized that DiffServ, unlike IntServ, does not offer performance guarantees. Even if packets are marked for preferential or priority treatment, it does not mean they will get it. For example, a router downstream that may be experiencing local traffic congestion will not hesitate to drop newly arriving packets no matter how they are marked. In addition, no QoS-related signaling provisions are provided in the protocol and consequently applications, which might need to adapt their behavior, depending on network status, do not have any means as to how go about implementing such a solution. However, it remains a simpler protocol computationally and structurally, and it is becoming a preferred method of working. Neither DiffServ nor IntServ can assign paths, though. The network equipment that is responsible for the traffic origination must assume that flows will run on the best-effort route that is assigned in the router regardless of priority.

Although the case of MPLS is a slight deviation from our subject, it is inside the same QoS realm. We will obviously not discuss MPLS in detail here. For more information, refer to the respective literature at the end of the chapter. However, it is interesting that in MPLS, specific bits are marked onto packets (or labeled), and explicit paths, which are called *label-switched paths* (LSPs), are assigned to packet flows by the network routing or switching equipment. This assignment is accomplished by the generation of corresponding packet labels. The appropriate label is contained in an MPLS header that now becomes a prefix to each IP packet. Policies that span the range of possible network applications must be established and then, based on these policies, labels are mapped/assigned to individual applications.

We cannot help but notice the striking similarity with the process of assigning applications to classes of traffic in a QoS realm as previously discussed. These labels will then be used by the *label-switched routers* (LSRs) in the MPLS network. More specifically, it is based on the content of the labels that the routers make on-the-fly and per-hop decisions as to how to appropriately forward each packet to the next downstream router along the stipulated path that is associated with the specific label.

Interestingly, MPLS can also work across network boundaries as long as network operators use the same version of MPLS on their routers.

From a QoS standpoint, MPLS has the important advantage of directing packet flows along specific paths. It therefore allows the implementation of sophisticated traffic engineering and VPNs. This consequently leads to the conclusion that the combination of MPLS with other protocols can potentially provide a sophisticated QoS platform for the future. It is important to note that traffic engineering, VPNs, and QoS are completely feasible without using MPLS, although MPLS is a convenient platform upon which the implementation of such capabilities is greatly facilitated. However, be aware of the potential impact that a VPN framework may have on the QoS mechanisms as there is no such thing as a free lunch.

In fact, MPLS has recently started to play a major role in the evolving QoS debate. Although it has been mostly regarded as a novel way for routing, the fundamental premise of MPLS is a good base for the differentiated QoS treatment of traffic packets. For example, MPLS allows a single IP backbone to deliver legacy services such as ATM and frame relay by offering similar QoS features to the ones previously offered by the other technologies. An example is IETF's Martini draft[1], which enables MPLS to effectively behave like a layer 2 protocol as it allows IP, which is a layer 3 network protocol, to actually carry layer 2 protocol traffic such as frame relay, *ATM Adaptation Layer 5* (AAL5), or Ethernet while providing a SONET circuit emulation service across an MPLS network and yet to preserve the same QoS guarantees usually associated with frame relay or other layer 2 traffic protocols.

Several new approaches are available, including the following, but nothing has been officially standardized yet.

- RSVP can be used to ask MPLS routers downstream about available resources and, based on their responses, an end-to-end MPLS tunnel can be configured.
- *RSVP Traffic Engineering* (RSVP-TE) can be used, which is a newer addition to the long list of possibilities.
- The *Constraint-Based-Routing Label Distribution Protocol* (CR-LDP) can be used, which is a variation/extension of LDP, which is currently widely used in MPLS networks.
- The *experimental* (EXP) bits in the MPLS shim header can be used. When taking such an approach, which is based on *EXP-inferred label-switched paths* (E-LSPs), eight service classes can be supported based on the DiffServ standard.

QoS is a technologically wide and deep field involving several requirements, protocols, and proposed solutions. Unfortunately, it has political and economical ramifications that are not immediately apparent to the casual onlooker. When the prospect of implementing cross-border SLAs that require monitoring, data collection, accounting, and policing even beyond country borders in some cases is considered, the subject becomes sensitive. The differentiation of service providers based on the quality of their network also stands to lose from such a prospect, because in a realm with one global common QoS, one set of traffic classes would be applicable worldwide. It is natural that providers resist embracing it.

On top of everything, the economics of network applications will dictate what is needed. Until massive futuristic consumer applications such as multiuser online gaming (with tight latency or jitter requirements), video on demand, or Internet support for wireless and mobile services become an absolute necessity, this may actually take some time.

We will not divert our attention deeper into the QoS debate as it is a giant subject that is not possible to cover here. We will therefore now return to our main topic: traffic managers. Hopefully, this discussion will explain how capable these chips must be in order to provide the network-processing platform upon which NEVs should be able to implement and manage QoS.

---

1. *www.ietf.org/internet-drafts/draft-martini-l2circuit-trans-mpls-10.txt.*

MAJOR TASKS AND ALGORITHMS

Based on this initial discussion, the significant tasks for traffic managers can be summarized as work that continuously pertains to one of the following categories:

- Statistics gathering.
- Traffic policing.
- Traffic shaping.
- Scheduling.
- Queuing and buffer management.
- Congestion avoidance and packet dropping.

STATISTICS

Modern network equipment systematically gathers a wealth of statistics, which can be accumulated at different levels of granularity from as fine as on a per-flow basis all the way to per-aggregates of flow. Traffic statistics can be used for mundane tasks such as network diagnostics and management and for more sophisticated goals such as billing purposes in a QoS realm. As mentioned in previous chapters, statistics can be gathered in different ways. Most platforms are based on the idea of using on-chip counters, which are incremented or decremented accordingly based on real-time traffic information. If these counters overflow at some point, then a control plane CPU will usually have to be interrupted for further assistance with appropriate code. Other systems offer different means for gathering statistics with special interface ports for external support. Some platforms simply do not have the traffic manager gather statistics by itself. In those cases, the network processor is designed to handle this chore instead.

We should clarify that although the network processor can do most of the statistics keeping, only the traffic manager will typically have visibility into congestion-related statistics and information. Likewise, if packets need to be marked differently based on the congestion that they have encountered, then the traffic manager must be able to support this kind of functionality.

TRAFFIC MARKING, SHAPING, AND POLICING

In earlier chapters, we discussed the details of the classification realm, where packets are assigned to specific flows. In the recent industry trend toward implementing SLAs, it is important that packet flows be policed. More specifically, it is imperative to ensure that whatever is specified in the corresponding SLA is what the flow is actually obtaining from the network. Packets are policed by being appropriately marked. Policing can occur at the domain level (as with protocols such as DiffServ) or they can be policed at the line-card level, usually at the ingress port.

This operation can be done by using several marking algorithms. The most common of these algorithms is the *three-color algorithm*. For example, in the implementation of such an algorithm, a packet that is within its flow's SLA can be marked green and the traffic manager will forward it if and when it is feasible. If the packet is out of the contract boundaries but it is still stipulated that the packet should be forwarded if at all possible, then the traffic manager will mark it yellow. Unless traffic congestion occurs, it will be forwarded when possible. Finally, a packet is marked red when it does not conform to the contract that manages that specific flow and it will be discarded (dropped). As an example, in the DiffServ AF PHB, a marking algorithm, such as the three-color approaches described by either the *two-rate three-color marker* (trTCM) or *single-rate three-color marker* (srTCM), establishes the packet-discarding precedence.

If policing is an ingress port task, then traffic shaping should be organized at the egress port. It is capable of handling the potentially bursty nature of some traffic by ensuring that it is released into the network in transmission rates that are commensurate with what is stipulated in the corresponding SLA. Several algorithms are used in this realm based on whether the underlying network is packet based and therefore connectionless (such as IP-based networks) or connection oriented (such as ATM). The *token bucket* approach is usually taken on packet-based networks, whereas the *leaky bucket* (LB) approach in its several variations is usually preferred on connection-oriented ATM-like networks.

The token bucket algorithm is based on the idea of a data structure called *bucket* where tokens are stored. Tokens are produced and thrown into the bucket at a constant rate. Each packet that is forwarded will consume one token. This means that the rate of consumption of the tokens is not constant (but it is limited). If no more tokens are in the bucket, the next packet is delayed and therefore buffered, and so will also be all subsequently arrived packets that await service/forwarding. This ensures that the rate of packet forwarding is effectively controlled, as the token-generation rate imposes a limit.

However, if several tokens are inside the bucket, they can potentially be consumed very fast by available packets. This means that through this approach, short bursts of traffic are possible and the depth of the bucket determines the length of potential burstiness that can be envisioned in the underlying traffic. When the algorithm is conversely used to determine whether a packet will be buffered or discarded, a straightforward scheme is used according to which if the accumulated tokens exceed a certain threshold value, then the next packet will be dropped.

The LB algorithm is used in ATM-like networks especially when a constant rate agreement must be enforced, as is the case with CBR traffic. For several reasons, ATM cells can arrive at a network node at an uneven rate. Each arriving cell contributes a token into a leaky bucket. Therefore, tokens are fed into the bucket at an uneven rate. Each departing cell will first need to consume a token from the bucket. As tokens are produced in a constant rate of leakage by design, the transmitted traffic rate is immediately pinned down to the desired level.

A creative variation of this algorithm is the *dual leaky bucket* (DLB) algorithm. With this algorithm, one bucket decides the peak rate and the other one sets the mean rate. An example of its efficient implementation is provided in Chapter 7, "Agere PayloadPlus™ Family of Network Processors."

The principle of the dual bucket approach is also used in policing. Two good examples can be found in RFC 2698 and RFC 2697, which define the trTCM and srTCM algorithms, respectively.

CONGESTION MANAGEMENT

We briefly mentioned the need to decide when to simply discard (drop) one or more packets. This radical decision is often the unfortunate result of a highly undesirable state called *congestion*. The reasons for congestion can be numerous and unpredictable. Fundamental textbooks on network traffic engineering discuss the problem in length. However, at a time when actual guarantees for specific levels of QoS are required and paid for by the customer, it is important to have robust mechanisms that ensure a fair approach toward making the decision of what gets dropped, when, and under what circumstances.

In fact, the traffic manager is responsible for action when it detects conditions of traffic congestion. Two choices of systems engineering are available:

- Decelerate the traffic arriving from upstream (this is similar to backpressure signaling, which we discussed in Chapter 14, "Switch Fabrics").

- Start dropping some packets locally. Originally, the brute-force method was used and only the latest arrivals were sacrificed. Because countless TCP sessions were by overwhelming proportions the direct victim of such an approach, the industry soon moved to levels of choice that reflect some choice based on a degree of fairness. The term *fairness* is not used here lightly or subjectively. Given the limited resources of a network, what is fair for one party may be completely unfair for the others. Therefore, some objectively defined criteria and agreed-upon techniques were required in order for the industry to enable such a regime.

One of the most sophisticated, but rarely used, techniques (similar to the approach taken by frame relay in the early 1990s) is the *Explicit Congestion Notification* (ECN) technique. In ECN, the packets that suffer due to congestion are marked appropriately and forwarded to their destination. Upon reception, the receiver notifies the sender of these packets, which slows down its transmission pace. A similar method used in connection-oriented ATM networks is the *Early Forward Congestion Indication* (EFCI).

The most common algorithms used to resolve traffic congestion are RED, WRED, and *RED with In and Out* (RIO) in IP networks (which may or may not be connectionless if MPLS is involved), and EPD and PPD in (connection-oriented) ATM and cell-based MPLS networks.

RED is a probabilistic method allowing the random discarding of packets as soon as the corresponding queue is filled beyond a prespecified threshold. As shown in Figure 15.1, this probability increases when packets fill the queue beyond the threshold level. The curve is composed of multiple points that depend on the actual application realm and network requirements. With RED, the same curve will be applicable if all the traffic has the same priority. If this is not the case and traffic has different priorities, then several RED curves will be applicable.

WRED curves do not have to be of the same geometric profile. For example, in Figure 15.2, we show a completely different profile of more aggressive action in a WRED environment that could be used between the profiles shown in Figures 15.1 and 15.2. If we follow a three-color marking example like the one we described earlier in the chapter, green packets can be discarded based on the profile shown in Figure 15.1 and yellow packets can be discarded based on the profile shown in Figure 15.2. Red packets can be simply dropped if no bandwidth is available. The ability to adapt thresholds and drop rates to the various traffic classes makes WRED more interesting for today's QoS needs in advanced IP networks.

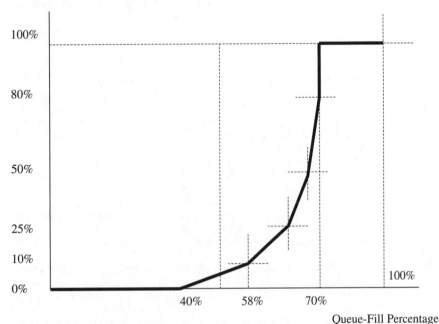

Probability of dropping a packet randomly

FIGURE 15.1 The principle of a RED curve tying the probability of randomly discarding packets based on how full a specific queue is

Probability of dropping a packet randomly

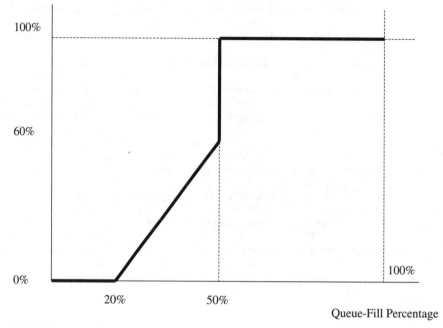

Queue-Fill Percentage

FIGURE 15.2 In a WRED environment, multiple RED curves are created for traffic of different priorities. This aggressive curve could be combined with the RED curve shown in Figure 15.1 for packets that are marked differently.

If the two algorithms are compared, it is apparent that RED is designed to prevent the tail-drop, which is caused by sudden bursts of traffic. Typical backbone routers can buffer up to around 100 ms worth of traffic per port. If a more significant burst of traffic occurs, buffers will overfill and oncoming packets will start to get dropped. For example, the phenomenon of tail-drop has affected TCP flows by causing them to decrease their rates only for them to increase later. In many occasions, undesirable oscillations can occur, which lead to instability in the application's performance and the network's traffic utilization.

RIO is a variation of the RED technique, as shown in Figure 15.3. In this example, two traffic classes are assumed. One is prioritized (also called *In*) and the other is nonprioritized (also called *Out*). The packet-discarding probability for nonprioritized traffic lies between the buffered queue sizes Min and Max, while the packet-discarding probability for prioritized traffic lies between the buffered queue sizes Max and Total. If the queue size of Max is exceeded, all arriving nonprioritized traffic packets will be dropped. The two shaded regions are called the *Graduated Dropping Regions* for the two classes of traffic. If the buffer queue size of Total is exceeded, all traffic will be dropped simply because buffers exhibit overflow.

The EPD and PPD algorithms are used in connection-oriented ATM networks, and their principle of operation is similar to the RED algorithm. The main difference between the two is that PPD will discard a complete frame from buffers. This frame may be composed of multiple cells, which will all be dropped. EPD will only discard the cells that are buffered in excess of a crossed threshold level. For example, when TCP/IP runs over ATM, with PPD, once a switch on the linking path between two communicating parties drops an AAL5 cell, all subsequent AAL5 cells belonging on the same *virtual*

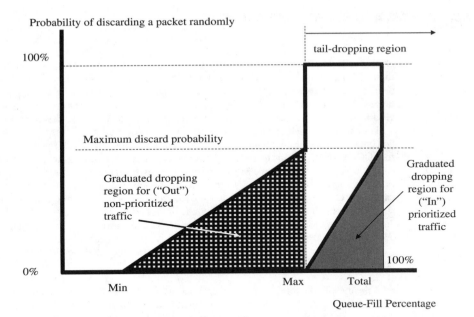

Probability of discarding a packet randomly

100%

tail-dropping region

Maximum discard probability

Graduated dropping region for ("Out") non-prioritized traffic

Graduated dropping region for ("In") prioritized traffic

0%

100%

Min

Max

Total

Queue-Fill Percentage

FIGURE 15.3   The RIO curve approach as a discarding mechanism (an example with two classes of traffic)

circuit (VC) will also be dropped until the switch sees the end of the AAL packet. PPD helps the overall throughput. However, all cells that happened to be transmitted before the specific dropped cell would definitely waste bandwidth.

With EPD, the idea is that the switch drops every cell belonging to the same IP datagram after its port buffers exceed some predefined safe threshold, but before they actually overflow. In other words, an IP datagram is dropped in its entirety, thereby preventing wasted bandwidth from the transmission of dead cells as would be the case with PPD.[2]

Congestion avoidance and management involves a wide area of activities that monitor the available resources and match them to specific traffic profiles that are simply desirable or that must be guaranteed by the network (in the case of SLAs). Based on multiple criteria, the network equipment may have to drop some packets. This is one of the tasks the traffic manager is called to handle. For example, the attainment of specific thresholds in some packet/flow queues may trigger the execution of congestion avoidance routines. Also as mentioned earlier, if the policing module of the traffic manager marks a packet with the appropriate color for demotion, then these packets stand a good chance of being discarded given the appropriate conditions.

---

2.   An interesting work on this subject is the paper by A. Romanow and S. Floyd, "Dynamics of TCP traffic over ATM Networks," *IEEE Journal on Selected Areas in Communications* (May 1995). This is available on the Internet at *ftp://ftp.ee.lbl.gov/papers/tcp_atm.ps*.

SCHEDULING AND BUFFER MANAGEMENT

In several examples presented already in the book as well as in the four-level framework of hierarchical traffic management, as discussed earlier in this chapter, designers must manage queues that are logically distinguishable and onto which usually flows are mapped. These streams of packets, cells, or frames are stored (buffered) into some sort of memory and piled up in queuing structures that are tapped in a systematic manner. This implies that these buffers must be properly managed in order to ensure that the traffic shaping matches the overall context that we have described so far.

For example, consider a *multiservice router* (MSR) operating at an edge network that happens to be connected upstream with an OC-192 link and downstream (through four different line cards) to four OC-48 cards. Assume that each card contains 16 ports of Fast Ethernet (100 Mbps each) and that each 100 Mbps Fast Ethernet carries 64 virtual ports with 10 active end-user systems per virtual port. For now, forget about redundancy for reliability and availability. This router contains four OC-48 cards, one OC-192 card, a master control CPU card, and a switch fabric card, in addition to power supply and extra backup features. Based on the figures shown here, there are $4 \times 16 \times 64 \times 10 = 40,960$ end users. If we assume 5 flows for each one of them, then we will have to manage 204,800 flows. With one queue per flow, the buffer management issue starts becoming clear.

So how can we go about organizing the problem of managing these 204,800 active queues simultaneously? We can perform one of the following actions:

- Logically split the traffic management function and distribute it to meaningful functional modules situated all over the MSR. More specifically, it can be distributed on the line cards as standalone traffic managers or as traffic manager work to be performed by the corresponding network processors.

- Concentrate it mostly on the central card (offering the OC-192 links in this case) that hierarchically dominates the network, and use commensurate standalone traffic manager chips.

In order to combine the problem of traffic management with the switch fabric choice and configuration, we must decide in this example whether traffic management must occur only on the egress (where traffic shaping usually occurs) or on both the ingress and egress sides of the switch fabric. Some trade-offs must be considered before making this decision. Creating these queues on the ingress side allows the implementation of *virtual output queues* (VOQs). Creating queues on the egress side usually allows the mapping of queues to multiple ports or channels for physical interface. Traffic management implemented at the egress side takes care of shaping since packets are retained in buffers and are only released to the network based on specific QoS and SLA criteria. These criteria are implemented around two fundamental families of algorithms—algorithms that handle *scheduling* and *buffer management*. The functions of queuing and scheduling are inter-related and are required for the determination of transmission of packets as well as for ordering them based on some priority schemes.

First, we will mention that the various scheduling algorithms are classified into two broad categories: *work-conserving schedulers* and *non-work-conserving schedulers* (see for example, a nice Motorola text at *http://e-www.motorola.com/collatoral/SNDF2002_N302.pdf*).

The distinction is straightforward, although the need for it may seem not obvious. A work-conserving scheduler is active when there are packets at hand, which are in need of its attention. The corollary is that when there is no traffic present that requires service, the work-conserving scheduler is not active. Conversely, a non-work-conserving scheduler may be inactive even if traffic is present which requires its assistance. Although it is counterintuitive, this is a highly desired behavior as it ultimately enables the deliberate instigation of some sense of predictability in the downstream traffic. This is accomplished by a decrease in the size of needed output queue buffers as well as in the delay jitter characterizing specific links. The reader may have noted that the function of the traffic shaping shown earlier is an example of scheduling done by a non-work-conserving method.

Now we turn our attention to schedulers. A number of different scheduling algorithms are used in network processing and it is not surprising that they are used in several areas, such as inside switch fabrics and in traffic managers. We will not discuss the internals or the merits of these algorithms, which are thoroughly covered by other more specialized sources[3] listed in the references.

Looking at data sheets of traffic manager chips requires a basic awareness of the most common algorithms, at least by name. The *priority queue* (PQ) or *strict priority* (SP) scheduler is based on a scheme of numerical priorities that are assigned to each queue. The highest-priority queue, which happens to contain packets, will always be serviced before others. This may sound good, but lower-priority queues may end up being starved as a side effect. The *fair queuing* (FQ) algorithm divides the available bandwidth by the number of available flows. By making the fundamental assumption that all flows occur in packets of the same size and by tackling one queue after the other in a sequential fashion, a certain degree of perceived fairness can be attained. The *round robin* (RR) algorithm, as the name implies, tackles one queue after the other by scheduling to transmit a certain amount of traffic from each queue.

Some interesting variations of these schedulers include the *weighted fair queuing* (WFQ) algorithm, which also takes into consideration the packet size that is associated with each flow. Based on a weight factor associated with each queue, it reallocates bandwidth between flows, especially in cases where significant differences in the packet size appear. A variant of WFQ is *guaranteed bandwidth WFQ* (GB-WFQ), which provides guaranteed bandwidth (such as CBR traffic in ATM) for a series of flows that are weighted differently. A similar approach is taken with a weighted variation of the RR algorithm known as *weighted round robin* (WRR). Traffic is forwarded essentially in a sequential fashion as in RR, but based on the weight assigned on each queue, some of these queues end up deliberately and predictably being serviced more often.

To address the possible complaints of unfairness associated with weighted queues and to look after the case of variable-length packet queues, two additional variations of the last two schedulers have come to existence called *deficit round robin* (DRR) and *deficit weighted round robin* (DWRR). DRR is aware of deficits accumulated in the frequency of visits of each queue at each round. It consistently tries to make it up to deficient queues as time goes by revisiting the lower-weight ones slightly more frequently for some periods of time. DWRR performs similar actions, but it also considers the size of the packets, which can be of variable length for some queues (regardless of the queue's weight). Another variation of RR is *frame-based deficit round robin* (FB-DRR), which uses a configurable quantum of service that ends up reducing the latency and jitter constraints associated with the DRR algorithm.

In general, the RR approach and its variations, including WRR and DRR, suffer from some unwanted side effects from traffic that is characterized by bursty or jittery time characteristics. As a result, some network session may end up being served twice when others have not even been served once.

However, some interesting proprietary variations are available, such as Agere's patented *smoothed-deficit weighted round robin* scheduling algorithm, which can reduce the service quantum to a single minimum-sized packet and thereby interleave service to different sessions. This is an interesting way to avoid this general RR drawback.

Another interesting, but less frequently used, approach is that of *class-based queuing* (CBQ), where the principle is similar to the one found in WFQ. The difference is that weights are now assigned to classes of traffic instead of being assigned to queues/flows. The effort of being fair is now made in the mathematical space of traffic classes. For some applications (as in ATM networks), some other algorithms are used. A good example is the *Earliest Deadline First* (EDF) scheduler, which makes a decision based on the time when a packet must be injected into the network in order to be able to meet stipulated latency specification requirements. Finally, as this is a very active domain of research in the industry and academia, many other algorithms have been developed. These include *Delay Earliest Due Date* (D-EDD), *Jitter Earliest Due Date* (J-EDD), *Rate-Controlled Scheduling*

---

3. For interested readers, Keshav's book (from references) is a very good place to delve deeper.

(RCS), and *Generalized Processor Sharing* (GPS). An exhaustive listing or description of these algorithms is beyond the scope of our coverage. For more information, refer to literature on this topic.

As an example of real-life applicability, we will mention that although the DiffServ EF PHB does not define the queuing algorithm, typical implementations are based on some combination of PQ and WFQ. In the AF PHB, although RED is used as an example in the IETF's RFC draft, WRED is preferred in practical systems. It is implemented in some combination with CBQ and WFQ scheduling.

In real-life designs, some system-design choices may require compensation through the engagement of complementary techniques. For example, some systems implement PQ (SP scheduling), but to avoid completely penalizing lower-priority traffic, they must do it in conjunction with some elaborate traffic policing that simultaneously controls the rate of the high-priority traffic coming in. These numerous blends of complementary functionality and an algorithmic approach are usually characteristics of product offerings through which network processing chip vendors or NEVs try to systematically differentiate their products.

We will now turn our attention to buffer management. The ultimate goal is to apply differentiated treatment on packets as a scalable means of offering sophisticated and scalable QoS. As shown in Figure 15.4, classified, policed, and shaped traffic is buffered (and therefore delayed in a controlled manner) before it gets scheduled. The appropriate buffering levels are usually decided from the corresponding SLA. Buffer management is a wide set of problems and techniques proposed as solutions. More specifically, in buffer management, a designer should be concerned about three things:

- Delineating and sharing the available buffering memory among a potentially large number of queues.
- A packet discard mechanism.
- A scheduling mechanism.

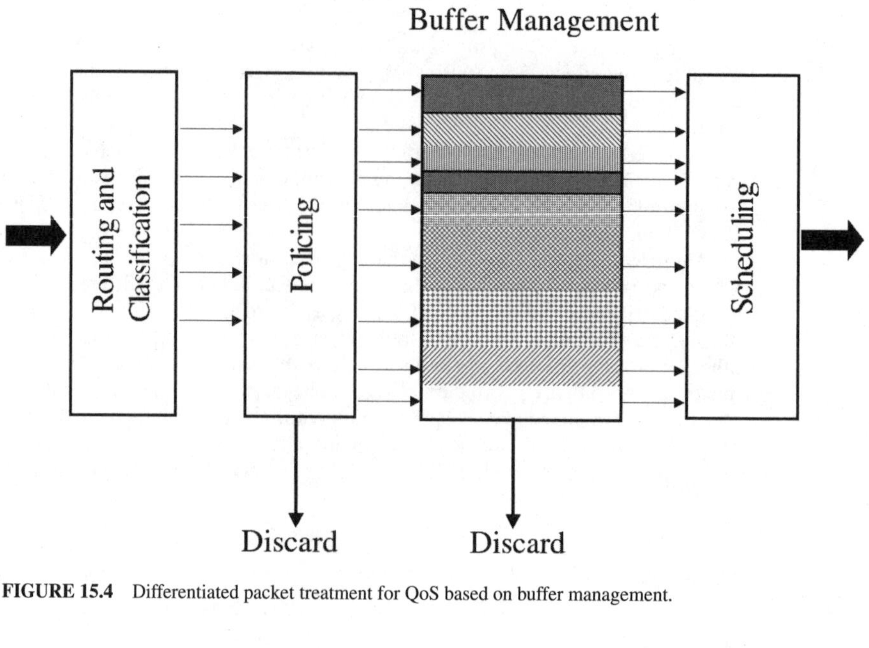

FIGURE 15.4 Differentiated packet treatment for QoS based on buffer management.

The last part is implemented upon the very same premises and techniques that we just discussed in the section on schedulers. Buffer space is split up among the multiple flows/queues along many approaches. The following three approaches are the most important:

- **Complete partitioning**   This is one of the two extreme approaches. With it, each queue is simply allocated a fixed amount of buffer space. This is easy to implement, but not very efficient as queues that are starved for more space cannot access the potentially empty parts of the overall buffer memory space.
- **Complete sharing**   This is the other extreme of the spectrum. Available buffer space is fully and equally shared among the queues. No space is wasted as may be the case with complete partitioning, but the attained degree of fairness is a problem, as any queue essentially can eat up the buffer space to the detriment of others.
- **Sharing with minimum allocation**   Somewhere between the top two extreme approaches lies the method of sharing with minimum allocation. This technique preallocates some fixed minimum space to each and every queue. The rest is then fairly and equally shared among all the queues. It is known for its simplicity, efficiency, and fairness.

Implementing QoS based on the differential treatment of packets is an interesting approach that tries to juggle computational resources (processing cycles and memory) to make a slew of decisions that allow a judicious compromise between the multiple and sometimes contradictory objectives of throughput maximization, the isolation of flows, rate guarantees, and the allocation of excess networking resources with some commonly acceptable notion of fairness.

The use of a *first-in first-out* (FIFO) approach to manage buffers with some sense of fair discarding of packets and of fair allocation of the excessive free headroom ends up being favorably compared to a major per-flow scheduling method such as WFQ. Ideally, multiple buffer pools enable a better traffic flow isolation. In terms of admissibility based on two criteria—namely, buffer availability and transmission bandwidth availability—buffer management with FIFO matches WFQ in bandwidth availability and outperforms WFQ in bandwidth availability. The technique is therefore more than worthy of attention. Some interesting work has also been done where the two methods are combined and concatenated. In other words, FIFO buffer management was followed by WFQ scheduling. Some systems also take this idea of differentiated packet treatment and expand it to the entire QoS toolkit, providing a more sophisticated approach to the problem.

TRAFFIC MANAGER CASE STUDIES

Multiple traffic manager chips have been discussed in the previous chapters at points where we presented the overall network-processing platform of each of the most prominent vendors in the industry. Refer to the appropriate chapters for specific case studies.

SUMMARY

In this chapter, we discussed the fundamental workings of typical standalone traffic managers. We reviewed fundamental concepts in QoS provisioning. We also discussed typical architectures of interfacing traffic managers with network processors and switch fabrics and, of course, algorithms used to implement traffic policing, shaping, queuing, scheduling, buffer management, and congestion avoidance. We also reviewed the major protocols involved in providing state-of-the-art QoS services in the evolving high-speed networks. Case studies of traffic manager chips are provided in previous chapters where specific vendor architectures have been discussed as network-processing platforms.

SUGGESTED REFERENCES

Basic policing, shaping, and scheduling algorithms are covered in classic computer networking text-books. The following is a good recommendation:

S. Keshav, "*An Engineering Approach to Computer Networking*," (Reading, Massachusetts: Addison-Wesley, 1997).

Among many sources on MPLS and traffic engineering, the following is a good source:

X. Xiao, A. Hannan, B. Bailey, and L. Ni, "Traffic Engineering with MPLS in the Internet," *IEEE Network Magazine* (March 2000).

The following is an informative work about issues encountered in the implementation of solid traffic management:

Jian-Guo Chen, David Sonnier, Robert Muñoz, and Ambalavanar Arulambalam, "Traffic Management in the Agere 10 Gigabits Network Processor," a white paper from Agere Systems, October 2002.

In addition to the IETF RFC documents mentioned in the text of the chapter, which can be easily downloaded from the IETF web site, the following are good sources for further information on the subject:

Roch Guerin, Sanjay Kamat, Vinod Peris, and Raju Rajan, "Scalable QoS Provision Through Buffer Management," *ACM SigComm* (1998).

A. Ioannou and M. Katevenis, "Pipelined Heap (Priority Queue) Management for Advanced Scheduling in High Speed Networks," *Proceedings of the IEEE International Conference on Communications (ICC 2001)* Helsinki, Finland (June 2001): 2043–2047. It is available online *http://archvlsi.ics.forth.gr/muqpro/heapMgt.html*.

Anthony Kam and Kai-Yeung Siu, "Linear-Complexity Algorithms for QoS Support in Input Queued Switches with No Speedup," *IEEE Journal of Selected Areas in Communications* 17, no. 6 (June 1999): 1040–1056.

R. LaMaire and D. Serpanos, "Two-Dimensional Round-Robin Schedulers for Packet Switches with Multiple Input Queues," *IEEE/ACM Transactions on Networking* 2, no. 5 (October 1994): 471–482.

Todd Lizambri, Fernando Duran, and Shukri Wakid, "Priority Scheduling and Buffer Management for ATM Traffic Shaping," NIST, Gaithersburg, MD (1999). It is also available online at *http://w3.antd.nist.gov/Publications/Hsnt/lizambri_1299.html*.

C. Ozveren, R. Simcoe, and G. Varghese, "Reliable and Efficient Hop-by-Hop Flow Control," *IEEE Journal on Selected Areas in Communications* 13, no. 4 (May 1995): 642–650.

D. Serpanos and P. Antoniadis, "FIRM: A Class of Distributed Scheduling Algorithms for High-Speed ATM Switches with Multiple Input Queues," *IEEE Infocom 2000 Conference,* Tel Aviv, Israel (March 2000).

M. Shreedhar and G. Varghese, "Efficient Fair Queuing Using Deficit Round Robin," *Proceedings of ACM SIG-COMM '95,* Boston (September 1995).

D. Stephens and Hui Zhang, "Implementing Distributed Packet Fair Queueing in a Scalable Switch Architecture," *IEEE INFOCOM'98 Conference*. It is also available online at *www-2.cs.cmu.edu/People/hzhang/publications.html* in ps.gz or in PDF format.

D. Stephens, J. Bennett, and Hui Zhang, "Implementing Scheduling Algorithms in High-Speed Networks," *IEEE Journal of Selected Areas in Communications* 17, no. 6 (June 1999): 1145–1158. It is available online as shown in the previous reference.

B. Suter, T. Lakshman, D. Stiliadis, and A. Choudhury, "Buffer Management Schemes for Supporting TCP in Gigabit Routers with Per-Flow Queueing," *IEEE Journal of Selected Areas in Communications* 17, no. 6 (June 1999): 1159–1169.

Tat Chee Wan and Swee Keong Joo, "Random Early Detection with In and Out (RIO) for Asymmetrical Geostationary Satellite Links," *Proceedings of the Joint International Conference IEEE MICC 2001 LiSLO 2001, ISCE 2001,* Kuala Lumpur, Malaysia (October 21–24, 2001).

Xi-Peng Xiao and Lionel Ni, "Internet QoS: A Big Picture," *IEEE Network Magazine* (March–April 1999): 8–18.

Hui Zhang, "Service Disciplines For Guaranteed Performance Service in Packet-Switching Networks," *Proceedings of the IEEE* 83, no. 10 (October 1995). It is available online at *www-2.cs.cmu.edu/People/hzhang/publications.html* in ps.gz or in PDF format.

The following are companies that provide technology documentation, white papers, and application notes involving their own standalone traffic manager chips:

Agere (*www.agere.com*)

AMCC (*www.amcc.com*)

Azanda (*www.azanda.com*)

EZchip (*www.ezchip.com*)

Internet Machines (*www.internetmachines.com*)

Mindspeed (*www.mindspeed.com*)

Motorola (*www.motorola.com*)

Sandburst (*www.sandburst.com*)

Teradiant (*www.teradiant.com*)

Vitesse (*www.vitesse.com*)

Xelerated (*www.xelerated.com*)

ZettaCom (*www.zettacom.com*)

Bay Microsystems offers embedded traffic management right inside their network processor. See *www.baymicrosystems.com*.

An excellent and biannually updated market and technology analysis report on traffic managers is available from the Linley Group. See *www.linleygroup.com/reports.html*.

P · A · R · T · IV

PUTTING EVERYTHING TOGETHER

CHAPTER 16
SYSTEMS ENGINEERING ISSUES

Throughout the book, we have not only discussed the most important components that comprise a network-processing platform, but we have also looked at the specific offerings of most important vendors in the field. We have examined network processors versus *net application-specific integrated circuits* (Net ASICs) versus configurable, multiprocessor-based custom designed ASICs. We have also discussed embedded or separate traffic managers, switch fabrics, search engines, *content-addressable memories* (CAMs), classification processors, and storage network processors. Security coprocessors are discussed in Chapter 17, "Security Coprocessors."

However, some issues cannot be easily categorized into one single area of the field as they span the entire problem space that designers of a *multiservice router* (MSR) confront. In this chapter, we attempt to combine the most important of these issues to wrap up the information provided throughout the book. We need to examine the soundness of the architectural choices made by the most important vendors to get a clear idea of the result of using approach A as opposed to approach B.

However, architecture is not the only issue. Software must also be developed. This often involves cost-related issues, which we will try to elucidate especially for newcomers into this field. Some experienced users may be surprised by our discussion as we will try to debunk some industry-wide myths. When making decisions, a designer must consider both visible and hidden costs.

For the sake of convenience, we will also clarify various memory subsystem technologies that may be encountered in this field. We will finish by taking a brief look at the feasibility and preliminary design analysis of a real-life case of a complex product—an MSR/multiservice switch. We will examine the trade-offs, concerns, options, ramifications, and compromises of the design. This case study also explains how to develop the conception of the architecture and subsequent systems design of such a major project.

MEMORY SUBSYSTEMS

In our review of numerous platforms for network processing, we have encountered a very broad collection of memory technologies that are involved in the overall picture. Different memory technologies are engaged by switching/routing systems designers at different places in a system's data paths and for different purposes. The abbreviated trade names of the various technologies have rightfully been characterized by one trade-journal editor as a "alphabet-soup of memory technologies." The rationale for the wide variety is that the system architect wants to maximize performance while minimizing cost. As different needs require memories with different characteristics, functionality, and price/performance ratios, an evolving palette of memory offerings must be assimilated continuously from multiple competing vendors. They must be mapped properly to the exact system application in order for a wise decision to be made. This decision can significantly affect the ultimate product differentiation.

We will provide a short overview of the most important memory technologies used in network-processing systems. We do not intend to explain the basics of memory chip operations. More information is available in the suggested references at the end of the chapter.

At a macroscopic level, the fundamental choices for data storage (whether it be packets, headers, parameters, statistics, or traffic descriptors) are *dynamic random access memory* (DRAM) and *static RAM* (SRAM), including all their variants. As junior engineering students learn early in school, SRAM is fast and very expensive. DRAM is slower, but more difficult to design due to the periodic need to refresh the retained data. It is also much less expensive to buy. Engineers have traditionally thought about arranging these two families into a hierarchy of memories to maximize performance while optimizing the cost budget. One result of such ingenuous approaches was the *caching* structure of traditional computers.

A small pool of very expensive and very fast SRAM memory operates next to the *central processing unit* (CPU) (often it is embedded right inside the CPU chip). This is where it retains a collection of the most recent and most frequently used blocks of data. Chances are the CPU will need one of these memory blocks for its next operation. Therefore, it is convenient to have it in the cache as opposed to initiating a memory *input/output* (I/O) cycle to go off-chip to the memory bank (usually made with DRAM) to fetch it, which would stall the processor's frantic pace of execution. This is caused by the fact processors are much faster than memories. This principle has also been carried over to multiple sophisticated levels of caching with the intention to maximize the performance while minimizing the cost.

We will not expand on these techniques here because traditional IT-computing environment memory hierarchies based on a cache approach do not work with network processors. Network processors need to search unpredictable and deep trees because both the spatial and the temporal locality of network traffic content data are radically different from traditional computer data. Therefore, designers are forced to use other means to solve storage and buffering problems.

Memory measures of merit for the network-processing arena include the work done per pin, the random cycle time, the capacity, the cost, the power, and the space.

DRAM Flavors

SDRAM stands for *synchronous DRAM*. Originally, DRAMs had been controlled asynchronously. A processor would present an address to the DRAM, and it would activate the row and column strobe signals. After waiting a certain amount of time called the *access time*, the DRAM either would present the corresponding data at its output for the processor to read or would write new data provided from the processor into that location, depending on the operation performed (read or write). However, as processors became faster, memory caused a bottleneck and the processor was forced to wait for the DRAM memory to deliver results. It was subjected to artificially inserted, idle, or wait states. Meanwhile, the processor could do other useful things if it was free.

By making DRAM synchronous (and therefore activated by an external clock), latches were inserted at the DRAM inputs and outputs so the information presented by a processor could be effectively latched on until the DRAM core was able to handle it. In the meantime, the processor was free to do something else. As soon as a read operation was finished, the results were latched on at the DRAM output for the processor to return and read it at its convenience. Of course, other architectural breakthroughs slowly appeared, such as the pipelining of addresses, the prefetching of data, and even the use of multiple modes (such as page mode, burst mode, and so on). The judicious combination of these breakthroughs has led to further improvements of DRAM performance in conjunction with the ever-increasing performance of processors. Ample information on this evolution is provided in the references at the end of this chapter.

Most network processors use DRAM in order to store packets. As packets must be written to and then read from memory, the memory system bandwidth must be twice as fast as the intended wire-speed performance. As a gauge of the cost and complexity involved, we will mention the obvious fact that if a system must be able to buffer (for safety) the equivalent of one second of full-capacity wire-speed traffic, we can determine the amount of memory that should be available for a 2.5 Gbps link (312.5MB), for a 10 Gbps link (1.25GB), and for a 40 Gbps link (5GB) system.

Using *single data rate* (SDR) SDRAM to transfer data at one edge of the clock signal is inexpensive, but very few companies still use this approach. The industry has moved over to *double data rate* (DDR) SDRAM memories. By transferring data on both edges of a clock signal (and therefore effectively at twice the clock rate) and in a manner that is synchronous to a data source, DDR SDRAM memories are more efficient for many applications.

For example, a DDR266 device with a clock frequency of 133 MHz has a peak data transfer rate of 266 Mbps or 2.1 GBps for a subsystem that is configured as a times-64 *dual inline memory module* (DIMM). This feat is accomplished by utilizing a two-times-prefetch architecture where the internal data bus is twice the width of the external data bus and data capture occurs twice per clock cycle. To provide high-speed signal integrity, the DDR SDRAM utilizes a bidirectional data strobe and interface with differential inputs and clocks.

DDR SDRAM has become the uncontested cost-per-bit capacity leader due to the vast PC market that commands unprecedented volumes for these same components. DDR SDRAM is also better in direct pin-count cost for the same I/O load than SDR SDRAM. Therefore, many companies use these memories on their network-processing platform for packet storage.

Although in some applications, such as the OC-48 realm, SDRAM bandwidth is sufficient for basic packet queuing at wire speed, some processors are characterized by small, embedded memory banks and therefore architectural features are needed to give them an upper hand. An example of such a processor is the IBM NP4GS3, which has less memory than AMCC, Agere, or even Motorola network processors for the same segment. However, the difference is that IBM's *network processing unit* (NPU) can access seven different external memories and one internal memory for table storage. IBM's approach excels in cases where many table accesses are required despite its apparent memory-related shortcomings.

In order to use DDR SDRAM and take advantage of the low cost of these memory components in 10 Gbps designs, the memory subsystem must be designed with a width of 128 bits for simplex 10 Gbps (OC-192) network processors and a width of 256 bits for full-duplex OC-192 network-processing systems. This means that the pin count required on the network-processing chips (especially if they are custom designed using cores, as shown in Chapter 10, "Alternative Approaches to Network Processing: Net ASICs and Designing with IP Cores") must be so high that the cost savings from moving to a commodity memory technology are completely wiped out by the spectacular increase in NPU packaging costs. In other words, we must always keep things in perspective.

The alternative is Rambus™-conceived RDRAM. This technology provides a lot of work per pin due to its significantly higher bandwidth. However, it requires its own signaling and coding and is expensive (four to five times more than SDRAM). RDRAM was designed for a cache line of products. Therefore, when it comes down to network processing, it unfortunately suffers from poor random cycle times, even when accessing the same bank of memory.

As the future of Rambus technology remains largely uncertain (since many companies are unhappy about the level of licensing rates), this randomness of bank access creates a real concern for network-processing designers. Some *original equipment manufacturers* (OEMs) are using it in their 2.5Gbps solutions (such as Vitesse with IQ2000 and 2200), but others have not been as convinced. Intel has also seemingly adopted it in their 10Gbps platform, although they have not yet adopted it in their 2.5Gbps solutions. This can be seen as a serious systems engineering shortcoming.

Rambus enables the use of many memory banks, but it is difficult to design a system around it because of high-speed signals. It is neither trivial nor within everyone's reach. Its shortcoming of the bank-access randomness can be overcome if adequate time is spent developing and testing highly specialized system software to randomly store information in different banks every time. This would minimize the impact of the access randomness by spreading it on all Rambus I/O operations. The problem with that approach is that a designer may be able to pick banks to write to cleverly, but he or she has no control over which bank information must be read from.

An interesting technology originally introduced by Toshiba and Fujitsu is *fast cycle DRAM* (FCDRAM) (sometimes also called FCRAM). Samsung calls this technology *network DRAM* (NDRAM). This technology is targeted for high-performance designs. Agere is using it heavily. It is characterized by a 25-nanosecond random cycle time (depending on the generation of the chip) back to back accessing the same bank. This means that FCDRAM is quite faster than DDR SDRAM and offers lower random access times than RDRAM.

FCDRAM can be electrically compatible with DDR SDRAM provided the corresponding memory controller is embedded in the NPU. The two solutions have different bandwidths and prices, but this can possibly lower the cost. Bay Microsystems and IBM are already doing it.

Another contender in the network-processing field is *reduced latency DRAM* (RLDRAM) from Infineon and Micron. This technology is competing with FCDRAM in pin bandwidth and latency.

SRAM Flavors

The basic cost/performance context of SRAM was expressed in the beginning of this section. High-speed contexts require SRAM, but designers do everything possible to minimize the use and the cost of it. However, the expected wire-speed performance of specific platforms may dictate the use of SRAM. For example, the Agere NPU 5 Gbps platform implies that the systems designer can get by without using external SRAM. This is because of the embedded 2.5MB of on-chip SRAM. If the same company's technology is used for a 10Gbps system, external SRAM is definitely required. As a memory technology, SRAM is indispensable, especially in situations where a large number of queues or a large number of linked lists must be traversed because of the multiple accesses needed to assemble working lists. SRAM is also used to monitor performance.

A common player on the high-speed network-processing stage is the *zero bus turnaround* (ZBT) *synchronous SRAM* (SSRAM). This high-speed SSRAM flavor is ideal for several current networking applications. This is because ZBT SSRAM offers superb bus utilization as buses can be used without any bus dead cycles, even when transitioning from read to write. In other words, ZBT SSRAMs can read or write every clock cycle for 100 percent bus efficiency. Clocked up to 166 MHz, ZBT SSRAM is available in chips of 2Mb to 18Mb. Micron, for example, seems to have a product roadmap for up to 72Mb.

DDR SRAM and *quadruple data rate* (QDR) SRAM are also available. The DDR SRAM approach accommodates transfers through different ports, so if an application is balanced in read/write operations, it turns out to be a superb choice. QDR SRAM was developed by Micron, Cypress, and IDT, and is useful in 10Gbps designs. The data inputs and outputs are separate and operate simultaneously in QDR SRAM. Because each data bus operates on two words of data per clock cycle, each bus effectively doubles its data rate. Since both of these buses operate in parallel, the QDR SRAM component operates on four bus widths of data per clock cycle. Therefore, the minimum set on which data can be operated on is two words—two times the component's bus width. QDR is ideal for balanced read/write workloads. Because of its common I/O, DDR is more pin efficient for workloads where read operations dominate write operations.

Some competition to DDR SRAM appeared from a newcomer technology called *Sigma SRAM* in early 2002. This technology was positioned mainly for the 5Gbps realm, but it does not seem to have attracted significant market acceptance yet.

CAM

CAM is a peculiar, but very interesting, memory technology. We discussed CAM and its role extensively in Chapter 12, "Search Engines," and Chapter 13, "Classification Processors." For more detailed information on CAMs, refer to those two chapters. We will summarize some systems-related issues and characteristics of this technology. The brute-force hardware lookups that CAMs provide require incurring both a significant cost and performing a worst-case power consumption analysis. Many CAM vendors do not happily provide their worst-case power needs. In a recent Communications

Systems Design conference in 2002, one of the major CAM vendors privately disclosed that their 9Mb CAM has a worst-case power consumption of 12 watts! Therefore, it is imperative that the correct numbers are unearthed from CAM vendors.

For the systems designer, CAMs are attractive in the following situations:

- When large bit patterns must be recognized using classification based on multiple bit fields because of the use of many pins. As we say in the industry, CAMs "do a lot of work per trip across the I/O pins." Approaches using conventional memory usually need to make many trips as the bit pattern that will be classified gets bigger.

- When table sizes are small, as storing large lookup tables in CAMs is often prohibitive, as measured both in direct chip cost and power consumption needs.

- When lookup latency is critical. However, latency can often be easily hidden with the use of architectural features such as pipelining and threads, especially when threads are combined with memory-based approaches.

An important point must be made regarding the first one of these three premises. When search-lookup-classification has to occur based on an algorithmic process such as a trie approach, then by necessity fewer pins are going to be used for I/O. In addition, the number of cycles involved in the execution of the algorithm significantly change the cost-performance comparison of a CAM versus other memory technology approaches.

Other network processors such as EZchip's, which enable the use of other memory technologies instead of CAM, also offer an important cost incentive to the intended user.

NPU ARCHITECTURE ISSUES

Throughout this book, we have examined the different approaches taken by multiple vendors. We have seen the scalar architectures offered by *complex instruction set computer* (CISC) and *reduced instruction set computer* (RISC) engines and where they fit. We discussed configurable computing, where new instructions can be easily generated that are specific to the application. This can be a successful means of dramatically enhancing the performance of such processors in the network-processing field. Most network processors offer multiple engines that can process packets simultaneously.

Four different computing philosophies exist in this realm:

- One approach is the so-called run-to-completion processors. They essentially hand a packet over to an engine among the multiple engines available in the chip. This engine will work on this specific packet until its tasks are completed. The packet will then be forwarded to the next step in its intended computation realm.

- Pipeline network processors with one or more packet engines per stage is another approach. Here different engines work on different packets and run different code at the same time. The efficiency of the pipeline is difficult to determine as bubbles are often created merely because some packets will require a different amount of processing than others. As they are paraded down the pipeline stages, they propagate these computational bubbles, which is a colloquial equivalent of unused or idle resources temporarily and/or locally (in time and space).

- Large-scale multiprocessing using some sort of computational symmetry is a third approach. This type of *symmetric multiprocessor* (SMP)-based parallel computing offers tremendous flexibility because it allows the allocation of packets and tasks to individual computing resources as well as determines the choice of software that will run on each of them. The downside of the SMP approach is that it becomes extremely difficult to optimize the allocation of the tasks to resources in real time. It also requires an elaborate fine-tuning of the runtime environment, which cannot be done every single time the application changes or the software is upgraded. This defeats the purpose of using network processors as a quicker way to the ever-evolving market for equipment vendors.

• The *simultaneous multithreading* (SMT)[1] approach is an interesting approach that seems to be a natural fit for the fast, heavy-load, and unpredictable network-processing field. It has not created network-processing products yet, although it marginally tried. It has created mainstream computing products as pushed by Intel and therefore deserves closer examination.

Of course, SMT should not be confused with the basic fundamental premise of multithreading that is characteristic of network-processor offerings, which we have discussed throughout the book. Computer architects devised multithreading as a way to hide from an application some undesirable but unavoidable effects, such as memory latency. Several combinations of these basic approaches can be made. For example, multiple parallel engines can be placed on each pipeline stage.

So although the architecture itself is very important and some vendors go out of their way to tout the advantages their approach bestows to the contemplated applications, take a step back and look at the overall picture before becoming emotionally involved with a specific choice.

SOFTWARE DEVELOPMENT ISSUES

Network-processing vendors approach the issue of software development from different angles. Aside from the Net ASIC solutions, which essentially provide a fixed-functionality nonprogrammable solution, the rest of the platforms require some sort of programming. This programming can be straightforward, tedious, efficient, or painful for those who have to actually do the coding. Further compounding the problem, the programming model offered by a specific network-processor platform may facilitate or actually inhibit the efficient programming that has to occur. Some architectures are pipelined, multithreaded, or even sequential. One software-engineering solution cannot be expected to fit all possible requirements. In fact, the primary concern of mapping wire-speed-performance software functionality onto the available computing resources (the number of computing engines and of processing cycles) exists on all architectures. The issue then becomes how efficiently each of these varied architectures handles the tasks at hand.

Let us look at the programming aspects of the problem. On one side of the spectrum, some vendors have been essentially preoccupied by the hardware prowess of their design and may have neglected and/or underestimated the need for superb software development tools to create and deliver wire-speed, feature-rich software. Text-based or *graphical user interface* (GUI)-based software development environments are available from different vendors trying to differentiate their offering. Other vendors propose a high-level language compiler, usually C, which is sometimes an optimizing compiler but sometimes it is not. Other vendors expect the users to write massive code in RISC assembly or, worse, in obscure NPU assembly. Neither case is ideal for full-fledged applications development. Some vendors also use proprietary languages with a certain value proposition.

Many vendors who have been preoccupied with offering a superb hardware solution have even resorted to offering free software implementation of protocols such as *Internet Protocol version 4 and 6* (IPv4/IPv6) and *Multiprotocol Label Switching* (MPLS) to their customers to entice them with one-stop shopping and a shortened time to market. In reality, this is usually the case from vendors whose software environments are not up to par in functionality with their competition or whose programming language compilers are not as efficient as those of others. As a result, vendors use this method

---

1. SMT was heavily researched by Dr. Mario Nemirovsky at the University of California Santa Barbara. It was then used in network processors he developed with Clearwater Networks. He is currently pursuing a similar approach with a new venture called Kayamba Inc. (*www.kayamba.com*). From a computation standpoint, SMT technology has been dubbed *hyperthreading* by Intel and recently introduced into products exhibiting improved performance throughput. An interesting article on this topic is "Hyperthreading Technology Architecture and Microarchitecture" by Deborah T. Marr et al. in the *Intel Technology Journal* Q1 (2002). This and other related papers can be also found online at *www.intel.com/technology/hyperthread/index.htm?iid5sr1hyper&*. Another interesting article is "Intel's Hyperthreading Takes Off" by Kevin Krewell in the *Microprocessor Report* (December 2, 2002). It is also available online to paid subscribers at *www.mdronline.com/mpr/h/2002/1202/164801.html*. SMT is researched at many schools with a notable example Prof. Susan Eggers' and Prof. Hank Levy's group at the University of Washington. See *www.cs.washington.edu/research/smt/index.html*.

to try to distract the customer's attention away from the real issues. Think about it. Why would someone offer sophisticated network engineering software for free? Be careful when free code is offered, regardless of which brand name is offering it.

Of course, typical customers look at this as an advantage if bug-free, tested, and validated software is obtained for free from a highly reputable NPU vendor. However, it is only helpful in the short term. Even if this solves some of the problems that the customer's product Release-One team confronts, the larger issue for the customer is how the Release-Two team will eventually deal with subsequent problems associated with modifying code that the customer has not written in the first place. Surprisingly, many companies do not think about this. This "hot potato" attitude is typical in some younger organizations where product Release-One teams just want to get rid of the current problem. This happens most often when teams are working under tremendous pressure to perform miracles within a short amount of time and with a limited budget. These teams ultimately do not exhibit much regard or sensitivity toward future problems down the line. However, this does not necessarily stem from ill intentions. It is usually a quasi-invisible by-product of the context we just discussed. Designers will pay a price for such a short-term advantage.

Let us look now at the problem from another angle—namely, how software can be mapped onto the underlying network-processing architecture. We have seen that there are sequential and parallel architectures. Parallel architectures exhibit some degrees of pipelining and multithreading. In order to develop software that performs at wire speed, a designer must ensure that very few computing cycles are wasted. Some software pertains to deciding what operations need to be performed on traffic packets, whereas other software deals with actually performing specific operations on the traffic packets.

If the programmer is required to have an intimate knowledge of the underlying NPU hardware architecture in order to set the functional module allocation and the performance optimization of the engines, then both of the following statements are true:

- This is very tedious work that few people can properly perform and that cannot be done on some platforms.
- This task makes the developed software completely dependent on the underlying architecture.

Perfectly and deterministically mapping the actual software on the available computing resources is not easy. In fact, it is only on rare occasions that it is possible. Therefore, it is not a discipline that we can expect to apply across the board. Most network processors are based on a single-image programming model, which completely shields the programmer from hardware intricacies and details.

This convenience results in immediate inefficiency as instruction cycles will be wasted at some point sooner or later. Typically, this implies that all processing engines within network processors execute the same code on various packets. Some pipelined designs offer several packet engines at each stage. A designer can have the convenience of different pieces of code running on different packets at the same time. This might bring up the subject of packet allocation to various computing resources. It is generally not desirable to extend this granularity of program allocation to the average software engineer for many reasons. Some applications might negate efficiency choices imposed by other programs. At the same time, some code must change as applications and protocols are upgraded. Reallocating the executable software modules usually becomes a continuous nightmare and a moving target.

The problem is even further compounded when the packet engines in network processors are expected to be multithreaded. The packet engines handle a packet, and whenever the execution thread stalls because something unforeseen occurs, such as a memory lookup or a longer calculation, the packet engine swaps the entire context and starts working on another packet that might have been kept aside temporarily. Meanwhile, the other packet is kept on hold pending the successful completion of the previous operation that caused the stall. A lack of adequate multithreading capabilities has been a major limitation of classical CISC/RISC processors in the network-processing field. SMT is an intriguing and highly promising field of computer architecture in this context. As we discussed previously, it has been the subject of intense research and development as well as passionate debate.

If the customer's organization must change software to reflect new functionality, techniques/ algorithms, and protocols (which is usually just a matter of time), the software must be massively rewritten. This is a nontrivial task. If the organization has to develop new product platforms on a new NPU (as will be the case if the NPU vendor goes out of business), the prospect of having to rewrite software is daunting. This shows why large *network equipment vendors* (NEVs) have on many occasions been reluctant to adopt NPU chips from startups that might not be around in a couple of years. This makes it nearly impossible to maintain and upgrade the equipment. The intertwined nature of so many seemingly unrelated problems could trigger a domino-like effect, if software development, which is their common denominator, is not addressed properly.

Yet another angle of looking at the problem is the choice of programming language. Writing low-level RISC code for each RISC or *very long instruction word* (VLIW) pico/microengine is a theoretical option, but it has little if any practical value. The task is daunting and the learning curve is very steep since few software engineers are competent enough to perform the task in a reasonable amount of time and with manageable quantities of bugs. As a result, the software will be difficult to maintain and upgrade. This is a huge price to pay just for the pleasure of efficiently utilizing the processing bandwidth of the NPU. On the other hand, high-level languages such as C/C++, which are widely known and used by engineers in the industry, also offer a series of undesirable trade-offs.

Traditional CPU devices do not have the specialized network-related machine instructions to perform the tasks that are required of them. Enhancing the instruction set architecture to tackle wire-speed requirements immediately hampers the chances that an optimizing compiler will be able to use these extensions transparently without the user noticing. It seems that what is gained in programming ease by abstracting away from the hardware is lost in performance-tuning capabilities of the application software.

The following example illustrates the important ramifications of these issues. Before this chapter went to press, IBM announced that its future generations of network processors will be based on engines that take advantage of the PowerPC instruction set, because the proprietary instruction set approach taken by the microengines in the NPUs such as the NP4GS3 seems to have run into a limited time span compared to the widely accepted PowerPC platform for which so many tools are available.

The problem becomes even more interesting when we consider the use of functional or descriptive languages in order to perform coding (such as Agere's *Functional Programming Language* [FPL] and Intel's *Network Classification Language* [NCL]) as opposed to procedural languages such as C/C++. However, we will not expand on this issue.[2]

SOFTWARE DEVELOPMENT COST

An interesting case (although not a very useful application from an everyday-life standpoint) is a benchmark test that Purdue University Professor Douglas E. Comer has documented in the web site that accompanies his network-processing book.[3] He calls the application "bump-in-the-wire." It is essentially a program that looks at incoming packet traffic, detects packets addressed to port 80 by properly parsing and classifying them, and then simply counts the number of packets that are addressed to port 80.

On his book's web site, Prof. Comer provides a mix of classical C code and assembly for the Intel IXP1200 NPU platform. The web site also supplies other submissions that describe Agere's implementation of the same application using FPL code. The numbers are quite telling. The Intel-based C-and-assembly language implementation requires 1,491 noncommented (counted using the NCSL tool) lines of source, whereas the Agere FPL approach was implemented in 46 lines of code. Counting lines of source code can often be an inaccurate task, so make sure oranges are compared with oranges. The same software engineering tool (NCSL) must be used. It does not take a rocket scientist to see

2. For more information, refer to the work covering this aspect in the white paper from Agere Systems "Network Processors Programming Models and Languages" written by Robert Muñoz in 2002.

3. The book we are referring to is *Network Systems Design Using Network Processors* by Douglas E. Comer (Upper Saddle River, New Jersey: Prentice-Hall, 2002). The author's web site with the comparative application source code can be found at *www.npbook.cs.purdue.edu/code.html*.

that a flagrant ratio of 1:30 in lines of code needed for an application written by two different methods creates significant and direct development cost differentials between the two methods.

A major problem associated with the software development process is the estimation of how much it costs to develop a solution with given resources within a certain amount of time and maintain it during its useful lifetime. Budgets, manpower, and other resource planning are essential by-products of such knowledge. This topic is extremely deep and we barely touch the surface here. Quite a few specialized references can be consulted for more information.[4]

Some very insightful ideas have been produced by proponents of functional (descriptive) languages (such as Agere's FPL) as opposed to those basing their approach on procedural languages (such as C and microengine assembly) and the quantified results are staggering. In fact, this turns out to be the case to such an extent that even if some of the publicly disclosed numbers (which can be easily replicated if the publicly available models and assumptions are similarly used by anyone else) are off by a wide percentage, the approach still deserves a meticulous consideration as it enables significant savings.[5]

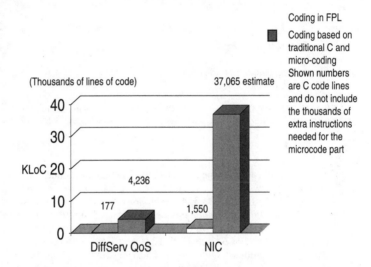

FIGURE 16.1   Based on an actual implementation of DiffServ-based *quality of service* (QoS) by programming in a functional language such as Agere's FPL as opposed to traditional procedural programming based on C, a scaling factor is calculated between the two approaches. It allows one to extrapolate to a first order of approximation the expected difference in program size when confronted with a new application, such as the development of software for a *network interface card* (NIC) of a third-generation *radio network controller* (RNC). The numbers still speak for themselves even if it is assumed they may be off from reality by a wide margin.   *(Source: Agere Systems)*

---

4.   Two classic textbooks on the subject are *Software Engineering Economics* by Barry W. Boehm, (Upper Saddle River, New Jersey: Prentice-Hall, 1982) and *Software Cost Estimation with COCOMO II* by Barry W. Boehm et. al. (Upper Saddle River, New Jersey: Prentice-Hall, 2000). Software engineers for 20+ years have been using the COCOMO model to make financial decisions, set project budgets and schedules, negotiate tradeoffs, plan to maintain or upgrade legacy products, and decide where to implement process improvement. The model accepts estimates of either logical lines of code or function points as the primary input parameter. One can also find in this source an implementation of the COCOMO model as well as coverage of emerging extensions such as object point data, application point data, the *phase schedule and effort model* (COPSEMO), dynamic COCOMO, the *RAD schedule estimation model* (CORADMO), the *commercial-off-the-shelf integration model* (COCOTS), the *quality estimation model* (COQUALMO), and the *productivity estimation model* (COPROMO).

5.   Modeling results and methodology were discussed in depth between the originators of this work and the author over a series of documented private communications in November 2002.

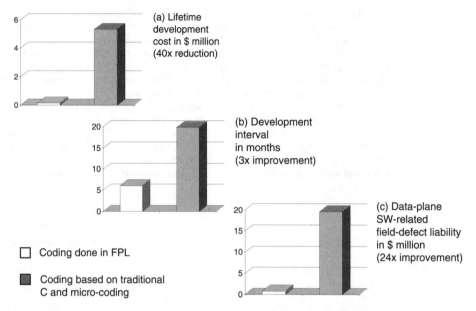

FIGURE 16.2 IP/MPLS router-based NPU software development cost comparison when coding in descriptive languages such as Agere's FPL, ASL, and classical procedural coding in C *(Source: Agere Systems)*

Figure 16.2 illustrates some of the results. The figure examines a case of an IP/MPLS router design that implements (in software) an impressive series of protocols. More specifically, it contains Ethernet *Digital Intel Xerox* (DIX), 802.3, *Asynchronous Transfer Mode* (ATM) signaling to host, ATM *operations, administration, and maintenance* (OAM) handling, IPv4 over *ATM Adaptation Layer Level 5* (AAL5), IPv6 over AAL5, *Point-to-Point Protocol* (PPP) over *Packet over SONET* (POS), *virtual local area network* (VLAN), MPLS, IPv4 and IPv6 routing, *Internet Control Message Protocol* (ICMP) to host, multicast, *Differentiated Services* (DiffServ), packet policing, and packet shaping.

As shown in Figure 16.2, the functional/descriptive coding approach has an intriguingly positive impact on the development cost as opposed to the classic approach taken based on coding in languages like C. These specific comparative results are based on the multiplicative scale factor inferred and calculated from the work shown in Figure 16.1.

It is therefore important for companies and organizations that contemplate developing software on network-processing platforms to realize that this is not business as usual, that it involves many hidden aspects, and that the choice of a platform and the choice of architecture it entails also implies several advantages or disadvantages regarding the timely delivery of functional, wire-speed-performance software that can be reused, upgraded, and maintained in the future. This cannot be stressed enough.

A REAL-LIFE CASE STUDY: DESIGN ISSUES WITH AN MSR

We will conclude this part of the book by looking at a complex real-life case where numerous choices are available. The systems architect who designs a cutting-edge multiservice switch/router is confronted with a dazzling array of different component technologies and possible configurations. We attempt to provide a sample of the challenges involved and the trade-offs confronted. This process is extremely complex so it is impossible to cover all the issues in a few pages, but we will try to give a clear idea of what it takes to come up with such a product.

Our analysis will serve as a tentative first iteration of a potential design. In a real-life environment, many teams of people are involved in researching, documenting, discussing, and debating issues before decisions are made. Here we will discuss a design process that is slightly less elaborate than one that an entire engineering department would undertake. That process would be much more involved than the traditional back-of-the-napkin designs for which some Silicon Valley or Greater Boston-based networking companies have become famous.

Task Definition

Assume that our designer's task is to build a complete, scalable, heavy-duty, edge network multi-service switch/router that can handle multiple 1 Gigabit Ethernet connections downstream connecting four IP LANs with a 10 Gigabit Ethernet uplink connection to some centrally located server. The designer must also have the capability to switch and route traffic to and from these two realms and to and from an OC-48c realm, which can be ATM and/or POS.

The desired system must be carrier-grade reliable, which implies the judicious redundancy of cards, and scalable in a typical 19-inch rack environment. This means the designer will need to consider choices that allow the original shelf to be upgraded eventually to a multishelf configuration without losing track of power consumption or the number of available slots in a chassis. The system will need to be upgradeable to potentially handle frame-relay and *time-division multiplexing* (TDM) traffic if required in the future. Multiple protocols will need to be supported and executed inside the MSR. We will also examine the most important task of eventually allocating these protocols on different parts of the platform.

Design Approach

Assuming the visible and hidden cost issues that we discussed earlier in the chapter will eventually be uncovered, our designer first looks at two broad categories of component choices: switch fabric scalability and network-processing platform. We anticipate that this context with four distinct 1 GbE LANs and a 10 GbE server connection in conjunction with double OC-48c traffic (ATM or POS) made redundant starts from 40 Gbps switching bandwidth. It can eventually scale to a massive 640 Gbps bandwidth with more 10 GbE server uplinks.

This fundamental requirement will steer the feasibility study team to choose an appropriate switch fabric chipset. Let's say that our designer is intrigued by Agere's PI40 switch fabric's scalability and implementation robustness. While considering the switch fabric from this vendor as well as from others, our designer also starts considering other systems-related issues. Let us assume that the team cannot afford to design ASICs, due to a lack of budget, time, and/or skills. Off-the-shelf components need to be chosen as much as possible.

The following items attract the designer's attention:

- A powerful network-processing platform from the same vendor that offers the switch fabric (always a good idea) for both the 1 GbE and the 10 GbE environments.
- The vendor's ability to offer chips that can handle TDM and ATM/POS traffic.
- The convenience of integrated *serialization/deserialization* (serdes) for the switch interfaces.
- *Physical* (PHY) interface/*Media Access Control* (MAC) capabilities in-house with the main vendor, which can provide one-stop shopping from an established vendor. All of these are very important for our hard-pressed designer.

The designer also notices that Agere's NPUs are based on a computational model and architecture that allow the optimal allocation of tasks to threads running on multiple pipelines. Unfortunately, the designer cannot count on getting application software for free like other vendors would try to propose to entice our designer to their platform. However, the solidity of the development environment and tools, especially the software engineering environment/language/tools and context whose metrics show undisputable economies of development, end up tilting the balance. Our designer tells his or her colleagues that code won't be written in plain vanilla C or C++ or even RISC assembly and that a short training will be needed in FPL and other advanced development tooling to make things happen. This seems like a reasonable investment with a significant and quantifiable return.

Preliminary Design Outlook

Requirements will need to be created for the various cards. If cards can be bought already with these specs, all the better. That saves time and money. Otherwise, the corresponding card design proposed by the vendor will have to be implemented by our designer's organization in new and perhaps custom-size cards that fit the available slots.

Our system designer first considers that the 19-inch rack will allow 12 slots. He or she can plug one full-size card or two half-size cards in each slot. The midplane and the backplane issues must also be considered as we advance in the conceptualization of the design. For proof-of-concept and evaluation steps, a designer would ask the vendor (in this case, Agere) if they have something to propose for this situation. The short answer is yes, because the company offers a full-fledged integrated development environment called Festino, which could be of help. In reality, however, this is not an optimized environment, and customers will want to ultimately design their own system that offers configurations that are not possible or available with Festino. We will not deviate from our subject though.

Our designer first considers a shelf with 12 slots. The designer starts doing a back-of-the-envelope calculation of what is needed.

- Slot 1 will be dedicated to the system host CPU card. Two half-size cards are needed for carrier-quality redundancy.
- Slot 2 will be the initial switch fabric card. Our designer first thinks of the 40 Gbps environment and will consider the 640 Gbps (or Tbps) realm later in a variation of the design. The initial concept will be based on the two PI-40SAX cards, which are half-size cards and can therefore be inserted into the same slot. Again two cards of each are needed for redundancy.
- Slots 3 and 4 will be occupied by two (for redundancy) full-size 10 GbE cards for the uplink connection to the server.
- Slots 5 and 6 will be dedicated to two double OC-48c cards for the ATM/POS connections.
- Slots 7 and 8 will each be devoted to two half-size cards. Each card will offer four 1 GbE connections for a total of eight 1 GbE ports per slot and for redundancy per shelf 16×1 GbE connections.
- This leaves slots 9 to 12 free for the moment.
- If necessary, the designer should expect to expand onto a new shelf.

We can consider using a frame-relay card or a TDM card as our system grows. Enough room is available to position a multiple-card switch fabric solution that expands the available bandwidth within one or even more than one shelf. The chassis will end up being quite overloaded so power per square foot will be an issue. Our designer thinks about two system-cooling choices: an upper fan tray on top that sucks the heat away from the space between the cards or a lower fan tray that blows the air away from the boards. He or she sighs with temporary relief. Not a hurdle yet.

The designer's attention turns now to the cards. First, the 4×1.25 GbE card must be able to provide connections occurring over copper or fiber optics. Both this card and the TADM line card (more about it here below) will interface with the OC-48 card using either the *Gigabit Media Independent Interface* (GMII) or the *System Packet Interface 3* (SPI-3) interface, which are carried on the same set of pins.

The 10 GbE card represents a challenge for our designer, but not an insurmountable problem. This is because in addition to the 10 Gigabit Ethernet NPU, Agere also provides their own in-house *Synchronous Optical Network* (SONET) framer. However, they do not provide the data engine, so a hybrid solution must be found.

Our designer does not need to tackle frame relay now, but he or she can think about solutions offered by an alternative source because Agere currently does not have high-density *High-level Data Link Control* (HDLC) framer and therefore cannot propose such a line card. This vendor could be someone like PMC-Sierra.

Cards then need to be found or designed. The first point to check is obviously the fabric and NPU vendor. Agere proposes multiple cards that could be used for this purpose but not all of them. The issues can be summarized as follows.

The designer notices that the network-processor aspects of the OC-48c card and the 4×1.2 GbE card are essentially identical; therefore, only the PHY/framer aspects of these cards will be different.

The designer notices that the OC-48 cards and the 4×1.25GbE cards are essentially identical as they are implemented around Agere's APP5xx chip; therefore, only the framer/PHY cards corresponding to these different interfaces will be different. The key component of the OC-48 ATM/POS framer/PHY card is Agere's TADM. Several versions are available of the TADM (the current one is the TADM042G5) and its sister-part the TDAT, which can handle the transmission convergence and SONET/*Synchronous Digital Hierarchy* (SDH) terminal/*add/drop multiplexing* (ADM) functionality in ring, linear, and mesh networks. The TADM can handle 155/622/2488 Mbps traffic and readily supports frame relay, POS, and ATM networks. It offers *low-voltage differential signaling* (LVDS) interfaces to the backplane for SONET/SDH ADM and crossconnect functions. It also supports either a synchronous or asynchronous generic microprocessor interface for control purposes. The TDAT is almost identical, but lacks the crossconnect and ADM functions of the full-featured TADM.

The TADM among other things provides for the encapsulation and de-encapsulation of packet and ATM streams into and out of SONET/SDH payloads. It interfaces with the NPU-based line card using standard *Universal Test and Operations PHY Interface for ATM* (UTOPIA) or enhanced UTOPIA interfaces (the latter is dubbed PLATO by Agere). As shown in Figure 16.4, the TADM interfaces with the fiber optics either directly to a transceiver, in the case of OC3/12 , or to a transponder, in the case of OC-48. For OC-12, the TADM can talk directly to four separate transceivers, interfacing with them with one differential pair for each optics at 622 MHz. For OC-3, the TADM can also talk directly to four separate transceivers, interfacing to them with one differential pair for each optic at 155 MHz. For OC-48, the TADM talks to the transponder with 16 bits operating at 155 MHz. The transponder then performs the serializer/deserializer functionality to interface to the 2.488 GHz optics.

Incidentally, our switch/router designer has also noticed that MAC controllers are included in the Agere NPUs, which ultimately makes the board design work much easier.

The approach just described based on the common use of the TADM chip is interesting as the communication will have to be handled properly at the midplane. The midplane will end up probably looking like a star and the backplane will need to contain serdes-based links that might run in the center height of the cards. At the same time, it seems obvious to our designer that the main physical challenge with the system design around the switch fabric card(s) is not going to be something like the power consumption per square foot that must be addressed for the line cards. Instead, the problem will be physically managing the access to and from the switch fabric for so many I/O lines.

More specifically, in terms of requirements of the 4×1.25 GbE cards,

- The 4×1.25 Gbps PHY I/O card must be able to interface to the APP550-equipped line card through connectors that attach to the midplane of the system platform. The I/O card probably should have a 6U form factor (233.35 mm×160 mm).

- A rough block diagram for the contemplated 4×1.25 Gbps PHY I/O card is provided in Figure 16.3. Each block is roughly described as follows:

  - It must be able to provide four *physical medium dependent* (PMD) interface connectors (RJ-45) for connecting *category 5* (CAT5) twisted-pair connections to a Gigabit Ethernet transceiver through the appropriate magnetics.

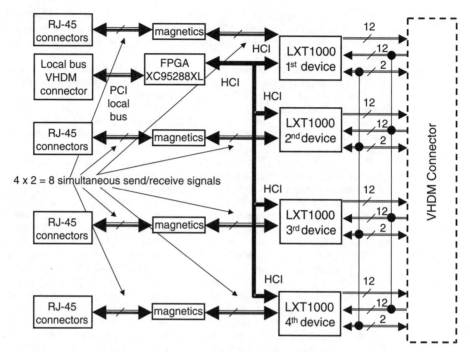

FIGURE 16.3 The PHY structure of a possible 4×1.25 GbE card based on the Intel LXT1000 *(Source: Agere Systems)*

- It must provide four Gigabit Ethernet PHY layer devices with a GMII/GPCS, which corresponds to layer 1 of the *Open System Interface* (OSI) model. The PHYs connect the media (CAT5 twisted-pair) to the MAC layer, which corresponds to OSI layer 2.

- It must allow for the appropriate connection of the GMII/GPCS and management/mode select interface signals to the backplane connector.

- It must provide programmable logic based on *field-programmable gate array* (FPGA) or *custom-programmable logic device* (CPLD) for the interfacing of the PHY *Hardware Control Interface* (HCI) to the local bus interface of the system platform.

- It must provide adequate logic analyzer connectivity as required for engineering debugging and/or other system test verification purposes.

- It must accept power from the port card through the backplane connector.

- It must provide sufficient software and drivers as required for configuration and control of the PHY I/O card.

- The PMD mentioned here can be four standard RJ-45 CAT5 twisted-pair connectors that interface to four of the Intel LXT1000's network interface through the appropriate magnetics.

- To support Gigabit Ethernet over copper twisted-pair connections, a very common choice is the Intel LXT1000 Gigabit Ethernet Transceiver. It will be used in our case as the PHY interconnect device. It supports Gigabit Ethernet over such a medium and supplies all of the PHY layer functions needed to interface to a Gigabit Ethernet controller. Four LXT1000 devices are required to interface to the four GMII interfaces of the Agere NPU. The physical connection between the LXT1000 and the GMII interfaces will be made through an industry-standard very-high-density VHDM backplane connector.

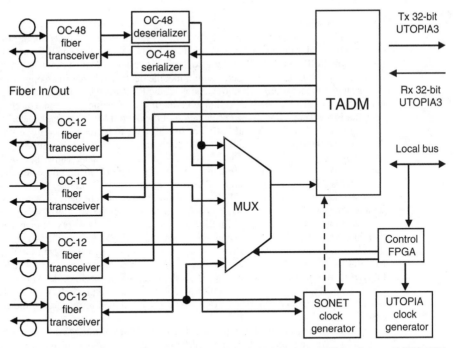

FIGURE 16.4   The principle of multiple-source use of the TADM chip in the PHY interfacing with POS, ATM, or frame relay, and OC-48 or slower links   *(Source: Agere Systems)*

- The LXT1000 uses a single common network interface to support 1000BASE-T, 100BASE-TX, and 10BASE-T. This physical interface consists of four signal pairs that are used for 1,000 Mbps transmission. Each signal pair consists of two bidirectional signals that transmit and receive at the same time.
- The LXT1000 also provides a *Management Data Input/Output* (MDIO) interface and an HCI. The MDIO enables upper-layer devices to monitor and control the state of the LXT1000. The HCI will be used to set configuration options and operational settings of the device. Both of these interfaces will connect to the local bus FPGA interface through the VHDM backplane connector for local control.
- The I/O card should provide a programmable logic device that will be used to interface the LXT1000's HCI interface to the *Peripheral Computer Interconnect* (PCI) local bus through the VHDM connector. The programmable logic should contain registers that can be accessed for reading or writing from over the PCI local bus. These registers will be used to control the signals associated with the HCI.
- The appropriate devices must be foreseen in the final design to provide support for the required clock frequencies, and the appropriate filtering and decoupling components must be used as required.

This seems like a reasonable list of requirements regarding this specific board for the engineering team, which must ultimately do the same thing for all components in the design and then put together full-fledged specifications on their intended design before starting the design of the board. We don't exhaustively design a system here. We just want to give a sample of the effort and diligence needed to come up with a system solution.

Switch Fabric

The system designer next starts conceptualizing some deeper insights about how the project will evolve. The vendor proposes the switch fabric PI-40SAX card that can be used in the lower configuration of the MSR, but as the designer also has in his or her mind the follow-on generation of the product, Agere's PI40C and PI40X cards sound like the desired solution in an upgraded system.

As shown in the Agere technology chapters, the PI40X is the crossbar and PI40C is the scheduler and queues. Agere has them as full-size cards in their Festino development environment, but our designer is preoccupied with the optimization of the overall MSR design. Hence, a design decision is tacitly made that the PI40X and PI40C cards will have to be redesigned to become half-size cards. Agere offers the design guidelines, but the designer's organization is responsible for producing its own (half-size now) PI40X and PI40C cards. (We are using the same names here but we will not repeat them from now on in this discussion; these names correspond to the in-house designed half-size switch fabric cards.) These must be combined as two of each (for redundancy) in our chassis. The designer must also keep in mind that the PI40X chips are unidirectional; hence, they must be doubled for a duplex solution. Also, one PI40C can be combined with up to four PI40X fabric chips.

Assuming that the redesigned PI40X cards are half-size and that two of them can occupy one slot, then in a combination of $1+5$ (meaning one scheduler PI40C and four crossbars PI40X), three slots must be filled—two that will individually contain two of the new half-size PI40X cards (for a total of four PI40X) and one that contains a half-size PI40C card. This means that for redundancy and reliability, this configuration must be doubled. Assume that six slots will be devoted to switch fabric cards. If the physical arrangement of the serdes links can be handled properly, this can possibly be brought down to five if the two redundant PI40C cards are combined in one slot of the chassis as half-size cards.

If the switching bandwidth requirements for the future of our MSR dictate a fifth PI40X, then a new PI40C will automatically be needed. However, the elegance of the approach starts appearing when our designer visualizes a second shelf in the same rack, where 12 new slots are available. A new PI40C or a combination of a PI40C and several PI40Xs can be put together in the new shelf and produce a multiple-shelf MSR that expands reliably and spectacularly.

It should also be clear that the switch fabric cards do not exhibit a density of integration that is anywhere close to the one encountered on line cards. As a result, power density is not as big a concern as it is with line cards.

This corroborates the designer's initial hunch that the switch fabric cards could and should be redesigned in half-size to save slots. However, the limiting factor for the tight integration will be the physical management of the numerous links interfacing these multiple fabric cards with the backplane.

System Considerations

The designer now starts thinking about the overall system functionality. First, he or she must consider what happens on the line cards.

Line Cards Line cards will be based on Agere's network processors. The APP550 will be used for each line card, as shown in Figure 16.5, handling the 4×1 GbE or OC-48c ATM/POS links, whereas the APP750NP/TM chipset will be used for the 10 GbE line cards. Depending on the applications that are expected to run, a further level of granular thinking is required. For example, the APP5XX family offers different NPU components optimizing the cost solution. The APP550 is a good choice for a full-fledged and varied MSR platform. The APP550TM offers the possibility of embedded *segmentation and reassembly* (SAR)ing and traffic management. The APP540 does not offer the possibility of extensive ATM reassembly functionality. Consequently, when the intended solution is purely IP-packet driven, APP540 may be the right choice. Another related component, the APP530, offers the possibility of cutting the performance speed in half, which may be the right choice for some

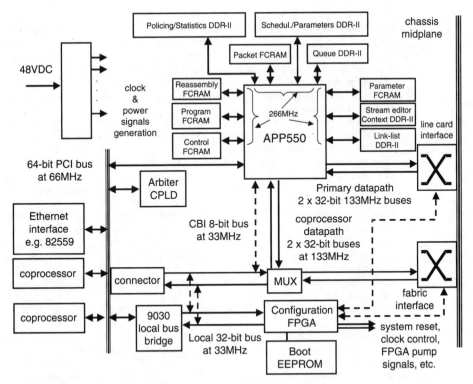

FIGURE 16.5 Typical 2.5 Gbps line card based on the APP550. The 10 GbE card of our case study would be structured along similar guidelines, but with the APP750 network processor and traffic manager chipset instead. *(Source: Agere Systems)*

traffic environments. Yet another member of this network-processing family, the APP520, supports only two as opposed to four Gigabit Ethernet interfaces.

Traffic management will be handled by the native Agere traffic managers for each wire-speed regime.

Some applications only require traffic management at the egress path. We will assume our system is complicated enough to require traffic management at both the ingress and the egress paths. Our designer therefore expects the 750TM on the 10 GbE cards on the ingress and egress paths. The same principle is also expected with the 550 on the 4×1 GbE and OC-48c cards.

The line cards will be computationally split into three planes of computing realms: the data plane, the control plane, and the management plane. Data plane tasks will be the responsibility of the network processor. Signaling protocols and some parts of routing protocols such as *Open Shortest Path First* (OSPF) and *Border Gateway Protocol* (BGP), or the signaling part of *Private Network-to-Network Interface* (PNNI) (central routing will be reserved for the system host CPU card) will be offloaded onto the line cards. They will all run on the control plane host CPU of each line card. The management plane work will combine configuration management of the various chips on each line card and exception processing. For our design, the control and management plane may be combined on the same host CPU running on each line card.

The designer goes back and forth in this iterative process, mentally scanning issues locally (per card) and globally (per system). Forward movement only occurs when consistency is ensured at both levels of granularity in his or her thought process. An issue that has now popped up in his or her mind is the granularity of the partitioning of each protocol among the host CPUs that are on the individual line cards and the central system host CPU. A few years ago a pioneering approach taken by some vendors involved distributing ATM protocols by putting the ATM signaling on the line cards.[6] This was a result of the poor performance during ATM call setup time. The same principle now can be used in routing protocol partitioning. The designer starts allocating protocols from an inventory of desired capabilities. Label distribution protocols such as MPLS or *Resource Reservation Protocol* (RSVP) will be allowed to run on the line card.

IPv4 and IPv6 will obviously be running on the line cards so the NPUs can handle classification and forwarding at wire speed. The routing tables will also be loaded into local memory residing on each line card; however, the routing table will be calculated centrally on the host CPU card. This is done because the centralized execution of the Dijkstra[7] algorithm, for example, or the policy aspects of protocols such as BGP, especially for a large network, will need to run on the host CPU card (here at slot 1). From this card, an updated copy will be periodically downloaded to the line cards for local use by the NPUs. These thoughts are crossing the designer's mind for a reason; they will soon guide him or her toward deciding the computational horsepower that must be made available on each and every card. The designer will then need to budget instructions per second and memory requirements to avoid any surprises. For the moment, however, he or she is just working through the problem.

A major concern is how to handle exceptions. For example, if a bizarre application such as massive and pervasive virus scanning is imposed onto the designer, the designer has two choices. Through its classification capabilities, the NPU will detect something unusual on the packet it processes. It will either flag that to the line card's control CPU for processing, which will realize that it has to be forwarded to the system CPU for the virus scanning, or (preferably) it will directly forward such a packet to the host CPU where the virus-scanning code is executed. The downside of this approach is that the designer does not know how often the system CPU will be interrupted by such requests.

If it is not virus scanning and say for instance that a denial of service attack is detected instead, the designer cannot afford to bring down the entire MSR with such a potential outburst of internal traffic. The application will therefore tell the designer whether it is acceptable to interrupt the system CPU often or whether it is preferable to provide the design with a dedicated coprocessor card (in another slot) where this type of processing can occur without penalizing the host CPU.

An example of such requirements could be some sophisticated firewall functionality. Typical firewall functionality can be described as a series of address and port-based filters that are expected to run on the NPU at each line card. *Access control lists* (ACLs) can be easily maintained by the NPU. However, more sophisticated firewalls adopt the stateful inspection method pioneered by CheckPoint Software. In such an environment, lots of session-related information must be retained in special (and numerous) data structures. Remember that Agere's approach is not optimized for the proper termination (including the setup and tear down) of TCP sessions. If high-speed stateful inspection functionality is desired, then not that many choices are available. Parts of the stateful-inspection-based application code will therefore have to run on a beefed-up system host CPU card or on a firewall-coprocessor card in a separate slot.

Another point of interest is *Network Address Translation* (NAT), which should run on the line cards at least partially. As soon as the start of an NAT session is detected by the NPU, it is flagged to the host CPU on the line card, which steps in and sets it up properly. An alternative is to have the translation table locally stored. The NPU would not only perform the classification, but it would also handle the address-bit-fields swapping and the forwarding.

It is also interesting to note that the network processor can handle all Ethernet address learning and aging on the wire-speed path, if necessary, as a design option. This can also be done on a control plane CPU. The former can be accomplished by stealing cycles from the engine that does the policing.

Similar considerations will have to be entertained mentally if other applications are required such as address learning based on multiple *virtual private networks* (VPNs) spanning more than one port.

6.   Cascade had originally done this on their ATM switches. Then Ascend acquired Cascade, but was eventually acquired by Lucent.

7.   The algorithm is described in depth in the book "Interconnections, 2nd Edition," by Radia Perlman, Addison-Wesley 2000.

Depending on the number and size of the VPNs as well as on the security policies in place, different approaches can be taken. It could be argued that the entire chassis is the customer's so it is safe to leave everything on the software to do and wherever it could run. Others may say that in a specific customer context (a sensitive government agency), the MSR software cannot be allowed to propagate addresses of new VPN members to members of other VPNs that share physical ports. On the other hand, even if this is not the case, the address learning is a minor computational chore that can be seen as a minute nuisance when the network is operating at steady state. Therefore, it can always be safely sent to the system CPU. The counterargument states that the problem is acute precisely when the network starts up with many users at the same time. This dilemma cannot be avoided, so for our purposes we will assume that the designer can live with the idea that his or her prospective users will be coming to work at random times during the day; therefore, this is not an issue. However, this shows how seemingly mundane events can drastically affect the design of such a complex product.

In terms of network management, an approach such as using the *Simple Network Management Protocol* (SNMP) will be required. Locally produced statistics memory should be accessible by GET commands to the centralized network manager that runs on the system CPU card. Things can run either by periodic polling of the line cards or in some cases by exception. In the latter case, an SNMP trap is thrown when a special event occurs and the manager catches it for logging and reporting and perhaps subsequent actions.

The last issue of interest to our designer is that the OC-48c card proposed by Agere can be configured with two different approaches in mind: to optimize performance or to optimize equipment density.

On the OC-48 card, two NPUs will be needed in the highest-performance setting—one for each direction (ingress and egress). Something such as Agere's VPP chip can also be placed at the coprocessor port to handle AAL2 traffic in a very dense configuration, if necessary. If security is needed, the possibility of a security coprocessor configuration must be evaluated.

System CPU Card The level of expected activity on the system CPU card tells our designer to orient his or her attention toward candidate boards with multiple CPUs. In addition to a native PowerPC, processors are available such as QED from PMC that embed multiple MIPS cores, which can handle many of these side chores that line cards conveniently offload onto the system card. To achieve the level of activity the designer expects on the host CPU where the routing protocols such as BGP and OSPF run usually requires a PowerPC of the caliber of a 7410. PMC is an example of a vendor that offers a multitude of processors that can fit on such cards. They will connect to the rest of the switch through the Compact PCI over the midplane, which also handles serdes-based connections between the cards.

The host CPU will run embedded Linux or VxWorks as the operating system. The control CPU on the line cards usually runs VxWorks as the operating system, although Linux is sometimes chosen on the line cards as well. It often becomes more of an issue about the development tools and environment that is available to design, package, and deliver a solution. However, for the moment, our designer does not need to make up his or her mind fully.

Control Plane CPUs on Line Cards The load of the 10 GbE cards will definitely justify something like a 7410 PowerPC as host CPU, whereas a PowerQUICC 3 class CPU or maybe a PowerPC 750 would usually be expected for the other lower-speed line cards (4×1 GbE and OC-48c). Incidentally, our designer realizes that such a PowerQUICC 2 or 3 CPU should also be placed ideally (but not necessarily) on the switch fabric cards to make a more complete design with diagnostics and handle front-panel *light-emitting diodes* (LEDs).

Resources Budget

The designer then starts adding up the available resources needed. Using the Agere simulator from the Festino development environment, the main data plane applications will be written, tested, simulated, and evaluated at typical and worst-case scenarios of traffic based on the available modeling.

Then the number of instructions that must be executed per second, as well as instruction memory and data memory requirements, will be tabulated for each line card's computational needs on the ingress and egress paths, depending on whether one or two NPUs are used on each card. The findings should corroborate his or her initial choices for CPU at each instance, or the choice will have to be revised.

The process will continue on the control plane CPU as well as on the system CPU based on the development tools that are available by the operating system of choice, such as Linux or VxWorks. A similar budget for control plane CPUs will be compiled and the appropriate choices of CPU and memory locally on each card will be settled upon.

This has been a very brief overview of how to tackle the issues of designing real-life equipment. For obvious reasons, we have not been able to expand on the task in depth, but the intention was to combine concepts and notions that we covered in numerous chapters into a meaningful stream of reasoning that shows how design architects try to squeeze system choices into the constraints imposed upon them by specification and product requirements.

SUMMARY

In this chapter, we tried to fill the most important gaps of material that remained uncovered in the rest of the book. We discussed memory technologies and looked deeper into software development costs. We uncovered some very important software ownership issues and discussed the costs and modeling of productivity to develop this software. We concluded by skimming the surface of a large-scale real-life design challenge—namely, that of an MSR—in order to show the magnitude and the complexity of the problem. We also put ideas and concepts into perspective, especially for newcomers in the field.

SUGGESTED REFERENCES

Memory manufacturer web sites have extensive literature, technical notes, and articles on the various memory technologies. These documents describe their functionality, principles of operation, systems design, and trade-offs. The following are a few good examples:

www.micron.com

www.fujitsu.com

www.infineon.com

www.rambus.com

www.samsung.com

www.toshiba.com

www.idt.com

A very good reference on numerous fundamental memory technologies is the following textbook:

Betty Prince, *High Performance Memories: New Architecture DRAMs and SRAMs Evolution and Function*, revised updated edition (New York: John Wiley, 1999).

Dr. Betty Prince's company Memory Strategies International is also a great source for more up-to-date memory-related advice, reference publications and material, and consulting. The company's web site is *www.memorystrategies.com*.

She provides a glimpse into the near future of memory technologies in her most recent book *Emerging Memories: Technologies and Trends* (Dordrecht, The Netherlands: Kluwer Academic Publishers, 2002).

RLDRAM is discussed in depth at *www.rldram.com*.

Regarding components mentioned in the MSR case study, pertinent data sheets and white papers can be found at Agere Microsystems' web site at *www.agere.com.*

An interesting article summarizing some issues regarding Network Processor architectures is "Steering Your Way Through Net Processor Architectures," by Scott Matheson from Silicon Access Networks, CommsDesign.com, July 24, 2002. Also availabe online at *www.commsdesign.com/ story/OEG20020724S0079.*

A very interesting article comparing Reduced Latency DRAM with CAM and SRAM in network processing contexts is: "RLDRAMs vs. CAMs/SRAMs, Part I" by Infineon's Eugene Chang, Bill Lu and Felix Markhovsky, CommsDesign.com, June 3, 2003, also available on line at *http://www.commsdesign.com/design_corner/OEG20030603S0007.* The second part, "RLDRAM vs. CAMs/SRAMs: Part 2," can be found online at *http://www.commsdesign.com/design_corner/ OEG20030609S0089.*

P · A · R · T · V

SECURITY COPROCESSORS

CHAPTER 17
SECURITY COPROCESSORS

NOTE

The subject of security coprocessors is directly related to the book's main theme of network processors. It would be an obvious omission if they were not somehow included in our coverage. In writing this chapter, our dilemma was how much we could assume that readers had an adequate cryptographic background that would enable them to understand the fundamental concepts of this chapter. After long thought and consultation with people in the industry, we decided that it would be better to include in this chapter some of the basic cryptography knowledge that is required. Therefore, this chapter is more or less self-contained for the nonspecialist reader. If the reader only wants to understand systems issues because cryptography may already be a familiar field, the chapter's introductory sections can be skipped. If the reader has no interest whatsoever in security, this chapter can be seen as an add-on at the end of the main book core. Some readers may not need it. However, others who do need to understand both the cryptography basics and the security coprocessor functionality and trade-offs in a *network processing unit* (NPU) system will find a lot of interesting information in this extensive chapter.

INTRODUCTION

Although the security of communications was originally a problem of government (predominantly of the military and intelligence) and of a few privileged powerful organizations/corporations that could afford expensive technology-based protection mechanisms, it is now one of the major concerns among individuals and corporations. Information-related crimes of all sorts are reported daily.

Identifying network and device vulnerabilities is a daily event and occurrences increase exponentially. Incidents of major fraud, petty theft, the misappropriation of identities, denial of service, virus infection, worm spreading, and even access to sensitive material about people, companies, and processes have skyrocketed. For example, corporate espionage is no longer a taboo expression, and organized crime in many nations is keeping its eyes on potentially vulnerable and lucrative targets. Unfortunately, the context has not escaped the attention of terrorist groups worldwide who are always out looking for opportunities to wreak havoc against their enemy societies by harming or destroying computer-based infrastructures and utility networks, thereby disrupting people's day-to-day lives. This can be done from a distance through the illicit use of networks.

Unfortunately, the techniques of committing these illicit acts have become widely known and applicable. Essentially anyone can find tutorials and recipes for this type of action either on the Internet, in easily available books/publications, and even at quasi-legitimate conventions where computer and network hackers gather to boast about their accomplishments. People do not need Ph.D.s in electrical engineering or computer science anymore to commit such cybercrimes. In many cases, high-school students with laptops are able to disrupt organizations and perpetrate these crimes. What is more worrisome in the current networks that offer global connectivity is that these crimes can happen in the privacy of someone's enterprise, office, or even home even though the perpetrator may have initiated the action from the other side of the planet. Because of the ease with which this trend has been spreading, the general consensus is that corporate (and often private) communications must be secured.

However, cybercrime does not just come from the outside. It can just as likely originate from inside the walls of an enterprise or organization through several ways and for several reasons, such as revenge from disgruntled employees. It can also occur through unauthorized access to privileged corporate information that can be used illegally to commit securities fraud, insider trading, and so on. Whether it is file downloading between sites or files exchanged between workstations on the same *local area network* (LAN), whether it is electronic purchasing or bill payment where a person wants to guarantee the confidentiality of the session, or whether it is corporate e-mail (or, more recently, full-fledged telephone conversations and videoconferencing), users want to rely on security.

However, confidentiality is only one aspect of the problem. Beyond matters of personal or corporate security, communications security is also mandatory in ordinary contexts. Users have become more mobile over the last decade and want easy access anywhere and anytime to published, audio, and/or video content. Companies can only safely distribute this material from their servers if the requesting user can be properly authenticated as a legitimate user with an account in good standing.

This overall context has therefore created a ubiquitous and ever-increasing need for communications security through the encryption of the transmitted content and the authentication of the parties involved, whether individuals are talking over the telephone, software programs are running on different computers that communicate with each other, or users are attempting to access a web site in order to carry out a legitimate electronic-purchase or payment transaction over the Internet.

Controlled access to specific resources, confidentiality, the authentication of communicating parties, and the nonrepudiation of financial transactions are all based on structured and often industry-wide standardized use of cryptographic technologies.

Robust cryptography in its multiple forms remains a set of heavily computation-intensive processes whose aggregate load only increases exponentially in some networked devices if the high speeds of today's networks and the plethora of communication sessions that exchange packets/frames are considered. For example, securing communications from a PC on an Ethernet does not imply the same load of computations in real time as securing hundreds or thousands of simultaneous sessions involving millions of packets entering and leaving a corporate LAN/*wide area network* (WAN) gateway. Because each one of these unrelated sessions is simultaneously encrypted with different keys under different algorithms and techniques, the security-related processing load can become staggering. It can easily drive the most powerful processors to their knees.

In this chapter, we will show how the fundamental technologies that enable communications security in the high-speed global network are implemented inside families of powerful and specialized chips that offload the computational workload that is associated with real-time encryption and authentication from ordinary network processors. These chips are collectively known in the industry as *security coprocessors*.

SECURE COMMUNICATIONS APPLICATIONS IN NETWORK PROCESSING

Many communication sessions involve material that has already been secured offline prior to its transmission. A typical example would be encrypting the content of a computer file or e-mail message and then transmitting it over an insecure link. The methods and techniques used for this purpose are also based on cryptography. As a result, they are intimately, but not completely, related to the realm described here. We will concentrate on communications security rather than information security. We will also limit our discussion to the high-speed processing environment of modern and future networks. We will discuss the applicability on network-attached devices, which support multiple parallel high-speed sessions over an insecure network such as the Internet or the WAN.

Three fundamental types of network applications require security-related processing:

- *Virtual private networks* (VPNs).
- Electronic transactions.
- Wireless communications.

VPNs

VPN is a generic term that describes several unrelated technologies and approaches toward building and managing a communication network between computers. This is done in a way that the users of the VPN have the continuous impression of communicating with each other over a closed, private, and secure network that is inaccessible to unauthorized outsiders. The word *impression* is a key term here as a VPN is not deployed over physically private links; it is deployed over a public and insecure network instead.

VPNs are created in two fundamental ways:

- By using network-induced methodologies and techniques such as the *Multiprotocol Label Switching* (MPLS) approach, where sets of labels are created for specific VPNs that the carrier sets up and manages for the customer. These labels are attached to all legitimate packets. On their way to a destination, these packets are switched and routed appropriately by the core network MPLS-enabled routers/switches in such a way that no physical access is given to packets originating from and destined to users who do not belong to a specific VPN.
- By applying cryptographic techniques, especially encryption and authentication, on the payload of packets. A security context is created that allows the generation of packets, which can still be routed by the core network like all regular packets (as their address headers are not encrypted and therefore are comprehensible by the routers), but whose payload is encrypted in such a way that only the intended recipient can decipher and therefore recover the original payload content.

The first technique among these two does not require cryptographic processing because the network infrastructure and the routers physically segregate access to and from the packets, as well as to and from the network nodes, according to strict policy and membership tables. These must be maintained for each VPN. This is an added-value function that carriers can offer to customers. The analysis of the potential vulnerabilities of MPLS routers or MPLS-based VPN schemes has nothing to do with our discussion; this topic would fall under the category of information security so we will not elaborate on it here. The second of these two techniques, however, does fall squarely on our lap. We discuss it in the section "IPsec" later in the chapter.

Conducting Secure Electronic Transactions

With the arrival of established cryptographic techniques, several new applications such as e-commerce have appeared during the last decade. People can now safely engage in actual financial transactions with a web-based server remotely over the Internet (or a similar network) using a bidirectionally authenticated and fully encrypted link that is set up for this purpose between the server and a workstation/PC. Browsing software allows the easy use of the link. The user obtains remote access to menus, services, or the possibility of safe and discrete data entry into online forms as if business was being directly conducted at a brokerage house or a bank. People can now safely make payments or transfer funds from one account to another. They can browse online catalogs before they decide what to purchase securely on the spot. They can also download sensitive material that is available online with restricted access without having to establish a special VPN first. *Secure Socket Layer* (SSL) has become the de facto standard protocol that devices and servers use to set up secure sessions that enable this type of secure communications environment.

Wireless Security

First it was the laptop, then it was the cellular phone, and now it is the *personal digital assistant* (PDA). Mobile users can connect to the world through a wireless carrier or through a *wireless LAN* (WLAN) or through a Wi-Fi hotspot on which they may be connected. They are then connected to the rest of the global network through some gateway. The expanded capabilities of mobile computing have now blurred the boundaries between the types of functionality available. People can access the global network from almost anywhere. Who could believe 10 years ago that cellular phones would be embedded with a color display and the capability of playing real-time video? The issues of security originally encountered by computers have now become everyday issues for all communicating devices. Secure access to servers, to the Internet for browsing or for conducting transactions is a key requirement for mobile devices that communicate through wireless links. From a network-processing standpoint, three fundamental types of security issues appear in this environment:

- End devices perform session authentication and, in some cases, encrypt and decrypt the two-way air interface, such as the part of the link between the wireless device and the base transceiver or repeater of the WLAN. The rest of the link to the other party with whom one communicates essentially remains insecure unbeknownst to the average user. This type of security (when available or when activated) is relatively low in computational requirements. It is easily handled by the embedded logic or software in the end device (a handset, laptop, and so on).

- Mobile devices that need to engage in secure browsing and electronic transactions are often based on the *Wireless Application Protocol* (WAP), which uses a security architecture called *Wireless Transport Layer Security* (WTLS) that is reminiscent of SSL. Figure 17.1 illustrates the structure. The WAP gateway between the mobile device and the server is needed to translate the transmitted web page content from the elaborate land-based *Hypertext Transfer Protocol* (HTTP) to the lean world of wireless microbrowsers where small displays cannot be clogged with undesirable banner-like advertisements. However, the same gateway that enables the Internet browsing in the first place is also the Achilles heel of the solution in terms of security. Although the two links are secured using encryption, uplink WTLS-based traffic coming from the mobile device over the air interface to the gateway is decrypted at the gateway and reencrypted for the subsequent link to the server. The same thing happens in the opposite direction (downlink). The gateway is the only place where the sensitive traffic can be intercepted while it is temporarily in the clear. Similar issues (albeit not based on WAP) surround the Wi-Fi and WLAN arenas, where wirelss security has been one of the major problems.

  This fact is slowly pushing the industry toward adopting end-to-end solutions. Several efforts are currently under way. New algorithms and techniques have been steadily introduced such as Kasumi (an offspring of Mitsubishi's MISTY cipher) for third-generation European cellular handsets and *Wi-Fi Protected Access* (WPA) for IEEE 802.11 networks. At the same time, some wireless equipment manufacturers use mainstream cryptography (such as *Data Encryption Standard* [DES] and *Advanced Encryption Standard* [AES]) to address some of the market's security concerns and a new sweeping standard IEEE 802.11i is being prepared. Not many people are fully aware of the issues surrounding the act of securing wireless links in the cellular industry such as reliable hand-off from cell to cell. We will not expand on these issues here. Refer to the book *Wireless Security: Threats, Models, and Solutions* by Randall K. Nichols and Panos C. Lekkas (New York: McGraw-Hill, December 2001) for a more in-depth discussion of all these issues.

- Base stations must be able to sustain several encrypted sessions with a plethora of devices. These sessions use different encryption keys, and entire cryptographic contexts must be maintained. This is the first indication of a potential need for a dedicated coprocessor where the main baseband processor of the base transceiver station will offload cryptographic computations in order to preserve its own capability of managing a cell or a family of cells in the first place.

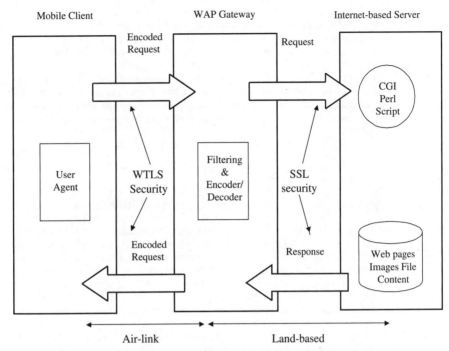

Mobile Client WAP Gateway Internet-based Server

Air-link Land-based

FIGURE 17.1 The security approach in a WAP environment based on a gateway that decrypts and reencrypts in both directions while operating between incompatible mobile and land-based network infrastructures. No end-to-end security can be guaranteed in this scheme.

CRYPTOGRAPHY: SOME BASIC NOTIONS

This book examines network processing. Therefore, we will not to spend too much time on the subject of cryptography. For more information on this topic, refer to the several outstanding texts we list at the end of the chapter. For the sake of convenience and because a significant portion of the readers most likely do not possess any special expertise on the subject, we will introduce some fundamental concepts and state some facts without any further elaboration. The notions we explain will also put the capabilities of security coprocessor chips into context, and the uninitiated reader will be able to appreciate the impact of what vendors integrate inside their chips and why.

At the transmitting station, *encryption* or *enciphering* is a deterministic mathematical process amounting to the controlled alteration of an input bit sequence called *plaintext*, which a user intends to send to the other party with whom he or she is in a communication session. The encryption process is controlled by the use of another bit sequence called *encryption key* in order to produce a new bit sequence called *cipher*, also known as *ciphertext* or *encrypted output*. Ciphertext is supposed to be illegible, if not unusable, by unauthorized parties. Describing the ciphertext as "unusable" implies that even if an eavesdropper copies it for subsequent analysis offline, it is not expected to yield any clues about the plaintext's content.

The inverse process is called *decryption* or *deciphering*. It is a mathematical process that receives the ciphertext as one of its inputs along with another bit sequence called the *decryption key* and produces the original plaintext.

The encryption and decryption processes map onto a series of logical and computational steps described in the corresponding *encryption* and *decryption algorithms* that implement the process in hardware or software. The traditional analysis of speed versus the cost of implementation dictates in every case what, if any, portion of a cryptographic algorithm should be implemented in hardware or software and why.

The encryption and authentication processes constitute the area of cryptographic operations. It can also be argued that *hashing* (discussed in a later section) is also part of that realm, but the boundary is somewhat blurred because some hashing is for security and other hashing is for other computational needs such as accelerating the indexing of database files. We will not discuss the latter form of hashing here.

Private- or Symmetric-Key Encryption

If the keys used for encryption and decryption are identical at the transmitter and receiver, respectively, we talk about *symmetric encryption*. In the case of devices communicating with each other from a distance, the encryption key that the transmitter intends to use must be somehow conveyed to the receiver in advance of any transmission. This poses some logistical security problems, which are thoroughly discussed in the references at the end of the chapter. For the moment, we will assume that the transmitting party manages to convey the encryption key to the receiver in a safe and timely fashion. The receiver, knowing the encryption algorithm, will be able to use the same key as its decryption key. Once it applies it to the received ciphertext, it will safely and reliably produce the original plaintext.

The security of the arrangement (assuming that both the algorithm and the key are considered cryptographically secure) requires that no unauthorized third party knows the key. Symmetric-key encryption is also called *private-key encryption* because of the secrecy required to protect the encryption/decryption key. In some private encryption schemes, the keys are not symmetric. A simple example would be a case where the same algorithm but two different keys (for example, K_1 and K_2) are agreed upon, where K_1 is used to encrypt traffic originating from party A and destined for party B, and K_2 is used to encrypt the traffic originating from party B and destined for party A. In this case, party A encrypts with K_1 and decrypts with K_2, whereas party B encrypts with K_2 and decrypts with K_1. Different algorithms can also be used for different directions of traffic, but systems synchronization issues usually become more difficult to manage.

Symmetric encryption is computationally fast, meaning that it can occur in real time on content that is unknown previously (such as live telephone conversation, streaming video transmission, high-speed data link, and so on). All securely transmitted live traffic today is encrypted using private/symmetric encryption algorithms. This does not necessarily include the offline-secured material such as a piece of e-mail (or other text file) that is first encrypted by software tools and then transmitted because this is material that can usually be secured with an acceptable delay in time. Therefore, depending on the availability of cryptographic tools, the user can often engage other cryptographic techniques and, more specifically, techniques that belong to the realm of *public-key cryptography*.

Public-Key Cryptography

The need to transfer the encryption and decryption key from the transmitter to the intended receiver securely has always been a weak point of symmetric encryption. The encryption algorithm may be powerful, but managing the key distribution is extremely tricky. Critical defense links have been compromised historically not by attacking the encryption algorithm or by trying to identify weaknesses in its design, but by compromising the safety of the courier that was entrusted to transport the encryption key to the recipient. If an adversary must absolutely decrypt an important communications session, it is infinitely easier to obtain a copy of the key physically (for example, from a careless secretary's drawer) rather than using supercomputers and a team of mathematical experts for years trying to crack the algorithm.

To solve this problem, the field of *public-key cryptography* was invented. The details and ramifications behind it are explained in any of the basic cryptography books listed at the end of the chapter. We will only describe the fundamental notions to set up our discussion on security coprocessors.

According to the public-key cryptographic model, every user A (a person, or a computer, or even a program) is issued a pair of bit sequences. One of them is called the user's *public key* (K_{pubA}) and the other is called the user's *private key* (K_{privA}). Similarly, user B who intends to communicate with user A is also issued a pair of public and private keys (K_{pubB} and K_{privB}). Several public-key cryptographic techniques are available in the industry and although the underlying mathematics may be different, they all share the same principles described here.

Once the pair of keys has been generated for a user, the public key can become publicly known without any constraint. It enhances security rather than compromises it. People freely share their public key with others without any fear. The user's private key, however, must remain private and secret so that only the user knows it. The robustness and beauty of the concept relies on the premise that the public key of user A or B is known by everyone, whereas the private key of user A or B is only known by user A or B respectively.

We deviate for a moment to clarify a concept. Earlier the term *issuance* was used for these pairs of keys. Public-key cryptography sometimes uses an overriding security officer (for example, at the enterprise level) that issues these keys to legitimate users of an organization. In other well-known cases, such as the famous *Pretty Good Privacy* (PGP) encryption environment, users generate their own public-private key pairs by clicking a few buttons from a menu-driven software program. We do not concern ourselves here with key management issues that pertain to information security. *Key management* is a generic term describing all issues related to generating, updating, discarding, recycling, verifying, storing, distributing, and managing cryptographic keys safely and reliably among a group of users in a specific group or community of interests according to applicable organizational policies. It is applicable to both private- and public-key cryptographic contexts, but it requires more work in public-key cryptographic contexts.

Let us return to the concept of public-key cryptography. In order for users A and B to securely communicate with each other based on such a method, the following process must occur.

User A starts by using user B's publicly known (or previously and openly shared) public key K_{pubB} and encrypts the plaintext content of the traffic that user A wants to transmit to user B. Once user B receives the ciphertext, it decrypts it by using its own private key K_{privB}, which is exclusively in user B's possession). This is the only way to decrypt traffic that is encrypted by user B's public key. This assures user A that only user B can decrypt the traffic. Inversing the flow direction, if user B needs to respond to user A using the same method, user B must use user A's openly known public key K_{pubA} in order to encrypt its message to user A and then user B transmits it. Only user A can decrypt this arriving ciphertext, as it is in possession of the private key K_{privA} needed to decrypt this traffic.

Because a completely different key is used by the transmitter for enciphering than the key that is used at the receiver for deciphering, the public-key cryptography realm is also known as *asymmetric-key cryptography*.

All currently known public-key cryptography techniques are extremely slow computationally. One public-key technique may have advantages over another, such as the size of the encryption keys or the computation time needed to perform operations. However, the common characteristic of all public-key cryptographic techniques is that they are much slower than symmetric encryption techniques. It is surprising how many people seem to be unaware of that fact.

This difference in speed is the fundamental reason a two-layer cryptographic approach is usually used in secure communication sessions:

- Private symmetric-key cryptography algorithms are used at the bottom layer to protect the transmitted content.

- Public-key cryptography algorithms are used at the top layer of this two-layered hierarchy to create a common shared secret bit-sequence, which can serve as the equivalent of a secure envelope for the safe and reliable exchange between the communicating parties of the *session key* that is usually generated by one of the two communicating devices. The two communicating parties usually use this session key as the initial encryption and decryption key of the underlying symmetric encryption algorithm.

The word *initial* was used in the previous paragraph. In older symmetric encryption systems (and in offline or mostly software-based cryptographic systems), the same key was used during the entire communications session. This is no longer the case with many state-of-the-art communications security designs, where systems engineering experience drawn from decades of military and intelligence-related networks has created the widespread know-how that allows systems engineers to design robust handshake protocols as well as methods and techniques that allow the encryption keys to be dynamically updated from both sides in a communication link at predetermined frequent, sparse, or even quasirandomly chosen points in time during the ongoing transmission. The advantage of such a cryptographic environment is that even if one of these symmetric keys is compromised, only a small portion of the overall transmission will be sacrificed. The disadvantage is that this increases the overall design complexity and key management.

The following are some interesting, but not obvious, points regarding this technique:

- Once a user has encrypted a bit sequence with someone else's public key, he or she cannot decipher the ciphertext back to plaintext, as the user is usually not in possession of the private key of the intended recipient. This is not the case with symmetric cryptography where both parties can produce cipher and plaintext, given the availability of the common key.

- The generation of the session key in the public-key cryptography arena and the symmetric key(s) used in the encryption of the actual content has nothing to do with the underlying cryptography. These are usually random or pseudorandom bit sequences that are generated by one of the parties based on *random number generation* (RNG) techniques. In some systems, a physically separate device generates the random numbers. This poses the danger of physically tampering with the system, if, for example, someone opens up the chassis and physically injects values into the encryptor chip that replace the intended RNG. The military and intelligence communities require tamper-resistant designs based on strict standards. In these systems, the slightest effort from an intruder to open a system chassis box or to even remotely try to physically access internal circuits will leave indelible traces and marks, and will zeroize internal registers, thereby prohibiting piracy and illicit use. The main processor (or encryption processor) requests a random bit sequence. The RNG then produces it and feeds it to the encryption processor, which can be a CPU running security algorithms in software or a security coprocessor chip.

In some other systems, an embedded RNG (many technologies allow this to be done inside a silicon die) inside the security coprocessor chip generates these random values. The subject of RNG is vast. The references listed at the end of the chapter can provide more information on the subject and its ramifications.

- We must distinguish between truly random and pseudorandom or quasirandom bit sequences. Several statistical tests correlating samples of larger windows rolling over a bit sequence can determine a series of criteria according to which scientists rate the degree of randomness in bit sequences. The *National Institute of Standards and Technology* (NIST) web site (*www.itl.nist.gov/fipspubs/ index.htm*) contains several test suites designed to check and rate the degree of randomness. Just because something looks random does not mean it is actually random. Very successful attacks have been staged against the source of randomness in many a cryptographic system. Random sources are ideal for the production of these values.

However, another environment may require a pseudorandom bit sequence instead. This is required for a family of encryption algorithms called *stream ciphers*, where a lot of additive key material must be generated that is not easy for a third party to guess. This material is deterministically created in synch between the transmitter and receiver. It is pseudorandom material that is usually created from *finite state machines* (FSMs) such as *linear feedback shift registers* (LFSRs) that are set up at both parties based on a commonly obtained session key and then advanced independently of each other in order to produce the same pseudorandom bit sequence at both communicating devices.

BLOCK CIPHERS, STREAM CIPHERS, AND CRYPTOGRAPHIC MODES

Symmetric encryption algorithms used for the traffic content protection are usually classified in one of two families: block ciphers or stream ciphers.

Block Ciphers

In this case, the encryption of the plaintext occurs block per block, where the block is a chunk of consecutive plaintext bits. Until recently, blocks for instance in the DES algorithm were 64 bits long. Most modern block ciphers use blocks that are 128, 192, or 256 bits long.

Block ciphers usually require a *key generation and scheduling* mechanism. Based on the original symmetric algorithm key, this mechanism applies specific steps to generate a series of subkeys, which will be scheduled for use subsequently by the algorithm and in a sequential order for each one of the multiple *rounds* that must be executed.

A block of plaintext is read in and combined with the first subkey. A series of processing steps has to be taken on it involving operations such as bit shifts, rotations, substitutions, and permutations. This is the first of several rounds. A *round* refers to the basic processing logic of the algorithm. The output of the first round is fed into the second round along with the second subkey and so on until the specified number of rounds and the required number of subkeys is exhausted. The number of rounds depends on the algorithm. For example, DES is specified with 16 rounds. Rijndael (AES) is specified with 9 rounds if both the key and block size are 128 bits long, 11 rounds if either the key or the block is 192 bits long and neither of them is longer than that, and 13 rounds if either the block or the key is 256 bits long. During the standardization process of the AES algorithm, it was settled that the official standard supports one block size only of 128 bits, whereas the key can be of 128, 192, or 256 bits. Rijndael supports all 9 possible combinations of key size versus block size, where each one them can be 128, 192, or 256 bits long. This issue of whether multiple key- and block-size combinations are possible is one of the many intercompatibility factors between communicating systems.

Figure 17.2 shows the evolution of a run of DES over a block of plaintext. K_i is the subkey used in each round. We use DES processing as an example because it is still the bulwark of the *Internet Protocol Security* (IPsec) arena, which applies strongly and predominantly on the network-processing field. All concepts we discuss in this chapter are directly related to the cryptographic reality; therefore, they can almost always be applied to other algorithms as well, such as Rijndael. In Figure 17.2, L and R denote the left half and the right half of the block-size reference, respectively. The 64-bit blocks would respectively correspond to the 32 MSB bits and the 32 LSB bits. The first block, called IP, represents an initial permutation where bits effectively are reshuffled according to a specific order. The last block is the inverse of that permutation and ensures that the same algorithm can be used for both encryption and decryption. Some DES implementations do not include these two steps. Although it has been proven that this does not alter the security characteristics of the algorithm, it does violate the standard specification, which is an issue when the intercompatibility of devices comes to play.

In order to give a better idea about how things work, Figure 17.3 shows the subkey generation and key-scheduling mechanism for DES, as well as the *f* function, which is involved in each DES round, as shown in Figure 17.2. The internal details of the substitution boxes (known as *S boxes*) and the permutation boxes (which are just small lookup tables with specific content from an implementation standpoint) can be found in any of the cryptography books listed at the end of the chapter. The folded modular structure based on the L and R approach in the implementation of the DES *f* function is a

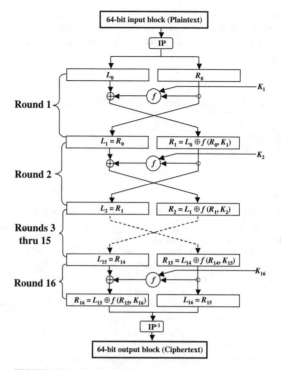

FIGURE 17.2 The flow of DES as an example of a typical block cipher in multiple rounds with subkeys scheduled for each round

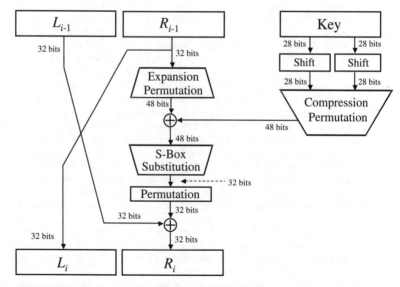

FIGURE 17.3 The right side shows the key-scheduling mechanism for DES; the left side depicts the Feistel network that is shown in Figure 17.2 with the DES *f* function.

classical cryptographic technique called *Feistel network*. This network is at the heart of many modern ciphers, such as Kasumi. One of the main characteristics of the Feistel structure is that working the properly scheduled subkeys at the receiver and rearranging the left and right sides ensures that decryption is a straightforward process. This property makes the same algorithm capable of encrypting and decrypting.

The encryption engine produces the ciphertext output that corresponds to the plaintext block it had previously read in only after all the specified rounds are completed. In the case of DES, it would mean that 64 bits of ciphertext are produced corresponding to the 64 bits (equal to one block length) of plaintext that it just finished processing. If more plaintext is available at the input, the process continues with a next block of input bits. If not, there is usually a specification as to what happens at the last block of plaintext—for example, padding with zeros to generate a block of input bits with some control characters to notify the receiver that no more input is available.

In some cases, higher-level protocols on the stack communicate the length of the transmitted bit sequence up front so by the time encryption occurs, the receiver is already set up with session counters to accept only the amount of bits it was told it was going to receive. Therefore, no cryptographic prowess is necessary to address the issue.

At decryption, the process is simply inversed. Based on the decryption key, the receiver starts by performing the same key generation and scheduling process. The list of subkeys is usually engaged with the reverse order, starting from the last one and moving on to the first one that was used at encryption.

Stream Ciphers

In this case, which is mostly (but not exclusively) favored by the military and intelligence communities, the plaintext is processed bit per bit with a quasirandom bit sequence (usually as long as the plaintext) that is known as key material. The key material can often be mapped through some lookup table or function mechanism onto the plaintext, or as more often is the case, it can be combined additively with the plaintext such as using an exclusive OR (XOR) logical operation in order to produce ciphertext. XORing plaintext traffic with good-quality random material makes the ciphertext extremely difficult for an adversary to tackle. An interesting property of the XOR function is that by XORing this ciphertext with the same additive material, the original plaintext is produced. In other words, a stream cipher has the following relationships:

Plaintext P $\oplus$ Additive key material K $\rightarrow$ Ciphertext C

Ciphertext C $\oplus$ Additive key material K $\rightarrow$ Plaintext P

Even if the same sequence that is generated at the transmitter is generated at the receiver for the production of the additive key material, ensuring that the correctly corresponding bit positions are used for XORing with the plaintext is not a trivial problem. The problem of synchronizing the additive key material sequence at the receiver with the one used at the transmitter has been notorious. It is known as the *problem of preservation of cryptographic synchronization* and has been one of the issues that have plagued stream ciphers. Refer to the book *Wireless Security: Threats, Models, and Solutions* by Randall K. Nichols and Panos C. Lekkas (New York: McGraw-Hill, December 2001) for a deeper discussion of the subject and for a description of a technique that solves this problem. The other major problem for stream ciphers has been the statistical quality of the key material, meaning the degree of randomness of the key-generation process. It falls under the same category of problems associated with random or random-like bit sequences.

Cryptographic Modes

Implementing encryption algorithms in communication systems brings along numerous systems engineering concerns that we cannot possibly cover in this short overview. They can range from efficiency

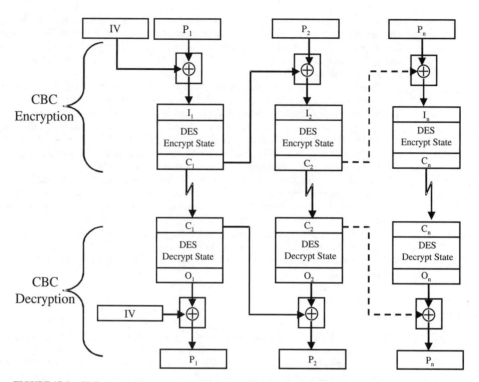

FIGURE 17.4   CBC mode applied on a block cipher like DES cipher and the associated need for an IV

and fault tolerance to capabilities to guard against the loss of synchronization all the way to resistance to specific types of attacks (such as chosen plaintext and error injection in the ciphertext). As the requirements are diverse based on each system implementation, the cryptographic community has developed several types of cryptographic modes of operation. These have been exhaustively analyzed and documented, and their behavior is well understood. A cryptographic mode is usually the combination of the underlying cipher algorithm along with some feedback mechanisms and other associated operations that produce a chain-like behavior.

The most widely known and used cryptographic modes are *electronic codebook* (ECB), *cipher block chaining* (CBC), *cipher feedback* (CFB) (also known as *cipher text auto key* mode [CTAK]), and *output feedback* (OFB) (also known as *key auto key* [KAK] mode. Many other less widely known modes such as propagating CBC mode, OFB with a nonlinear function, and plaintext feedback are available, but we will not elaborate on those here. Bruce Schneier's book *Applied Cryptography: Protocols, Algorithms, and Source Code in C* (New York: John Wiley, 1996) examines the pros and cons of many cryptographic modes. The important point for our discussion is that different cryptographic modes are available and the choice of which one to use in a system or session may very well decide whether two devices can communicate. As mentioned earlier, the interoperability of cryptographic systems has traditionally been a problem. Just saying that two systems support the same encryption algorithm does not mean they can exchange information even if they posses the encryption key.

The basic modes, except for ECB, which is essentially the fundamental mode of encrypting block per block until the plaintext has been exhausted, require the use of *initialization vectors* (IVs). These are also sometimes referred to as *seed values*.

For example, let us examine the case of the CBC mode as applied for example to DES (but it will be the same situation no matter what algorithm is used). As shown in Figure 17.4, the cipher output C_{n-1} from processing every block is used to XOR the following incoming plaintext block P_n prior to encryption.

So how is the first pass started since no previous ciphertext is available? The answer is by using a random bit sequence (equal in length to the block size) and by proceeding accordingly. The IV can be any bit sequence of that same length. In fact, cryptographic systems openly communicate the chosen IV to their communications partner. IVs are generated by the same RNG module that we discussed earlier. In some systems, it can come from different sources, such as a timestamp, an internal clock, or a combination of some internal states as read from registers.

It makes absolutely no difference for the algorithm's security whether the IV is known. In fact, an eavesdropper who is monitoring a link will always be seeing the $C_1, C_2, \ldots C_n$ ciphertext chunks (which are used as XOR inputs in the following block processing). These are all publicly available so they are irrelevant for the easy breaking of the cipher. Schneier's book elaborates on this subject.

So why was a mode like CBC needed in the first place? In the traditional ECB mode (block-per-block encryption with the same key schedule), the same plaintext will always generate the same ciphertext output, which sometimes opens other ways for an attacker to crack the code (by guessing the key). For example, all payment checks have the phrase "Pay to the order of" at the same position. By correlating this type of plaintext and ciphertext information, an attacker can guess the encryption key with higher probability. When the algorithm is operating in the CBC mode, the same plaintext will never be the same at the input of the encryption process, as it is always XORed first with the ciphertext that was produced from the previous block processing. Identical plaintext blocks will always produce different ciphertext; therefore, CBC is a much more secure mode of operation.

Keep in mind, however, that there is no free lunch in technology. Despite its beauty, CBC inevitably has shortcomings. For example, a single-bit error in the CBC-produced ciphertext output can be shown to always cause a whole block at the receiver to decipher to garbage and a single-bit error at the corresponding position of the original error of the recovered plaintext from the subsequent block. This drawback is called *error extension*, and it can be detrimental in some applications. After these two errors (first a block and then a bit are deciphered incorrectly), the system recovers automatically, and CBC is said to be self-recovering. However, this is only true as long as the block-level synchronization is maintained. If a bit is lost or an extra bit is inserted into the ciphertext due to some interference, the receiver will continue decrypting traffic to garbage and the algorithm will be unable to recover. This is a good example of a classic case where higher-layer communication protocols coupled with *error correction codes* (ECCs) must be combined at the right place to preserve the integrity of CBC-based cryptographic operations.

In the IP network world, this can mean that the security requirements of a streaming audio or video application that runs with the *Real-Time Protocol* (RTP) over *User Datagram Protocol* (UDP) (which is less reliable than *Transmission Control Protocol* [TCP]) may actually negate the use of underlying encryption if the latter is configured to run in an undesirable cryptographic mode, whereas file transfer or e-mail sent simultaneously with the former may have absolutely no problem with it, as they run using TCP's reliability of connection.

Some people are surprised to suddenly find out that the choice of encryption technique is often not dictated by security concerns and requirements only, but by the apparently irrelevant network protocol stack and software infrastructure, which seem completely unrelated factors at first glance.

IMPORTANT CRYPTOGRAPHIC CONSIDERATIONS IN COMMUNICATIONS

Several techniques are being systematically used in the implementation of cryptographically equipped communication systems. Some of these techniques have become mainstream ideas, whereas some remain less well known to users. We will mention several of these techniques to give an idea of what issues the systems architect must consider in order to design a robust system and how all of these capabilities weigh in adversely when they must be performed under the proverbial gun of high-speed links that require real-time attention.

Many of these techniques originated from defense networks and were initially implemented and perfected on civilian/industrial networks by being embedded inside protocol-sensitive encryptors that operated at the data link layer on X.25 and frame-relay networks in the early 1990s. The positive experience acquired from those networks was successfully taken advantage of at the network layer when the *Internet Engineering Task Force* (IETF) designed IPsec, which adapts most of these methods to the IP layer.

Weak Keys

When some of the possible keys used in an algorithm are known or found to be less secure than others, this specific cryptographic algorithm is characterized by a *nonlinear* key space. Conversely, when this is not the case, the algorithm's key space is *flat*.

For example, the DES algorithm contains weak, semi-weak, and possibly weak keys:

- Four DES keys that are composed only of 1s or only of 0s, or keys where one half is composed of 1s whereas the other half is composed only of 0s, are *weak*. The DES algorithm and key expansion structures cause these keys to suffer from the highly undesirable effect that all rounds of DES are encrypted by the same subkey.

- Some pairs of DES keys encrypt plaintext to the same ciphertext. Instead of generating 16 subkeys (for the 16 rounds of DES), these keys only generate two different subkeys, each of which is used eight times during encryption or decryption. The result is that in any of these pairs, if one of the keys is used for the encryption, the other key from the pair can be used for decryption and the plaintext will be recovered. These keys are *semi-weak keys* and are listed in any good basic cryptography book, such as Schneier's book.

- Some DES keys are known to produce only four subkeys. Each subkey is used only four times in a DES run (16 rounds). They are not as weak as the semi-weak keys, but they are not as strong as typical keys; hence, they are *possibly weak keys*.

- A lot of cryptanalysis has been done on the safety of using *complement keys* (replacing 1s with 0s and 0s with 1s) and determining what can occur if plaintext is encrypted with a key and subsequently with its complement. Under these conditions, specific relationships can be identified between ciphertext and its logical complement. This can lead to specific types of attacks. It is recommended that the use of complement keys be avoided.

A sophisticated security coprocessor that uses internal random number sources to generate symmetric keys must have the ability to verify whether the random bit sequences that it generates satisfy any of these unfortunate and dangerous profiles. If they pass the test, they can be used and are formally issued and transmitted to the other party. If they fail, the RNG must be consulted again for a new random bit sequence. As the number of weak keys is usually not large, it just takes a small lookup table (such as in the case of DES with 64 entries with 64 bits in each slot) and a small checkup routine.

For complement keys, the system should keep a copy of all previously generated keys in a buffer and check if any new generated key during the session happens to be the complement of a previous one. Incidentally, this is obviously not a requirement when the same symmetric key is used throughout a session (in other words, when no rekeying occurs).

The cost of implementing this kind of verification is negligible, but the cost of loss due to an attack can be significant if it is not implemented. It is surprising how many DES-encrypting systems in the market do not check for weakness when they produce a new key.

Protocol-Sensitive Encryption

The first use of communications encryptors on digital lines was applied to point-to-point links, where all the transmitted bits including the useful information, the headers, and the trailers of frames (in the case of old BSC/SDLC protocols) are encrypted. These encryptors were called *link encryptors* and did not care what protocol was used on the link they purported to secure. Assuming that the receiver was always listening to the channel and that it was in possession of the symmetric key used for the encryption, decryption was not a problem. For obvious reasons, people applied this technique predominantly on private leased lines or on wireless radio links (in land-mobile or space communications).

However, with the arrival of packetized networks such as X.25 and frame relay, as well as with the subsequent *asynchronous transfer mode* IP/ATM explosion, the act of encrypting indiscriminately all traffic on a link bit stopped. Instead, only the payload of packets or frames would be encrypted by leaving the headers (which contain source and destination addresses) untouchable so that the encrypted packets or frames could still be routed and switched by the network infrastructure without problems. That provides the benefit of packet-switched networks to customers who needed to move away from the old regime of leased lines for financial reasons while providing them with the ability to operate and exchange information with other sites in a more secure environment than before.

From a security standpoint, the idea was that decryption would only be possible at the intended destination receiver, and any other party accidentally or intentionally intercepting the packets or frames could not decipher its contents. Of course, the CRC codes of the new packets/frames would have to be regenerated after the payload encryption. The CRC process result is usually attached as part of the packet's trailer, so in this case, a new trailer would have to be calculated and attached instead. However, that was difficult. This marked the beginning of a new era in communications security—the era of protocol-sensitive encryption. The IPsec effort decided to implement this idea, taking security one layer higher and securing IP packets.

Hashing

The term *hashing function* in cryptography denotes a one-way mathematical procedure that takes an input bitstream of arbitrary length and produces a fixed-length output called the *hash*. Hashing functions are also known in the industry by several other names, such as *message digest* (MD), compression or contraction function, cryptographic checksum, fingerprint, *message integrity check* (MIC), and *manipulation detection code* (MDC). The term *one-way* means that hashing functions cannot be inverted—in other words, given the hash output, the input bitstream that produced it cannot be found.

Good hashing functions have several spectacular characteristics. First, their collision probability is very low. In other words, given an input and its hash, it is extremely difficult and highly unlikely that another bitstream input that produces the same hash output will be found. Second, given a bitstream input of arbitrary length and its corresponding hash, even a subtle change of only one bit at the input will produce a new hash by changing on average around half of the bits of the original hash. Third, hashing algorithms are supposed to be public knowledge and should not be kept secret.

The use of hashes is an outstanding mechanism that can ensure the integrity of transmitted information. By calculating the hash of an underlying piece of information and by attaching the hash to the transmitted information, the recipient is given an opportunity to verify the integrity of the data transmission. This can be done locally by calculating the hash of the received data and by comparing the received hash with the locally calculated one. If no one tampered with the transmitted data, the hashes should be identical.

Of course, if the information must be kept confidential, encryption must be used in addition to hashing. This eliminates the potential danger that someone may intercept the transmission, alter the actual content, calculate a new hash to replace the old one, attach the new hash to the new altered transmission, and then forward the new data-plus-hash combination to the unsuspecting recipient.

The converse argument also works. Even if the data are not confidential, both encryption and hashing must be used to protect against the stipulated attack. For example, payment clearinghouses receive millions of short messages daily from financial institutions defining different combinations of payments of specific amounts from a debit account to a credit account. An attacker does not necessarily intend to steal information or money. In this example, he or she can simply wreak havoc on the operation by completely destroying the reliability of the transmission where amounts of money are systematically erred and accounts are messed up without the attacker having direct monetary advantage. Therefore, hashing must be intimately combined with encryption even when confidentiality and secrecy are not required.

Message Authentication Codes (MACs)

By appending a secret key that is mutually known to both communicating parties next to the information bitstream input of a hashing function, a hash is created (also known as *keyed hash*) that can only be verified by the recipient if he or she is in possession of this secret key. This technique is called *message authentication code* (MAC) or *data authentication code* (DAC). In the case of an entire bitstream of substantial length, the stream is usually fragmented into reasonable-size chunks, which are key-hashed to produce the corresponding MACs that become part of the new bitstream and are sent along with the information bits at predetermined locations (for example, at the end of each one of the stream chunks of information).

A special type of keyed MAC is the *hashed MAC* (HMAC), which is defined in RFC 2104. It is actually a keyed hash inside another keyed hash and is used by IPsec for the authentication of all messages. The exact definition of this function can be found in any basic cryptography book.

Digital Signatures

A spectacular by-product of public-key cryptography is the ability to generate a *digital signature*. This is produced if the fundamentals of public-key cryptography are engaged in an inverse scenario. More specifically, if user A encrypts a message using its own private key K_{privA} and transmits it to user B, then user B can look up user A's public key K_{pubA} in a public directory or over the Internet at some key server. User A may have also already openly communicated it to user B along with other users. The retrieved public key K_{pubA} is used to decrypt the message.

As the pair of keys (K_{privA} and K_{pubA}) allow the encryption and decryption in either direction and no other key can be used for either of these roles, the message could have been encrypted only by user A—hence, the equivalent notion of digitally signing. If instead of just signing the message digitally, user A also wanted to keep it confidential, user A should encrypt the information with the intended recipient's (in this case, user B) public key as discussed earlier in addition to digitally signing.

Public-key operations take a long time. For example, on the same hardware platform, *Rivest-Shamir-Adleman* (RSA) encryption is about 1,000 times slower than DES encryption. To ensure that digital signatures are completed in a reasonable amount of time, it is now common practice on all digital signature algorithms that a hashing function is used to produce a hash of the bitstream input (which can be long for a communications session). Then the digital signature algorithm is invoked on the hash itself. As the hash is usually a few bytes long, the process should now take an acceptable amount of time to calculate.

At the receiver, it works as follows. The receiver receives information. If it is encrypted, it is decrypted. Then the hash is locally generated from what was received and the signature is verified on the hash. If it is correct (that is, the decrypted hash matches the locally produced one), the information is accepted as legitimately signed.

Session Key Exchange

Public-key cryptography also enables a session key to be easily generated, which can be used by the communicating parties for many different purposes, such as a secure envelope for the secure exchange of the first symmetric key to be used for the content encryption or a mutual seed of pseudorandom LFSR structures to be used in some systems to generate specific random-looking streams as in stream ciphers.

The beauty of the session-key idea is that two (or more) parties can engage in a remote handshake protocol to produce it. This takes a maximum of a few seconds in most real-life settings. At the end of the handshake, a common bit sequence has been generated that is essentially impossible for any eavesdropper who monitored the handshake exchanges to guess. This now becomes a common shared secret, which the parties can use in any form and fashion to authenticate themselves to each other.

Digital Certificates

In the example given in the previous section, when user A signed a message and sent it to user B, if user B was not already in possession of user A's public key K_{pubA}, it would have to retrieve it from a server or from some other source. A legitimate fear is that what is perceived as the legitimate public key K_{pubA} of user A may actually not be the real public key of user A. This is especially a concern when the physical distance between the parties or other logistical issues inhibit the easy key exchange or when a lesser degree of acquaintance and confidence in the content of a key exchange is involved. An imposter user C could impersonate user A and send digitally signed messages presenting itself as user A.

The only way to know true from false here is the use of *digital certificates*. These are small digital data structures, which are attached to the transmission content that certify the signer's bona fide thereby verifying the content's validity. The certificate is acceptable to the receiver if it is digitally signed by a *certificate authority* (CA) that is acceptable to both parties. It is common that certificates of certificates are required in many cases where parties must communicate securely but don't know or fully trust each other. This often culminates to the absurd situation of hierarchical chains of certificates dangling behind the actual information message of importance. Stripping each level of certification at the receiver may require that special directories be looked up. A CA may have to be contacted in order for the recipient to verify the signature involved and before the recipient's security management system proceeds with the following layer of security. The process concludes when the actual payload has been authenticated and deciphered, and can be accepted for subsequent local processing.

The process can be intricate and involves algorithmic and computational aspects. Issues of scalability are of paramount importance in network-processing equipment so the systems architect must be always aware of what is absolutely necessary and what is superfluous.

We conclude the short overview of this subject by clarifying that two computational aspects are associated with digital signatures: generating digital signatures and verifying digital signatures. Both are needed for a system to handle authentication; however, the former is computationally slower than the latter. When a vendor says that their equipment handles digital signatures in X milliseconds, find out which one of the two operations they refer to.

Embedded Sequence Counters

We often take for granted that in packetized networks higher-level protocols generate counters and map their content into header fields. The advantage of this is that the receiver knows upon reception of a packet that this is the kth packet out of a total of m packets. If packets arrive out of order, which is extremely probable in a routed world as they are traversing many different routes on their way to the same destination, the appropriate intelligence will reorder the packets at the receiver in the correct sequence, stripping away the headers and trailers. The reordered recovered payload is then presented to the higher layers that await to process it (the presentation and application layers).

To ensure that unintended receivers (eavesdroppers) do not acquire any knowledge about how many packets intended for a destination are parts of a specific transmission sequence, protocol-sensitive encryptors (and now IPsec) embed internal counters inside the secure payload with a sliding window over the bitstream. The receiver upon decryption will know that this is the kth packet from a transmission of m consecutive packets that should be accepted according to the current window. If the counter is connected to a well-thought-out timestamping process, it protects against playback attacks, even if someone knows the cryptographic context and tries to alter the packets. In simpler words, we cannot be receiving the 141st packet for payment clearance as all the packet-related counters in today's sessions should be in the 10,000s.

Address Tunneling

Before the word *firewall* became a household name in the 1990s, protocol-sensitive hardware encryptors operating at the data link layer and positioned at gateway positions between LAN and the WAN pioneered the concept of masking addresses to protect end stations. The concept was easily expanded to the IP layer with IP-level encryption and has by now been institutionalized by IPsec in one of its operational modes. Instead of encrypting just the payload of the packet, the addresses are also encrypted, so the real source and destination are not publicly visible to the routers and switches in the network infrastructure. New default addresses are created and prefixed (as opposed to appended) as new headers to the encrypted packet. The appropriate new trailers are calculated and attached.

In addition to encryption, this process therefore encapsulates the original packet (along with its headers and trailers) inside a new packet, which behaves like an envelope and can be easily routed by the IP network routers without a problem.

The encapsulating packet can have its payload encrypted (a second superposed time) by the encapsulating mechanism, and the original packet can be tunneled inside the externally encrypted process. Upon arrival at the destination, the encapsulation is stripped away, the top-level packet is decrypted, and the new packet that appears beneath it from the tunnel with the intended destination addresses is steered accordingly to the correct destination by the use of translation tables that map, for instance, external to internal addresses, which must remain invisible to the WAN.

Timestamp (Nonrepudiation)

As mentioned earlier, events such as counters must sometimes be associated with specific windows in time. Some transmissions also require a good indication of the moment in time that they were initiated or executed. *Timestamping* is a generic term that embeds a reading of time in a time-sensitive message. The significance of this message depends on the requirements of the application. Time can be read from an internal clock or from a time server, which may be next door or at the other side of the planet. The concepts of how this time server is trusted and how its answer is kept untampered are related to the information security discipline, which is amply discussed in the references. The idea is however, that if a payment instruction to a financial institution is timestamped, hashed, and digitally signed, then upon signature verification, no discussion can occur about whether the order was sent or whether it was actually originated by the person who digitally signed the order. Cryptography therefore enables nonrepudiation and in some cases this can be an issue of very serious importance.

Rekeying

Sophisticated hardware-based encryption systems are capable of changing the underlying symmetric encryption key several times during the same session. Some systems can do it periodically, whereas others can do it at quasirandom points in time. This dramatically enhances the security of the communication, but it causes a significant increase in the cost of implementation. This is because specialized signaling protocols (usually working in-band) are needed to safely and reliably halt the exchange of live traffic with the necessary buffering and without loss of information, to renegotiate or generate a new symmetric key, to initialize the appropriate internal states at both the transmitting and receiving portions of the communicating parties, and to resume traffic from the correct point using the new cryptographic context and key.

Security Associations (SAs)

In a network-processing environment, where numerous sessions are set up and maintained, it is vitally important to preserve the cryptographic context of each session in such a way that no mixing of keys or parameters can occur between sessions. This capability is handled by IPsec through *security associations* (SAs).

COMMON CRYPTOGRAPHIC ALGORITHMS

Many encryption algorithms of variable capabilities and commercial success are available. The symmetric encryption world has been so far dominated by the use of DES during the last 25 years and the use of the Triple DES derivative in the last 5 years or so. Other widely used algorithms are RC4, IDEA, etc. Details about the internal workings of these algorithms can be found in any good fundamental cryptography textbook.

Triple DES is a composite algorithm, which is usually based on a two-key approach and extremely rarely on a three-key approach. Because of mathematical and efficiency reasons, a discussion of the cryptographic efficiency merits between the two approaches goes beyond this short overview. See the references at the end of the chapter for more information. In the two-key approach of Triple DES, the plaintext P is encrypted with a key K_1. Then the ciphertext is decrypted with a key K_2 and the new result is reencrypted with the first key K_1. The length of the bit concatenation of K_1 and K_2 creates a new effective composite key that is twice as long as single-key DES. This is the base of the increased security afforded by Triple DES. Instead of being 64 bits long, the keys are now 128 bits long, which is much more secure. This method is called *encrypt decrypt encrypt* (EDE). EDE has other variations such as using three different keys (as explained in Schneier's book). Figure 17.5 illustrates the principle.

As our emphasis is on network-processing environments, readers with an interest in chip architectures may want to contemplate the fact that 1 run of DES implies 16 rounds, each of which contains several operations bit shifts, logical XORing, table lookups, substitutions, and permutations. A Triple DES environment requires $3 \times 16 = 48$ rounds, with different subkeys scheduled for each round. As there is no free lunch in technology, from a hardware design standpoint, two extreme choices are available. One choice is to have a small efficient DES module that implements the computations of one round and keeps the internal state in buffers, so the triple DES data flow actually implies 48 passes through this same module (a solution that is inexpensive and slow). The other choice is to have a huge pipeline of 48 replicas of a DES encryption round module that allows wire-speed performance, but with a huge expense in silicon and power consumption.

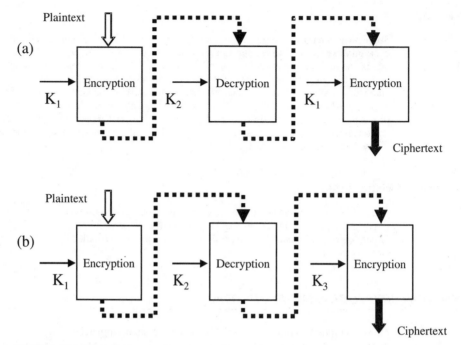

FIGURE 17.5 Triple DES encryption using the EDE configuration: (a) using more often two keys and (b) rarely three keys

As fears about DES security grew over the last few years, the U.S. government moved through a public bid and scrutiny process to establish the new *Advanced Encryption Standard* (AES), which ended up being the Belgian algorithm Rijndael. Several good descriptions of the Rijndael algorithm can be found in the references at the end of the chapter. A good source to start with is located at J. Savard's web site at *http://fn2.freenet.edmonton.ab.ca/~jsavard/crypto/comp04.htm*. The NIST web site (*www.itl.nist.gov/fipspubs/index.htm*) has extensive high-quality documentation of the entire process that led to the scrutiny and the acceptance of the new standard. Numerous papers analyze the algorithm and discuss its implementation in software and in hardware in either *field-programmable gate array* (FPGA) or in *application-specific integrated circuit* (ASIC) form. The inventors of the algorithm, Dr. Joan Daemen and Dr. Vincent Rijmen, also have lots of interesting information on their web site[1], which is maintained at the cryptographic research group of the *Katholieke Universiteit Leuven* (KUL). Their site includes code and links to other sites that have Rijndael implementations in C, Lisp, Ada, Visual Basic, Java, and other languages, as well as even a Matlab-based simulator of the algorithm. Daemen and Rijmen have also recently published a book called *The Design of Rijndael: AES—The Advanced Encryption Standard*, which explains the design philosophy and the algorithm functionality in depth. All these sources are listed in the references at the end of the chapter.

For the sake of clarity, we will just mention a few basic characteristics here to provide a first impression of the Rijndael algorithm compared to DES.

---

1. *http://www.esat.kuleuven.ac.be/~rijmen/rijndael/index.html*

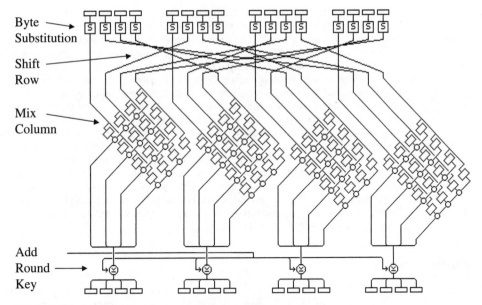

FIGURE 17.6 Operations during one round of Rijndael processing

Rijndael uses multiple rounds of a series of essentially repetitive operations except for a couple of minor variations at the beginning and the end of the execution. Figure 17.6 shows the Rijndael operations. They are as follows:

- *Byte substitution*, where each byte of the input block is replaced according to a substitute found in a specified way in an S box.
- *Shift row*, where the input block (for example, 128 bits) is composed of bytes ordered from 1 to 16. These byte orders are arranged vertically from 1 to 16 in 4 columns of 4 rows, each creating a 4×4 table. Then the rows of this table are made to rotate horizontally by a specific number of positions depending on the row number. For example, the first row stays intact, the second is rotated by one position, the third is rotated by two positions, and the fourth is rotated by three positions. As a result, a new 4×4 table is produced that reorders the bytes.
- *Mix column*, where this last result is now multiplied by the 4×4 matrix:

$$
\begin{matrix}
2 & 3 & 1 & 1 \\
1 & 2 & 3 & 1 \\
1 & 1 & 2 & 3 \\
3 & 1 & 1 & 2
\end{matrix}
$$

This is done using Galois field multiplication techniques in $GF[2^8]$, which means that the bytes are treated as polynomials and not as numbers. If more than 8 bits are produced by some multiplication, the result is XORed with Rijndael's generating polynomial of the chosen Galois field—that is, the bit sequence 100011011.

- *Add round key*, where the subkey used in the current round is XORed with the result of the previous operation.

We will not discuss the exact key schedule generation for Rijndael here as it can be found in the suggested references. Rijndael has been designed to take advantage of the latest microprocessor architectures. It can be easily and inexpensively implemented to run even on small processors such as in smart cards. Several publicly disclosed hardware designs enable the algorithm to sustain a variety of multigigabits-per-second-throughput (albeit not in smart cards).

So far our discussion has presented the types of questions that *network equipment vendor* (NEV) designers should ask security co-processor vendors when probing about the internals of an implementation. Duplicate implementations are available to support the encrypting and decrypting of full-duplex operations. Designs are also available where one fast cryptographic module that is alternatively switched between both directions is used on a time-shared basis. The trade-off will be the cost of implementation as opposed to the speed of operations and throughput. In high-speed network-processing designs, duplicate (if not multiple) cryptographic modules are expected for the symmetric algorithm. Different cryptographic modes and combinations of different key sizes (and ideally also block sizes) must be supported. Support for the capability to appropriately seed the IVs in the CBC implementation must also be available. The latency of the key-scheduler switching overhead should be considered, whereby supporting different cryptographic contexts should not be excessively long to avoid penalizing the aggregate performance.

Diffie-Hellman (DH)

The most fundamental and definitely the most famous technique for the generation of a common shared secret is the *Diffie-Hellman* (DH) algorithm, which is described and analyzed at length in all fundamental cryptography books. We provide a brief description here for the sake of convenience.

Two (or more with some small protocol extensions—see Schneier's book) communicating parties (A and B) agree in public on a specific large prime number p, which is called the *modulus* of the operation, and a (mathematical group) *generator* g that they will use for the ensuing calculations. The generator must be chosen in such a way that for any integer $Z<p$, a number W exists so that $g^W \bmod p = Z$. In other words, g can produce or generate all numbers from 1 to $p-1$.

The DH protocol is a four-step process:

1. First, the two parties independently generate a secret large random number (bit sequence), which is denoted here with the lowercase letters a and b for users A and B, respectively. These random numbers must be smaller than p.

2. Then they both proceed with the mathematical process of calculating the modular exponentiation of the chosen group generator g raised to a power equal to the respective random number *modulo* the large prime p that has been previously agreed upon. In other words, divide by p and retain the remainder of the operation. The result is called the *DH public key* of each party. It is shown here with the uppercase letters A and B. The public key and the public key from public-key cryptographic schemes such as RSA are irrelevant to each other:

$$A = g^a \bmod p$$

$$B = g^b \bmod p$$

3. User A then sends over the public insecure network its own public key A to user B, and user B sends its own public key B to user A:

Public key A

User A $\longrightarrow$ User B

Public key B

User A $\longleftarrow$ User B

4. They both then independently modularly exponentiate the public key, which they just received from the other party, by raising it to the power of their own privately generated random number, a and b, respectively. In other words, user A, which has received the public key B, which is equal to g^b mod p, now raises it to the secret power a that only user A knows; therefore, user A now privately obtains a new sequence:

$$S = B^a = (g^b \bmod p)^a = g^{ba} \bmod p \qquad (17.1)$$

Likewise, user B, which has received the public key A, which is equal to g^a mod p, now raises it to the secret power b that only user B knows; therefore, user B now privately obtains a new sequence:

$$S = A^b = (g^a \bmod p)^b = g^{ab} \bmod p \qquad (17.2)$$

The right-hand sides of Equations 17.1 and 17.2 are identical. Therefore, the two parties have just independently generated a common bit S sequence that no third party that may have been eavesdropping on their session can guess or calculate. The interesting characteristic of the DH protocol is that although the associated computations are quite straightforward (albeit difficult to conceive on the back of an envelope), the inverse problem that an attacker confronts is extremely difficult to solve. This is called the *discrete logarithm* problem. In this example, it would mean that an attacker who knows g and p (as they are not secret) and who monitored the line and therefore knows A and B, cannot easily guess or calculate the S sequence.

This S sequence, or some derivative of it, can be used as the session key. In some systems, the first device can do the following:

• Generate randomly a symmetric key K_s for the underlying content encryption.

• Encrypt this secret key K_s using the session key S as the encryption key.

• Send it to the other party.

Because the second device also generated the same session key S, it will be able to decrypt the message and recover the underlying symmetric key K_s. They are now both in position to engage into secure communication using their algorithm of choice, such as Triple DES, Rijndael, and so on.

The possibly of a *man-in-the-middle attack* must be eliminated, which in some exaggerated and highly improbable but quite possible cases can make user A believe that it is communicating with user B when user A is actually communicating with attacker C instead and make user B believe that it is communicating with user A when in fact user B is communicating with attacker C instead. The idea is shown in the following, where users A and B are led to believe that they are communicating with each other when they are actually not.

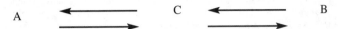

The danger in this scenario is that the attacker negotiates a different DH session with each user. To thwart such an attack, users A and B must digitally sign the public keys A and B that they send out to the other party. Because attacker C cannot forge the signatures, the attack fails. The DH protocol handshake will go through only if the signatures are verified by the respective recipients.

From a computational standpoint, we will simply mention that on a specific hardware platform, doubling the DH key size from 512 to 1,024 bits almost quadruples the time needed per operation. A significant difference appears in performance when the protocol is executed in hardware with assist units or purely in software. For example, modular exponentiation is done as a series of successive multiplications. The most reputable algorithm for modular multiplication is the *Montgomery algorithm*. Numerous descriptions of circuit architectures that implement Montgomery arithmetic are described in the annual proceedings of the *Cryptographic Hardware and Embedded Security* (CHES) conference. For more information, consult either Çetin Kaya Koç's web site (*http://islab.oregonstate .edu/koc/*) or Christof Paar's web sites (*www.ece.wpi.edu/faculty/cxp.html* and *http://crypto.ruhr-uni-bochum.de*). Many security coprocessors have an embedded specialized hardware unit that carries out Montgomery exponentiations in hardware, thereby tremendously accelerating the performance.

COMMON PUBLIC-KEY CRYPTOGRAPHY ALGORITHMS

Many hashing and keyed-hashing algorithms are available. However, the most common ones to be aware of are the *Secure Hashing Algorithm* (SHA-1), which is very strong as it was designed by the *National Security Agency* (NSA) to preserve the integrity of messages. This has become a key ingredient of the *Digital Signature Algorithm* (DSA). The establishment of the new AES standard has launched an ongoing discussion about expanding the same hashing principle to larger block sizes that are commensurate with the higher performance afforded by the more modern and powerful algorithms such as Rijndael and other contestants of the AES. These discussions regarding SHA-256 and even longer along with the latest proposals are documented at the NIST cryptographic web site.

Other widely known hashing algorithms include MD5 and RIPEMD. They are all documented in detail in essentially every good cryptography textbook.

In the public-key cryptography arena, the undisputed leading technology in terms of years of use, several deployed solutions, and worldwide acceptance is RSA technology, which is based upon the difficulty of factoring very large numbers that are the product of two large primes. The steps taken in RSA-based security are as follows:

1. One first comes up with a number n that is the product of two large primes p and q, where p and q must remain secret and are eventually discarded at the end of an operation.

2. A number e is agreed upon, which is relatively prime to (that is, has no common factors with) the product $(p-1)(q-1)$. Different sources recommend different values, but e is typically one of the following: 3, 17, and more often 65537 (= 2^{16} + 1). The RSA public key is the pair (n, e).

3. The RSA private key is then the number $d = e^{-1} \bmod [(p-1)(q-1)]$.

4. Either key can be used for encryption and the other one will be able to decrypt the cipher.

5. Encryption on a message m to produce cipher c is $c = m^e \bmod n$.

6. Decryption of the cipher c to recover the message m is $m = c^d \bmod n$.

Implementations for the decryption are usually based on the *Extended Euclid Algorithm* to find the inverse of a number in modular arithmetic. However, to accelerate the private-key operations, some cryptographic coprocessors offer the possibility of using a special module that implements the famous *Chinese Remainder Theorem* (CRT). In that case, the values of p, q, and values close to them, such as $d \bmod (p-1)$ or $d \bmod (q-1)$, are saved and others are easily calculated from them for subsequent use. Both theorems are fundamental pieces of knowledge in elementary number theory and can be found in any good textbook on cryptography or number theory.

Several sources of benchmarking information are available to compare the time it takes between various operations. Without intending to scrutinize the efficiency of algorithms or code implementations, we show some results publicly available from Wei Dai's Crypto++ benchmarking work in Table 17.1 (see *www.eskimo.com/~weidai/benchmarks.html*). This is provided solely to compare the impact of some typical cryptographic operations on systems. The absolute values of these items are irrelevant to our discussion. What is important here is the impact on the performance from the choice of algorithm and the key size.

TABLE 17.1   Time Comparisons of Performance on the Same Platform of Several Typical Public-Key Cryptographic Operations

Public-key based cryptographic operation description	Key size	Time spent on an Intel Celeron 450 platform
RSA signature generation	512 bits	0.4 msec/operation
RSA signature verification	512 bits	0.3 msec/operation
RSA signature generation	1,024 bits	27 msec/operation
RSA signature verification	1,024 bits	0.7 msec/operation
Digital Signature Standard (DSS) signature generation	512 bits	4 msec/operation
DSS signature generation with precomputation	512 bits	2 msec/operation
DSS signature verification	512 bits	5 msec/operation
DSS signature generation	1,024 bits	15 msec/operation
DSS signature generation with precomputation	1,024 bits	5 msec/operation
DSS signature verification	1,024 bits	18 msec/operation
RSA encryption	512 bits	0.2 msec/operation
RSA decryption	512 bits	4 msec/operation
RSA encryption	1,024 bits	1 msec/operation
RSA decryption	1,024 bits	27 msec/operation

The longer the key, the more computationally heavy the work. Notice the time difference (for specific key lengths especially when they are getting longer) between actually generating a digital signature and simply verifying one. Public-key decryption takes more time than public-key encryption. This is because in order to enhance security, the private key is usually longer than the public one. As a result, the modular exponentiation operations involved with public-key decryption (the equivalent of digital signature generation) are computationally more involved. This is why security coprocessors sometimes possess extra assist units (such as the implementation of the CRT or Montgomery exponentiators to implement the associated arithmetic efficiently) or why they use other design tricks such as the precomputation of several powers of the modulus that accelerate calculations and keep them readily available in a lookup table.

If a potential user wants to further scrutinize the internals of a security coprocessor, he or she may want to find out how the quality of primes is ensured. Naïve users usually think that in order to create that effect, large random numbers are generated and then the system tries to factor them in order to ensure that a generated number (intended to be a prime) is a prime. Many different probabilistic primality testing algorithms are available and a system at configuration time or at design time was put together in such a way that it allows the setting of a threshold of acceptable failure risk in the generation of such numbers. The embedded cryptographic system will then generate large random numbers and apply one or more primality tests to them. These test are run with a varying degree of confidence based on the risk threshold stipulated by the user. The system will make a judgment as to whether the generated number is indeed prime. The subject has fascinating ramifications, but it goes way beyond the coverage of our chapter.

The following are a few other well-known public-key techniques:

- The unpatented El-Gamal system, which is similar in approach to the DH and RSA processes, but in terms of encryption, the ciphertext is twice as long as the plaintext.
- The DSS, which is standardized as FIPS-186.

- The lattice-cryptography-based NTRU cryptosystem, which is starting to make a significant inroad in the wireless market due to its speed and modest computational resource requirements. The company's web site contains a series of white papers and excellent tutorials (see *www.ntru.com*).

- *Elliptic-curve cryptography* (ECC) and some more recent variations of its principles such as hyper-elliptic-curve-based cryptography. In the last few years, ECC public-key cryptography has become a potential large-scale contestant, as it offers similar levels of protection to RSA but with significantly shorter keys (an order of magnitude shorter), which translates into bandwidth savings. A quick series of tutorials on fundamentals of ECC can be found at the Certicom web site at *www. certicom.com*. A series of standards have been already compiled using ECC cryptography to implement session key exchanges (such as *elliptic-curve-based Diffie-Hellman* [EC-DH]) and digital signatures (such as EC-DSA, which is the standard *elliptic-curve-based Digital Signature Algorithm* [EC-DSA]).

Both the *Institute of Electrical and Electronics Engineers* (IEEE) and *American National Standards Institute* (ANSI) have been actively pursuing standards along several dimensions of these technologies, although many of these technologies (such as Rijndael) have not yet received the IETF's blessing and therefore do not appear in implementations such as IPsec for the moment—a situation that is most likely to change in the not-so-distant future.

In general, a designer must weigh the pros and cons between execution speed and security. Longer keys provide higher security, but they take computationally longer to complete. Dedicated security coprocessors do not penalize the main CPU or NPU; therefore, the dilemma goes away. The issue then is to decide how flexible or powerful the security coprocessor is and how easy it will be to interface with the NPU at hand.

STANDARDIZED SECURITY PROTOCOLS

By now, it should be clear that in order for devices to communicate safely and securely with each other, a series of issues needs to be settled at communication setup time, including the following:

- Which algorithms will be used among the many that may be implemented inside a system?
- What will be the session key generation mechanism?
- What if the other party supports only some but not all the available techniques?
- How do both parties fall back on a common denominator?
- What if the parties are able to set up a session according to multiple options? Who gets to decide what will be used and based on what criteria?
- How will the symmetric key be generated?
- What will decide the size of the symmetric key?
- How will IVs be generated?
- For some algorithms, what will decide the size of the block?
- What mode of the symmetric algorithm will be used?
- What wider provisions must be set up at both communicating parties in order to allow the possibility of safe recovery of the session if cryptographic errors or losses prohibit the correct decryption of some block? For instance, will one have to go back and reconstitute the stack buffers with prior ciphertext and previously scheduled subkeys in order to recover the session (and if so, up to what extent) or should a system drop a session altogether?
- How is all of this communicated to and enforced on the other party?
- What will be the session key exchange protocol?

- How will the communicating parties ensure that none of the generated symmetric keys are weak?
- How will the authentication be handled?
- Will it be unidirectional or bidirectional?
- Will it be necessary to certify either party's credentials?
- Will it be necessary to perform integrity checking and other security mechanisms? If so, which ones?
- Will it be necessary to rekey the underlying symmetric algorithm periodically? If so, how often will rekeying have to be reinitiated? Based on which protocol?
- How are rekeyed IVs generated and communicated to other parties every time?
- How are IVs generated and communicated to the other party every time?
- Which one of the communicating parties will initiate it? Who enforces the mechanism?
- Will the communicating parties spend the rest of their lives trying to handshake a multitude of protocols in order to set up a meaningful session or will they eventually time out and drop the effort? Under what conditions does this occur and what event logging takes place?

The list can go on and on. The intention here is not to be exhaustive, but simply to put things into the overall computational-load context and to show the magnitude and importance of the interoperability problem.

IPsec

IPsec is the current de facto standard in the security of IP-protocol-based networks as it provides multiple mechanisms that ensure confidentiality, integrity, and authentication. The book *IPsec: The New Security Standard for the Internet, Intranets, and Virtual Private Networks* by Naganand Doraswamy and Dan Harkins (Upper Saddle River, New Jersey: Prentice-Hall, 1999) is an outstanding source of tutorial information. It also contains numerous references to the multiple RFC documents that detail all aspects of IPsec. Here we will limit ourselves only to some generalities to put things into a network-processing context.

IPsec establishes security relationships between end devices, between end devices and gateways (such as firewalls), or between gateways connecting two LANs from different sites with each other over the insecure WAN. It is designed to function in one of two modes (*transport* and *tunneling*), and it allows the use of two basic packet-forming protocols in either of these two modes with the intention of securing IP datagram traffic: *Authentication Header* (AH) and *Encapsulating Security Payload* (ESP). The landscape is also complemented with a sophisticated key-generation-and-sharing handshake protocol, which is called *Internet Key Exchange* (IKE).

Several cryptographic techniques are used in the heart of IPsec's protocols (along the directions we briefly described earlier). Support is available for several symmetric (such as DES, Triple DES, IDEA, and RC4), asymmetric (such as RSA and DSS), and hashing (such as SHA-1 and MD5) algorithms all operating with different key sizes, which are configurable to suit security and performance requirements. The fundamental idea behind the IETF's effort that created IPsec was that IPsec-compliant devices designed by different manufacturers that desired to communicate with each other securely should be able to engage in a handshake mechanism at session establishment time, communicate their cryptographic capabilities and user preferences to each other, and automatically negotiate transparently for the user the setup of a session that is configured along a greatest common denominator of capabilities as well as security requirements.

The IPsec transport mode is used when upper-layer protocols need to be protected such as UDP or TCP in end-to-end sessions, whereas the tunnel mode is used when complete IP datagrams need to be protected, as is the case in gateway-to-gateway communications where one does not want LAN-internal destination addresses or even specific sessions to be monitored outside while they traverse the WAN. In transport mode, the IPsec header is inserted between the IP header and the upper-layer

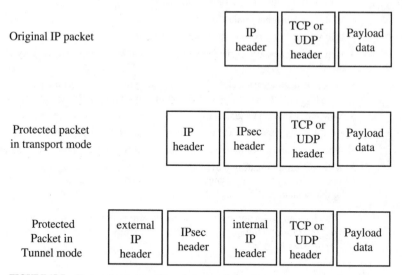

FIGURE 17.7   IPsec restructuring of IP packets according to the transport and tunnel modes

protocol header. In tunnel mode, the entire packet to be protected is encapsulated inside another datagram, and an IPsec header is inserted between the outer and inner headers. Figure 17.7 illustrates this.

We mentioned earlier the concept of security associations (SAs), which allow the structured handling of multiple secure sessions. SAs are unidirectional; therefore, a different SA is used for inbound as opposed to outbound traffic. In the case of IPsec, they are identified by an index called *Security Parameter Index* (SPI), which is inside each packet and ultimately enables the parties to map the security services they require on specific links based on established security policies (as contained in a policy database) as well as on multiple cryptographic contexts (keys, sequence numbers, destination and source addresses, and the choice of protocol used).

The AH protocol provides data integrity, data source authentication, and protection against playback (replay) attacks. As it does not provide confidentiality, it involves a new packet header but not a new trailer.

The ESP protocol provides confidentiality, data integrity, and data source authentication of IP packets as well as protection against replay attacks. We saw earlier that a new ESP header is created. Part of the ESP packet payload is encrypted and a portion is in the clear to allow the recipient to properly process the packet. All the operations needed to create this new material obviously require a new trailer. The result is a new combination of plaintext-ciphertext-and-authenticated material, as shown in Figure 17.8.

It goes without saying that as both confidentiality and authentication are ensured by the ESP protocol, the SPI, which identifies an SA, refers to a combination of cipher and authenticator algorithms. In addition to using block ciphers in CBC mode, communication of the necessary IV is also foreseen.

The AH and ESP protocols can be used independently or jointly in some sessions. Security and computational issues are involved depending on which one is performed first and on which mode IPsec is configured to perform (transport or tunnel) in that specific case. The pros and cons go beyond our scope and the reader is referred to the literature.

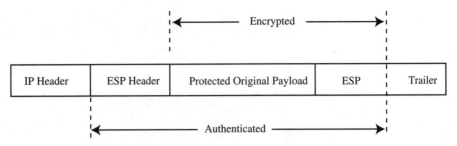

FIGURE 17.8   The packet structure after application of the ESP protocol

IKE is used to establish shared security parameters and authenticated keys, which amounts to establishing SAs between peer systems that are engaged into an IPsec session. More specifically, IKE is a hybrid of the *Oakley* (from which it has adopted multiple modes) and SKEME (from which it has adopted sharing and rekeying techniques) protocols. It operates in the *Internet Security Association and Key Management Protocol* (ISAKMP) framework, which fully defines packet formats, retransmission timers, and message construction requirements. As a result, IKE defines its own way of deriving authenticated keying material and negotiating shared policy.

A few words are appropriate here about the module names that we just mentioned. First, Oakley was designed by University of Arizona cryptographer Hilarie Orman. It is a free-form protocol that essentially enables parties to advance the state of the protocol at their own pace. It allows an authenticated key exchange. IKE actually borrowed the idea of some of its modes from Oakley and codified them into concretely specified handshake exchanges.

SKEME is a key exchange protocol that was originally designed by cryptographer Hugo Krawczyk. It defines a mechanism of authenticated key exchange, whereby the communicating parties use public-key encryption to authenticate each other and then they share some randomly generated components of the exchange, which the parties exchange among themselves using public encryption. It may be combined with a regular DH handshake.

ISAKMP is a framework that was developed originally by the NSA and defines how peer devices engage into a secure communications session. This includes how the exchanged messages are constructed and formatted, and the exact state transitions that have to be undergone in order to secure the communication session. It provides means that allow a device to properly authenticate a peer, to exchange information needed for a key exchange, and to negotiate security services. Although this covers the whys and hows, it does not cover the whats of the exchanges. The details of what exactly must be negotiated is left to another type of specification document, which in the case of IPsec is called the *Domain of Interpretation* (DOI). The exact bit field positions reflect the exact significance of all parameters involved with algorithm choices and configurations.

In addition to encryption and authentication, IPsec addresses several other interesting aspects of operation. One of them is network-layer compression. The payload content is compressed prior to being encrypted. In fact, this is exactly how it should always be, because if it is done the other way around, the process ends up being worthless because compression is impossible or, as more often is the case, because compressing an encrypted stream is not a reversible operation and the receiver cannot decrypt traffic. IPComp is the protocol used for IPsec compression. It creates a type of SA known as the IP *Payload Compression Protocol* (PCP). Other techniques of compressing are available in this same IP realm such as the *Lempel-Ziv standard* (LZS) and Deflate, but we will not spend any more time on them here. Other issues of interest are also how IPsec behaves in a multicast environment and how the *Network Address Translation* (NAT) protocol can be consolidated within the same context where IPsec security may be required. The textbook by Doraswamy and Harkins covers all of these issues in detail.

As a conclusion, the overriding point we want to make is that the consolidated need for significant computational power justifies offloading all IPsec operations onto an accelerating coprocessor.

SSL

SSL is a layered handshake protocol that was originally designed by Netscape and through lots of bumpy industrial quasi-consensus and intervention by the IETF, it has become a formal standard now covered by RFC 2246 and it is known as *Transport Layer Security* (TLS). Its so-called record layer lies on top of reliable transport protocols such as TCP.

The principles on which SSL is built have already been discussed so nothing should be surprising. If a client device wants to establish a secure communication session with a server, three steps must be taken during the *handshake phase*:

- After the server has been authenticated, the two parties must agree on the cryptographic algorithms to be used in the session.
- The keys to be used by the agreed-upon algorithms must be derived and established.
- The handshake may optionally authenticate the client.

Once the handshake phase has been successfully concluded, the *data transfer phase* begins.

To understand the computational load needed at session setup time, let's look at the handshake process a little closer. For more details, refer to the excellent book *SSL and TLS: Designing and Building Secure Systems* by Eric Rescorla (Reading, Massachusetts: Addison Wesley, 2000).

1. The client first sends a list of algorithms it can support and a random number to the server. Both parties will use this random number for subsequent calculations.

2. The server chooses the cipher it prefers from the proposed list and sends it back to the client along with a digital certificate, which contains the server's public key. Other information is also located in the certificate that the client can potentially use to further authenticate the server, if necessary. The server also sends another random number that will also be used by the parties as part of the key-generation calculations.

3. The client verifies the server's certificate and extracts the server's public key. The client will then generate a random secret string called the *premaster secret*. It encrypts it with the server's public key and sends it to the server.

4. From the premaster secret, the two parties will independently calculate the MAC and encryption keys. This is done by both parties using the same *Key Derivation Function* (KDF) in order to derive the same *master secret*, which will subsequently be used to generate the actual cryptographic keys.

5. The handshake process is concluded by first having the client calculate a MAC of all the handshake messages and then send it to the server.

6. The server calculates and sends a MAC of all the handshake messages to the client.

The load is heavier at session setup time than during the actual data transfer phase where encryption/decryption acceleration is desirable. SSL accelerators are therefore needed where it makes sense to introduce them—for example, at servers or gateways.

The SSL protocol sometimes allows the use of a significantly accelerated variant called *resumption* or *resume handshake*, where the parties agree to use the same master secret from a previous session and therefore forego the time spent starting cryptographic calculations again from scratch. According to Rescorla, this has been shown to accelerate the session setup by almost a factor of 20 in the case of session keys that are 512 bits long. This special *resume handshake* should only be used in cases where the convenience of some types of frequent connections for specific client devices outweighs the security concerns; therefore, it should only be engaged judiciously.

SECURITY COPROCESSORS: A CLASSIFICATION

Security coprocessors come in many forms and shapes. Users can offload computationally intensive cryptographic operations to specialized chips. This has been the case from the original DES encryptor chips in the early 1990s from companies like VLSI Technology (now Philips Semiconductor) to the latest coprocessors like SafeNet's SafeXcel-2141 coprocessor, which has been designed together with Analog Devices (and also contains a powerful embedded and user-programmable *digital signal processor* [DSP] core), as shown in Figures 17.9 and 17.10, which are used in computational environments of workstations or small gateways with aggregate throughputs less than 1 Gbps. Figure 17.9 shows how the coprocessor fits into a system, and Figure 17.10 shows the internal structure of the chip. The cryptographic modules are shown in gray for clarity.

However, the trend in the network-processing arena, where wire speeds of several gigabits per second is the norm and where every packet flowing by belongs to a completely different session and must be treated in a different way involving different ciphers and keys, places a tremendous stress on traditional encryption chips. These classical encryption chips contain special hardware units that will calculate all modular exponentiations with hardware assistance. On top of that are usually embedded modules that will calculate DES (and/or AES in the latest designs) in the appropriate mode, as well as hashes such as MD5 and SHA-1. The main systems processor therefore hands out the data to be secured along with the parameters required to the security coprocessor, which will calculate all that is required and pass the results of the cryptographic calculations back to the main CPU. The main CPU during that time can use its power on other tasks.

In our case, we emphasize two types of security coprocessors: IPsec accelerators and SSL accelerators. The products appear in chip and board format.

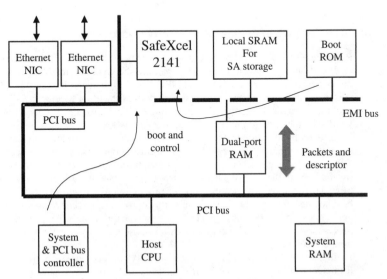

FIGURE 17.9   The systems configuration of a SafeNet-2141 security coprocessor based on the *Peripheral Computer Interconnect* (PCI) bus   *(Source: SafeNet)*

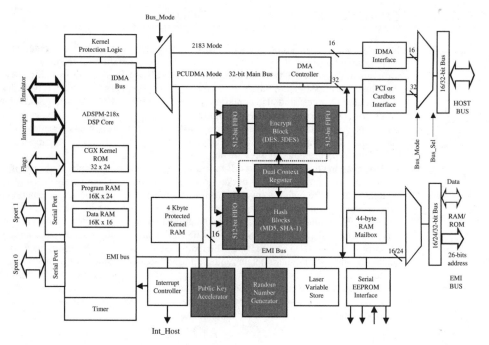

FIGURE 17.10 The internal structure of the Analog Devices/SafeNet-2141L security coprocessor chip *(Source: Analog Devices)*

Systems Considerations when Engaging a Security Coprocessor

Regardless of their speed of processing, security processing can be conceived either in a *look-aside* or *flow-through* architecture. The principles are shown in Figures 17.11 and 17.12, respectively, where the corresponding cases of a typical NPU-based high-speed line card is depicted. Some interesting systems-related issues are associated with both of these architectures.

In the look-aside architecture, incoming traffic(1) that needs decryption or decapsulation is forwarded by the NPU to the security coprocessor. After local processing, it is sent back(2) to the NPU for further classification and forwarding. Outgoing traffic(3) that needs encryption, hashing, or encapsulation is forwarded by the NPU to the security coprocessor. After processing, it is sent back to the NPU for subsequent scheduling(4) and transmission over the line interface. As soon as the host CPU (or the NPU in the case of a network-processing piece of equipment) receives a packet, it will determine whether it requires security-related processing. If it does, it will forward the packets through the host bus (usually PCI, but in some security coprocessors, it can also be PCIX, which just doubles the performance of the bus, or HyperTransport, which unfortunately is not natively supported by network processors) to the security coprocessor for subsequent processing, whether this is hashing, encryption, or decryption.

In the flow-through architecture, the security coprocessor sits in front of the network processor so it intercepts all incoming traffic. If it requires decryption or decapsulation, it can handle these tasks independently, in which case it will ultimately hand over decrypted traffic to the NPU for subsequent classification and forwarding. Likewise, outgoing traffic coming out of the NPU will be encrypted and/or encapsulated prior to its transmission over the line interface. As shown in Figure 17.12, bidirectional traffic at points 1 and 4 is encrypted, whereas it is decrypted at points 2 and 3.

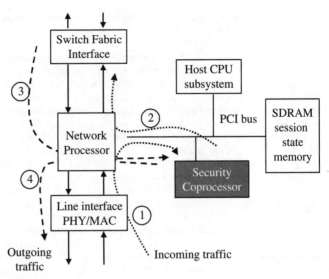

FIGURE 17.11 A security coprocessor sitting next to an NPU in a look-aside architecture

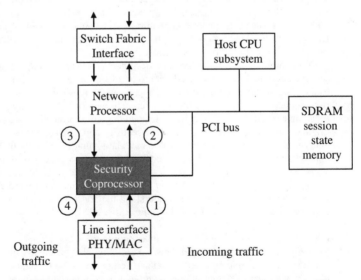

FIGURE 17.12 A security coprocessor sitting next to a network processor in a flow-through architecture

The security processor will need access to the data associated with the secure session such as the keys, policies, and parameters. This usually requires a storage capacity per session of up to a few hundreds of bytes. In many cases where the number of sessions is limited, if the security coprocessor chip contains on-chip memory this information can be stored on-chip where it is both convenient and relatively safe. However, when several sessions (tens of thousands, if not hundreds of thousands of sessions) need to be maintained, some sort of massive external memory will be required to preserve the session states. In these cases, *synchronous dynamic random access memory* (SDRAM) is usually used.

The obvious bandwidth issue associated with the frequent duplex use of the PCI/host bus to send traffic back and forth between the network processor and the security coprocessor indicates that the look-aside solution can only be sustained for relatively low-end systems where line speed is OC-3 or OC-12 and not more than 1 Gbps. The flow through is extremely scalable to line speeds to the extent that the security coprocessor has been designed to sustain the wire-speed processing. The system-related downside of the flow-through approach is that some comprehensive information about the state of various applications that are running at higher layers must be made systematically available to the security coprocessor, so that it can make the correct decisions about what to do on incoming or outgoing packets without disturbing the network processor or other CPUs in a complete system.

In terms of the security ratings of a system, several levels of certification of how secure a device is deemed to be are available. The *Federal Information Protection Standard* (FIPS)-140 standard describes the requirements a device must satisfy to be rated at a specific level of security needed (in this case) by the U.S. federal government. As an example of the implications, we will mention the danger of having the security coprocessor access an external block of SDRAM to obtain the necessary keys and other session-related information. If the system is not designed to be tamper-proof, essentially anyone can open the chassis and gain physical access to the memory pins, thereby reading the sensitive information simultaneously with the security coprocessor in order to subsequently attack a system or a link.

If the system is designed according to the FIPS-140 standard and the security coprocessor is housed in the same tamper-proof cryptographic module with the session-state memory in such a way that when it detects intrusion and a specific trigger is created, then a specific pin can be activated (when available) on some security coprocessor chips, and that event will zeroize all internal registers and storage so no sensitive internal material can be obtained by an intruder. Of course, the box design (if it has been designed according to the same standard guidelines) will also leave physical traces and undeniable proof of the physical tampering.

Yet another approach that systems designers can take is for the security coprocessor to write and retrieve encrypted session-related data to and from the session memory only when it obtains access to the session-state memory directly (bypassing the network processor or host CPU). This way even if someone obtains access to that session-state memory, the illicitly retrieved session-state data are meaningless. The price for such an elevated degree of protection is a significant decrease in cryptographic throughput performance as the security coprocessor now has to meticulously encrypt and decrypt all the session's cryptographic metadata—that is, the data that it needs in order to actually encrypt and decrypt user traffic on behalf of the network processor or host CPU.

Be extremely cautious with public-key cryptographic performance ratings, as companies unfortunately report results partially and in arbitrary units that are extremely difficult to compare. The prospective user should therefore understand the constraints of the technology and ask vendors the appropriate questions until the full picture is disclosed.

IPsec Acceleration Some manufacturers quote RSA operations or DH operations, but in addition to being vague, they are not even close to defining what is actually obtained by acquiring their product. IPsec has multiple parameters that can define the computational context. It is therefore important to know what parameter compares with what. It is important first and foremost to know the following items as an IPsec accelerator spends most of its time encrypting and reformatting payloads of packets:

- How it performs with a typical (such as RC4) and worst-case cipher (such as the combination of Triple DES with SHA-1 hashing).
- How it performs with different key sizes (and, when applicable, when combined with different block sizes).

- How performance is affected when IPsec is configured respectively with AH, ESP, and potentially AH/ESP protocols.
- What happens when IPsec works in transport and, more importantly (in terms of workload that it then has to churn), in tunnel mode.
- How the link performance is degraded when the link is sustained as simplex or as full duplex.
- What happens under a specific and pertinent key exchange protocol. The exact context of the exchange environment must be defined, so that oranges are compared with oranges, so to speak. For instance, one may want to know what happens when IKE is used with a DH operation, an RSA (or DSA) signature generation, and an RSA (or DSA) signature verification where the key sizes are standardized. For example, some DH cases can use 180-bit exponents and RSA keys can be stipulated to be 1,024 bits (typically good security) or 2,048 bits (much more conservative security) and so forth.

Do not neglect finding out how the IPsec accelerator behaves in a more subtle situation where it must be decided whether the consecutively flowing packets belong to different SAs. Many vendors would assume that many of the consecutive packets belong to the same SA and therefore the configuration data are readily available in a lookup table. This is one extreme case that suits the vendor. A question at the other extreme of the argument should be asked instead—namely, how is the performance affected if all parading packets during the test belong to different security associations and the cryptographic context has to be completely updated after each packet has been processed? The truth of a typical case would then lie somewhere in the middle.

With this fundamental set of overall IPsec performance parameters more or less delineated, the potential user should then go even deeper into the relevant worth of actual hardware-assist units and probe deeper into the vendor claims in order to ultimately bring out the real substance of performance numbers, which are relevant to one's own design requirements and targets. The public-key-related questions that must be answered by a prospective vendor are listed in the section "SSL Acceleration."

On the cipher side, the prospective user must also keep an eye on the advertised throughput numbers. If an IPsec accelerator or a security coprocessor can handle multiple algorithms such as Triple-DES and AES, be aware of the extra header bytes that the algorithm may require. Triple DES can require up to 57 extra bytes to be added to a packet, whereas Rijndael/AES with its longer IV requirements can add up to 65 bytes to the size of an original packet. If the packets used in a throughput test are long enough (longer than 1KB), this extra overhead may be conveniently ignored. If an environment is tested where packets are very short, as is the case in some instances where packets can be as short as 64 bytes, then the extra crypto overhead suddenly occupies half the bandwidth available. Therefore, when speaking about throughput, define exactly what is measured.

The wire-speed bit rate, packets per second, and actual cumulative payload (such as the data content of packets after headers and trailers are stripped away) divided over time are all very distinct entities and all denote bit rates. Yet they are somehow usurped concepts that some security chip vendors will not hesitate to use liberally to present a favorable view of their product. It is up to the user to discern the truth.

These pointed questions toward a vendor not only formulate a set of expected answers that reflect the environment of a potential user, but they necessarily lead to results that are more meaningful when mapped onto an environment.

SSL Acceleration   From what we discussed, it should be obvious that SSL performance is more an issue at session setup time. An SSL server has a very different workload from a client device in that case, so be careful about what is compared. The server first signs its certificate, which it sends to the client, and then hashes some pieces of information, which it also sends to the client. Signing is a euphemism for encrypting something with a private key. In the public-key arena, people usually talk loosely about encryption when they refer to the operation of encrypting content with someone's public key. In the time rating that vendors typically report to characterize such processes, they quote RSA sign operations per second.

To start with, this number is not clear at all. In fact, all vendors do not use it similarly:

- Some manufacturers include the time that their chip takes to produce hashes, in which case some new questions naturally arise: Which hash algorithm was used? How will the performance change if a different hash algorithm were used instead? What will the performance look like without the hashing?
- Some manufacturers do not include time in the time quotes for these public-key operations to account for hashes because they say that digitally signing takes much more time than producing the hash, so they argue that it can be neglected (or so they think).

Even if we focus on the RSA encrypt/decrypt operations, which in a digital signature environment directly correspond to the processes of verifying and generating a digital signature, respectively, we will need to know the size of the keys if a hardware-assist unit is used, such as for the CRT calculation. In Table 17.1, where we show some qualitative relationships between operations, we used the term *operations* to denote the modular exponentiation of the signature generation or verification process.

Some vendors talk in terms of sessions per second, transactions per second, keys per second, or exponentiations per second. Some vendors also quote figures for the shorter resume-handshake protocols when they should actually be quoting their performance on the full-fledged SSL session setup handshake. It is therefore critical for a prospective user who is comparing security coprocessors to submit the prospective vendors to the rigors of a thorough questioning so that no misunderstandings will occur. The best way to go about doing this is to leave no stones unturned and to insist on figures that show the full-fledged SSL handshake protocol capabilities, where hashes must be calculated, signatures must be generated, and signatures must be verified.

The result of a performance inquiry with a manufacturer should not just yield a couple of numbers as many vendors would have the users believe. It should produce a three-dimensional matrix of numbers that will say a lot to the overall systems architect about how a specific coprocessor might work in a new systems design that he or she is trying to sketch out and budget.

Last but not least, we need to point out that unfortunately several security coprocessor vendors are not fully prepared to easily share this type of objective detail with customers; therefore, be persistent.

SUMMARY

In this chapter, we discussed issues related to funtionality of security coprocessors operating in a network-processing environment. We looked at the fundamentals of most aspects surrounding private- and public-key cryptographic algorithms and operations as required in communications systems. We also discussed the most commonly used security protocols from a systems standpoint. We explained how to compile elaborate educated lists of criteria that consolidate the best practices of cryptography into ways that allow an individual to probe deeper into vendors' claims in order to identify the true substance of a security coprocessor product. Finally, we reviewed the architectures of typical security coprocessors and identified systems engineering techniques of introducing them into larger switching/routing systems next to network processors. We also discussed the trade-offs that are associated with their respective use.

SUGGESTED REFERENCES

Information, documentation, and white papers for products in the areas discussed in this chapter can be found from the web sites of the following companies:

Analog Devices (*www.analogdevices.com*)

Broadcom (*www.broadcom.com*)

Cavium (*www.cavium.com*)

Certicom (*www.certicom.com*)

Corrent (*www.corrent.com*)

Hifn (*www.hifn.com*)

Layer N (*www.layern.com*)

Motorola (*www.motorola.com*)

NTRU (*www.ntru.com*)

NetOctave (*www.netoctave.com*)

RSA Security (*www.rsasecurity.com*)

SafeNet (*www.safenet.com*)

Tehuti Networks (*www.tehutinetworks.com*)

J. Savard's web page is a good source on the internals of several algorithms. See *http://fn2.freenet. edmonton.ab.ca/~jsavard/crypto/comp04.htm*.

All FIPS standards can be found at the corresponding U.S. government's NIST web site at *www.itl.nist.gov/fip-spubs/index.htm*.

Wei Dai's benchmarks using his Crypto++ 3.1 (2000) package can be found at *www.eskimo.com/~weidai/ benchmarks.html*.

Information about the CHES conferences can be found at *http://islab.oregonstate.edu/ches/*.

The proceedings of the CHES workshops are published by Springer-Verlag in its *Lectures Notes in Computer Science* series *www.springer.de/comp/lncs/index.html*.

An astounding number of good books have been written on all aspects of communications security and cryptography. The following sources should be good starting points for discussing the various types of security technology mentioned or discussed in this chapter:

Uyless D. Black, *PPP and L2TP: Remote Access Communications* (Upper Saddle River, NJ: Prentice-Hall, 1999).

David M. Bressoud, *Factorization and Primality Testing* (New York: Springer-Verlag, 1989).

Johannes A. Buchmann, *Introduction to Cryptography* (New York: Springer-Verlag, 2001).

Joan Daemen and Vincent Rijment, *The Design of Rijndael: AES—The Advanced Encryption Standard* (New York: Springer-Verlag, 2002).

Bruce S. Davie and Yakov Rekhter, *MPLS: Technology and Applications* (San Francisco: Morgan Kaufmann Publishers, 2000).

Naganand Doraswamy and Dan Harkins, *IPsec: The New Security Standard for the Internet, Intranets, and Virtual Private Networks* (Upper Saddle River, NJ: Prentice-Hall, 1999).

Sheila Frankel, *Demystifying the IPsec Puzzle*, Artech House Computer Security Series (Boston: Artech House, 2001).

Jim Guichard and Ivan Pepelnjak, *MPLS and VPN Architectures: A Practical Guide to Understanding, Designing, and Deploying MPLS and MPLS-Enabled VPNs* (San Jose, CA: Cisco Press, 2000).

Neal I. Koblitz, *A Course in Number Theory and Cryptography*, Graduate Texts in Mathematics, no. 114 (New York: Springer-Verlag, 1994).

Peter Loshin, *Big Book of IPsec RFC's: Internet Security Architecture* (San Francisco: Morgan Kaufmann Publishers, 1999).

Michael G. Luby, *Pseudorandomness and Cryptographic Applications* (Princeton, NJ: Princeton University Press, 1996).

Alfred J. Menezes, Paul C. Van Oorschot, and Scott A. Vanstone, ed., *Handbook of Applied Cryptography*, CRC Press Series on Discrete Mathematics and Its Applications (Boca Raton, FL: CRC Press, 1996).

Randall K. Nichols, *ICSA Guide to Cryptography* (New York: McGraw-Hill Professional, 1998).

Randall K. Nichols and Panos C. Lekkas, *Wireless Security: Threats, Models, and Solutions* (New York: McGraw-Hill, December 2001).

Randall K. Nichols, Daniel J. Ryan, Julie J. C. H. Ryan, and Arthur W. Coviello, Jr., *Defending Your Digital Assets from Hackers, Crackers, Spies, and Thieves* (New York: McGraw-Hill, 2000).

Eric Rescorla, *SSL and TLS: Designing and Building Secure Systems* (Reading, MA: Addison-Wesley, 2000).

Michael Rosing, *Implementing Elliptic Curve Cryptography* (Greenwich, CT: Manning Publications, 1999).

Bruce Schneier, *Applied Cryptography: Protocols, Algorithms, and Source Code in C*, 2nd ed. (New York: John Wiley, 1996).

G. Seroussi, Nigel P. Smart, and Ian F. Blake, *Elliptic Curves in Cryptography* (New York: Cambridge University Press, 2000).

Richard Shea, *L2TP: Implementation and Operation* (Reading, MA: Addison-Wesley, 1999).

William Stallings, *Cryptography and Network Security: Principles and Practice*, 2nd ed. (Upper Saddle River, NJ: Prentice-Hall, 1998).

Douglas Stinson, *Cryptography: Theory and Practice*, 2nd ed. (Boca Raton, FL: CRC Press/Chapman & Hall, 2002).

Michael Welschenbach, *Cryptography in C and C++* (Berkeley, CA: APress, 2001).

LIST OF ACRONYMS

AAL	ATM Adaptation Layer
ABR	Available Bit Rate
ACK	Acknowledgement signal (of reception)
ACM	Association for Computing Machinery
ADM	Add/drop multiplexer
ADPCM	Adaptive differential pulse code modulation
ADSL	Asymmetric Digital Subscriber Line
AES	Advanced Encryption Standard
AF	Assured Forwarding (for example, AF-PHB)
AFB	Absolute Fairness Bound
AH	Authentication Header
ALU	Arithmetic logic unit
AMAT	Average memory access time
API	Application programming interface
ARP	Address Resolution Protocol
ASIC	Application-specific integrated circuit
ASN	Abstract Syntax Notation
ASSP	Application-specific standard product
ATM	Asynchronous Transfer Mode
AUI	Attachment Unit Interface
BB-DCS	Broadband Digital Crossconnect System
BCAM	Binary content-addressable memory
BECN	Backward Explicit Congestion Notification
BER	Bit-error rate (also basic encoding rules)
BGA	Ball grid array

BGP	Border Gateway Protocol
BIOS	Basic input/output system
BIST	Built-In Self-Test
BLEC	Business/Building Local Exchange Carrier
BNR	Backus-Naur Form
BR	Boot record
BSC	Boundary Scan Cell
BSR	Boundary Scan Register
BST	Boundary Scan Test
CA	Certificate authority
CAS	Column address strobe
CASE	Computer-aided software engineering
CAM	Content-addressable memory
CBC	Cipher block chaining (mode)
CBQ	Class-based queuing
CBR	Constant bit rate
CBR	CAS-before-RAS (double row refreshing mode)
CDMA	Code Division Multiple Access
CDR	Clock and data recovery
CFG	Context-free grammar
CFP	Classification and forwarding processor
CIDR	Classless InterDomain Routing
CIF	Caltech Intermediate Format
CISC	Complex instruction set computer
CLEC	Competitive Local Exchange Carrier
CLIP	Classical IP (over ATM)
CLNP	Connectionless Network Protocol
CML	Current Mode Logic
CO	Central office
CoS	Class of service
CPE	Customer premises equipment

CPI	Clock cycles per instruction
CPP	Control plane processor
CPU	Central processing unit
CRC	Cyclic redundancy check
CRL	Certificate revocation list
CR-LDP	Constraint-based Routing Label Distribution Protocol
CRLSP	Constraint-based Routed Label-Switched Path
CS	Chip select signal
CSIX	Common Switch Interface (specification from the NPF)
CSMA/CD	Carrier sense multiple access with collision detection
CTI	Clock Tree Insertion
CTS	Clear-to-Send signal
DA	Destination address
DARPA	Defense Advanced Research Project Agency
DCA	Digital Certificate Authority
DCS	Digital Crossconnect System
DCT	Discrete Cosine Transform
DDR	Double data rate
DDS	Digital data system
DEDD	Delay Earliest Due Date
DES	Data Encryption Standard
DFB	Distributed feedback
DFT	Design for testability
DFT	Discrete Fourier Transform
DFV	Design for verification
DH	Diffie-Hellman protocol
DHCP	Dynamic Host Configuration Protocol
DiffServ	Differentiated Services
DLB	Dual leaky bucket
DLC	Digital loop carrier
DLCI	Data Link Circuit Identifier

DLEC	Data Local Exchange Carrier
DMA	Direct memory access
DMT	Discrete Multitone
DNS	Domain Name Server
DOD	U.S. Department of Defense
DP	Destination port
DPCM	Differential pulse code modulation
DPP	Data plane processor
DPRAM	Dual-port random access memory
DRAM	Dynamic random access memory
DRC	Design rule checking
DRR	Deficit round robin
DRU	Data recovery unit
DS	Differentiated Services
DS0	Digital Signal Level 0
DS1	Digital Signal Level 1
DSA	Digital Signature Algorithm
DSCP	Differentiated Services Code Point
DSL	Digital Subscriber Line
DSLAM	Digital subscriber line access multiplexer
DSM	Distributed shared memory
DSM	Deep submicron (such as processing or technology)
DSP	Digital signal processing (or processor)
DSS	Digital Signature Standard
DTAU	Digital test access unit
DTE	Data Terminal Equipment
DTMF	Dual Tone Multiple Frequency
DVMRP	Distance Vector Multicast Routing Protocol
DWDM	Dense wavelength division multiplexing
DWRR	Deficit weighted round robin
ECB	Electronic codebook (mode)

ECC	Elliptic-curve cryptography
EC-DH	Elliptic-curve-based Diffie-Hellman protocol
EC-DSA	Elliptic-curve-based Digital Signature Algorithm
ECL	Emitter-Coupled Logic
ECN	Explicit Congestion Notification
EDA	Electronic Design Automation
EDF	Earliest Deadline First
EDFA	Erbium-doped fiber amplifier
EDGE	Enhanced Data Rates for GSM Evolution
EDIF	Electronic Data Interchange Format
EDO/FPM	Extended data-out/fast-page mode
EEPROM	Electrically erasable programmable read-only memory
EF	Expedited Forwarding (for example, EF-PHB)
EFCI	Early Forward Congestion Indication
E-LSP	Experimental-bits-inferred LSP
EPD	Early Packet Discard
EPROM	Erasable programmable read-only memory
EGP	Exterior Gateway Protocol
EOT	End of transmission
ESP	Encapsulating Security Payload
ETSI	European Telecommunications Standards Institute
EU	European Union
FA	Fairness algorithm
FC	Fibre Channel
FCC	Federal Communications Commission
FCFS	First-come first-served algorithm
FCS	Frame checksum (or sequence)
FDDI	Fiber-Distributed Data Interface
FEC	Forward Equivalency Class
FEC	Forward error correction
FECN	Forward Explicit Congestion Notification

FEP	Front-end processor
FF	Flip-flop
FFP	Fast facility protection
FFT	Fast Fourier Transform
FIFO	First-in first-out
FIPS	Federal Information Protection Standard
FPA	Floating point accelerator
FPU	Floating point unit
FPGA	Field-programmable gate array
FR	Frame relay
FSM	Finite state machine
FT	Fault tolerant
FTP	File Transfer Protocol
FTTC	Fiber To The Curb
FTTH	Fiber To The Home
GbE	Gigabit Ethernet
GBR	Guaranteed bandwidth (algorithm)
GCRA	Generic cell rate algorithm
GMII	Gigabit Media-Independent Interface
GPR	General Purpose Register
GPRS	General Packet Radio Service
GPS	Generalized processor sharing
GRE	Generic Route Encapsulation
GSM	Global System for Mobile communication
GUI	Graphical user interface
HDL	Hardware Description Language
HDLC	High-level Data Link Control
HDSL	High-bit-rate Digital Subscriber Line
HDTV	High-definition television
HFC	Hybrid fiber/coax
HLA	High-level architecture

HOL	Head of Line
HSCSD	High-speed circuit-switched data
HTML	Hypertext Markup Language
HTTP	Hypertext Transfer Protocol
IANA	Internet Assigned Numbers Association
IAB	Internet Architecture Board
IAP	Integrated Access Platform
IC	Integrated circuit
ICE	In-Circuit Emulation
ICMP	Internet Control Message Protocol
ICP	Integrated communications provider
IDE	Integrated Development Environment
IDRP	Interdomain Routing Protocol
IEEE	Institute of Electrical and Electronics Engineers
IESG	Internet Engineering Steering Group
IETF	Internet Engineering Task Force
IGMP	Internet Group Membership Protocol
IGP	Interior Gateway Protocol
IKE	Internet Key Exchange
ILEC	Incumbent Local Exchange Carrier
IMA	Inverse multiplexing for ATM
INMS	Integrated Network Management System
IntServ	Integrated Services
I/O	Input/output
IOP	Interoperability
IP	Intellectual property (for example, in IP cores)
IP	Internet Protocol
IPsec	Internet Protocol Security
IPCS	IP over channelized SONET
IR	Instruction register
ISAKMP	Internet Security Association and Key Management Protocol

ISDN	Integrated Services Digital Network
ISIS	Intermediate-System to Intermediate-System protocol
ISO	International Standards Organization
ISP	Internet service provider
ISV	Independent software vendor
ITU	International Telecommunications Union
IV	Initialization vector
IVR	Interactive voice response
IXC	Interexchange carrier
JEDD	Jitter Earliest Due Date
JTAG	Joint Test Action Group
KS	Key schedule
LAN	Local area network
LANE	Local Area Network Emulation
LAI	Look-aside interface
LB	Leaky bucket algorithm
LDAP	Lightweight Directory Access Protocol
LDC	Long-distance carrier
LDP	Label Distribution Protocol
LEC	Local Exchange Carrier
LFIB	Label Forwarding Information Base
LFSR	Linear feedback shift register
LIB	Label information base
LIS	Logical IP subnet
LMDS	Local Multipoint Distribution System
LOF	Loss of frame
LOS	Loss of synchronization
LPM	Longest-prefix match (algorithm)
LSP	Label-switched path
LSR	Label-switched router
LSSD	Level-sensitive scan design

LTE	Line terminating equipment
LU	Lookup
LUT	Lookup table
LVDS	Low-voltage differential signaling
LVS	Layout versus schematic checking
LZ	Lempel-Ziv (compression algorithm)
LZS	Lempel-Ziv standard
MAC	Media Access Control
MAC	Message Authentication Code
MAN	Metropolitan area network
MAPOS	Multiple Access Protocol over SONET
MAU	Medium attachment unit
MBS	Mirror backup site
MD	Message digest
MGCP	Media Gateway Control Protocol
MIB	Management information base
MII	Medium Independent Interface
MIPS	Millions of instructions per second
MIPS	Microprocessor without interlocked pipelined stages
MIMD	Multiple-instruction multiple data
MLD	Multicast listener discovery
MLPS	Millions of lookups per seconds
MMFA	Min-max fair allocation
MMU	Memory management unit
MP-BGP	Multiprotocol Border Gateway Protocol
MPEG	Motion Picture Expert Group
MPLS	Multiprotocol Label Switching
MPOA	Multiprotocol over ATM
MPPS	Millions of packets per second
MSDP	Multicast Source Discovery Protocol
MSPP	Multiservice providing platform

MSR	Multiservice router
MTBF	Mean Time Between Failures
MTTR	Mean Time to Repair
MUX	Multiplexer
MZAP	Multicast Zone Allocation Protocol
NAS	Network-attached storage
NAT	Network Address Translation
NB-DCS	Narrowband Digital Crossconnect System
NE	Network element
NEBS	Network Equipment Building Standards
NGN	Next-generation network
NHRP	Next-Hop Routing Protocol
NIC	Network interface card
NIST	National Institute of Science and Technology
NLRI	Network layer reachability information
NPF	Network Processing Forum
NP	Network processor
NPU	Network processing unit
NRZ	Nonreturn to zero
NSA	National Security Agency
NTP	Network Time Protocol
NTU	Network termination unit
NVRAM	Nonvolatile random access memory
OA	Optical amplifier
OADM	Optical add/drop multiplexer
OAM&P	Operations, Administration, Maintenance, and Provisioning
OC	Optical carrier
OCH	Optical channel
OCS	Optical crossconnect system
OE	Output enable
O/E	Optical to electrical

OEM	Original equipment manufacturer
OFB	Output feedback (mode)
OFC	Optical Fiber Conference
OFDM	Orthogonal frequency division multiplexing
OIF	Optical Internetworking Forum
OSI	Open System Interface
OSPF	Open Shortest Path First
OXC	Optical Crossconnect
P&R	Placement and routing
PBX	Private Branch Exchange
PC	Program counter
PCB	Printed circuit board
PCI	Peripheral Computer Interconnect bus
PCM	Pulse code modulation
PCS	Personal communications service
PDH	Plesiochronous Digital Hierarchy
PDU	Protocol data unit
PECL	Positive Emitter Coupled Logic
PGPS	Packet-per-packet Generalized Processor Sharing
PHB	Per-hop behavior
PHY	Physical layer (of the OSI model)
PIC	Presubscribed Interexchange Carrier
PIM	Protocol Independent Multicasting
PKI	Public Key Infrastructure
PLD	Programmable logic device
PLO	Prefix-length ordering
PLL	Phase-locked loop
PNNI	Private Network-to-Network Interface
POM	Parallel optical module
POP	Point of presence
POS	Packet over SONET

POTS	Plain Old Telephone Service
PPD	Partial Packet Discard
PPP	Point-to-Point Protocol
PQ	Priority queue
PRI	Primary Rate Interface
PRNG	Pseudorandom number generator (or generation)
PROM	Programmable read-only memory
PSTN	Public Switched Telephone Network
PTT	Post, Telephone, and Telegraph
PVC	Permanent virtual circuit
QDR	Quad data rate
QoS	Quality of service
RAID	Redundant array of inexpensive disks
RAM	Random access memory
RARP	Reverse Address Resolution Protocol
RAS	Row address strobe
RBOC	Regional Bell Operating Company
RDMA	Remote DMA
RDRAM	Rambus Dynamic Random Access Memory
RED	Random Early Detection
RF	Radio frequency
RFB	Relative Fairness Bound
RFC	Request for Comments
RIP	Routing Information Protocol
RISC	Reduced instruction set computer
RLDRAM	Reduced-latency dynamic random access memory
ROM	Read-only memory
ROR	RAS only Refresh
RMON	Remote Network Monitoring
RNG	Random number generator (or generation)
RPC	Remote Procedure Call

RPF	Reverse-path forwarding
RR	Round robin
RSA	Rivest-Shamir-Adleman (inventors of the original algorithm)
RSVP	Resource Reservation Protocol
RTBI	Reduced Ten-Bit-Interface
RTCP	Real-Time Control Protocol
RTL	Register Transfer Language
RTOS	Real-time operating system
RTP	Real-Time Protocol
RTS	Request-to-Send signal
R/W	Read/write signal
Rx	Receive signal
RZ	Return to zero
SA	Security association
SA	Source address
SAN	Storage area network
SAR	Segmentation and reassembly
SCE	Search and classification engine
SCP	Service control point
SCP	Security coprocessor
SCSI	Small Computer System Interface
SCTP	Streaming Control Transmission Protocol
SDF	Standard Delay Format
SDRP	Source Demand Routing Protocol
SF	Switch fabric
SFF	Small form factor
SDCC	SONET data communication channel
SDH	Synchronous Digital Hierarchy
SDLC	Synchronous Data Link Control
SDM	Space division multiplexing
SDR	Single data rate

SDRAM	Synchronous dynamic random access memory
SDSL	Symmetric Digital Subscriber Line
SERDES	Serialization and deserialization
SIF	SONET Interoperability Forum
SIMD	Single-instruction stream multiple data streams
SIP	Session Initiation Protocol
SIR	SONET-IP Remote
SLA	Service-level agreement
SLDRAM	Synchronous link dynamic random access memory
SLIC	Subscriber Loop Interface Circuit
SMDS	Switched Multimegabit Data Services
SME	Small medium enterprise
SMP	Symmetric multiprocessor
SMS	Short Message Service
SMT	Simultaneous Multithreading
SNA	Systems Network Architecture
SNMP	Simple Network Management Protocol
SNP	Storage network processor
SNR	Signal-to-noise ratio
SOC	System-on-a-chip
SOHO	Small office/home office
SOI	Silicon on insulator
SONET	Synchronous Optical Network
SP	Source port
SPF	Standard Parasitic Format
SPE	Synchronous payload envelope
SPI	System Packet Interface specification from the OIF
SPMD	Single program multiple data
SRAM	Static random access memory
srTCM	Single-rate three-color marker
SSRAM	Synchronous static random access memory

SSL	Secure Socket Layer
SSP	Superscalar pipeline
STA	Static Timing Analysis
STM	Synchronous Transfer Mode
STP	Shielded twisted pair
STS	Synchronous Transfer Signal
SYN	Not an abbreviation but used as one. A flag in TCP header requesting handshake.
SVC	Switched virtual circuit
TBI	Ten-Bit Interface
TCA	Traffic conditioning agreement
TCAM	Ternary content-addressable memory
TCM	Three-color marker
TCP	Transmission Control Protocol
TDD	Time division duplexing
TDM	Time division multiplexing
TDMA	Time Division Multiple Access
TDMoIP	Time Division Mpultiplexing over Internet Protocol
TE	Traffic engineering
TLB	Translation look-aside buffer
TLS	Transport Layer Security
TLV	Type/Length/Value encoding
TM	Traffic manager (or Management)
TMU	Traffic management unit
TOE	TCP Offload Engine
TOS	Type of Service
TRNG	Truly random number generator (or generation)
trTCM	Two-rate three-color marker
TSI	Time slot interchange
TTL	Time to Live
Tx	Transmit signal
UBR	Unspecified bit rate

UDP	User Datagram Protocol
UMTS	Universal Mobile Telecommunication System
UNI	User-to-Network Interface
URL	Uniform Resource Locator
USB	Universal synchronous bus
UTOPIA	Universal Test and Operations PHY Interface for ATM
UTP	Unshielded twisted pair
VBR	Variable bit rate
VC	Virtual channel
VCI	Virtual Connection Identifier
VCSEL	Vertical cavity surface emitting laser
VCX	Virtual Channel Crossconnect
VHDL	Very High-Speed IC Hardware Description Language
VLAN	Virtual local area network
VLIW	Very long instruction word
VLSI	Very large scale integration
VoD	Video on demand
VoFR	Voice over frame relay
VoIP	Voice over Internet Protocol
VP	Virtual path
VPI	Virtual Path Identifier
VPN	Virtual private network
VPX	Virtual path switch
VSR	Very short reach converter integrated circuit
WADM	Wavelength add/drop multiplexer
WAN	Wide area network
WAP	Wireless Application Protocol
WB-DCS	Wideband Digital Crossconnect System
W-CDMA	Wideband Code Division Multiple Access
WDM	Wavelength division multiplexing
WE	Write enable
WEP	Wired Equivalence Protocol

WFQ	Weighted fair queuing
WLAN	Wireless local area network
WML	Wireless Markup Language
WRED	Weighted Random Early Detect
WRR	Weighted round robin
WSP	Wireless service provider
WTLS	Wireless Transport Layer Security
XFER	Transfer
ZBT	Zero bus turnaround

OVERVIEW OF NETWORK-PROCESSOR PRODUCTS AND PLATFORMS

Vendor	Gigabit Ethernet	2.5 Gbps OC-48	10 Gbps OC-192	40 Gbps	Traffic Manager: Integrated (I) or External (X)	Classification (I/X)	Packet Modification (I/X)	Switch Fabric (Y/N)	PHY and MAC	NPU	Net ASIC	CPU Core	Availability Status
Agere Systems	APP520	FPP, RSP, and ASI chipsets, or APP5XX family	APP750 network processor and traffic manager chipset	—	Integrated inside the NPU or part of chipset	I	I	PI40X, PI40C, and PI-40SAX	N	Y	N	N	OC-48 shipping and fabric chips in production
AMCC	nP3400 or nP3404	nP7250	nP7510	—	GbE (I), 10 Gbps (X) nP5700, and 40 Gbps (X) nP5700 and 5710/5720 pair	I	I	nPX5800 through VIX-v3 to serdes	N	Y	N	N	34XX in production, 72XX samples, 57XX samples, 5800 samples, and 75XX samples (end of 2002)
Bay Micro-systems	—	—	Montego InP™	—	I	I	I	SPI-4 and CSIX	N	Y	N	N	Production
Broadcom	BCM1250	BCM1250	—	—	X	I	I	N, connectable through Hyper-Transport, but with lots of glue or an FPGA bridge	GbE and MAC	Y, but more suited for the control plane	N	N	Production
Cognigine	—	—	VISC INP	—	I	I	I	SPI-4.2	N	Y	N	N	Production

Vendor	Gigabit Ethernet	2.5 Gbps OC-48	10 Gbps OC-192	40 Gbps	Traffic Manager: Integrated (I) or External (X)	Classification (I/X)	Packet Modification (I/X)	Switch Fabric (Y/N)	PHY and MAC	NPU	Net ASIC	CPU Core	Availability Status
EZchip	—	—	NP-1 and QX-1 set	NP-2 and QX-2 set	I	I	I	N, CSIX-L1	MAC	Y	N	N	10G production and 40G samples postponed
IBM	NP2G	NP4GS3 and NP4GS4	—	—	I and X, needed for DiffServ	I	I	IPowerPRS Q64G, DASL, or CSIX	MAC	Y	N	N	Production
Intel	IXP1200 and IXP2400	IXP2400 and IXP2800	IXP2800	—	X and I	I	I	IX bus, UTOPIA 1,2 or 3, SPI-3, and CSIX	1200 (N) and 2400 (Y) with ATM/POS PHY or Ethernet MAC	Y	N	N	Production of 1200 and 2400, and samples of 2800
Internet Machines	—	NPE10-TMC-10 chipset	—	—	Part of chipset	I	I	Y, SE200 serdes	N, SPI-4.2	Y	N	N	Samples
Motorola (C-Port)	C-3e and C-5	C-5e, Q-5, and M-5, set of three chips	C-10 and Q-10 chipset	—	Q-5 and Q-10 chipset	I	I	I	N	N	Y	N	N C-3e, C-5, and C-5e in production, and 10 Gbps set for 2003

Vendor	Gigabit Ethernet	2.5 Gbps OC-48	10 Gbps OC-192	40 Gbps	Traffic Manager: Integrated (I) or External (X)	Classification (I/X)	Packet Modification (I/X)	Switch Fabric (Y/N)	PHY and MAC	NPU	Net ASIC	CPU Core	Availability Status
Paion	GEP4C04	GEP4C04	XEP1010	—	Integrated but rudimentary	I, only on ingress	I	Y, GES0032 and GMII/TBI	GbE and MAC	Y	N	N	Samples
PMC-Sierra	—	—	—	—	—	ClassiPI		TT1 fabric chip	A	N	N	N	Production
Raqia/SafeNet	—	—	RQDP-C10G	—	X	I	I	N	N	Part Y	N	N	Samples
Silicon Access	—	iFlow™ chipset of iPP, iCL, iAC, and iAP	iFlow™ chipset of iPP, iCL, iAC, and iAP	—	Now X with an FPGA and later iTM	iCL	iPP	N, SPI-4.2	N	Y	N	N	Samples
Tran-Switch (Onex)	—	OMNI	—	—	I	I	I	Y, serdes	OC-48 framer and no MAC	Y	N	N	Production
Vitesse	IQ2000 and IQ2200	IQ2200	—	—	I	I	I	N	GbE and MAC	Y	N	N	Production
Wintegra	WinPath™	—	—	—	I	I	I	N	MAC	Y	N	N	Samples
Xelerated	—	—	x10s, x10d, and x10q chipset	x40 and t40 chipset	I	I	I	X, SPI-4.2 or SPI-5	N	Y	N	N	Samples

APPENDIX II

TYPICAL TRAFFIC LOAD (IN MILLIONS OF PACKETS PER SECOND) CORRESPONDENCE AT VARIOUS LINK SPEEDS AND PACKET SIZES

Link	Aggregate Throughput	Type of Link	Equivalent Traffic Load (354-Byte Packets)	Equivalent Traffic Load (84-Byte Packets)	Equivalent Traffic Load (40-Byte Packets)
1 GbE	2 Gbps	Full duplex	0.7 Mpps	2.98 Mpps	6.26 Mpps
1 GbE	1 Gbps	Simplex	0.35 Mpps	1.49 Mpps	3.13 Mpps
OC-12	1.25 Gbps	Full duplex	0.44 Mpps	1.84 Mpps	3.88 Mpps
OC-12	622 Mbps	Simplex	0.22 Mpps	0.92 Mpps	1.94 Mpps
OC-48	2.5 Gbps	Simplex	0.88 Mpps	3.72 Mpps	7.81 Mpps
OC-48	5 Gbps	Full duplex	1.76 Mpps	7.44 Mpps	15.62 Mpps
OC-192	10 Gbps	Simplex	3.53 Mpps	14.88 Mpps	31.25 Mpps
OC-192	20 Gbps	Full duplex	7.06 Mpps	29.76 Mpps	62.5 Mpps
1×10GbE	10 Gbps	Simplex	3.53 Mpps	14.88 Mpps	31.25 Mpps
1×10GbE	20 Gbps	Full duplex	7.06 Mpps	29.76 Mpps	62.5 Mpps
OC-768	40 Gbps	Simplex	14.12 Mpps	59.52 Mpps	125.0 Mpps
OC-768	80 Gbps	Full duplex	28.24 Mpps	119.04 Mpps	250.0 Mpps

APPENDIX III
STANDARDIZATION EFFORTS IN NETWORK PROCESSING

In every new industry, some prominent vendors and hopeful startups introduce pioneering products with features and characteristics that the vendors believe are unique, desirable, and necessary. Sooner and later competition with new differentiable arrivals in the product arena as well as systems integration issues widely faced by the business management of customer organizations (regardless of how enamored or disinterested the customers may ultimately be regarding choices and the adoption of new technologies) lead to a shake-out in the industry. Industry body forums propose, discuss, scrutinize, negotiate, and formulate new specifications. These industry bodies end up establishing guidelines for future development work. This is how industry standards are created. Products must usually conform to a series of complementary standards to ensure that customers appreciate the product capabilities and that interoperability in a customer's systems design is (more or less) straightforward.

In the area of network processing, several major forums address multiple aspects of the technology standardization to ensure the connectivity and (mostly seamless) interoperability of a chip with chips from other vendors, and an objective performance measurement that reflects real-life behavior. This appendix does not intend to delve into the details of the work of these standards. Instead, it provides sources where the nonexpert reader can find more information. The work of these bodies is usually extremely fluid, given the adolescent level of maturity of the technology involved, so many changes are bound to occur. For more information, individuals should systematically monitor the evolution of work produced by these bodies or even join several of these bodies.

NETWORK PROCESSING FORUM (NPF)

The *Network Processing Forum* (NPF) (*www.npforum.org*) is the preeminent body in this industry. It was set up to foster and standardize many aspects of a telecommunications systems design that is based on network-processing technologies. It was built upon previous successful work from two other industry bodies: the *Common Switch Interface Consortium* (CSIX), which was predominantly hardware oriented, and the *Common Programming Interface Forum* (CPIX), which was correspondingly software oriented. The NPF combined the two approaches with new working groups, which also cover benchmarking and education. The NPF's current working groups are discussed in the following sections.

Hardware Working Group (HWG)

The task of this working group according to the NPF is to "define, promote, and deliver hardware interfaces for the packet data path through a network processor (Streaming Interface) as well as for components attached to a network processor, but not in the data path (Look-Aside Interface)." Its work on the Streaming Interface (referred to a little later in this appendix) is largely based on the previous 2.5 Gbps CSIX-L1 specification. This work takes the view that a reduced pin-count approach to a 10 Gbps interface is highly desirable between network processors and other ancillary needed chips such as framers and switching fabrics. The NPF's working group has tried as much as possible to use the work previously accomplished within the *Optical Internetworking Forum* (OIF) *System Packet Interface Level 4* (SPI-4) phase 2 interface. The Look-Aside Interface (again referred to later in this appendix) formalizes the interface with lookup and address-mapped devices such as search engines, classifiers, and encryption coprocessors that operate using a request/response model.

Software Working Group (SWG)

This working group is responsible for the definition, promotion, and delivery of standardized software interfaces comprising initially a framework, namespace, and management *application programming interfaces* (APIs) that are validated through a top-level command set. The overriding goal is to permit a seamless integration between multiple software applications and network processors.

Benchmarking Working Group (BWG)

With all the hype that accompanies newly introduced products and with the usually biased performance rating of products by the vendors who designed and built them, the industry requires a consensus on the methods, tools, and results reporting of tests and benchmarking procedures. The goal of the BWG is to create such a context.

Technical Education and Marketing Working Group (TEMWG)

This working group creates a forum for the continuing education of the industry on network-processing technologies and uses several tools from seminars and presentations during major industry events, congresses, and trade shows to publications and programs of awareness.

Implementation Agreements (IAs)

The sets of interrelated specifications that the NPF produces are known as *Implementation Agreements* (IAs). Finalized and published specifications are free for anyone to download from the NPF's web site. Specifications under development are only available to NPF members. At the time of this writing, the finalized and published NPF specifications are as follows.

From the NPF Hardware IAs,

- Streaming Interface (NPSI) (September 2002) *www.npforum.org/techinfo/HWStreamingIA.pdf.*
- Look-Aside Interface LA-1.0 (June 2002) *www.npforum.org/techinfo/npf2001.114.14a.pdf.*

From the NPF Software IAs,

- API Framework Lexicon (August 2002) *www.npforum.org/techinfo/LexiconIAv1.pdf.*
- Software API Conventions (August 2002) *www.npforum.org/techinfo/FoundationsIAV1.pdf.*

- API Software Framework (August 2002) *www.npforum.org/techinfo/NPF-SwAPI-Software_ FrameworkIAv1.pdf.*
- Interface Management APIs (August 2002) *www.npforum.org/techinfo/InterfacesIAV1.pdf.*

From the NPF Benchmarking IAs,

- NPF *Multiprotocol Label Switching* (MPLS) Application Level Benchmark, Annex, and Template (March 2003) *www.npforum.org/benchmarking/licenseagm_MPLS.shtml.*
- NPF Fabric Benchmarking Traffic Models (March 2003) *www.npforum.org/benchmarking/ licenseagm_Traffic.shtml.*
- NPF Fabric Benchmarking Performance Metrics (March 2003) *www.npforum.org/benchmarking/ licenseagm_Performance.shtml.*
- NPF Switch Fabric Benchmarking Framework (September 2002) *www.npforum.org/techinfo/ BMFabric_IA.pdf.*
- *Internet Protocol version 4* (IPv4) Forwarding (July 2002) *http://www.npforum.org/techinfo/ IPv4.shtml.*

A related agreement from CSIX is the CSIX-L1 specification (August 2001), which is now available at *www.npforum.org/techinfo/CSIX.shtml.*
The following sections discuss other organizations of related interest.

OPTICAL INTERNETWORKING FORUM (OIP)

The OIF (*www.oiforum.com*) has the charter to "foster the development and deployment of interoperable products and services for data switching and routing using optical networking technologies." Plenty of useful information and white papers are available on their web site. For some applications (such as high-speed switch fabric interfaces with backplane and *physical* [PHY] layer interfaces with fiber optics), the content is directly helpful to network-processing professionals or interested readers.
The Electrical Interface IAs from the OIF include the following:

- **OIF-SPI3-01.0**   *System Packet Interface Level 3* (SPI-3): OC-48 System Interface for Physical and Link Layer Devices. It is available online at *www.oiforum.com/public/documents/OIF-SPI3-01.0.pdf.*
- **OIF-SFI4-01.0**   *Serdes Framer Interface* (SFI-4): Proposal for a Common Electrical Interface Between SONET Framer and Serializer/Deserializer Parts for OC-192 Interfaces. It is available online at *www.oiforum.com/public/documents/OIF-SFI4-01.0.pdf.*
- **OIF-SPI4-01.0**   *System Packet Interface Level 4* (SPI-4) Phase 1: A System Interface for Interconnection Between Physical and Link Layer, or Peer-to-Peer Entities Operating at an OC-192 Rate (10 Gb/s). It is available online at *www.oiforum.com/public/documents/OIF-SPI4-01.0.pdf.*
- **OIF-SPI4-02.0**   SPI-4 Phase 2: OC-192 System Interface for Physical and Link Layer Devices. It is available online a *www.oiforum.com/public/documents/OIF-SPI4-02.0.pdf.*
- **OIF-SPI5-01.1**   *System Packet Interface Level 5* (SPI-5): OC-768 System Interface for Physical and Link Layer Devices. It is available online at *www.oiforum.com/public/documents/OIF-SPI5-01.1.pdf.*
- **OIF-SFI5-01.0**   *Serdes Framer Interface Level 5* (SFI-5): Implementation Agreement for 40Gb/s Interface for Physical Layer Devices. It is available online at *www.oiforum.com/public/ documents/OIF-SFI5-01.0.pdf.*

• **OIF-SxI5-01.0** *System Interface Level 5* (SxI-5): Common Electrical Characteristics for 2.488 to 3.125 Gbps Parallel Interfaces. It is available online at *www.oiforum.com/public/documents/OIF-SxI5-01.0.pdf*.

ATM FORUM

The *Asynchronous Transfer Mode* (ATM) Forum (*www.atmforum.org*) has long been the standard-bearer for ATM technologies and standards. The organization's web site provides ample technical material for both the ATM and the MPLS world. It also provides links to other industry bodies that have mostly coherent objectives and interests, such as the *Digital Subscriber Line* (DSL) Forum (*www.adsl.com*), the *European Telecommunications Standards Institute* (ETSI) (*www.etsi.org*) in Europe, and the Telecommunications Technology Committee in Japan (*www.ttc.or.jp/*).

INSTITUTE OF ELECTRICAL AND ELECTRONICS ENGINEERS (IEEE)

The various IEEE 802.3 groups have standardized several aspects of internetworking technology up to the recent 10 Gigabit Ethernet standard (IEEE 802.3ae-2002), and they can all be found at *www.ieee802.org/3/*. Of course, the IEEE web site also gives access to all other IEEE 802 work.

10 GIGABIT ETHERNET ALLIANCE

The 10 Gigabit Ethernet Alliance (*www.10gea.org*) is an industry group that strives to establish a common forum for the further development and acceptance of 10 Gigabit Ethernet technology in different settings and contexts such as data center (storage networks and server farms), *local area network* (LAN), *metropolitan area network* (MAN), and *wide area network* (WAN).

METRO ETHERNET FORUM (MEF)

The MEF (*www.metroethernetforum.org*) is another industry group that works toward the accelerated acceptance and deployment of 10 Gigabit Ethernet technology in metropolitan networks.

INTERNET ENGINEERING TASK FORCE (IETF)

The IETF's web site is located at *www.ietf.org*. All *Request for Comments* (RFC) documents can be accessed for free from their web site.

Other relevant standards include the InfiniBand™, RapidIO™, and HyperTransport™ technologies.

INFINIBAND

InfiniBand architecture is a new interconnect technology defined for servers, which drastically changes the way data centers are built, deployed, and managed. Based on a centralized *input/output* (I/O) fabric, InfiniBand enables greater performance while enhancing reliability and scalability. The technology is based upon an industry-standard, channel-based, switched fabric, point-to-point architecture. Because the I/O subsystem is the biggest bottleneck of computing systems performance, InfiniBand architecture offers three levels of link performance at 2.5 Gbps, 10 Gbps, and 30 Gbps. It enables low-latency communication within the fabric, enabling higher aggregate throughput than traditional standards-based protocols. This gives InfiniBand architecture an edge as an I/O interconnect for data centers.

The technology is considered complementary to Fibre Channel and to Gigabit Ethernet as devices on such networks can have access to InfiniBand-based computational resources that they can share. As this technology is able to support tens of thousands of nodes within a single subnet, it is expected to lead to a better balance of I/O and processing resources within an InfiniBand fabric.

RAPIDIO

Another interesting industry trend that also happens to enjoy the prominent backing of Motorola is the RapidIO technology, which is formally promoted by the RapidIO Trade Association (*www. rapidio.org/home*). RapidIO technology specs cover both parallel and serial types of interfaces using minimal pins and differential current mode signaling. The technology was originally designed for embedded systems, primarily for networking and communications equipment, enterprise storage, and other high-performance embedded markets. It is a data communications standard for interconnecting chips on a circuit board and circuit boards using a backplane. The RapidIO architecture, which is intended to be used in processor and peripheral interfaces, especially in cases where bandwidth and low latency are crucial, is partitioned into a three-layer hierarchy of logical, transport, and physical specifications. This approach enables scalability and future enhancements while maintaining compatibility.

The RapidIO interconnect provides a common connection architecture for general-purpose *reduced instruction set computer* RISC processors, *digital signal processing* (DSP) processors, communications processors, network processors, memory controllers, peripheral devices, and bridges to legacy buses. This efficient architecture enables users to curtail their costs and reduce their time to market and the complexity of designs.

HYPERTRANSPORT

HyperTransport (*www.hypertransport.org*) is a high-speed, high-performance, point-to-point link used for interconnecting integrated circuits on a motherboard. Among its most important characteristics, it can be significantly faster than a *Peripheral Computer Interconnect* (PCI) bus for an equivalent number of pins. It was introduced by AMD, who is one of the major driving forces behind it. In network-processing applications, it finds a niche in the interface between packet-processing *application-specific integrated circuits* (ASICs) and control-path RISC processors as well as between control-path *central processing units* (CPUs) and fast memory.

The following are some interesting articles that will help put these multiple efforts into context: "Intel Reveals Details of Arapahoe/3GIO After SIG Approval" by Mark Hachman, *ExtemeTech* (August 3, 2001), which is available at *www.extremetech.com/article2/0,3973,118922,00.asp*; "Networking Bus Wars" by Linley Gwennap, *Nikkei Electronics Asia Online* (February 2002), which is available at *http://neasia.nikkeibp.com/nea/200202/cmpu_167414.html*; and "Renamed 3GIO Interface Moves Towards Standardization" by Rick Merritt, *EE Times* (April 18, 2002), which is available at *www.eetimes.com/story/OEG20020418S0062*.

PERFORMANCE BENCHMARKING

In terms of objectively and realistically evaluating the performance of network processors, several efforts are noteworthy. Besides the benchmarking efforts of the NPF, The Linley Group has also recently introduced a proposal for a performance test that it calls LinleyBench™. The benchmark is available for licensing directly from the company at a nominal license fee of $1,000. More information can be obtained at *www.linleygroup.com/benchmark/linleybench.html*. This test suite is based on a functional baseline from the NPF's IPv4 forwarding benchmark IA. The LinleyBench 2002 test suite, however, goes beyond basic IPv4 forwarding, which only tests a very limited set of modern network-processor capabilities. As a result, *Differentiated Services* (DiffServ) classification and marking have been added to reflect a more real-life-like environment. Some optional ATM-IPv4 interworking tests are also available. Network-processor vendors are free to choose whether they will self-publish results, adhering to strict disclosure rules, or whether they will have their results certified by The Tolly Group (www.tollygroup.com), a respected and well-known network-equipment performance testing and certification organization.

Another effort toward the standardization of benchmarks for the assessment of computing performance is the *Embedded Microprocessor Benchmark Consortium* (EEMBC) benchmark suites available from EEMBC (www.eembc.org). These tests are favored by CPU vendors because they show the traditional computational performance of processors. However, the EEMBC benchmark suites for the moment lack the testing depth needed for the performance evaluation and the assessment of computational architectures used in real-time diversified networking applications, as they do not contain any flavor of packetized traffic mix based on different *class of service* (CoS) or *quality of service* (QoS) requirements. This comment, however, is applicable only for the moment and it may very well change in the future.

INDUSTRY FORUMS

Some interesting forums for the monitoring of the evolution of the network-processing industry, technology, and network-processing community, in addition to the web sites of the major vendors, which we have covered in several places in this book, include the specialized conferences and trade shows:

- Network Processors Conference (East/West) (*www.networkprocessors.com*)
- Networld+Interop (*www.interop.com*)
- SuperComm (*www.supercomm2002.com/*)
- Microprocessor Forum (*www.mdronline.com*) with its Embedded Processor Forum
- Next-Generation Networks (*www.bcr.com/ngn/default.asp*)
- Communications Design Conference (*www.commdesignconference.com/*)

INDUSTRY ANALYSTS

Leading groups offering substantive industry analysis for the network processing community include the following:

- The Linley Group (*www.linleygroup.com*).
- Microprocessor Forum (*www.mdronline.com*).
- Gilder Technology Report (*www.gildertech.com*).
- Allied Business Intelligence (*www.alliedworld.com/servlets/Home*), among other areas with an emphasis on wireless infrastructure.

INDEX

Note: Boldface numbers indicate illustrations and tables.

100BaseT networks, 33
10Base-T, 13, 15, 17
3090 IBM mainframe, 6
68000 microprocessors, 20
80286 microprocessors, 21
80386 microprocessors, 21

A

abacus switches, 290
acceleration chips, 39
access control lists (ACL)
    classification processors and, 239
    iFlow chipset, 138–141, **139**
    PayloadPlus family network processors and, 115
access networks, 45, **46**, 49–50, 77
    CPort family network processors and, 123
    PowerNP (NPRGS3) and, 62
accelerators, ClearSpeed Technologies and, 170
accountability and billing information, classification
    processors and, 239
acronym list, 397–413
active queue control, traffic managers and, 316
active queue management (AQM), CPort family network
    processors and, 129, 130
active redundancy, 279
Adaptec, 206–209
add/drop multiplexer (ADM), 47, 347
address resolution protocol (ARP), 197
address tunneling, 376
addressing, 5, 53, 56
admission control, 265
advanced encryption standard (AES), 91, 362, 367–369, **368**,
    378, 389, 393
Advanced Interactive Executive (AIX), 8
Advanced Micro Devices, 20
advanced mobile phone system (AMPS), 16
agents, SNMP, 12

Agere, 33, 342
    PayloadPlus (*See* PayloadPlus family network processors)
Agere system interface (ASI), PayloadPlus family network
    processors and, 105, 106, 107, 112–114, **113**
Agere Systems PI40 family chips in, 298–301, **299–301**
Agere's Scripting Language (ASL), PayloadPlus family
    network processors and, 116
aggregate route based IP switching (ARIS), 24, 26
Aho-Corasick algorithm, 254
AIX, 10
Alacritech, 209
Alcatel, 33
Algorithms on Strings, Trees, and Sequences, 254
allocation, traffic managers and, 329
Altera, 177
Ambegaonkar, Prakash, 9
AMCC nP family network processors, 33, 93–103
    architecture, 93–95, **94**
    bandwidth control in, 99
    block structure, **95**
    class of service (CoS) and, 101
    compiler, assembler, debugger for, 97
    fifth generation technology and, 102–103
    interfaces in, 94
    MANs and SANs using, 100–101
    master slave systems, 100
    memory and storage in, 99, 100
    multicasting and, 97
    network optimized instruction set computing (NISC) in, 93, 94
    pipeline bubble and, 96–97
    policy engine in, 95
    quality of service (QoS) and, 97, 99, 100
    serialization and deserialization (serdes), 100, 101
    snooping mode in, 98
    software development using, 96–97
    switch fabrics and, 99–101, **99**
    systems design using, 101–102, **102**
    traffic engineering/management and, 96, 97–99, **98**
    ViX interconnect bus in, 100, 101
    weight arrays in, 95
AMD, 425
Amdahl, 3

American National Standards Institute (ANSI), 18–19, 384

Analog Devices Co., 21

AnyMapping, Montego and InP family and, 142

Apollo, 6

APP550 (*See also* PayloadPlus family network processors), 106

APP550 (EX-INP5) processor, 116–118, **117**

APP750NP (EX-NP10) and APP750TM (EX-TM10) chipset in, 114–116, **115**

AppleTalk, 9

application programming interfaces (API), 422, 423
    AMCC nP family network processors and, 96
    ClearSpeed Technologies and, 172
    CPort family network processors and, 129, **132, 133**
    PowerNP (NPRGS3) and, 67
    storage network processors (SNPs) and, 209

application specific integrated circuit (ASIC), 27, 31, 34, 51, 52, 56, 335, 425
    classification processors and, 245
    content addressable memory (CAM) and, 222
    CPort family network processors and, 128
    design constraints and, 36–38
    design time required for, 37–38
    encryption and, 378
    flexible length instruction set (FLIX) and, 182–184
    iFlow chipset and, 140
    IXA family network processors and, 78
    Net ASICs, 159–160, **161**
    PayloadPlus family network processors and, 106
    PowerNP (NPRGS3) and vs., 63
    programmability constraints in, 37
    speed of processing using, 36–38
    switch fabrics and, 272
    Vitesse IQ/IQ2000 and, 153

Applicon, 6

Applied Cryptography, 370

arbitrated crossbar switch fabrics, 285–287, **285, 286**

arbitration algorithms switch fabrics, 283

ARC cores, 162, 184–186, **185**

architecture issues for NPU, 4, 339–340

area based quad tree (AQT), 250

arithmetic logic unit (ALU), 164
    iFlow chipset and, 139
    PayloadPlus family network processors and, 108
    PowerNP (NPRGS3) and, 62

ARM, 21, 162, 169, 173

Art of Computer Programming, The, 245

assemblers, AMCC nP family network processors and, 97

assembly languages, 20

associated data memory, 53, 220

assured forwarding, 150, 319

asymmetric digital subscriber line (ADSL), 33

asymmetric key encryption, 365

asynchronous transfer mode (ATM), 13, 15–16, 24, 31, 40, 45, 47, 48, 49, 51, 56, 160, 265, 266, 277, 344, 345, 424, 426
    AMCC nP family network processors and, 98, 100
    classification processors and, 240, 243
    Cognigine and, 145
    congestion control in, 324–325
    CPort family network processors and, 125, 126, 128, 129, 133
    DiffServ and, 320
    Fibre Channel and, 197
    IXA family network processors and, 83, 87, 90
    Montego and InP family and, 141, 142, 144
    PayloadPlus family network processors and, 106, 111, 112, 114, 118, 119
    quality of service (QoS) and, 28, 320
    switch fabrics and, 270, 271, 272–273, 280
    Wintegra and, 154

AT&T Bell Labs, 6, 14

ATI, 177

ATM adaptation layer (AAL), 16, 18, 56, 344
    IXA family network processors and, 90
    Montego and InP family and, 142
    PayloadPlus family network processors and, 106
    Wintegra and, 154

ATM Forum, 15, 24, 424

ATM switches, 265

authentication, 26, 27

authentication header (AH), 28, 56, 385–389

automotive industry, 7

availability of network, 47, 50

available bit rate (ABR), 316

Avici, 45

B

backbones, 13, 15, 16, 24, 34, 35, 43–46, 55

backplane, 10, 25, 275–277, 348, 423

backpressure (in-band/out-of-band), switch fabrics and, 274

Baeza-Yates, R., 254

ball grid array (BGA), 140, 258

bandwidth, 12, 13, 15, 22–23, 34–35, **35**, 47
    AMCC nP family network processors and, control of, 99
    Gigabit Ethernet and, 17–18
    traffic managers and, 314
    wide area network (WAN), 48

bandwidth provisioning, traffic managers and, 314

Banyan based switches, 288–289, **289**

Batcher-Banyan switches, 289–290

Bay Microsystems (*See also* Montego and InP family), 141–145, 177

behavior aggregates (BAs), DiffServ and, 319

Bell Labs, 4

benchmarking, 40, 422, 423, 426
    PowerNP (NPRGS3) and, 72
    Tensilica and, 178–180, **178, 179, 180, 181**

Benchmarking Working Group (BWG), 422

best efforts forwarding, DiffServ and, 319
beta elements, switch fabrics and, 269–270, **269**
"big iron", 4
billing information, classification processors and, 239
binary CAM (BCAM), 53, 220–221
binary trees, 245
bit level parallelism, 249–250
Black, Ulysses, 9
block ciphers, 367–369, **368**
blocking, 265, 268–269
boot sequence, IXA family network processors and, 80
border gateway protocol (BGP), 11, 65, 232, 241
Boyer-Moore algorithm, 254
bridges, 5, 7, 8, 9, 87, 141
broadband networks, 19–20, 35, 45
Broadband Packet Switching Technologies, 282
broadcast switches, 282
Broadcom, 33, 39
browsers, 10
    micro-, 16
brute force approach, 254
BSC protocol, 7, 10
bubbles, 339
buffer management unit (BMU), CPort family network
    processors and, 125, 127
buffered crossbar switch fabrics, 285
buffers, 55, 283–284
    switch fabrics and, 273–274
    traffic managers and, 326–329
bugs, 34
Burroughs, 3
burst tolerance (BT), 114
bus architectures, 10
bus function model (BFM), 177
buses
    CPort family network processors and, 127
    IXA family network processors and, 83, 88
    PayloadPlus family network processors and, 107
business management, 39
byte slicing, 287
byte substitution encryption, 379

C

C/C++, 7, 342
CAB interface coprocessor, PowerNP (NPRGS3) and, 68–70, **68**
cable broadband networks, 45
Cabletron, 13
cache, 36
campus networks, 13
capacity, traffic managers and, 313
card punch computers, 3–4
carrier sense multiple access with collision detection
    (CSMA/CD), 5

carriers, 8
categories of network processors, 39–40
Cavium, 39
cell delay variation tolerance (CDVT), 114
cell loss priority (CLP), PayloadPlus family network
    processors and, 114
cell phones, 16
cell switched router (CSR), 24, 26
cells, 31, 57, 266, 273
cells, ATM, 15
cellular digital packet data (CDPD), 17
central processing units (CPUs), 4, 6, 10, 20, 21, 22, 27,
    28, 33, 38, 51, 52, 55, 56, 163, 353, 425
    AMCC nP family network processors and, 93
    bandwidth demand and, 34–35, **35**
    CPort family network processors and, 123
    IXA family network processors and, 79
    off-the-shelf, design using, 34–36
    packet processing and, 34
    PayloadPlus family network processors and, 107
    performance restraints of, 34–36, **35**
    PowerNP (NPRGS3) and, 62
    switch fabrics and, 276
certificate authority (CA), 375
channel processors (CPs), CPort family network processors
    and, 125
Chao, H. Jonathan, 282
Cheapernet, 13
Checksum coprocessor, PowerNP (NPRGS3) and, 68–70, **68**
checksums
    Montego and InP family and, 142
    PayloadPlus family network processors and, 108
Chinese remainder theorem (CRT), 382
Chiussi, F., 282
cipher block chaining (CBC) mode, 370, 371
cipher feedback (CFB) mode, 370
cipher text auto key mode (CTAK), 370
ciphertext, 363
circuit emulation services (CES), Wintegra and, 154
circuit-switched networks, 8–9, 11, 265
Cisco Systems, 10, 24, 26, 45, 156, 209
class based queuing (CBQ), 327
class of service (CoS), 57, 426
    AMCC nP family network processors and, 101
    classification processors and, 238
    iFlow chipset, 138–141, **139**
    Montego and InP family and, 141, 142
    PayloadPlus family network processors and, 105
    PayloadPlus family network processors and, 109, 115
    search engines and, 220
    switch fabrics and, 277
    traffic managers and, 311, 315
class selector, 150
classification engines, 219–220
classification of packets, 52–53, 55–56

classification processors, 237–261, 335
  access control lists (ACLs) and, 239
  accountability and billing information in, 239
  Aho-Corasick algorithm in, 254
  algorithms and structures to support, 245–247
  area based quad tree (AQT) in, 250
  asynchronous transfer mode (ATM) and, 240, 243
  bit level parallelism in, 249–250
  Boyer-Moore algorithm in, 254
  brute force approach to, 254
  classes of IP addresses and, 240–241, **241**
  classless interdomain routing (CIDR), 241
  content addressable memory (CAM) vs., 252–255, **253**
  cross producting, 249–250
  deep packet classification and, 238, 247–249, **248**
  dictionary and, 245
  differentiated services (DiffServ) and, 238
  DNA matching sequences and, 253–254
  Ethernet and, 243
  fat inverted segment (FIS) tree in, 250
  field programmable gate array (FPGA) and, 255
  forwarding and, 238, 239, **239**, 240–244, **241**, **243**
  forwarding tables, 238
  hash buckets in, 247
  hidden Markov model (HMM) algorithm in, 254
  hierarchical intelligent cuttings approach in, 250
  Horspool algorithm in, 254
  implementation of, 251–252, **252**
  integrated services (IntServ) and, 238
  integrated vs. standalone systems in, 256
  intrusion detection systems (IDS) and, 256–259
  IPv6 and, 241
  Knuth-Morris-Pratt (KMP) algorithm in, 254
  leaf pushed binary trees in, 245
  longest prefix match (LPM), 240, 245, 251
  lookup and classification in, 238, **239**, 240–244, **241**, **243**
  lookup and search defined for, 237–239
  lookup tables in, 244–245
  Lulea scheme in, 246–247
  media access control (MAC) and, 255
  memory vs. search time in, 246
  multiple-field classification in, 249–251
  multiprotocol label switching (MPLS) and, 240, 243
  Needleman-Winsch algorithm in, 254
  NetLogic Microsystems and, 251
  network address translation (NAT) and, 241
  network ID (netID) and, 240–241
  network processing elements in, 256
  packet classification and, 237
  Patricia trie in, 245
  prefix length ordering (PLO) in, 244–245
  programmable logic device (PLD) and, 255
  programmable state machine (PSM) and, 253
  quality of service (QoS) and class of service (CoS) in, 238
  radix trie in, 245, 246
  Raqia regular expression classification coprocessor in, 256–259, **257**
  recursive flow classification (RFC) in, 250, **251**
  Smith-Waterman algorithm in, 254
  SSRAM and, 254
  Sunday algorithm in, 254
  ternary CAM (TCAM) in, 240, 243, 244, 251
  traffic management and, 239
  traffic shaping and, 239
  trie in, 245–246, **246**, 249
  type of Service (TOS) and, 238
  virtual connection identifiers (VCI) and, 240
  virtual private network (VPN) and, 238, 242
classless interdomain routing (CIDR), 52, 229, 241
ClearConnect, 170, 234
ClearSpeed Technology, 162, 164–173, **165**, **167**, **168**, 233–235
Clearwater Networks, 137
client-server model, 7–8, 12, 13, 36
clock speed (*See* speed, clock speed)
coaxial cable, 7, 13, 23
code division multiple access (CDMA), 16
Cognigine, 145–148, **146**, **148**
Comer, D.E., 342
committed information rate (CIR), 114
common lisp object system (CLOS), 7
Common Programming Interface Forum (CPIX), 421
Common Switch Interface (CSIX), 153, 421
compilers
  AMCC nP family network processors and, 97
  ClearSpeed Technologies, 171
  IXA family network processors and, 86
  PowerNP (NPRGS3) and, 72
complement keys, 372
complementary metal oxide semiconductors (CMOS), 34, 73, 118
complex instruction set computers (CISC), 20, 31, 51, 52, 339, 341
compressor chips, 39
computer aided design/computer aided manufacturing (CAD/CAM), 5, 6, 7
computer assisted software engineering (CASE), 22
computers, evolution of, 3–29, **29**
ComputerVision, 6
Conexant, 33
configuration bus interface (CBI), PayloadPlus family network processors and, 107
congestion management, 15
  switch fabrics and, 281
  traffic managers and, 322–325, **323**, **324**, **325**
connectivity, 7

constant bit rate (CBR), 99, 130, 316
constraint based routing label distribution (CR LDP), 320
content addressable memory (CAM) (*See also* search engines), 52, 53–54, 220, 335, 338–339
  application specific integrated circuit (ASIC) and, 222
  cautions against use of, 230–232
  classification processors and vs., 252–255, **253**
  classless interdomain routing (CIDR) and, 229, 241
  developments in, 233–234
  DNA matching sequences and, 253–254
  double data rate (DDR) bus and, 229
  global mask registers in, 222
  iFlow chipset, 138–141, **139**
  IXA family network processors and, 82
  learning in, 223
  longest prefix match (LPM) and, 229, 240, 245, 251
  lookup tables and, 224–228, 224, **224**
  mask registers in, 222
  maximum search capability of, 224–225, **225**
  PayloadPlus family network processors and, 107, 117
  PowerNP (NPRGS3) and, 69
  prefix length ordering (PLO) in, 244–245
  programmable state machine (PSM) and, 253
  search engines and, 219, 220–234, **222**
  search keys and, 229
  skip bits in, 227
  sort-free type, 233
  speed of, 223–225, **225**, 229, 230
  structure of, 221–226, **222**
  systems engineering issues of, 227–230, **228**
  tag bits in, 223, **223**, 227
  TOPcore (EZchip) and, 152
  validity bits in, 227
  Xelerated packet devices and, 155
  See also associated data memory.
Control Data, 3
control plane processing, 50–51, 353
control point (CP) processors, PowerNP (NPRGS3) and, 64
converged networks, 16, 19, 35
cookies, 46
cooperative computing (*See also* client-server model), 7
coprocessors and assist hardware, PowerNP (NPRGS3) and, 68–70, **68**
core language processor (CLP), PowerNP (NPRGS3) and, 70
core networks, 50
core router, 45, **46**, 95
Corrent, 39, 134
cost of software development, 342–344, **343**, **344**
counter coprocessor, PowerNP (NPRGS3) and, 68–70, **68**
CoWare, 177–178
CPort family network processors, 123–135, **124**
  application programming interfaces (APIs) and, **132**, **133**
  architecture of, 124–127, **125**

buffer management unit (BMU) in, 125, 127
buses in, 127
C Ware for, 131–134
channel processors (CPs) in, 125
executive processor (XP) in, 125, 127
fabric processor (FP) in, 125, 126–127
HiTCE CBGA package for, 124
interfaces and, 123–124
M-5 Channel Adapter and, 126, 127
memory and storage in, 126, 128, 129
packet data units (PDU) in, 127
pipelining in, 125–126
PowerQUICC architecture and, 123
Q-5 TMC in, 127–131, **128**, **129**, **132**
quality of service (QoS) and, 126, 129, 131
queue management unit (QMU) in, 125, 126
serial data processors (SDP) in, 124–125
software development kit (SDK) for, 131–135
speed, throughput, clock speed in, 124, 125
switch fabrics and, 135
systems design using, 134–135
table lookup unit (TLU) in, 125, 126
traffic engineering/management and, 123
traffic management coprocessor (TMC) in, 125
traffic management interface (TMI) in, 128, 130
very long instruction word (VLIW) engines in, 124
credit table mechanisms, 274
cross producting, 249–250
crossbar switch fabrics, 266, **266**, 285–287, **285**, **286**
  (*See* arbitrated crossbar)
  (*See* buffered crossbar)
crossconnects, 276
crosspoint switch, 267–268, **267**, **268**
Crypto++, 382
cryptography (*See* encryption)
CSIX, 275, 316
current mode logic (CML), 276
custom programmable logic device (CPLD), 177, 348
customer network, 44
customer premises, 44
customer premises equipment (CPE), 44, 49, 50, 77
cut through designs, 55
CWare, CPort family network processors and, 131–134
cyclic redundancy check (CRC), 8, 55, 56
  CPort family network processors and, 125
  IXA family network processors and, 77, 90
  Montego and InP family and, 142
  PayloadPlus family network processors and, 108
  Tensilica and, 174
Cypress coprocessor, 134

D

Dai, Wei, 382

data aligned serial links (DASLs), PowerNP (NPRGS3) and, 65–68, **66**

data encryption standard (DES), 362, 367–369, **368**, 372, 377–378, **378**, 385, 389, 393

Data General, 4, 6

data link layer, 9

data plane processing, 50–51, 64

data store coprocessor, PowerNP (NPRGS3) and, 68–70, **68**

datagrams, 270

debuggers, 97, 348

decapsulation, storage network processors (SNPs) and, 205

decryption/deciphering, 363

DECtalk, 9

Deegan, J.J., 282

deep packet classification/inspection, 205, 238, 247–249, **248**

Defense Advanced Research Projects Agency (DARPA), 9

deficit round robin (DRR), 130, 327

deficit weighted round robin (DWRR), 327

delay, switch fabrics and, 281

delay earliest due date (DEDD), 327

dense wavelength division multiplexing (DWDM), 35, 142

descriptor buffer, CPort family network processors and, 130, 131

deterministic design, 160

dictionary
   addresses, 56
   classification processors and, 245
   Cognigine and, 146

differential mode positive emitter coupled logic (DM-PECL), 276

differentiated services (DiffServ), 27, 28, 52, 53, 55, 57, 318–320, 344, 426
   classification processors and, 238
   iFlow chipset, 138–141, **139**
   IXA family network processors and, 88
   Montego and InP family and, 141, 142
   PowerNP (NPRGS3) and, 66, 72
   TOPcore (EZchip) and, 150
   traffic managers and, 315, 328
   Vitesse IQ/IQ2000 and, 153

Diffie-Hellman (DH) encryption, 380–382, 384

DiffServe code point (DSCP), 318

digital certificates, 375

Digital Equipment Corporation (DEC), 4, 5, 6, 7

Digital Intel Xerox (IDX), 344

digital processing, 21

digital signal processing (DSP), 21, 22, 105, 162, 425
   security coprocessors and, 389
   Wintegra and, 154

digital signals, 11

digital signature algorithm (DSA), 382, 393

digital signature standard (DSS), 383

digital signatures, 374

digital subscriber line (DSL), 35, 45, 97, 424

digital subscriber loop access multiplexers (DSLAMs), 63, 123

direct memory access (DMA), 170
   IXA family network processors and, 79
   PayloadPlus family network processors and, 112, 113
   storage network processors (SNPs) and, 209
   Tensilica and, 174

directly attached storage (DAS) devices, 193

DNA matching sequences, classification processors and, 253–254

domain of interpretation (DOI), IPSec and, 387

Doraswamy, N., 385

double cycle deselect (DCD), iFlow chipset and, 139

double data rate (DDR) bus, 117, 337
   content addressable memory (CAM) and, 229
   CPort family network processors and, 128
   iFlow chipset and, 140
   IXA family network processors and, 82
   PowerNP (NPRGS3) and, 63

DS-0, AMCC nP family network processors and, 98

dual leaky bucket (DLB)
   Montego and InP family and, 142
   PayloadPlus family network processors and, 113, 114
   traffic managers and, 322

dual round robin matching (DRRM), 283

dual tone multiple frequency (DTMF), 87

dyadic processor units (DPPUs), PowerNP (NPRGS3) and, 64

dynamic RAM (DRAM), 73, 170–171, 336–338
   AMCC nP family network processors and, 99
   iFlow chipset, 138–141, **139**, 138
   PayloadPlus family network processors and, 116
   search engines and, 221
   TOPcore (EZchip) and, 149, 152
   Xelerated packet devices and, 156

E

E1, 48

E3, 48

earliest deadline first (EDF), 327

early forward congestion indication (EFCI), 323

early packet discard (EPD), 57, 324–325
   AMCC nP family network processors and, 99
   Montego and InP family and, 142
   PayloadPlus family network processors and, 109

eBusiness, 13

eCommerce, 13

economic climate and network processing, 137

edge label switched routers (Edge LSRs), 25

edge networks, 50

edge routers, 45, **46**, 47, 123

egress enqueue/dequeue scheduler (E-EDS), PowerNP (NPRGS3) and, 65–68, **66**
egress points, 33
El Gamal encryption, 383
electrical interfaces, 423–424
electrical wavelength routers, 25
electrically erasable PROM (EEPROM), 209
electronic codebook (ECB) mode, 370, 371
Electronic Design Automation (EDA), 6, 177
elliptic curve cryptography (ECC), 384
email, 49
embedded microprocessor benchmark consortium (EEMBC), 426
Embedded Multiprocessing Benchmarking Consortium (EEMBC), Tensilica, 178–180, **178, 179, 180, 181**
embedded powerPC complex (ePPC), PowerNP (NPRGS3) and, 65–68, **66, 67**
embedded processor complex (EPC), PowerNP (NPRGS3) and, 65–68, **66, 67**, 69, 70
Embedded Processor Forum, 165
embedded sequence counters, 375–376
encapsulating security payload (ESP), 28, 56, 385–388, **387**
encapsulation, 26, 28, 56
enciphering, 363
encrypt decrypt encrypt (EDE), 377
encryption, 26, 27, 28, 361, 362, 363–384
   address tunneling and, 376
   advanced encryption standard (AES)), 367–369, **368**, 378, 389, 393
   algorithms for, 377–384
   asymmetric key, 365
   block ciphers in, 367–369, **368**
   byte substitution in, 379
   Chinese remainder theorem (CRT), 382
   ciphertext in, 363
   Crypto++, 382
   data encryption standard (DES), 367–369, **368**, 372, 377–378, **378**, 385, 389, 393
   decryption/deciphering in, 363
   Diffie-Hellman (DH), 380–382, 384
   digital certificates and, 375
   digital signature algorithm (DSA), 382, 393
   digital signature standard (DSS) in, 383
   digital signatures and, 374
   El Gamal, 383
   elliptic curve cryptography (ECC), 384
   embedded sequence counters in, 375–376
   enciphering in, 363
   encrypt decrypt encrypt (EDE) in, 377
   error correction codes (ECC) in, 371
   error extension in, 371
   extended Euclid algorithm in, 382
   Galois field multiplication in, 379
   hashing, 373–374

IDEA, 385
initialization vectors and, 370
IXA family network processors and, 91
key generation and scheduling in, 367
lattice based, 384
MD5, 382, 385, 389
message authentication code (MAC) in, 374
modes for, 369–371, **370**
Montgomery exponentiation in, 383
pretty good privacy (PGP), 365
private key/symmetric key, 364
protocol sensitive, 373
public key, 364–367, 382–384
random number generator (RNG) in, 366
real time protocol (RTP) and, 371
rekeying and, 377
Rijndael algorithms in, 378–380, **379**, 384
RIPEMD, 382
RSA Technology, 382, 383, 385, 393, 394
S boxes in, 367
secure hashing algorithm (SHA), 382, 385, 389, 392
security associations (SAs) and, 377
seed values and, 370
session key exchange and, 375
stream ciphers in, 369
timestamps (nonrepudiation) and, 376
transmission control protocol (TCP) in, 371
triple DES, 377–378, **378**, 385, 392
user datagram protocol (UDP) in, 371
weak keys in, 372
enhanced data rates for GSM evolution (EDGE), 17
Enqueue coprocessor, PowerNP (NPRGS3) and, 68–70, **68**
enterprise network, 44, 47
erasable PROM (EPROM), 34
error correction code (ECC)
   CPort family network processors and, 127
   encryption and, 371
   iFlow chipset and, 139
   IXA family network processors and, 77
error extension, encryption and, 371
EtherChannel, storage network processors (SNPs) and, 209
Ethernet, 5, 7, 10, 12–13, 15, 17–18, 23, 44, 46, 47, 48, 51, 55, 271, 275, 344, 347, 348, 424
   AMCC nP family network processors and, 98, 100, 101
   classification processors and, 243
   Cognigine and, 145
   CPort family network processors and, 123, 125, 129
   Fibre Channel and, 197
   iFlow chipset and, 139
   IXA family network processors and, 86, 87, 90
   Montego and InP family and, 142, 144
   PayloadPlus family network processors and, 106, 116, 118

serialization and deserialization (serdes) module, 277
storage area networks (SANs) and, 195, 204, 205
storage network processors (SNPs) and, 209
TOPcore (EZchip) and, 149, 150, 151
Wintegra and, 154
EtherType, 64
European Telecommunications Standards Institute (ETSI), 424
evolution of network processing, 48–49
evolution of network technology, 3–29, **29**
excess bandwidth weighted fair queuing (EB-WFQ), 130
executive processor (XP), CPort family network processors and, 125, 127
EX-INP5 (*See* APP550 [EX-INP5] processor)
EX-NP10/EX-TM10 (*See* APP750NP [EX-NP10] and APP750TM [EX-TM10] chipset)
EXP inferred label switched paths (E LSPs), 320
expedited forwarding, 150, 319
experimental (EXP) bits, multiprotocol label switching (MPLS), 320
explicit congestion notification (ECN), 323
extended Euclid algorithm, 382
extensible markup language (XML), 119
exterior gateway protocol (EGP), 11
extranets, 13–14
EZchip (*See also* TOPcore), 138, 148–152, 165

F

fab and fabless, 22, 34
fabric processor (FP), CPort family network processors and, 125, 126–127
fair queuing (FQ), 327
fairness concept, 322
fast cycle DRAM (FCDRAM), 338
fast cycle RAM (FCRAM), 73, 115, 116, 117
Fast Ethernet, 12–13, 17, 18, 44
IXA family network processors and, 86–90, **89**
PowerNP (NPRGS3) and, 62, 63, 64
fast forwarding, 55–56
fast path processing, 33, 50–51, 108–109, 272
fast pattern processor (FPP), PayloadPlus family network processors and, 105, 106–109, **108**
FastChip, 251
fat inverted segment (FIS) tree, 250
fat pipe, PowerNP (NPRGS3) and, 63
fault tolerance, 50
FCIP, storage network processors (SNPs) and, 199, 204
Federal Information Protection Standard (FIPS), 392
Festino (*See also* PayloadPlus family network processors), 119
fiber distributed data interface (FDDI), 12–13
fiber optics (*See also* Fibre Channel; optical networks), 9, 19–20, 18–19

Fibre Channel, 18, 196–199, **198**, 425
CPort family network processors and, 125
FCIP and, 204
IP storage and, 199–200
iSCSI bridging and, 204
security issues and, 214–215
serialization and deserialization (serdes) module, 277
storage area networks (SANs) and, 195–196, 195
Fibre Channel arbitrated loop (FCAL), 19
Fibre Channel Industry Association (FCIA), 197
Fibre Channel over IP, 44
field programmable gate array (FPGA), 40, 54, 177, 348
AMCC nP family network processors and, 101
classification processors and, 255
encryption and, 378
IXA family network processors and, 86–87
fifth generation technology, AMCC nP family network processors and, 102–103
file transfer protocol (FTP), 10, 72, 314
finite state machines (FSM), 173, 270
firewalls, 27, 63, 376
FIRM, 283
first in first out (FIFO), 28, 57, 88, 130, 276, 329
five-tuple lookup, 52
flash memory, 22, 34, 80
flat keys, 372
flexible length instruction extension (FLIX), 173, 182–184
floating point processors, 6
flow control, PowerNP (NPRGS3) and, 63
flow through, traffic managers and, 312
FOCUS connect, Vitesse IQ/IQ2000 and, 153
forward equivalency class (FEC), 25
forward error correction (FEC), 23
forwarding, 55–56, 72, 205, 238, 239, **239**, 240–244, **241**, **243**, 319, 423
forwarding policies, 28, 52
forwarding tables, 238
fractional T1, 47, 98
frame based deficit round robin (FBDRR), 130, 327
frame relay, 9, 10, 26, 28, 45, 48
CPort family network processors and, 125, 129
Montego and InP family and, 141, 142
PayloadPlus family network processors and, 105, 112
framers, 105
frames, 8, 9, 51, 266, 273
framing of packets, 51
Fujitsu, 33
full match (FM) algorithm, 69
functional blocks in networking equipment, 32–33, **32**
functional bus interface (FBI), PayloadPlus family network processors and, 107, 109
functional programming language (FPL), 106, 114–115, 118–120, 342

G

Galois field multiplication, encryption and, 379
gateways, 5, 45
general data handler (GDH), PowerNP (NPRGS3) and, 71
general packet radio service (GPRS), 17
general powerPC handler (GPH), PowerNP (NPRGS3) and, 71
general purpose processors (GPP), PowerNP (NPRGS3) and, 70
general table handler (GTH), PowerNP (NPRGS3) and, 70
generalized processor sharing (GPS), 328
generic cell rate algorithm (GCRA), PayloadPlus family network processors and, 112, 113
Gigabit Ethernet, 13, 15, 17–18, 33, 40, 44, 45, 46, 47, 48, 50, 275, 347, 348, 424, 425
    AMCC nP family network processors and, 93, 100, 101
    CPort family network processors and, 125
    Fibre Channel and, 198
    IXA family network processors and, 86, **89**
    Montego and InP family and, 144
    PayloadPlus family network processors and, 106, 116, 118
    PowerNP (NPRGS3) and, 65–68, **66**, 72
    storage area networks (SANs) and, 195, 204, 205
    storage network processors (SNPs) and, 209
    TOPcore (EZchip) and, 149, 151, 152
Gigabit Ethernet, 24
gigabit media independent interface (GMII), 51
    CPort family network processors and, 125
    PayloadPlus family network processors and, 116
    PowerNP (NPRGS3) and, 65–68, **66**
    TOPcore (EZchip) and, 149
GigaStream, Vitesse IQ/IQ2000 and, 153
global bus, CPort family network processors and, 127
global mask registers, content addressable memory (CAM) and, 222
global semaphore unit, 171
global system for mobile communication (GSM), 16
graceful degradation, switch fabrics and, 278
graduated dropping regions, traffic managers and, 324
graphical user interfaces (GUI), 340
graphics, 6
guaranteed bandwidth weighted fair queuing (GBWFQ), 130
guaranteed bandwidth WFQ (GB WFQ), traffic managers and, 327
guided frame handler (GFH), PowerNP (NPRGS3) and, 70–71
guided frames, 64
Gupta, P., 250
Gusfield, Dan, 254
Gwennap, Linley, 426

H

Hachman, Mark, 426
handlers, PowerNP (NPRGS3) and, 70–71
handshakes, SSL, 388
hard disk drives, 6
hardware control interface (HCI), 348, 349
hardware description language (HDL), Vitesse IQ/IQ2000 and, 153
Hardware Working Group (HWG), 422
Harkins, D., 385
hash buckets, 247
hash unit, IXA family network processors and, 79
hashed MAC (HMAC), 374
hashing, 373–374
HCL, 135
head of line (HOL) blocking, 269
helix switches, 291
Hewlett-Packard, 4
hidden Markov model (HMM) algorithm, 254
hierarchical intelligent cuttings approach, 250
hierarchy of networks, **44**
HiFn, 39, 215
high level data link control (HDLC), 347
    CPort family network processors and, 125
    IXA family network processors and, 87
    PowerNP (NPRGS3) and, 62
    Wintegra and, 154
high speed serial (HSS) links, 275
high touch, ClearSpeed Technologies and, 173
history of networking (*See* evolution of network technology)
Hitachi, 169
HiTCE CBGA packages, CPort family network processors and, 124
Horspool algorithm, 254
host bus adapters (HBAs), storage network processors (SNPs) and, 199, 200
host processing, 33
hot pluggable, 47, 50
hot swapping, Fibre Channel and, 197
hub and spoke architectures, 13
hubs (*See also* repeaters), 13
Huitema, Christian, 9, 240
Hurricane accelerator, CPort family network processors and, 134
hybrid wavelength routers, 25
hyper task chaining, IXA family network processors and, 78
hypertext markup language (HTML), 10
hypertext transfer protocol (HTTP), 10, 52, 259, 362
HyperTransport, 424, 425

I

IB, 33

IBM, 3–10, 13, 20, 21, 24, 26, 33, 34, 138, 148, 156, 162, 182, 284, 287
  PowerNP (*See* PowerNP), 61–75, **62**

IBM PowerPRS, 291–298, **293–297**

IDEA, 385

IDT, 54, 101

IEEE 488, 10

IEEE 802.11, 17

IEEE 802.16, 17

IEEE 802.2, 18

IEEE 802.3, 5, 90

IEEE 802.3ae, 17

IEEE 802.4, 7

IEEE 802.5, 5

iFlow chipset, 138–141, **139**

implementation agreements (IAs), 422–423

implementation specific instruction set (ISIS), ClearSpeed Technologies and, 172

Improv Systems, 162, 186–187, **186**

industry analysts, 427

industry forums, 426

Infineon, 22

InfiniBand, 424, 425

information technology (IT), evolution of, 3–4

ingress enqueue/dequeue scheduler (I-EDS), PowerNP (NPRGS3) and, 65–68, **66**

ingress points, 33

initialization vectors, 370

InP family (*See also* Montego), 141–145

input- and output-buffered switches in, 283–284

input/output (I/O), 6, 10, 50
  CPort family network processors and, 125
  InfiniBand, 425
  IXA family network processors and, 88
  PowerNP (NPRGS3) and, 65
  RapidIO, 425
  switch fabrics and, 266

Institute of Electrical and Electronics Engineers (IEEE), 5, 384, 424

instruction set, PowerNP (NPRGS3) and, 62, 174

instruction set simulators (ISS), 177

integrated services (IntServ), 27, 28
  classification processors and, 238
  traffic managers and, 317–318

integration, storage network processors (SNPs) and, 205

Intel, 5, 20, 21, 33, 34, 138, 342

intelligent cuttings approach, classification processors and, 250

Intelligent Network Processor (Cognigine), 145

interface management, 423

interfaces
  AMCC nP family network processors and, 94
  CPort family network processors and, 123–124
  switch fabrics and, 275

Intergraph, 6

interior gateway protocol (IGP), 10

Interconnections, Bridges, Routers, Switches and Internetwork Protocols, 9, 24

Internet, 9–11, 13, 45

Internet control message protocol (ICMP), 344

Internet Engineering Task Force (IETF), 10, 11, 16, 24, 26, 27, 241, 316, 372, 424

Internet exchange architecture (IXA) (*See also* IXA family network processors), 77

Internet Key Exchange (IKE), 385, 387, 393

Internet Packet Exchange (IPX), 9, 26

Internet protocol (IP/IPv4) (*See also* IPv6; TCP/IP), 9, 10, 13–14, 16, 18, 26, 35, 43, 160, 168, 265, 340, 344, 423, 426
  Cognigine and, 145
  CPort family network processors and, 125
  iFlow chipset and, 139, 141
  Montego and InP family and, 141, 142
  PayloadPlus family network processors and, 115
  storage area networks (SANs) and, 195–196
  switch fabrics and, 271
  TOPcore (EZchip) and, 151, 152
  traffic managers and, 312
  Wintegra and, 154
  Xelerated packet devices and, 155

Internet Security Association and Key Management Protocol (ISAKAMP), 387

Internet service providers (ISP), 14–15, 24–25, 43, 45, 63

Internet Telephony, 9

Internetworking Development System (IDS), Montego and InP family and, 144, **145**

Intranet Resources Kit with CD-ROM, 9

intranets, 13–14, 25–26

intrusion detection systems (IDS), classification processors and, 256–259

inverse multiplexing for ATM (IMA), Wintegra and, 154

IP addresses, 17, 52–53, 240–241, **241**

IP core based network processing, 21, 22, 160–162
  search engines and, 234

IP networks, 13–14, 44, 48

IP over ATM, 15, 56, 154

IP over Ethernet, 101

IP over SONET, 63, 72, 119

IP Security (IPSec), 26, 27, 28, 39, 56, 367, 385–388, 392–393
  IXA family network processors and, 82
  Vitesse IQ/IQ2000 and, 153

IP storage, storage network processors (SNPs) and, 199–200

IP switching, 24, 26

IP telephony, 11, 14–15

Ipsilon, 24, 26
IPv6, 340, 344
    classification processors and, 241, 242
    Montego and InP family and, 142, 144
    PayloadPlus family network processors and, 115
    TOPcore (EZchip) and, 152
    Xelerated packet devices and, 155
IQ2000, 152–153
iSCSI, storage network processors (SNPs) and, 199, 200,
        202–204, **203**, 214
iSNAP architecture for, 211–214, **212**, **213**
iterative round robin with matching (IRRM), 283
iterative round robin with SLIP (iSLIP), 283
IXA family network processors, 77–92
    architecture of, 78–83, **79**
    boot sequence in, 80
    buses in, 83, 88
    compiler for, 86
    interfaces for, system packet versions, 81–82
    IXP2850 follow-on processor for, 91
    memory and storage in, 78, 79, 80, 82
    microengine technology in, 78–83, **81**, 84, 85, 86, 90–91,
        90, **91**
    platform development kit (PDK) for, 85
    portability framework for, 78, 84, 86
    scalability and, 86–87
    security and encryption in, 91–92
    software and systems development using, 85–86
    software architecture in, 84–85, **84**, **85**
    software developer's kit (SDK) for, 84, 86
    speed of, 77, 78
    speed, clock speeds, 80, 86
    StrongARM processor core and, 80, 84
    switch fabrics and, 77
    symmetric multiprocessing (SMP) in, 81
    threading in, 80–81
    trade-offs when designing with, 86–92
    XScale technology in, 78, 84

J–K

Java, 7, 8, 144
Jazz VLIW, 186
jitter
    storage network processors (SNPs) and, 205
    switch fabrics and, 281
    traffic managers and, 314
jitter earliest due date (JEDD), 327
Joint Test Action Group (JTAG), 78
Jupiter, 45
just a bunch of disks (JBOD), storage network processors
        (SNPs) and, 193
just in time (JIT), 13–14, 39

Kasparov, Gary, 7
Katholieke Universiteit Leuven (KUL), 378
Kawasaki LSI (KLSI), 54, 222
Keshav, Srinivasan, 130, 273
key auto key (KAK) mode, 370
key generation and scheduling, encryption and, 367
knockout switches, 290
Knuth, Donald E., 245
Knuth-Morris-Pratt (KMP) algorithm, 254
Kwok, T., 282

L

label distribution protocol (LDP), 25
label edge router (LER),
    ClearSpeed Technologies and, 171
    Montego and InP family and, 142
label switched points (LSP), DiffServ and, 319
label switched router (LSR), 25
    ClearSpeed Technologies and, 171
    DiffServ and, 319–320
    Montego and InP family and, 142
Lam, C.H., 282
lambda switching, 24–25
LAN emulation (LANE), 15–16
languages, 7, 10, 20, 22, 37, 342
languages, 22
large scale integration (LSI), 162
latency, 55
    lookup and searches, 54
    storage network processors (SNPs) and, 205
    traffic managers and, 314
lattice based encryption, 384
layer 1 switching, 49
layer 2 switching, 10, 16, 23–24, 26, 35, 40, 46, 49
layer 2 tunneling protocol (L2TP), 115
layer 3 switching, 9, 15, 16, 23–24, 26, 35, 46, 49, 56
layer 7 switching, TOPcore (EZchip) and, 149
leaf pushed binary trees, 245
leaky bucket (LB) algorithm
    PayloadPlus family network processors and, 112, 113, 114
    traffic managers and, 322
learning, content addressable memory (CAM) and, 223
Lekkas, P.C., 362, 369
Lempel-Ziv standard (LZS), 387
Lexra, 162
licensing, 40
line cards, 272, 276, 350–353, **351**, **352**
link availability, traffic managers and, 314
link encapsulation, 18
link encryptors, 373
link redundancy, switch fabrics and, 275

Linley Group, 426
LinleyBench 2000 benchmark, 72, 426
Linux, 354
    ClearSpeed Technologies and, 171
    IXA family network processors and, 86
    PayloadPlus family network processors and, 120
    PowerNP (NPRGS3) and, 71
liquid crystal display (LCD), 16
Litchfield Communications, 49
load balancers, 46
    CPort family network processors and, 123
    Cognigine and, 145
    TOPcore (EZchip) and, 151
    Wintegra and, 154
load sharing redundancy, 279
local area networks (LANs), 5–6, 7, 8, 17, 18, 44, 46, 77, 424
    CPort family network processors and, 123
    IXA family network processors and, 88, **89**
    PowerNP (NPRGS3) and, 63
    switched, 12–13
    switching and routing in, 23–24
logic blocks, megacells, 21, 22
logical link control (LLC), IXA family network
    processors and, 90
longest prefix match (LPM), 52, 56, 69, 126, 154, 229,
    240, 245, 251
look aside interface, 312, 422
Look Aside Task Group, 232
lookup and classification, 238, **239**, 240–244, **241**, **243**
lookup elements, 234–235
lookup tables, 5, 10, 28, 52, 53, 54
    classification processors and, 244–245
    content addressable memory (CAM) and, 224–225, **224**,
        226–228
    CPort family network processors and, 126
    PowerNP (NPRGS3) and, 64
    TOPcore (EZchip) and, 151
    update management algorithms for, 244–245
low voltage differential signaling (LVDS), 82, 140, 276, 347
Lucent Technologies (*See also* Agere), 105
Luderer, G.W.R., 282
Luleå scheme, 246–247
LXT1000, 349

M

M-5 Channel Adapter, CPort family network processors
    and, 126, 127
MAC addresses, 53
mainframe computers, 3–7
management data input/output (MDIO), 349
management information base (MIB), 12, 72, 209

management of networks, 11–12
management path interface (MPI), PayloadPlus family network
    processors and, 107
management plane, 51
manipulation detection code (MDC), 373
manufacturing automation protocol (MAP), 7
mapping, Montego and InP family and, 142
markup languages, 10
Marvell, 154, 177
mask registers, content addressable memory (CAM) and, 222
massively parallel branch accelerator (MPBX), iFlow chipset
    and, 140
master-slave networks, 40, 100
maximum burst size (MBS), 114
maximum transmission unit (MTU), 209
McGinnis, Evan, 12
McKeown, N., 250
mcommerce, 16
MD5, 382, 385, 389
media access control (MAC), 5, 51, 345
    AMCC nP family network processors and, 101
    classification processors and, 255
    CPort family network processors and, 123, 126
    iFlow chipset, 138–141, **139**
    IXA family network processors and, 78, 86
    PayloadPlus family network processors and, 116
    PowerNP (NPRGS3) and, 62
    search engines and, 220
    switch fabrics and, 275
media independent interface (MII), IXA family network
    processors and, 87, 88
media, storage, 193
medium scale integration (MSI), 162
megacells, 21
memory, 6, 10, 22, 34, 53–54, 425
    AMCC nP family network processors and, 99, 100
    classification processors and, 246
    Cognigine and, 147
    CPort family network processors and, 126, 128, 129
    IXA family network processors and, 78, 79, 80, 82
    PayloadPlus family network processors and,
        107, 109, 115, 117
    PowerNP (NPRGS3) and, 69, **70**, 73
    storage network processors (SNPs) and, 209
    TOPcore (EZchip) and, 149
    Vitesse IQ/IQ2000 and, 153
    Wintegra and, 154
memory subsystems, 335–339
MEPG4, 17
message authentication code (MAC), 374
message digest (MD), 373
message integrity check (MIC), 373
metering, traffic managers and, 314
Metro Ethernet Forum (MEF), 18, 424

metropolitan area network (MAN), 13, 17, 45, 47, 424
  AMCC nP family network processors and, 100
  PowerNP (NPRGS3) and, 63
  storage network processors (SNPs) and, 204
MIB object identifiers, 12
MIB variables, 12
microbrowsers, 16
microengine technology, 78–86, 90–91, **90**, **91**, 342
midrange computers, 7–8
MIPS Technologies Inc., 21, 162, 163
MISTY cipher, 362
Mitsubishi, 362
mobile computing, 22
Mobile IP, 17
modem banks, 49
modems, 45
Modern Information Retrieval, 254
modes for encryption, 369–371, **370**
modification of packets, 56
modular design, 47
MontaVista, 86
Montego and InP family, 141–145, **143**, 177
Montgomery exponentiation encryption, 383
Motorola, 20, 21, 154, 162, 274
  C-Port (*See* CPort family network processors)
MP3 music players, 17
multicasting, 33, 72, 344
  AMCC nP family network processors and, 97
  CPort family network processors and, 133
  switch fabrics and, 281, 282, 287–288
  traffic managers and, 316
multimedia, 23
multiple instruction multiple data (MIMD), 164
multiple instruction single data (MISD), 143
multiple storage protocols, storage network processors (SNPs) and, 205
multiple-field classification, 249–251
multiplexing, 8, 45
multiply and accumulated (MAC) operations, 21, 164
multiprotocol label switching (MPLS), 15, 24–25, 27, 35, 47, 50, 55, 56, 165, 168, 344, 423
  AMCC nP family network processors and, 98
  classification processors and, 240, 243
  ClearSpeed Technologies and, 171
  Cognigine and, 145
  CPort family network processors and, 125, 126, 135
  DiffServ and, 319–320
  iFlow chipset and, 141
  Montego and InP family and, 141, 142, 144
  PayloadPlus family network processors and, 115
  PowerNP (NPRGS3) and, 63, 64, 72
  quality of service (QoS) and, 320
  search engines and, 220
  security coprocessors and, 361

TOPcore (EZchip) and, 151
traffic managers and, 315–316
Vitesse IQ/IQ2000 and, 153
Xelerated packet devices and, 155
multiprotocol lambda switching (MPLms), 25
multiprotocol over ATM (MPOA), 15–16, 24
multiprotocol switches, storage area networks (SANs) and, 204, 205
multiservice access platforms (MSAP), CPort family network processors and, 123
multiservice providing platform (MSPP), 47
multiservice router/switches (MSR), 43, 47, 57 265, 271–275, **273**, 326, 335
  design issues and, case study, 344–354
  traffic managers and, 326
multiservice systems, traffic managers and, 315
multistage switches, 288–290
multithreaded array processing (MTAP), 164–173, **165**, **167**, **168**
multithreading, 340
  Cognigine and, 146
  IXA family network processors and, 80–81

N

National Committee for Information Technology Standards (NCITS), 193
National Institute of Standards and Technology (NIST), 366
National Security Agency (NSA), 382
NEC, 177
Needleman-Winsch algorithm, 254
Net ASICs, 159–160, **161**
NetGx, 177
NetLogic Microsystems, 54, 101, 251
NetMark, 180
netOctave, 215
Netplane, 135
network address translation (NAT), 387
  classification processors and, 241
    TOPcore (EZchip) and, 151
    Vitesse IQ/IQ2000 and, 153
network ASIC (Net ASIC), 335
network attached storage (NAS), 18–19, 86, 194
network classification language (NCL), 342
network database search engine, CPort family network processors and, 134
network DRAM (NDRAM), 115, 338
network equipment building standards (NEBS), 50
network equipment vendor (NEV), 64, 97, 156, 164, 342
network ID (netID), classification processors and, 240–241, **241**
network interface cards (NIC), IP storage and, 199–200

network manager, 12

network optimized instruction set computing (NISC), 93, 94

network processing elements, classification processors and, 256

Network Processing Forum (NPF), 40, 72, 82, 100, 116, 232, 256, 316, 421–423

network processor application services (NPAS), PowerNP (NPRGS3) and, 71–72

network processor specialized assembler (NPASM), PowerNP (NPRGS3) and, 71

network processing forum streaming interface (NPFSI), 275

next neighbor registers, IXA family network processors and, 83

NextWARE, Montego and InP family and, 144

Nichols, R.K., 362, 369

Nokia, 24, 26

nonblocking switches, 269

nonlinear key spaces, 372

nonrepudiation, 376

nonsteady state processing, PowerNP (NPRGS3) and, 64

nonwork conserving schedulers, 326

Novell, 9

NP2G network processor (*See also* PowerNP), 62

NP4GS3 network processor (*See also* PowerNP), 61, 156

NP4GX (*See also* PowerNP (NPRGS3)), 73

nPcore (*See also* AMCC nP family of network processors), 93

NPe405 network processor (*See also* PowerNP), 61–62

NP-Ic, TOPcore (EZchip) and, 152

Nucleus PLUS, 177

O

Oakley, 387

object identifiers, MIB, 12

OC-3

    CPort family network processors and, 125

    Wintegra and, 154

OC-12, CPort family network processors and, 124–126

OC48, 45, 48, 54, 345

    AMCC nP family network processors and, 93–103

    CPort family network processors and, 126

    Montego and InP family and, 144

    PayloadPlus family network processors and, 105, 116

    PowerNP (NPRGS3) and, 63

    Vitesse IQ/IQ2000 and, 153

OC-192, 45, 47, 48

    AMCC nP family network processors and, 93–103

    PayloadPlus family network processors and, 105

    TOPcore (EZchip) and, 149

OC-768, 48

    Cognigine and, 145, 147

    Xelerated packet devices and, 156

offloading, storage network processors (SNPs) and, 205

Oie, Y., 282

Oki, E., 282

Omni Service Processor, 177

one plus one redundancy, 279

Onex, 177

online transaction processing (OLTP), storage network processors (SNPs) and, 194

open shortest path first (OSPF), 10, 64, 65, 71

Open Software Foundation (OSF), 8

open system interface (OSI) model, 8, 9, 348

open systems, 8

operations administration and maintenance (OAM), 344

Optical Internetworking Forum (OIF), 103, 146

Optical Networking Forum (OIF), 422, 423–424

optical networks (*See also* fiber optics), 19–20, 45

    Cognigine and, 146

    CPort family network processors and, 128

    IXA family network processors and, 87

optical wavelength routers, 25

original equipment manufacturers (OEM), 6

output feedback (OFB) mode, 370

overhead, ATM, 16

overspeed, switch fabrics and, 281

overview of network processor products and platforms, 415–418

P

packet classification, 237–261, 237

    search engines and, 219–220

packet data unit (PDU)

    CPort family network processors and, 127

    PayloadPlus family network processors and, blocking and scheduling in, 109–111, **110**, 118–120

Packet Instruction Set Computing (PISC), Xelerated packet devices and, 155

packet loss, traffic managers and, 314

packet over SONET (POS), 51, 344, 345

    AMCC nP family network processors and, 100

    CPort family network processors and, 125, 126, 129

    IXA family network processors and, 87, 88

    Montego and InP family and, 142, 144

    PayloadPlus family network processors and, 106, 112, 113, 116, 119

    PowerNP (NPRGS3) and, 64

    Raqia regular expression classification coprocessor and, 258

    Wintegra and, 154

packet processing, 33, 34, 43–57

    classification of packets in, 52–53, 55–56

    content addressable memory (CAM) and, 53–54

    encapsulation of packets in, 56

    forwarding and, 55–56

    framing, 51

modification of packets in, 56

packet parsing and, 54–55

pattern search in, 52–53

search engines and, 54

switching in, 57

traffic management and, 57

packet switches, 265

packets, 31, 270

packet-switched networks, 8–9, 11

Palo Alto Research Center (PARC), 5

parallel computing, 339

parallel iterative matching (PIM), 283

parsing of packets, 54–55

partial packet discard (PPD), 142, 324–325

partitioning, traffic managers and, 329

Patricia trie, 245

pattern matching, 54

PayloadPlus family network processors and, 106–107, 108

pattern processing engine (PPE), PayloadPlus family network processors and, 108

pattern search control blocks (PSCBs), PowerNP (NPRGS3) and, 69

pattern searching, 52–53

payload bus, CPort family network processors and, 127

payload compression protocol (PCP), 387

PayloadPlus family network processors, 105–121

Agere system interface (ASI) in, 105, 106, 112–114, **113**

Agere's Scripting Language (ASL) and, 116

APP550 (EX-INP5) processor in, 106, 116–118, **117**

APP750NP (EX-NP10) and APP750TM (EX-TM10) chipset in, 114–116, **115**

architecture of, 105–107, **106**

buses in, 107

configuration bus interface (CBI) in, 107

dual leaky bucket (DLB) algorithm in, 113, 114

fast pattern processor (FPP) in, 105, 106, 107–109, **108**

functional bus interface (FBI) in, 107, 109

functional programming language (FPL) for, 106, 114–115, 118–120

management path interface (MPI) in, 107

memory and storage in, 107, 109, 115, 117

packet data unit (PDU) blocking and scheduling in, 109–111, **110**, 118–120

pattern matching optimization in, 106–107

pattern processing engine (PPE) in, 108

quality of service (QoS) and class of service (CoS) in, 105, 109, 115

routing switch processor (RSP) in, 105, 106, 107, 109–111, **111**

serialization and deserialization (serdes) in, 116

software development environment (SDE) for, 119

software development kit (SDK) for, 119

stream editor in, 111, 117

switch support package (SSP) and, 120

systems and software development using, 118–120

TMS system for software development using, 120, **120**

traffic engineering/management in, 109–111

very long instruction word (VLIW) architecture in, 109, 111, 119

voice packet processor (VPP) in, 106, 107

peak cell rate (PCR), 114

Pentium, 21, 33, 120

per hop behavior (PHB), TOPcore (EZchip) and, 150

performance, 9, 28, 40, 423, 426

CPUs, 34–36, **35**

PowerNP (NPRGS3) and, 72–73

switch fabrics and, 281

Tensilica and, 178–180, **178**, **179**, **180**, **181**

peripheral component interconnect (PCI), 349, 425

CPort family network processors and, 127, 128

iFlow chipset and, 140

IXA family network processors and, 79

PayloadPlus family network processors and, 105

peripheral network processors, 39–40

Perkins, David, 12

Perlman, Radia, 9, 24

permanent circuits, 8

personal computers (PC), evolution of, 6–7, 10, 20

personal digital assistants (PDAs), 17, 362

physical layer (PHY) interface, 5, 32–33, 39, 51, 345, 348, 423

AMCC nP family network processors and, 94

CPort family network processors and, 125, 126

IXA family network processors and, 86

PayloadPlus family network processors and, 107, 116

PowerNP (NPRGS3) and, 65–68, **66**

switch fabrics and, 275

TOPcore (EZchip) and, 150

physical MAC multiplexer (PMM), PowerNP (NPRGS3) and, 65–68, **66**

PI40 family chips, switch fabrics and, 298–301, **299–301**

picocode, PowerNP (NPRGS3) and, 70–71

picoengines, 342

pipeline bubble, 36, 96–97

pipelining, 36, 339

CPort family network processors and, 125–126

Montego and InP family and, 144

Tensilica and, 175–176, **176**

TOPcore (EZchip) and, 150

Xelerated packet devices and, 156

plain old telephone service (POTS), 45

platform development kit (PDK), IXA family network processors and, 85

platform network processors, 39–40

platforms, overview of network processor products and platforms, 415–418

plesiochronous digital hierarchy (PDH), 45, 105

point-to-point protocol (PPP), 51, 72, 344
Cognigine and, 145
CPort family network processors and, 125
iFlow chipset and, 141
Montego and InP family and, 142
PayloadPlus family network processors and, 119
Wintegra and, 154
point to point protocol over Ethernet (PPPoE), PayloadPlus
family network processors and, 115
policing, policer modules, traffic managers and, 314, 315,
321–322, 344
policy coprocessor, PowerNP (NPRGS3) and, 68–70, **68**
policy engine, AMCC nP family network processors and, 95
polling, 12
portability framework, IXA family network processors and,
78, 84, 86
post office protocol (POP), 49
power consumption, 49
PowerNP (NPRGS3), 61–75, **62**
architecture of, 62–65, **62**, **63**
compiler for, 72
connectivity options, 64
control point processor in, 64
coprocessors and assist hardware for, 68–70, **68**
external switch fabric and CPU in, 63
handlers in, 70–71
input/output lines in, 65
major functional blocks in, 65–68, **66**
memory and storage for, 69, **70**, 73
migration issues and, 62
NP2G network processor and, 62
NP4GS3 network processor and, 61
NP4GX and, second-generation of, 73
NPe405 network processor and, 61–62
performance of, 72–73
scalability in, 63
software and systems development (toolkits) using,
71–73, **71**
software architecture and picocode for, 70–71
speed of, 73, 74
switch fabric and, 69
trade offs when designing with, 73–74
traffic engineering/management and, 72, 74
tree search engine (TSE) in, 62, 69
PowerPC, 33, 120, 162
PowerPRS, 127, 152, 291–298, **293–297**
PowerQUICC architecture, CPort family network processors
and, 123
prefix length ordering (PLO), 244–245
pretty good privacy (PGP), 365
Prime Computer, 4, 6
printed circuit boards (PCB), 5, 162
priority queues (PQ), 327

private key encryption, 364
probes, 12
processing elements (PEs), 164–173
processing paths, 33, 50–51, 108–109
processor generator, Tensilica and, 174
processor interface (PIF), Tensilica and, 174
programmable logic controller (PLC), 7
programmable logic device (PLD), classification processors
and, 255
programmable read only memory (PROM), 22, 127
programmable state machine (PSM), 253
programmed I/O (PIO), 170
protocol data units (PDU), PayloadPlus family network
processors and, 107–109
protocol sensitive encryption, 373
protocol stacks, 44
protocols, 10, 11, 35
provider networks, 45
pseudorandom number generator (PRNG)
IXA family network processors and, 83
PowerNP (NPRGS3) and, 62
public key encryption and, 364–367, 382–384
public switched telephone network (PSTN), 45, 48, 86
pulse code modulation (PCM), 8, 87, 154
PXF chip, Cisco, 156

Q

Q-5 TMC, CPort family network processors and, 127–131,
128, **129**, **132**
quad data rate (QDR), 73, 82, 115, 156, 338
quality of service (QoS), 10, 11, 22–25, 27–28, 31, 35, 47, 52,
55, 57, 265, 426
AMCC nP family network processors and, 97, 99, 100
asynchronous transfer mode (ATM), 15, 16, 320
classification processors and, 238
ClearSpeed Technologies and, 173
congestion management in, 322–325, **323**, **324**, **325**
CPort family network processors and, 126, 129, 131
differentiated services (DiffServ) and, 318–320
Fibre Channel and, 197
iFlow chipset, 138–141, **139**
integrated services (IntServ) and, 317–318
IP telephony, 14–15
IXA family network processors and, 83
Montego and InP family and, 141, 142
multiprotocol label switching (MPLS) and, 319–320
PayloadPlus family network processors and, 105, 109, 115
PowerNP (NPRGS3) and, 63, 64
resource reservation protocol (RSVP) and, 317, 320
search engines and, 220
SONET and, 320

switch fabrics and, 277, 279, 284
TOPcore (EZchip) and, 149, 150
traffic managers and, 311, 313, 314, 315, 317–320, 329
Vitesse IQ/IQ2000 and, 153
queue management unit (QMU), CPort family network
    processors and, 125, 126
queuing, traffic managers and, 316
QX-1, TOPcore (EZchip) and, 150

R

radix trie, 245, 246
Rainier (*See* NP4GS3; PowerNP), 61
Rambus, 337
random access memory (RAM), 22
random early detection (RED), 66, 323–325, **323**, **324**, **325**
    AMCC nP family network processors and, 99
    CPort family network processors and, 130
    PayloadPlus family network processors and, 109
    traffic managers and, 316
    Vitesse IQ/IQ2000 and, 153
random number generator (RNG), encryption and, 366
RapidIO, 424, 425
Raqia Networks, 251
Raqia regular expression classification coprocessor in,
    256–259, **257**
rate controlled scheduling (RCS), 327–328
RC4, 392
RDRAM, 153, 337
read only memory (ROM), 22, 34, 80
real time control protocol (RTCP), 87
real time operating systems (RTOS), IXA family network
    processors and, 80
real time protocol (RTP), 14, 87, 371
rearrangeably nonblocking (RNB) switches, 269
reassembly, storage network processors (SNPs) and, 205
receiving of information, CPUs and, 32–33
reconfigurable communications units (RCUs), Cognigine and,
    145, 146, **148**
recursive flow classification (RFC), 250, **251**
Red Hat Linux, PowerNP (NPRGS3) and, 71
RED with In and Out (RIO), 57, 323–325, **325**
reduced instruction set computing (RISC), 6, 7, 20, 21, 31, 33,
    34, 38, 51, 52, 163, 164, 169, 170, 339, 341, 342, 425
    AMCC nP family network processors and, 93, 94
    classification processors and, 251
    CPort family network processors and, 124, 125
    iFlow chipset and, 140
    IXA family network processors and, 78
    PayloadPlus family network processors and, 106
    PowerNP (NPRGS3) and, 62
    Tensilica and, 173, 174

reduced latency DRAM (RLDRAM), 156, 338
redundancy of switch fabrics, 278–279, **278**, **279**, **280**
redundant array of independent disks (RAID), 18, 193
    Fibre Channel and, 197
    storage network processors (SNPs) and, 205
reflector mode pathways, IXA family network processors
    and, 83
regular expression classification coprocessor in, 256–259, **257**
rekeying, encryption and, 377
reliability of network, 17, 47, 50
remote monitoring (RMON), 12
    PayloadPlus family network processors and, 112
    PowerNP (NPRGS3) and, 65
    storage network processors (SNPs) and, 209
repeaters (*See also* hubs), 13
request for comments (RFC), 424
requirements for network equipment, 49–50
Rescoria, Eric, 388
research & development, 20
resource reservation protocol (RSVP), 14, 27, 55, 315,
    317, 320
resources budget, systems engineering and, 353–354
restructuring, 13
return on investment (ROI), 37
reverse address resolution protocol (RARP), 197
Ribeieo-Neto, B., 254
Rijndael encryption, 91, 378–380, **379**, 384
ring buffers, IXA family network processors and, 83
ring bus, CPort family network processors and, 127
ring networks, 13
RIPEMD, 382
roots, in FPL, 118
round robin (RR), 99, 283, 284, 327
round robin greedy scheduling (RRGS), 283
route aggregation, 241–242
routers and routing, 9–11, 23–24, 43, 44, 45, 56, 265
    classification of packets and, 52–53
    Cognigine and, 145
    PowerNP (NPRGS3) and, 63
    switch fabrics and considerations and, 279–281
    wavelength type, 24–25
Routing in the Internet, 9, 240
routing information protocol (RIP), 10, 65
routing protocols, 10
routing switch processor (RSP), PayloadPlus family network
    processors and, 105, 106, 107, 109–111, **111**
RS/6000 IBM supercomputer, 7, 10, 21
RS-232, 10
RSA Technology, 382, 383, 385, 393, 394
RSVP Traffic Engineering (RSVP-TE), 320
run-to-completion processors, 339

S

S boxes, encryption and, 367
S/370 IBM computers, 6
SafeNet, 256, 389, **389**, **390**
SafeXcel, 389, **389**, **390**
SAN Protocol Processor (SPP), 210–211, **210**
scalability, 13
  IXA family network processors and, 86–87
  switch fabrics and, 277
scalable processor architecture (SPARC), 162
schedulers, 57
  PowerNP (NPRGS3) and, 63
  switch fabrics and, 274, 275, 284
  traffic managers and, 316, 326–329
Schneier, B., 370
scratch pad memory, IXA family network processors and, 79
SDLC, 7, 10
search engines (*See also* content addressable memory),
    53, 54, 219–236
  alternatives to CAM in, 234–235
  associated memory, DRAM, and SRAM in, 220
  binary CAM (BCAM) and, 220–221
  classification engines in, 219–220
  content addressable memory (CAM) and, 219, 220–234,
      222, **233**
  developments in, 233–234
  IP cores and, 234
  lookup elements in, 234–235
  MAC layer and, 220
  multiprotocol label switching (MPLS) and, 220
  packet classification context of, 219–220
  quality of service (QoS) and class of service (CoS) in, 220
  search keys and, 220, 229
  sort free CAM in, 233
  table lookup engine (TLE) and, 234–235
  ternary CAM (TCAM) and, 220–221
  TOPcore (EZchip) and, 149
  trie algorithm and, 234
  virtual private network (VPN) and, 220
search keys, 53, 220, 229
second generation technologies, 17
secure electronic transactions, 361–362
secure hashing algorithm (SHA), 382, 385, 389, 392
secure sockets layer (SSL), 206, 361–362, 388, 393–394
security, 17, 23, 26–27
  coprocessors for, 26–27
  IXA family network processors and, 91–92
  storage network processors (SNPs) and, 214–215
  traffic managers and, 312
security associations (SAs), 377
security coprocessors, 359–394
  classification of, 389
  digital signal processing (DSP) and, 389

encryption and, 361, 363–384
  Federal Information Protection Standard (FIPS) and, 392
  firewalls and, 376
  IP Security (IPSec) and, 385–388, **386**, 392–393
  multiprotocol label switching (MPLS) and, 361
  need for security and, 359–360
  synchronous dynamic RAM (SDRAM) and, 392
  secure electronic transactions, 361–362
  secure sockets layer (SSL) and, 361–362, 388, 393–394
  security parameter index (SPI) in, 386
  standardized protocols for, 384–388
  systems considerations when engaging, 390–394, **391**
  timestamps (nonrepudiation) and, 376
  transport layer security (TLS) and, 388
  virtual private network (VPN) and, 361
  wireless security and, 362, **363**
security parameter index (SPI), IPSec and, 386
seed values, 370
segmentation and reassembly (SAR)
  AMCC nP family network processors and, 97
  CPort family network processors and, 126, 133
  IP storage and, 200
  IXA family network processors and, 83, 90
  Montego and InP family and, 142
  PayloadPlus family network processors and, 106, 116
  traffic managers and, 316
self-routing switch fabrics and, 283
semaphore coprocessor, PowerNP (NPRGS3) and, 69–70, **68**
semaphores, 171
semiconductors, evolution of, 4, 9
serial data processors (SDP), CPort family network processors
    and, 124–125
serialization and deserialization (serdes), 51, 345
  AMCC nP family network processors and, 100, 101
  Cognigine and, 146
  PayloadPlus family network processors and, 116
  switch fabrics and, 277, 280
Series/1 IBM computers, 7
server adapters, PowerNP (NPRGS3) and, 63
service level agreements (SLAs), 25, 28, 57, 101, 129, 265
  PayloadPlus family network processors and, 115
  traffic managers and, 312, 315
service provider networks, 47
session key exchange, 375
Session Layer Interface Card (SLIC), 209
SH5, 169
SHA-1 encryption, 91, 382, 385, 389, 392
shared memory switches, 287–288, **288**
sharing, traffic managers and, 329
shock absorber random early detection (SRED), PowerNP
    (NPRGS3) and, 63
SiberCore, 54, 229, 231
signaling, traffic managers and, 315
Silicon Access Networks, 138–141, **139**

Silicon Graphics, 163
Silverback Systems iSNAP architecture for, 211–214, **212**, **213**
simple mail transfer protocol (SMTP), 49
simple network management protocol (SNMP), 11–12, 64, 314, 353
   IXA family network processors and, 87
   PowerNP (NPRGS3) and, 71
simultaneous multithreading (SMT), 340
single instruction multiple data (SIMD), 164–173, **165**, **168**
single rate three color marker (srTCM), 321
SKEME, 387
skip bits, content addressable memory (CAM) and, 227
slow processing paths, 33, 50–51, 108–109, 272
small computer system interface (SCSI), 18, 193
Small computer system interface (SCSI) over IP, 44
small medium enterprises (SME) networks, IXA family network processors and, 77
small office/home office (SOHO), 34, 77
Small talk, 7
SMII interfaces, 51
   CPort family network processors and, 125
   PayloadPlus family network processors and, 116
Smith-Waterman algorithm, 254
smoothed deficit weighted round robin (SDRR), 327
SNMP, SNMPv2..., 12
snoop mode, AMCC nP family network processors and, 98
sockets, 8
software, 5, 6
   IXA family network processors and, 84–85, **84**, **85**
   PowerNP (NPRGS3) and, 70–71
software developer's kit (SDK)
   CPort family network processors and, 135
   IXA family network processors and, 84, 86
   PayloadPlus family network processors and, 119
software development, 340–342
   costs of, 342–344, **343**, **344**
   CPort family network processors and, 131–134
   IXA family network processors and, 85–86
   management issues for, 341
   PayloadPlus family network processors and, 118–120
   PowerNP (NPRGS3) and, 71–73, **71**
software development environment (SDE)
   CPort family network processors and, 135
   PayloadPlus family network processors and, 119
software managed tree (SMTs) algorithm, 69
Software Working Group (SWG), 422
Solaris, ClearSpeed Technologies and, 171
Solidum, 251
SONET, 15, 16, 45, 47, 48, 51, 105, 275, 347
   AMCC nP family network processors and, 93
   CPort family network processors and, 123, 125
   iFlow chipset and, 141
   IXA family network processors and, 87
   Montego and InP family and, 141

PowerNP (NPRGS3) and, 63, 64
   quality of service (QoS) and, 320
   switch fabrics and, 271
   traffic managers and, 314
sort free CAM, 233
spanning tree, PowerNP (NPRGS3) and, 65
speculative execution, 107
speed, clock speed, throughput, 34, 36, 46, 49
   ClearSpeed Technologies, 171
   Cognigine and, 145, 146
   content addressable memory (CAM) and, 229, 230, 223–225, **225**
   CPort family network processors and, 124, 125
   iFlow chipset and, 140
   IXA family network processors and, 77, 78, 80, 86
   Montego and InP family and, 144
   PowerNP (NPRGS3) and, 73, 74
   switch fabrics and, 281
   Tensilica and, 178–180, **178**, **179**, **180**, **181**
   TOPcore (EZchip) and, 149
   traffic load and, chart of, **419**
SPI 3, 51
split mode operation, IXA family network processors and, 81
Srinivasan, V., 249
SSL and TLS, 388
Stallings, William, 12
standardization efforts in network processing, 10–11, 421–427, 421
   security coprocessors and, 384–388
static RAM (SRAM), 53, 54, 170–171, 336, 338
   Cognigine and, 147
   CPort family network processors and, 126, 129, 135
   iFlow chipset, 138–141, **139**
   IXA family network processors and, 79, 80, 82
   PayloadPlus family network processors and, 112, 115, 117
   PowerNP (NPRGS3) and, 63, 69
   search engines and, 220
   TOPcore (EZchip) and, 152
   Xelerated packet devices and, 156
statistics collection, CPort family network processors and, 133
statistics module, traffic managers and, 316
steady state processing, PowerNP (NPRGS3) and, 64
STM, PayloadPlus family network processors and, 105
STMicro, 169
storage area networks (SANs), 18–19, 44, 194–196, **195**
   AMCC nP family network processors and, 100–101
   Cognigine and, 145
   Ethernet and Gigabit Ethernet in, 195
   Fibre Channel and, 195–199, **198**
   Gigabit Ethernet and, 204, 205
   Internet protocol (IP) networks and, 195–196
   iSCSI in, 202–204, **203**, 214
   multiprotocol switches in, 204, 205
   PowerNP (NPRGS3) and, 63

storage area networks (SANs) (*continued*)
  redundant array of independent disks (RAID), 205
  serialization and deserialization (serdes) module, 277
  storage network end systems and, 205
  TCP/IP and, 195–196, 205
storage media, 193
storage network end systems, 205
storage network processors (SNPs), 177, 193–217, 335
  application programming interfaces (APIs) and, 209
  applications for, 204–205
  directly attached storage (DAS) devices and, 193
  Ethernet and Gigabit Ethernet in, 209
  FCIP in, 199, 204
  Fibre Channel and, 195–199, **198**
  Fibre Channel to iSCSI bridging in, 204
  host bus adapters (HBAs) and, 199, 200
  IP storage and, 199–200
  iSCSI in, 199, 200, 202–204, **203**, 214
  just a bunch of disks (JBOD) and, 193
  maximum transmission unit (MTU) in, 209
  media for, 193
  memory and, 209
  multiprotocol switches in, 204, 205
  network attached storage (NAS) in, 194
  network interface cards (NIC) and, 199–200
  online transaction processing (OLTP) and, 194
  redundant array of independent disks (RAID) and, 193, 205
  requirements for, 205
  SCSI bus and, 193
  security issues in, 214–215
  Session Layer Interface Card (SLIC) and, 209
  Silverback Systems iSNAP architecture for, 211–214, **212**, **213**
  storage area networks (SANs) in, 194–196, **195**
  storage network end systems and, 205
  storage protocol accelerator (SPA) in, 208–209
  storage virtualization and, 200–202
  TCP offloading and, 205, 206–209, **207**, **208**
  TCP termination engines and, 206–209, **207**, **208**
  TCP/IP and, 206–209, **207**, **208**, 215
  Trebia Networks SAN Protocol Processor (SPP), 210–211, **210**
  virtual private network (VPN) and, 215
Storage Networking Industry Association (SNIA), 199
storage networks, 18–19
storage protocol accelerator (SPA), 208–209
storage virtualization, 200–202
store and forward architectures, 55
straightline code, 168
stream ciphers, 369
stream editor, PayloadPlus family network processors and, 111, 117
stream I/O (SIO), 170
streaming data, 11

streaming interface, 422
strict priority (SP), 130, 284, 327
strictly nonblocking (SNB) switches, 269
string copy coprocessor, PowerNP (NPRGS3) and, 68–70, **68**
stripping, switch fabrics and, 287
StrongARM processor core, IXA family network processors and, 80, 84
subnets, 23–24
subnetwork access protocol (SNAP), IXA family network processors and, 90
Sun Microsystems, 6, 7, 21
Sun Solaris, PowerNP (NPRGS3) and, 71
Sunday algorithm, 254
sunshine switches, 290–291
sustained cell rate (SCR), PayloadPlus family network processors and, 114
switch cell interface (SCI), PowerNP (NPRGS3) and, 65–68, **66**
switch data mover (SDM), PowerNP (NPRGS3) and, 65–68, **66**
switch fabrics, 33, 39, 57, 105, 263–309, 350–353, **351**, **352**, 350, 423
  abacus switches in, 290
  Agere Systems PI40 family chips in, 298–301, **299–301**
  AMCC nP family network processors and, 99–101, **99**
  application specific integrated circuit (ASIC) and, 272
  arbitrated crossbar, 285–287, **285**, **286**
  arbitration algorithms in, 283
  architectures for, 281–288, 281
  asynchronous transfer mode (ATM), 271, 272–273, 277, 280
  asynchronous vs. synchronous operation in, 276
  backplane description and, 275–277
  backpressure (in-band/out-of-band) in, 274
  Banyan based switches in, 288–289, **289**
  Batcher-Banyan switches in, 289–290
  beta elements in, 269–270, **269**
  blocking and, 265, 268–269
  broadcast switches in, 282
  buffered crossbar, 285
  buffers and, 273–274, 283–284
  cells in, 273
  Cognigine and, 145
  congestion and, 281
  considerations for, 279–281
  CPort family network processors and, 135
  CPU controller in, 276
  credit table mechanisms in, 274
  crossbar switch in, 266, **266**
  crossconnects in, 276
  crosspoint switch in, 267–268, **267**, **268**
  current mode logic (CML) in, 276
  datagrams and, 270

definition of, 264

differential mode positive emitter coupled logic (DM-PECL) in, 276

finite state machines (FSM) and, 270

frames in, 273

graceful degradation in, 278

head of line (HOL) blocking and, 269

helix switches in, 291

high speed serial (HSS) links and, 275

IBM PowerPRS, 291–298, **293–297**

input and output buffers in, 266

input- and output-buffered switches in, 283–284

interfaces in, 275

Internet protocol (IP), 271

IXA family network processors and, 77

knockout switches in, 290

line cards and, 272, 276

link redundancy and, 275

low voltage differential signaling (LVDS) in, 276

media access control (MAC) and, 275

multicasting and, 281, 282, 287–288

multiservice router/switches in, 271–275, **273**

multistage switches in, 288–290

nonblocking switches and, 269

overspeed in, 281

packet routing switches (PRS) in, 291–298, **293–297**

packets, 270

performance and standards in, 281

physical (PHY) layer and, 275

platforms for switching, 271

PowerNP (NPRGS3) and, 64, 69

quality of service (QoS) and class of service (CoS) in, 277, 279, 284

rearrangeably nonblocking (RNB) switches in, 269

redundancy of, 278–279, **278**, **279**, **280**

scalability of, 277

schedulers and, 274, 275, 284

self-routing, 283

serialization and deserialization (serdes) module, 277, 280

shared memory switches in, 287–288, **288**

slow and fast path processing in, 272

SONET/SDH and, 271

strictly nonblocking (SNB) switches in, 269

stripping or byte slicing, 287

sunshine switches in, 290–291

switching application specific integrated circuit (ASIC)s and, 264–268

switching elements in, 269–270

time division multiplexing (TDM) and, 277

time slot interchanger (TSI) in, 267

time space (TS) switching in, 269

time space time (TST) switching in, 269

TOPcore (EZchip) and, 152

traffic managers and, 316

virtual output queues (VOQs), 283–284, 286

virtual queues and, 280

switch interface (SWI), PowerNP (NPRGS3) and, 65–68, **66**

switch support package (SSP), CPort family network processors and, 120, 135

switched LANs, 12–13

switching (*See also* switch fabrics), 9, 15, 44, 57, 264–268

symmetric key encryption, 364

symmetric multiprocessing (SMP), 81, 339

synchronous digital hierarchy (SDH), 15, 16, 45, 105, 347

CPort family network processors and, 123, 125

iFlow chipset and, 141

IXA family network processors and, 87

PowerNP (NPRGS3) and, 64

switch fabrics and, 271

traffic managers and, 314

synchronous dynamic RAM (SDRAM), 337

Cognigine and, 147

CPort family network processors and, 128

IXA family network processors and, 79, 80

PayloadPlus family network processors and, 107, 111

PowerNP (NPRGS3) and, 63, 69

security coprocessors and, 392

storage network processors (SNPs) and, 209

TOPcore (EZchip) and, 149

Wintegra and, 154

synchronous optical network (*See* SONET)

synchronous static RAM (SSRAM), 338

AMCC nP family network processors and, 101

classification processors and, 254

Cognigine and, 147

iFlow chipset and, 139

PayloadPlus family network processors and, 107

PayloadPlus family network processors and, 112–113, 112

Raqia regular expression classification coprocessor and, 258

system CPU cards, 353

System Network Architecture (SNA), 8, 9, 10

system on a chip (SOC), 21, 22, 162

ARC cores, 184–186, **185**

flexible length instruction set (FLIX) and, 182–184

Improv Systems, 187, **186**

Tensilica and, 174, 175

Trebia Networks SAN Protocol Processor (SPP), 210–211, **210**

system packet interface (SPI), 275, 316, 42

IXA family network processors and, 81, 82

PayloadPlus family network processors and, 116

Raqia regular expression classification coprocessor and, 258

systems development and engineering, 335–355

AMCC nP family network processors and, 101–102, **102**

architecture issues in, 339–340

content addressable memory (CAM) and, 227–230, **228**

control planes in, 353

CPort family network processors and, 134–135

systems development and engineering (*continued*)
    design approach in, 345–346
    IXA family network processors and, 85–86
    line cards in, 350–353, **351**, **352**
    memory subsystems and, 335–339
    multiservice switch/router (MRS) in, 344–354
    PayloadPlus family network processors and, 118–120
    PowerNP (NPRGS3) and, 71–73
    preliminary design outlook in, 346–349
    resources budget and, 353–354
    security coprocessors and, 390–394, **391**
    software development costs and, 342–344, **343**, **344**
    software development issues and, 340–342
    switch fabric in, 350–353, **351**, **352**
    system CPU cards in, 353
    task definition in, 345
systems engineering (*See* systems development and
    engineering)
Systems Network Architecture (SNA), 4

T

T1, 47, 48
    AMCC nP family network processors and, 98
    Wintegra and, 154
T3, 47, 48, 154
table lookup engine (TLE), 170–171, 234–235
table lookup unit (TLU), CPort family network processors
    and, 125, 126
tag bits, content addressable memory (CAM) and, **223**, 227
tag switching, 24, 26
Tality, 135
task definition, systems engineering and, 345
task optimized processing (TOP), TOPcore (EZchip) and,
    149, 150
TCP offloading, storage network processors (SNPs) and,
    205, 206–209, **207**, **208**
TCP termination engines, storage network processors (SNPs)
    and, 206–209, **207**, **208**
TCP/IP, 9, 52–53
    FCIP and, 204
    Fibre Channel and, 197
    IP storage and, 199–200
    iSCSI in, 202–204, **203**
    IXA family network processors and, 87
    Montego and InP family and, 144
    Raqia regular expression classification coprocessor and, 259
    storage area networks (SANs) and, 195–196, 205
    storage network processors (SNPs) and, 206–209, **207**,
    208, 215
Technical Education and Marketing Working Group
    (TEMWB), 422
Tehuti Networks, 215

Telecommunications Technology Committee (Japan), 424
telephony (*See also* IP telephony), 14, 45, 87
10 Gigabit Ethernet Alliance, 18, 424
Tensilica, 162, 172, 173–182, **175**, **176**
Tensilica Instruction Extension (TSE) language, 176–177
terabit routers, 45
TeraStream, Vitesse IQ/IQ2000 and, 153
ternary CAM (TCAM) (*See also* content addressable
    memory), 53, 54
    classification processors and, 240, 243, 244, 251
    iFlow chipset, 138–141, **139**
    IXA family network processors and, 82
    Montego and InP family and, 142
    prefix length ordering (PLO) in, 244–245
    search engines and, 220–221, 220
    Xelerated packet devices and, 155, 156
Texas Instruments, 21, 22
third-generation technologies, 17
threading, 340
    Cognigine and, 146
    IXA family network processors and, 80–81
    multithreaded array processing (MTAP), 164–173,
        165, **167**, **168**
three-color algorithm, traffic managers and, 321
throughput (*See* bandwidth; speed)
time division multiple access (TDMA), 16
time division multiplexing (TDM), 45, 48, 49, 164,
    265, 267, 345
    AMCC nP family network processors and, 100
    Cognigine and, 145
    IXA family network processors and, 86
    Montego and InP family and, 141
    PowerNP (NPRGS3) and, 65–68, **66**
    switch fabrics and, 277
    traffic managers and, 314
    Wintegra and, 154
time in market, 38
time multiplexing, 8
time slot exchange (TSI), 87, 267
time space (TS) switching, 269
time space time (TST) switching, 269
time to live (TTL) fields, PowerNP (NPRGS3) and, 64
time to market, 38
timestamps (nonrepudiation) and, 376
TMS system for software development, PayloadPlus family
    network processors and, 120, **120**
Tobagi, F.A., 282
TOE (*see* TCP offloading and TCP termination engines)
token buckets, traffic managers and, 322
token ring, 5, 13, 46
toll quality, 14
Tolly Group, The, 72, 426
tool control language/toolkit (TCL/Tk), PowerNP (NPRGS3)
    and, 72

toolkits, PowerNP (NPRGS3) and, 71–73
TOPcor (EZchip), 148–152
topologies, 4
Tornado for managed switches (TMS), 135, 177
Toshiba, 24, 26
traffic engineering and management, 24, 27, 47
    AMCC nP family network processors and, 96, 97–99, **98**
    classification processors and, 239
    CPort family network processors and, 123, 128, 130
    packet processing and, 57
    PayloadPlus family network processors and, 109–111
    PowerNP (NPRGS3) and, 64, 72, 74
    traffic managers 311–331
    Wintegra and, 154
    Xelerated packet devices and, 155
traffic load, chart of, **419**, 397
traffic management coprocessor (TMC), CPort family network
    processors and, 125
traffic management interface (TMI), 128, 130
traffic managers, 39, 311–331, 335
    active queue control in, 316
    bandwidth or throughput and, 314
    bandwidth provisioning in, 314
    buffers in, 326–329
    capacity and, 313
    class based queuing (CBQ) in, 327
    congestion management in, 322–325, **323**, **324**, **325**
    deficit round robin (DRR) in, 327
    deficit weighted round robin (DWRR) in, 327
    definition and purpose of, 311
    delay earliest due date (DEDD), 327
    differentiated services (DiffServ) and, 315, 318–320, 328
    dual leaky bucket (DLB) algorithm in, 322
    earliest deadline first (EDF) in, 327
    early forward congestion indication (EFCI) in, 323
    early packet discard (EPD)), 324–325
    explicit congestion notification (ECN) in, 323
    fair queuing (FQ) in, 327
    fairness in, 322
    first in first out (FIFO) and, 329
    flow through, 312
    frame based deficit round robin (FBDRR) in, 327
    generalized processor sharing (GPS) in, 328
    graduated dropping regions in, 324
    guaranteed bandwidth WFQ (GB WFQ) in, 327
    integrated services (IntServ) and, 317–318
    Internet protocol (IP) and, 312
    jitter and, 314
    jitter earliest due date (JEDD), 327
    latency and, 314
    leaky bucket (LB) algorithm in, 322
    link availability and, 314
    look aside, 312
    major tasks and algorithms in, 321

metering in, 314
multicasting and, 316
multiprotocol label switching (MPLS) and, 315–316
multiservice routers (MSR) and, 326
multiservice systems and, 315
packet loss and, 314
partial packet discard (PPD), 324–325
partitioning, sharing, allocation in, 329
policies and, 314
policing, policer modules in, 315, 321–322
priority queues (PQ) in, 327
quality of service (QoS) and class of service (CoS) in, 311,
    313–320, 329
queuing and, 316
random early detection (RED) in, 316, 323–325, **323**,
    324, **325**
rate controlled scheduling (RCS), 327–328
RED with In and Out (RIO) in, 323–325, **325**
requirements of, 312, **313**
resource reservation protocol (RSVP) and, 315, 317, 320
round robin (RR) in, 327
schedulers and, 316, 326–329
security and, 312
segmentation and reassembly (SAR) in, 316
service level agreements (SLAs) and, 312, 315
signaling in, 315
smoothed deficit weighted round robin (SDRR) in, 327
standalone chips for, 311–312
statistics module in, 316, 321
strict priority (SP) in, 327
switch fabric interfaces and, 316
three-color algorithm in, 321
token buckets in, 322
traffic engineering and, 316
traffic management and, fundamental concepts in, 312–316
traffic marking in, 321–322
traffic shaping and, 314, 321–322
virtual output queues (VOQs) and, 326
virtual private network (VPN) and, 316
weighted fair queuing (WFQ) in, 327, 329
weighted random early detection (WRED) in, 323–325, **323**,
    324, **325**
weighted round robin (WRR) in, 327
traffic marking, traffic managers and, 321–322
traffic shaping, 314, 344
    classification processors and, 239
    PowerNP (NPRGS3) and, 65–68, **66**
    traffic managers and, 321–322, 321
transactors, IXA family network processors and, 85
transfer registers, IXA family network processors and, 79
transmission control protocol (TCP), 8, 9, 28, 46, 49, 72, 371
    Tensilica and, 173
    traffic managers and, 315
transmission of information, CPUs and, 32–33

transport layer security (TLS), 388
transport mode, IPSec and, 385
traps, 12
Trebia, 177, 215
Trebia Networks SAN Protocol Processor (SPP), 210–211, **210**
tree search engine (TSE), PowerNP (NPRGS3) and, 62, 69
trie, 56, 234, 245–246, **246**, 249
triple DES, 91, 377–378, **378**, 385, 392
TSE coprocessor, PowerNP (NPRGS3) and, 69–70, **68**, 69
TSMC, 34, 118
tunneling, 26, 27, 28, 141, 376, 385
twisted pair, 13
2.5 generation technologies, 17
two-rate three color marker (trTCM), 321
type of service (TOS)
    classification processors and, 238
    traffic managers and, 313

U

Ultra3, 18
Ultrix, 6
Understanding SNMP MIBs, 12
Unibus, 7
unicasting, 72
uniform resource locator (URL), 35, 46, 52, 95
universal test and operations PHY interface for ATM
        (UTOPIA), 51, 275, 347
    AMCC nP family network processors and, 101
    CPort family network processors and, 125, 126, 127
    PayloadPlus family network processors and, 109, 116
    Wintegra and, 154
University of California Berkeley, 6, 20
University of California Stanford, 20
UNIX, 4, 6, 7, 8, 9, 11
UNIX BSD, 6
unshielded twisted pair (UTP), 13
user data memory, 53
user datagram protocol (UDP), 87, 371

V

Vaidya, A.K., 282
validity bits, content addressable memory (CAM) and, 227
value proposition for network processors, 38–39
variable bit rate (VBR), 99
variable first rate real time (VBR-rt), 109
Variable Instruction Set Communications Architecture, 145
variables, MIB, 12
variable bit rate (VBR), 316
VAX, 5, 6
VAX/VMS, 4

vector facility (VF), 6
Verilog, 37, 176
Verilog code, 87
Verizon, 14
vertical data processing (VDP), Montego and InP family
        and, 143
vertical instruction processing (VIP), Montego and InP family
        and, 143, **143**
very large scale integration (VLSI), 33, **167**
very long instruction word (VLIW), 164, 342
    CPort family network processors and, 124
    flexible length instruction set (FLIX) and, 182–184
    Improv Systems, 186–187, **186**
    PayloadPlus family network processors and, 109, 111, 119
    Tensilica and, 173
VHDL, 37
video, 45
video over IP, 45
Virtex, 177
virtual channel identifier (VCI), CPort family network
        processors and, 126
virtual channels (VCs), CPort family network processors
        and, 129
virtual circuits, 8
virtual component interfaces (VCIs), 170
virtual connection identifier (VCI), 112, 240
virtual instruction machine (VIM), ClearSpeed Technologies
        and, 172
virtual LANs (VLANs), 24, 344
    CPort family network processors and, 126
    iFlow chipset and, 141
    PowerNP (NPRGS3) and, 64, 66, 72
    Wintegra and, 154
virtual machine simulator, ClearSpeed Technologies and, 172
virtual output queues (VOQs), 283–284, 286, 326
virtual path identifier (VPI)
    CPort family network processors and, 126
    PayloadPlus family network processors and, 112
virtual private network (VPN), 24, 25–26, 47, 55
    classification processors and, 238
    CPort family network processors and, 128, 134
    DiffServ and, 320
    iFlow chipset and, 139
    IPv6 and, 242
    IXA family network processors and, 88
    Montego and InP family and, 141
    search engines and, 220
    security coprocessors and, 361
    storage network processors (SNPs) and, 215
    traffic managers and, 316
virtual queues, 280
virtualization of storage, 200–202
VISC Architecture, 145–148, **147**
Vitesse, 33

Vitesse IQ, 152–153
ViX interconnect bus, AMCC nP family network processors and, 100, 101
voice, 11, 14–15, 23, 43, 45, 47–48
    CPort family network processors and, 127
    IXA family network processors and, 87
    Montego and InP family and, 141
    PayloadPlus family network processors and, 106, 107
voice over IP (VoIP), 45, 48, **89**
    AMCC nP family network processors and, 95, 101
    CPort family network processors and, 127
    IXA family network processors and, 86
    switch fabrics and, 281
    traffic managers and, 314
voice packet processor (VPP), PayloadPlus family network processors and, 106, 107
VxWorks, 177, 354
    IXA family network processors and, 86
    Montego and InP family and, 144
    PayloadPlus family network processors and, 120
    PowerNP (NPRGS3) and, 71

W

wave division multiplexing (WDM), 16, 19–20
wavelength routers, 24–25
weak keys, encryption and, 372
web pages, 10
web switches, 46
    CPort family network processors and, 123
    TOPcore (EZchip) and, 149
weight arrays, AMCC nP family network processors and, 95
weight round robin (WRR), 284
weighted fair queuing (WFQ), 284, 327, 329
    AMCC nP family network processors and, 97, 101
    Montego and InP family and, 144
    PowerNP (NPRGS3) and, 66
    Vitesse IQ/IQ2000 and, 153
weighted random early detection (WRED), 57, 63, 66, 323–325, **323**, **324**, **325**
    AMCC nP family network processors and, 99
    CPort family network processors and, 130
    PayloadPlus family network processors and, 109, 119

    Montego and InP family and, 142, 144
    Vitesse IQ/IQ2000 and, 153
    Xelerated packet devices and, 156
weighted round robin (WRR)
    AMCC nP family network processors and, 97
    traffic managers and, 327
    Vitesse IQ/IQ2000 and, 153
Wellfleet/Bay Networks, 13
Wi Fi Protected Access (WPA), 362
wide area network (WAN), 13, 17, 34, 43, 45, 47, 49, 55, 424
    bandwidth in, 48
    IXA family network processors and, 87, 88, **89**
    PowerNP (NPRGS3) and, 63
    storage network processors (SNPs) and, 204
Wideband code division multiple access (WCDMA), Tensilica and, 177
Wind River, 71, 86, 120, 177
Windows
    ClearSpeed Technologies and, 171
    PowerNP (NPRGS3) and, 71
WinPath, 154
Wintegra, 154–155
wireless application protocol (WAP), 16, 362
wireless LAN (WLAN), 22–23, 17, 123, 362
wireless networks, 16–17
Wireless Security, 369
Wireless Security Threats, 362
wireless security, 362, **363**
wireless transport layer security (WTLS), 362
work conserving schedulers, 326
workgroup switching, 46
workstations, evolution of, 6–7
world wide web (WWW), 10

X

X.25 networks, 8, 9, 10, 26
Xelerated Packet Devices, 155–156
Xelerator, 155–156
Xenix, 6
Xerox, 5
XScale technology, IXA family network processors and, 78, 84
Xtensa, 172, 173–182, **175**, **176**

ABOUT THE AUTHOR

Panos C. Lekkas has been intensely active in the industry for more than 20 years and is currently the Founder and President of Xstream Technologies LLC in Boston, Massachusetts. Lekkas is involved in advanced technology and business development in the areas of network processing, broadband optical and RF wireless communications and networks (3G, Wi-Fi, Wi-Max, etc.), advanced signal processing, optical recording, communications security, and neural computing. Lekkas is known for his expert technology advisory role for both government and leading hi-tech companies, as well as for top-of-the-line investment banks and venture capital companies. His combined experience in both technology and business development worldwide has been applied to projects involving introduction of new technologies, performing due diligence process and technology evaluation for clients, as well as conducting valuation and sale of companies and technology assets to prospective corporate acquirers. His company has also been developing a series of patents and intellectual property that it licenses and it also provides turnkey solutions in projects involving communications systems analysis/simulation/development, VLSI/SOC architecture design, FPGA prototyping, and development of embedded software.

Lekkas started his career as a VLSI engineer with Silvar-Lisco and he rose to supervise the company's European applications engineering group. He joined IBM in the early 1980s. Among several positions he was Leading Architect & Systems Engineer in Austin, Texas, in charge of processor and memory management architecture and he has been instrumental in IBM's successful worldwide introduction of the RISC architecture, which ultimately evolved to become the heart of the renown IBM RS/6000 supercomputers. Lekkas has held several positions in advanced technology development and technical marketing management with IBM in both the United States and Europe.

After he left IBM and before starting Xstream Technologies, Lekkas has held positions as CTO & Technology Division General Manager of THLC in Marlboro, Mass., a fabless semiconductor company in the area of high-speed communications security where he built the engineering division by a series of mergers and acquisitions while hands-on leading the development of the company's highly complex ASIC product in collaboration with IBM Microelectronics; Co-Founder, President/CEO of wirelessEncryption.com Inc., a Burlington, Mass. fabless semiconductor company where he invented and started developing a patented streaming communications security technology that culminated to a multimillion dollar IPO; VP Engineering at ACI, a Hudson, Mass. a fabless semiconductor company designing advanced communications ASICs; Director of Business Development with TCC, in Concord, Mass. where he further built and supervised the in-house cryptography team, participating in industry-standards bodies, and having led the product definition for systems destined for military and intelligence communications including link- and protocol-sensitive encryptors for the industry; Director of International Technical Sales & Applications Engineering with Galileo Corp. where he pioneered the introduction of their WDM (wavelength division multiplexing) and praseodymium-doped-fluoride fiber telecom amplifier communications technology and where he increased by 10 times the Japanese business of the company within two years. At Galileo he led the establishment of the company's advanced electro-optics technologies as a de facto standard worldwide in the fields of lithography, scanning electron microscopy, and surface analysis for semiconductor equipment manufacturers and he was instrumental in the effort to diversify the fiberoptics-based coherent-imaging business division from heads-up avionics displays to medical markets that revolutionized the ways minimally invasive surgery is conducted. Before that, Lekkas had also set up and run successfully with full P&L responsibility Galileo's European branch working closely with manufacturers of image intensifier tubes (military night-vision), large scientific/analytic instrumentation, and medical imaging systems, as well as with particle accelerator labs, space agencies and nuclear weapons labs of friendly countries.

Lekkas has done his graduate research in quantum electronics at Rice University, in Houston, TX where he was a student of Nobel-laureate Professor Robert Curl. He has two graduate degrees in electrical engineering, one specialized in VLSI and semiconductor technology, and one on wireless communications with emphasis on RF microelectronics and microwave antennas. He has done his MBA work in Corporate Finance at the (KUL) Katholieke Universiteit Leuven, in Belgium and he is a Professional Engineer in the European Union. Lekkas has invented and authored several US and foreign patents in the areas of communications transmission, coding, and security and he is a member of the IEEE, of the American Mathematical Society and of the Mathematical Association of America.

Together with now George Washington University's Professor Randall Nichols who was a key member of one of his previous R&D teams, Lekkas has coauthored another major textbook titled Wireless Security, which was published worldwide by McGraw-Hill in December 2001. It contains a foreword written by Admiral Michael McConnell, former DIRNSA (head of the U.S. National Security Agency).

Lekkas has worked extensively in N. Europe, Japan, Asia, Middle East and Latin America and he can fluently speak and write in more than a dozen of European, Asian and Middle Eastern languages.